Small Ruminant Production in India

STRATEGIES FOR ENHANCING

Editors

S.A. KARIM

M.K. TRIPATHI

O.H. CHATURVEDI

A.K. SHINDE

2008

SSPH

SATISH SERIAL PUBLISHING HOUSE

403, Express Tower, Commercial Complex, Azadpur, Delhi-110033 (India)
Phone : 011-27672469, 27672852, Fax : 91-11-27672046
E-mail : ssp@ndf.vsnl.net.in, hkjain1975@yahoo.com
Website : www.satishserial.com

Published by :

SATISH SERIAL PUBLISHING HOUSE
403, Express Tower, Commercial Complex, Azadpur,
Delhi-110033 (India)
Phone : 011-27672469, 27672852, Fax : 91-11-27672046
E-mail : info@satishserial.com, hkjain1975@yahoo.com
Website : www.satishserial.com

First Published 2008

ISBN 81-89304-45-3

Printed at

SALASAR IMAGING SYSTEMS, DELHI-35

DR T. J. RASOOL

Assistant Director General
(Animal Sciences)
Indian Council of Agriculture Research,
New Delhi-110 114

Foreword

The contribution of livestock sector to agricultural growth has been ubstantial and is linked with the rural economy and nutritional ecurity of our country. Sustainable livestock production can be ensured rough improvement in per animal productivity rather than increasing eir number. Increasing human population and concomitant increase demand for food of animal origin and shrinkage of grazing area arrants newer approaches to meet our future requirement.

India has vast resources of small ruminants and it is presently nked second in the world in goats and third in sheep population. he capacity of sheep and goats to survive and maintain their oduction under diverse environmental conditions has made it possible r the large segment of landless and marginal farmers of our country rear them for their livelihood under different agro-climatic nditions. They produce 4.2% of the total milk and 15% of the total eat produced in the country. However there are many gaps in sheep roduction that can be improved by appropriate technological terventions. The lambs and kids in their active phase of growth hardly hieve 40-50 g average daily gain and weigh 20-22 kg when marketed r slaughter at around 9-12 months of age. In the present scenario e demand for meat in India has increased rapidly and there is ortage of meat for domestic consumption besides demand for export. he productivity of these two small ruminant species can be further creased using modern tools in the areas of breeding, feeding and production.

One of the major limitations of our small ruminant production stem is poor nutrition, which results in low carcass yield. Feeding is e of the critical components in small ruminant production system d accounts to 65% of the total cost of production. There is a need for increasing awareness among the scientific and developmental agencies for boosting commercial meat production and value addition through integration of technologies. It is in pursuance of this objective that the Central Sheep and Wool Research Institute, Avikanagar is organizing a Winter School on *"Small Ruminant Feeding System for Commercial Mutton Production and Value Addition to Augment Income Generation"* from November 20 to December 10, 2007. I appreciate the efforts of the Editors, Dr. S.A. Karim, Dr. M.K. Tripathi, Dr. O.H. Chaturvedi and Dr. A.K. Shinde on this occasion in compiling and publishing of invited lectures on the subject contributed by national experts engaged in research, teaching and development in small ruminant production system in the form of a book entitled *"Small Ruminant Production in India: Strategies for Enhancing"*. I hope that this book will serve as a valuable repository of information on the focal theme of the Winter School for the benefit of researchers and teachers engaged in small ruminant production.

November 27, 2007

(T.J. Rasool)

Preface

Sheep and goat production play an important role in livelihood security and economic sustenance of large number of small, marginal and landless laborers in harsh climatic zones of the country where agriculture farming is always a gamble due to uncertainty of rainfall and topography. Small ruminants depend on grazing resources, which include wasteland, fallow and, community land and stubbles. These resources ensure better nutrition to sheep and goats during monsoon season but with the progress of season, scarcity of feed resources compel them to compromise the production. Such condition in desertic environment compels them to migrate to neighboring states for meeting their forage requirements.

Meat is an important produce contributing 70 % to earning of sheep and goats farmers amounting to 20 % of the total red meat production of the country. At present per capita availability of meat is about 4.6 kg as against the recommended intake of 11 kg. There is need to increase meat production in the country to ensure intake of 25 g/day animal protein from present intake of 9 g. In the recent past, incidence of diseases in human beings from consumption of infected meat has raised issues of food safety of consumers. International and National standards have been strengthened for sanitary and phyto-sanitary measures for hygienic meat production. Country has made rapid strides in export market of meat. To further increase export and compete in international trade strict rule of food safety and cold chain is required to be strengthened in the country.

In the country in addition to growing sheep and goats, large numbers of spent animals are slaughtered to meet the demands. The meat of such animals is poorly accepted by the consumers owing to toughness of muscle fiber. The processing of meat and by-products at present stage is not geared up commercially for wider range of marketable value added products. There is ample scope of value addition of meat and by-products to ensure better price to producers and safe and quality produce to consumers.

The Institute has been working on enhancing mutton production from sheep and goats by integrated approaches of feeding, breeding and health management since last 30 years and made strides in achieving targeted slaughter weights of 30-33kg at 6 month of age in native breeds with dressed meat yield of 15-16kg, lean 60-65%, bone 20-22% and fat 16-18%. Similarly large number of processed products utilizing meat and by-products has been developed for improving keeping quality, value addition and consumer acceptability.

The Institute is organizing winter school on "*Small ruminant feeding system for commercial mutton production and value addition to augment income generation*" to extend knowledge and information to participants for popularization of mutton production technology and meat processing to meet ever increasing demand of mutton in the country. On this occasion a book on "*Small Ruminant Production in India Strategies for Enhancing*" of lecture notes is being published. We hope present book would be useful to all those who are involved with small ruminant production and meat and by-product processing. We express our sincere thanks to all chapter contributors for their timely submission of lecture notes and participation in the winter school for delivering the lectures.

Editors

Contents

1

Perspectives of Small Ruminant Production for Augmenting Rural Prosperity

S.A. KARIM

Central Sheep & Wool Research Institute, Avikanagar (via-Jaipur), Rajasthan - 304 50

Introduction

The livestock sector is in a massive transformational state to meet increased demands of animal foods for increasing human population. Progressive urbanization and income growth has further increased demand of food from animal origin. The animal sector's challenge is to fulfill requirement of food from animals without increasing animal population as it is realized that the land is already overstocked with the existing livestock population. Sheep and goat rearing is traditional occupation of small and marginal farmers and land less labors of the country. These two species are predominantly reared under extensive range management on community rangeland employing self or family labor under low input and sustainable output system of production management. Management of small ruminants does not require specialized skill hence the surplus family labors are gainfully employed for grazing management and upkeep of the animals. Although cost of feed input alone accounts for more than 70 % in organized livestock rearing, the expenditure of the sheep and goat farmer on this account is negligible. Moreover the rangeland feed resources of semiarid and arid region of the country wither off during summer months compelling the farmers of western and sub Himalayan region to migrate with their stock to neighboring states and high altitude alpine ranges, respectively seeking greener pasture. The farmers adopt traditional selection criteria in identifying the breeding rams and apply the health technology, specifically drenching with anthelmintics to some extent, still the application of the technology requires considerable improvement. However, organized programs of breeding, feeding management, husbandry practices, health coverage and marketing structure for the live animals and their produce are ineffective. Although the individual production traits of Indian sheep and goats are poor still they have sizable contribution in form of meat, wool, milk, skin and manure for sustenance of rural poor and GDP mainly due to higher population figures.

Sheep and goat rearing is less capital intensive and an appropriate technology for Indian economy where the capital is always a limiting factor for any type of entrepreneurship. The sheep and goat farmers belong to weaker segment of society with little potential for providing inputs. Hence, the *in vogue* production system with little or marginal input and sustainable output is the appropriate type of technology for them. The management parlance of high input and high output (HIPHOP) and low input and low output (LIPLOP) has little relevance for small ruminant rearing under Indian context. In addition the trade provides round the year employment to rural poor and his dependant family members. The existing production system does not require specialized skills, expensive infrastructure, organized breeding plan and feeding management as the farmers adopt age-old traditional skills of sheep rearing passed over the generations. Additionally, the small ruminants provide employment to rural artisans who depend on the produce from these animals to manufacture specialty products. The animal produce viz. meat, wool, hide, offal and slaughterhouse by products serve as raw material for organized industry.

Population and Production Trends

The country possesses 23.5 % of world sheep and goat population ranking first in goat and fifth in sheep population. The sheep and goat population during the Quinquennial period extending from 1951 to 2002 is presented in Table 1. Considering that about 32 and 37 % of sheep and goats are slaughtered annually, the production potential of mutton/chevon in the country is presented in Table 2. Together the two species contributed 633 million kg mutton and chevon to the GDP. The slaughter rate of Indian goats is reported to be much lower than even our neighboring countries like China (49.4 %) and Pakistan (56.6 %) still these countries have maintained higher population growth of sheep and goats at 7.8 and 3.6 % (Kondaiah and Agnihotri 1995) whereas the growth rate of the two species in India during last half century is hardly 0.7 and 5.6 %, respectively (Table 1). One of the major contributing factors to lower slaughter rate of Indian sheep and goats could be comparatively lower reproductive efficiency (65- 70 %) of small ruminants under field conditions while for success full mutton production program higher prolificacy of the animals with twining and triplication is an essential trait. Additionally, the real assessment of slaughter rate is difficult in India as majority of animals are slaughtered under unorganized sector without any record. Further, the average body weight of Indian sheep and goats is considered to be 22- 25 kg and the carcass weight 10 kg as against the production level of 25 kg carcass weight in developed countries. Under farmers *in vogue* feeding management without organized inputs, the lambs and kids in active phase of growth hardly achieve 40- 50 g average daily gain (ADG) and marketed for slaughter around 9- 12 months of age weighing about 20- 22 kg. Three decades of concentrated research under intensive feeding program at Central Sheep and Wool Research Institute improved the growth rate of native sheep by 300 % with 18- 20 % feed conversion efficiency (Karim, 2000). The developed technology has been demonstrated successfully on farmers' field while large-scale adoption of the technology is not effective due to poor economic background of the sheep farmers.

Table 1. Population trend of sheep and goats in India (in millions)*.

Year	Sheep	Goats
1951	38.96	47.14
1956	39.28	54.42
1961	40.22	60.84
1966	42.01	64.48
1971	39.99	67.50
1977	40.91	75.62
1982	48.76	95.26
1987	44.84	110.21
1992	46.00	115.30
1997	47.61	121.21
2002	49.28	128.48
2006	62.58	124.0

Table 2. Mutton/chevon production potential of India (Million kg).

Year	Sheep	Goats
1951	124.7	174.4
1956	125.7	201.3
1961	128.7	225.1
1966	134.4	238.6
1971	128.0	249.7
1977	130.9	279.8
1982	132.0	352.5
1987	162.0	407.8
1992	167.0	426.6
1997	222.0	448.5
2002	233.0	475.4
2005	238.8	475.0

The sheep and goat rearing regions are deficient in energy and protein supplements while in feed supplement rich regions small ruminants are not traditionally reared. In order to bridge the gap in demand and supply of meat in the country, it is suggested that the weaner lambs from dominant small ruminant rearing region can be procured by entrepreneurs and reared/fattened near some cities or exit port by procuring feed supplements from agriculturally developed regions and market the finisher lambs profitably (Karim *et al.* 1998). Although there is considerable margin of profit in organized meat industry still the business houses and entrepreneurs avoid investing in the industry due to social taboos. The weaker segment of society involved in the trade neither have resources of their own to establish industry nor are financially solvent to furnish collateral to the financial institution for drawing loans.

Economic Contribution of Mutton/Chevon to GDP

It was reported that the contribution of sheep and goat meat to total meat produced in the country was 5.4 and 13.9 %, respectively in the year 2000 AD (Bhattacharya *et al.* 1992). The contribution of meat from small ruminants has decreased over the years (Sheep 14 to 5 %: goat 30 to 13 %) while the contribution of buffalo meat has increased during the same period to 30 % (Kondaiah and Agnihotri 1995). The trends in mutton production clearly indicate that the mutton and chevon production could not meet the demand hence as an alternate source, the use of buffalo meat has increased.

Sheep husbandry is an important enterprise from temperate to tropical ecology of India characterized by sparse vegetation, marginal land and a high incidence of poverty. Sheep are reared mainly by the poorest people in the lower strata of society and serve either as the main or supplementary source of income in this category. Perhaps it is the most appropriate species of livestock for utilization of available sparse vegetation and owing to their multi facet utility (wool, meat, skin, milk and manure) constitute an important part of rural economy. Usually sheep are raised as an extra investment without a major labor input, based on road side and fence vegetation, which provides a sizable portion of the energy consumed by them. This species is a low-investment sustainable enterprise yielding reasonably high rates of return and continuously fulfilling the demand of huge human population in terms of valuable products relatively at cheaper cost.

The mutton and chevon production in the country was 152.3 and 450.4 million kg in the year 1997 with projected to increase to 157.7 and 475.4 million kg by the year 2002 (Table 2). Considering minimum selling price of Rs.120/kg both for mutton and chevon, the sheep and goats contributed Rs. 15235 and 45036 million, respectively totaling to a sum of Rs. 15685 million to GDP. Additionally, the slaughterhouse byproducts viz. edible offal, head and hooves, skin and viscera fetching an average price of Rs. 200/animal contributed Rs. 120546 to GDP. Hence the total contribution of sheep and goat meat and byproducts to GDP amounted to Rs. 136231 million. It is generally accepted that the return from edible offal, head and hooves, skin and by products is the net profit of the butcher while the return from the dressed carcass is the purchase price of the animals which in turn is shared by the primary producer and the middle man.

Marketing Structure

Marketing of live animal is unorganized while wool marketing is partly organized with a chain of intermediaries and commission agents linking the primary producer to industrial houses/end users. Leaving aside the metropolitan cities where organized slaughter housed exist in Government sector, the slaughter of sheep and goats are mostly under taken by the local butcher as back yard enterprise. The meat from the dressed animals is sold to the consumers on per kg basis while the concept higher price for choice cuts, as prevalent in developed counties, is not applicable under our conditions. The edible offal viz. liver, kidney, depot fat and pancreas are included in the dressed meat and sold on per kg weight basis while viscera, lungs, head and hooves and skin are sold as unit.

Under the prevailing marketing structure, the disposal of live animals involve many agencies, middle man and commission agents hence it is difficult to assess their exact number and role in operational chain (Bhasin, 1988; Krishna, 1990). Earlier studies also revealed that organized marketing structure is lacking in the country for small ruminant produce and no government, semi-government or cooperative society are involved in live animal trade (Bhattacharya and Khan, 1988). Traditionally, the middle man involved in live animal trade loan out required money to the farmers for meeting their social obligation and family emergencies. While giving the loan the trader generally accept verbal commitment of the farmer that he will sell off his animals to the trader, at a later date, in the price fixed on that date. In such transaction benefits accrued by occasional upwards trend in market price is pocketed by the middleman while the slump in price is directly passed on to the farmers by lowering purchase price of the animals. Butchers involved in the trade are also largely benefited by the slump in market price of the live animals as the market price of meat does not follow the downward trends in consonance with the fall in live animal cost (Karim, 1996).

Sizeable number of sheep and goats are marketed in weekly *hats*/markets owned by private trusts, local bodies or even individuals. These markets lack in the facilities of shelter, fodder and even sometimes the basic need of water. The animals from the region are assembled in these weekly markets and purchased by local traders or representative of business houses from cities. In most of these markets the trading is based on body score examination of the animals and its probable meat yield. Although both the seller and buyer involved in the trade are generally illiterate, the final deal is based on quantity and quality of the probable meat yield. The major limitation of such type of trading are periodic nature of the markets, large numbers of agents involved, lack of infrastructure for holding the animals and above all absence of pricing based on unit weight. The animals collected from the villages are transported to city markets in Lorries. The major operator in the city retains the best animals for export while other animals are sold to local butchers for end utilization.

The spent sheep and goats are sold by the farmer to the traders/butchers at about 50 % of the prices fetched by the finisher lambs/kids. The traders usually mix the spent ewes/does and the finishers lambs/kids and offer them as a lot for disposal in the cities. The butchers after slaughtering the animals however sell the meat at a uniform price irrespective of whether it comes from finisher males or spent ewes/rams. Thus in such type of trading both the small ruminant producer and the consumer of meat are cheated while the middleman and the butchers are benefited. It is therefore felt that the meat from spent animals after tenderization and processing can be converted to value added ready to eat finished products to avoid such malpractice and cheating: the required infrastructure and facilities has to be created. Moreover the farmer can harvest better price for his finisher lambs if sold as separate entity. The awareness on this aspect has to be created by the development agencies.

In summary it is felt that an efficient marketing system with back up facilities for transport, storage/holding, processing, market intelligence and floor price support is a must to ensure reasonable return to the farmers and appropriate price of the commodity for the consumers (Seetharaman, 2002). The existing marketing structure for animal produce, specifically for meat and meat products, should be upgraded targeting the private sector to invest in the venture and harvest profitable returns.

Export Potential of Live Animals and Dressed Carcass

India has the advantage of close proximity to Middle Eastern market. The promising market for Indian produce is approachable by 1 and 3 hrs flight time from Mumbai and New Delhi, respectively. Further by sea route, the market can be reached in 8 hrs from West coast of the country (Kandla exist port). Presently the Middle Eastern countries are meeting their domestic requirement of mutton and chevon by importing live animals from developed countries like Australia/ New Zealand and Africa/West Asia, respectively. These countries are distantly located requiring 20 to 30 days transit time by sea route to reach the destination. The developed countries mostly land their consignment at the entry port of Ras al Khaima from where the animals are distributed throughout the Middle East by specially designed Lorries. The native population prefers fresh and quality mutton from sheep/goat slaughtered by local butchers while the migrant population from Indian sub continent usually subsists on dressed meat imported under ice.

India's share in world meat export is negligible i.e. hardly 1 % while the commodity has sizable export contribution from developed countries viz. USA

(13 %), Netherlands (12.4 %), France (10.0 %) and Australia (8.4 %). At the present sheep and goat meat export is hardly 2 % of total production in the country whereas the country has substantially higher export potential. Export of sheep and goat meat has increased by 18.1 % during the period 1993- 94 to 1995- 96 while by 2000- 2001 India exported 25 % more meat than the previous year. During last decade the meat export is one of the fastest growing sectors as assessed by APEDA. The major meat export houses in the country are Al Kabeer, Allana Sons, Arabian Exports, Hind-Agro Foods, MK Overseas and Fair export. Lately Egypt has emerged as one major importers of meat from India accounting for more 50 % of total meat exported from the country. The others major buyers of Indian meat are Malaysia, Philippines and Saudi Arabia: the monthly meat exports to these countries are 14000, 4000, 3500 and 3000 tones, respectively. The export potential will further increase in near future with establishment of export oriented abattoirs meeting the international specification and eradication of contagious diseases like Rinderpest, Blue Tongue etc. The advantage of India in export market is its huge sheep population, comparatively leaner carcass and close proximity to promising Middle East market. The limitation of hygienic slaughterhouses meeting international specifications and the organized campaign by developed countries as regards to poor quality of Indian meat has to be tackled in appropriate forum.

Prospects of Meat Based Convenience Products

The meat from 6- 12 months old male lambs/kids, surplus to the breeding requirement are preferred by the consumers while 30- 40 % slaughtered animals are spent ewes/does contributing tough and fibrous meat. Since tenderness is on of the sought after trait of the meat, development of technology for improving texture of such meat assumes greater importance. Therefore to harvest maximum return from such meat it is essential that the meat from spent animals be converted to ready to eat products after tenderization. Presently about 3 % (0.1 m MT) of meat produced in the country is processed into ready to eat meat products as against 80 % in Finland and 50 % in Russia. There are about 200 licensed meat processing units in the country producing 50 % cured products, 20 % sausages, 20 % canned products, and 10 % products of miscellaneous nature. It is projected that by the year 2010 about 2 m MT processed meat will be available for domestic and export market.

Problems and Opportunities for Export of Meat and Meat Products

Agriculture in India as in any other developing country is a crucial sector and key to growth of core sectors like industry, transport and other nonagricultural activities. Moreover, the livestock sector contributes nearly 25 % to total Ag GDP: in that meat from small ruminants is one of the major produce. No longer is agriculture production and marketing shaped by intuitive decision of the farmers based on age old practices and experience but increasingly becoming a complex policy decision governed by agro-economic, environmental and global considerations. However, till date the small ruminant produce from poor sheep and goat farmer is not effectively covered by such consideration rather governed by traditional marketing structure.

After remaining a protected economy for long, the country has opted for reforms in economic and financial sectors during last decade, ushering in a free market economy. India was among the 23 founder members of General Agreement on Tariff and Trade (GATT) established in 1948 and has participated in protracted Uruguay Round Table discussion leading to formation of World Trade Organization (WTO) in 1995 with 168 member nations. Consequent to globalization, the unification of markets is fast taking shape and forces of global competition are already shaping a single market concept. With the liberalization, the effect of WTO regime is being felt by all sectors of the economy, both in import and export scenario. Presently the demand and supply of meat and meat products in the country is balanced because a sizable population is either facultative meat eaters or not adapted to meat eating. Moreover the purchasing power of Indian consumers is relatively poor. However it is envisaged that with improvement in economic status and awareness for healthy and balanced food, the demand for meat and meat products in the domestic sector will increase.

Conclusion and Recommendations

Sheep and goats are poor men's livestock mostly reared by the weaker section of the society in the regions of country inhospitable for conventional crop production. The economic contribution of the two species constitutes about 15 % of total livestock production while their share in terms of developmental inputs are marginal. The marketing structure of sheep and goat produce are not organized rather exploitive as a result, the middleman and traders pocket major share of return without involving themselves in animal rearing. Past experience of state concern in establishing cooperative societies/ marketing federations to facilitate the farmers in harvesting real value of his produce died their natural death due to rampant corruption. However, considering the high margin of profit in the venture, of late, some of organized business houses have shown interest in live animal/ meat trade . With the liberalization, the effect of WTO regime is being felt by all sectors of the economy, both in import and export scenario. Presently the demand and supply of meat and meat products in the country is balanced because a sizable population is either facultative meat eaters or not adapted to meat eating. Moreover the purchasing power of Indian consumers relatively poor. However it is envisaged that with improvement in economic status and awareness for healthy and balanced food, the demand for meat and meat products in the domestic sector will increase. The trade still has future because of increasing demand of meat both for domestic consumption and export market.

Table 3. Annual Export details of animal products

Item	1993-94		1994-95		1995-96		1996-97		1997-98		1998-99		1999-00		2000-01		2001-02	
	Qty	Value	Qty	Value	Qty	Value	Qty	Value	Qty	Value	Qty	Value	Qty	Value	Qty	Value	Qty	Value
Animal Products																		
Buffalo Meat	101665.58	280.75	116137.57	324.37	159703.91	555.00	157574.00	619.04	176328.68	729.30	153956.15	691.29	167291.41	706.42	288027.84	1375.04	243355.58	1144.42
Sheep / Goat Meat	11022.16	62.16	10819.35	66.03	8612.80	56.42	8691.19	71.04	7546.60	62.66	8648.45	78.48	12419.44	90.89	11902.85	78.16	3915.06	33.07
Poultry Products	3002.97	11.08	27665.24	8.87	10067.42	26.21	16682.91	81.44	11040.46	88.84	7390.89	51.73	8367.93	54.25	15836.12	86.18	19876.02	130.07
Dairy Products	2105.91	12.47	8957.06	40.11	4712.08	32.56	2018.49	15.36	2378.24	13.39	2568.42	13.65	6134.42	37.21	11068.86	83.90	24774.13	182.45
Animal Casings	306.57	6.56	334.97	7.71	325.14	0.08	446.38	12.27	371.22	11.95	382.06	13.59	352.65	11.69	573.70	12.29	464.28	9.63
Processed Meat	182.41	0.97	192.92	1.27	476.64	4.07	511.99	4.44	264.77	2.22	297.96	2.99	815.77	4.58	130.69	1.58	267.13	1.29
Total for Animal Products	**118285.60**	**373.99**	**164107.11**	**448.36**	**183897.99**	**674.34**	**185924.96**	**803.59**	**197929.97**	**908.36**	**173243.93**	**851.73**	**195381.62**	**905.08**	**327540.06**	**1637.16**	**292652.20**	**1500.93**

Source: www.apeda.com

Table 4. Export of Sheep and goat meat to different countries

Country	1999-00		2000-01		2001-02	
	Quantity (Kg)	Value (Rs.)	Quantity (Kg)	Value (Rs.)	Quantity (Kg)	Value (Rs.)
Lamb carcass (fresh)						
Bahrain	-	-	1950	186438	27190	1398572
China	-	-	25000	541140	-	-
Spain	-	-	-	-	1500	182125
Germany	20000	273905	-	-	-	-
U.K.	2700	250188	-	-	-	-
French Guiana	-	-	15653	1378602	-	-
Jordan	25000	928774	-	-	-	-
Kuwait	25650	964277	42629	2403546	-	-
Lebanon	800	74712	-	-	-	-
Sri Lanka	6900	370339	-	-	-	-
Malaysia	25000	1282741	-	-	-	-
Oman	57155	3151087	185461	12928162	52468	3838992
Qatar	3500	291168	45950	2602738	2500	189507
Saudi Arabia	57437	4072119	94034	6987181	-	-
U.A.E.	939421	71027550	820484	67341802	126834	10724056
Ukraine	4650	412764	-	-	-	-
U.S.A.	4750	285261	700	70190	-	-
Product total	1172963	85850033	1231861	94439799	210492	16333252
Sheep Carcasses (Fresh)						
Bahrain	4600	340566	4750	447399	4350	438750
Bahamas	1000	76632	-	-	-	-
China	-	-	25997	707912	-	-
Germany	5000	263376	-	-	-	-
Jordan	-	-	-	-	7100	697924
Japan	-	-	-	-	1050	106099
Kuwait	-	-	46000	1842245	-	-
Sri Lanka	18300	946358	148519	6253392	34851	1545250
Maldives	1580	73704	-	-	-	-
Oman	176115	10657367	202389	13324555	163082	15290651
Qatar	21830	1598863	57610	4093259	80545	6456304
Saudi Arabia	430582	31752408	205950	16160772	4975	547322
Sudan	1200	145470	-	-	-	-
U.A.E.	1121887	76226765	1352400	106515854	1852526	159815695
U.S.A.	2320	92200	1070	101481	1700	134356
Product total	1784414	122173709	2044685	149446869	2150179	185032351

Contd....

Table 4. Contd....

Sheep meat with bone (Fresh)						
Bahrain	1500	118344	-	-	1000	79311
Brazil	17160	1572305	-	-	-	-
Cameroon	-	-	-	-	1000	102735
Hong Kong	-	-	1000	75235	-	-
Jordan	-	-	-	-	2512	320427
Japan	-	-	-	-	1000	120207
Kuwait	51989	2909364	-	-	500	54831
Maldives	24	1521	40	4738	44	5362
Oman	37810	2982919	56380	4786160	22213	2167513
Qatar	4910	305981	9800	999962	10920	1103457
Saudi Arabia	351638	11613712	80514	8168341	4280	469460
U.A.E.	697243	44833768	310920	29902559	244744	23000781
U.S.A.	-	-	1554	150831	-	-
Product total	1162274	64337914	460208	44087826	288213	27424084
Boneless sheep meat (Fresh)						
Argentina	1300	123710	-	-	-	-
Bahrain	17000	1747895	4050	382109	-	-
Bahamas	850	91648	-	-	-	-
Brazil	25992	2591532	-	-	-	-
Barbados	-	-	6550	622252	-	-
Canada	-	-	-	-	35000	2554731
France	9200	599590	11085	1046034	4000	458155
U.K.	-	-	1000	58359	-	-
Kuwait	1302	139961	52352	4546127	-	-
Sri Lanka	-	-	-	-	17001	725400
Maldives	-	-	24	1910	44	6044
Mali	40	5485	-	-	-	-
Nepal	-	-	1950	89142	-	-
Oman	240197	17304518	119307	9228008	38914	3487358
Qatar	36082	3981169	28279	2301129	19520	1747851
Saudi Arabia	300841	29413305	203727	19100419	1695	181175
Singapore	-	-	330	39400	-	-
U.A.E.	1593050	129373472	782806	71334057	302143	26866080
U.S.A.	26102	2495063	1400	135745	-	-
Yemen Arab Republic	-	-	9760	917338	-	-
Product total	2251956	187867348	1222620	109802029	418317	36026794

Contd....

Table 4. Contd....

Lamb carcass (Frozen)						
Bahrain	2525	167556	-	-	-	-
Kuwait	24010	1320860	-	-	-	-
Maldives	153	11054	-	-	-	-
Oman	2700	166974	50207	2719415	-	-
Qatar	3540	262636	7200	523109	-	-
Saudi Arabia	2700	259705	141822	12509419	-	-
Switzerland	-	-	-	-	2000	173463
U.A.E.	168128	8466987	2179914	10246263	42000	2986243
Product total	203756	10655772	2379143	25998206	44000	3159706
Sheep Carcasses (Frozen)						
France	-	-	2500	245215	-	-
Japan	8000	786452	2000	117102	-	-
Sri Lanka	-	-	16801	663375	-	-
Oman	145966	9059201	99315	6137244	16000	1456926
Saudi Arabia	44194	3890894	50004	4671137	-	-
U.A.E.	290366	20151525	452358	31393997	267242	20319521
Product total	488526	33888072	622978	43228070	283242	21776447
Sheep meat with bone (Frozen)						
Bahrain	77950	10159637	47097	4869834	38669	3472877
Kuwait	-	-	-	-	24949	1088125
Sri Lanka	10080	554880	-	-	17850	723075
Maldives	300	24471	4422	457736	3133	384323
Oman	87207	6496659	154275	10959902	68916	5302224
Saudi Arabia	4000007	275953771	2006110	161769457	-	-
U.A.E.	634325	58012325	542603	52172443	198347	17894128
U.S.A.	25000	2281307	-	-	3800	276573
Product total	4834869	353483050	2754507	230229372	355664	29141325
Boneless sheep meat (Frozen)						
Bahrain	240	21672	-	-	-	-
France	-	-	-	-	4000	467403
U.K.	1500	133212	-	-	-	-
Jordan	-	-	-	-	2000	174855
Kuwait	-	-	53200	3619385	-	-
Lebanon	-	-	25999	1244731	-	-
Sri Lanka	-	-	10000	856950	-	-
Maldives	-	-	50	7516	-	-
Malaysia	268	22334	-	-	-	-

Contd....

Table 4. Contd....

Oman	47983	3579593	26664	2198548	16992	1497186
Philippines	-	-	100000	4508111	-	-
Saudi Arabia	288614	31706965	185690	18722663	2000	168622
Singapore	500	52514	-	-	-	-
Slovenia	-	-	-	-	18000	740125
U.A.E.	99565	8510166	364513	17140115	69140	6045412
Yemen Arab Republic	-	-	-	-	26584	916943
Product total	438670	44026456	766116	48298019	138716	10010546
Goat Meat (Chevon)						
Bahrain	210	18940	-	-	-	-
Japan	-	-	23	2000	-	-
Sri Lanka	16822	514423	-	-	-	-
Maldives	-	-	73	8455	228	31610
Oman	14973	1149196	59017	3908001	-	-
Saudi Arabia	50002	5019788	242300	23281173	-	-
U.A.E.	-	-	119315	8888103	26008	1718094
Product total	82007	6702347	420728	36087732	26236	1749704

Source: DGCIS, Annual Volume

Table 5. Production potential of sheep and goats under extensive range management

Traits	Sheep	Goats	Dressing yield (%)
Birth weight (kg)	3.0	2.9	-
Weaning weight (kg)	12	11	40- 45
Six months weight (kg)	18	15	45- 48
Nine months weight (kg)	20	18	45- 48
12 months weight (kg)	22	20	45- 48
Spent ewes/does (kg)	27	24	40
Spent breeding rams/bucks (kg)	35	30	45- 48

Table 6. Production potential of sheep and goats on protected rangeland

Traits	Sheep	Goats	Dressing yield (%)
Birth weight (kg)	3.5	3.0	-
Weaning weight (kg)	13.0	11.5	40- 45
Six months weight (kg)	21.0	16.2	43- 45
Nine months weight (kg)	25.0		43- 45
12 months weight (kg)	30.0	21.5	43- 45
Spent ewes/does (kg)	30	30.0	40
Spent breeding rams/bucks (kg)	35.0	35.0	45- 48

Table 7. Production potential of lambs/kids on cultivated pasture

Traits	Lamb	Kid	Dressing yield (%)
Birth weight (kg)	3.2	3.2	
Weaning weight (kg)	13.9	15.4	
Six months weight (kg)	20.6	18.0	43- 45
Nine months weight (kg)	24.0	20.0	43- 45
12 months weight (kg)	27.0	24.0	43- 45

Table 8. Production potential of lambs/kids on silvipastures

Traits	Lamb	Kid
Weaning weight (kg)	15.2	12.8
Six months weight (kg)	20.8	21.9
Nine months weight (kg)	24.0	-
12 months weight (kg)	27.0	-

Table 9. Production potential of lambs under grazing with supplementation

Traits	1.5 % of BW	2.5 % BW
Weaning weight (kg)	11.9	11.9
Six months weight (kg)	22.6	27.2
ADG (g)	118	170
FCR	5.5	4.7

Table 10. Production potential of lambs under intensive feeding

Traits	R:C 50:50	
	Native	CB
Weaning weight (kg)	17.2	19.8
Six months weight (kg)	30.8	35.7
ADG (g)	151	178
FCR	9.1	8.3

Table 11. Export trends of mutton/chevon

Year	Sheep/goat meat export (000 MT)	% of production	% of total meat export	Value of meat export (Rs. In million)
1980- 81	2.195	0.54	4.97	59.0
1984- 85	13.384	2.68	27.90	268.9
1990- 91	8.327	1.41	12.20	312.9
1995- 96	8.613	1.13	4.82	564.3
1996- 97	8.691	1.27	4.74	710.3
1997- 98	7.547	1.11	3.87	626.6

Table 12. Sheep population in different states of India (In 000)

States	1972	1977	1982	1987	1992	1997	2003
Andhra Pradesh	8343	7064	7519	6872	7787	9743	21376
Assam	0051	0059	0046	0067	0150	0084	0170
Bihar	0983	1121	1322	1557	2027	1956	0382
Gujrat	1722	1592	2357	1557	2027	2158	2062
Haryana	0459	0541	0758	0890	1044	1275	0633
Himachal Pradesh	1040	1055	1090	1112	1079	1080	0926
Jammu & Kashmir	1073	1210	1909	2493	2947	3170	3411
Karnataka	4662	4536	4792	4727	5432	8003	7256
Kerala	0010	0003	0007	0030	0029	0003	0004
Madhya Pradesh	1009	0968	0959	0835	0836	0657	0546
Maharastra	2128	2636	2671	2872	3074	3368	3094
Manipur	0002	0002	0014	0016	0014	0008	0006
Meghlaya	0018	0020	0026	0015	0023	0017	0018
Orissa	1369	1432	1990	1840	1840	1765	1620
Punjab	0388	0493	0611	0508	0526	0436	0220
Rajasthan	8556	9938	13431	9933	12497	14585	10054
Sikkim	-	0016	0011	0011	0016	0005	0006
Tamil Nadu	5393	5289	5537	5881	5849	5259	5593
Tripura	0002	0003	0005	0003	0005	0006	0003
Uttar Pradesh	1956	2059	2307	2181	2404	1905	1437
West Bengal	0793	0793	1365	1460	1488	1462	1525
Total	**39957**	**40830**	**48765**	**44851**	**50799**	**57494**	**61469**

Table 13. Biomass yield, carrying capacity and production performance of sheep on different types of pastures

Pastures	Biomass Yield (q DM/ha)	Carrying capacity (sheep/ha)	Growth rate (g/day)	Carcass yield (%)	Annual fleece (g /head)
Degraded range land	4.0-4.5	1	35-45	35-37	800-900
Protected rangeland	8.0-10.0	2	50-60	40	950-1050
Cenchrus/Sewan pastures	16-20	4	70-80	42.5	1200-1300
Grass-legume pastures	22-25	6	90-100	45	1500
Two tier silvipastures	22-25	6	90-100	45	1500
Three tier silvipastures	30-35	7-8	100-120	47.5	1500-1600
Cenchrus/ Sewan pasture + Concentrate Supplementation	Grazing 25%of adlib 50%of *adlib* 75%of *adlib* *Ad lib*		66 90 100 110 125	40 43 48 49 51	620 800 900 1200 1250

Table 14. Economic ration formulations and mutton production

Ration	Preparation	Target wt kg	Days	ADG	FCE	Dressing %
Tree leaves/ legume hay + conc.	50:50	25	130	162	16	49-50
Tree leaves/ legume hay + conc.	40:60	25	130	179	18	50-51
Tree leaves/ legume hay + conc.	30:70	30	130	204	18	50-51
Supplementary feeding	1.0% 1.5% 2.0% 2.5%	19 22 23 25	90 90 90 90	83 108 123 149	15 17 18 19	48 48 50 50
Free grazing with *ad libitum* concentrate supplementation	-	33	90	170	20	50- 51

REFERENCES

Bhasin, N.R. 1988. Marketing of goat meat. Proceeding: IDRC Workshop on goat meat production in Asia (Ed. C. Devendra), Tendojan, Pakistan, pp. 212- 219.

Bhattacharya, N.K. and Khan, B.U. 1988. Goat meat production in India. Proceeding of Workshop on Goat meat production in Asia, Tendojan, Pakistan, pp. 125- 134.

Bhattacharya, N.K., Agnihotri, M.K. and Pal, U K. 2000. Estimated production of meat by 200 AD in India and its impact on defense logistics. Proceedings of National Seminar on "Latest trends in equine breeding/husbandry practices with critical approach to changed role of RVC in modern scenario. 28- 29 November, New Delhi, pp. 42- 49.

Karim, S.A. 1996. Market trends in live animal transaction. Proceedings, Advances in sheep production, CSWRI, Avikanagar, pp. 186- 195.

Karim, S.A., Shinde, A.K. and Sankhyan, S.K. 1998. Feeding and grazing management of sheep under different production system. Golden Jubilee Seminar on Sheep, goat and rabbit production and utilization. Jaipur, Rajasthan pp. 102- 107.

Karim, S.A. 2000. Feeding small ruminants for meat and wool production. Lead paper: National Seminar in "Animal Nutrition in Health and Disease. Department of Animal Nutrition, Veterinary College, University of Agricultural Science, Bangalore, 17- 19 February.

Kondaiah, N. and Agnihotri, M.K. 1995. Strategies for processing and marketing of meat from small ruminants. Lead paper: 3rd National Seminar on sheep and goat production and utilization, Central Sheep and Wool Research Institute, Avikanagar, 8- 10 April, pp. 50- 71.

Krishna, A.Y. 1990. 1990. Small ruminant production in India by the year 2000: Strategies for development (Ed. G. V. Raghavan, M. R. Reddy and N. Krishna), Tirupatti, 13- 17 November, pp. 226- 227.

Rawat, P.S. and Riyazuddin. 1997. A report on socio-economic survey of households of village Bheepur. Institute publication, CSWRI, Avikanagar.

Riyazuddin, Karim, S.A. and Rawat, P.S. 1998. Role of small ruminants in rural economy. Lead paper: Golden Jubilee Seminar on sheep, goat and rabbit production and utilization, Jaipur, Rajasthan, 24- 26 April, pp. 139- 143.

Seetharaman, M.K. 2002. Policy prospective in Indian agriculture. Seminar on "Leveraging the Indian agribusiness- emerging roles for corporates and other stake holders", 26- 27 July, New Delhi.

❑❑❑

2

Goat Rearing for Chevon Production: National Perspective

N.P. SINGH

Central Institute for Research on Goats, Makhdoom, Mathura- 281 122

Introduction

World's human population is around 5910 million. India possesses about 17.50 % of the World human population. Two-third of the world's poor live in Asia below nationally defined poverty lines and 65 % of them are poor livestock keepers who derive a large part of their household from domesticated animals. Underprivileged families account for about one-fourth of the population and contribute 70-80 of the total livestock produce. The rapidly changing patterns of demand for livestock and livestock products point to livestock production being an increasing component of the agricultural economy in India. The extent to which the rural poor will benefit from these changes depends on how livestock can be integrated into developing markets and whether cheaper livestock products benefit the rural poor as consumers as well as producers. There is scope for small ruminants to play an important role for smallholder farmers in accessing these new markets. Their significance, which is now being exploited in several countries, is that they are small livestock in high demand and can thrive on low inputs and local resources. Of the two small ruminants - sheep and goat, the goat is much more important in most parts of the region.

Poor men's cow - the Goat is thought to have been the earliest domesticated ruminant and of all the species of domesticated animals except dog, has the widest ecological range. Originating in Asia, goats have spread over all the continents and inhabit almost all-climatic zones from arctic- circle to the equator. They have distinct social, economical, managerial and biological advantages over other livestock species. They significantly contribute to the agrarian economy and play a very vital role in the livelihood security of the small and marginal farmers and landless labourers especially in arid, semi-arid and mountainous regions of the country. Goats have several advantages over other animals. They are widely adapted. They thrive well and reproduce in tropical, cold, humid as well as dry regions. Their small size compared to cattle and buffaloes permit them to be maintained on a limited area. The breeding animals are inexpensive. Goats consume a wide variety of grasses, weeds, forbs, bushes, shrubs, tree leaves and crop residues that would otherwise go waste and cause pollution. They are gentle and easy to control. Their small size makes them suitable for home slaughter and the meat can be consumed by the family before it spoils in warm climate where no refrigeration facilities are available. The goats can be milked any time of the day and are therefore named as the moving refrigerators. Goat milk is prescribed for children, old and sick as it is easily digestible and has medicinal value. Goat meat (chevon) is preferred over other meats because it is leaner and there are no religious taboos against its consumption.

Present Scenario

World's current population of cattle, buffaloes, sheep and goats is around 1355, 174, 1081 and 808 million respectively. Asian region possesses about 33.61, 96.88, 42.29 and 64.33 % and India 13.65, 56.31, 5.79 and 14.87 % of the total world population of the four respective livestock species (FAO, 2005). The developing countries of the world have about 94.5 % of the goats and the developed ones only the rest 5.5 %. India possesses about 124 million goats making 14.87 % of the world population and stands second to China. Although the population of all livestock species has shown increasing trend since 1951 the goat population has increased at a much faster rate than others in India. The growth rate of goats in India has varied from 0.94 to 5.10 % with an average of 3.05 % during 1951-2005 in spite of about 41 % slaughter and about 15 % natural annual mortality. The goats around the world contributed 12438.4 TMT of milk, 4562.1 TMT of meat and 985.9 TMT of fresh skins per annum. The Asian region contributed 54.08 % of the milk, 75.34 % of the meat and 79.91 % of the fresh skins of the world production of goats. India produced 2700 TMT of milk, 475 TMT of meat, 130 TMT of skins, 8.5 MT of Pashmina and 400 TMT of manure which make 21.71 % of the milk, 10.47 % of the meat and 13.15 % of the fresh skins of the world goat production. The estimated value of different types of goat produces works out to about Rs. 1, 06, 335 million per annum. The goat sector also generates about 5 % rural employment and about 20 million families belonging to small and marginal farmers and landless labourers are engaged in goat keeping.

Per capita meat consumption in India is less than 5 kg per year as against 13.7 kg in Pakistan, 38.6 kg in China and 58.6 kg in Brazil. Estimated annual demand for meat is 7.7 m MT against the availability of 5.6 m MT. Goats contribute 8.5% of the total meat production of the country. Goats contribute about 3.0 % to the total milk produced in the country. India exported 3 lakh live goats, 280 Metric Ton goat meat and 25 Metric Ton goat skins and earned over 4 m US Dollars during 2003. A great export potential exists for live goats, goat meat, goat skins and their products. Around 250 commercial goat farms have been established in different parts of the country. The rapidly changing patterns of demand for livestock and livestock products point to goat production being an important component of the agricultural economies of India. Lack of awareness to adopt improved technologies due to poverty, illiteracy and little or no say in decision making process is the weakness of the goat owners. Non-availability of elite breeding males, effective vaccines, grazing lands and feed resources is the other weaknesses. The goat farmers are exploited by the middle men in marketing of surplus animals for meat markets. Sufficient modern abattoirs for hygienic slaughter are not available and a large number of goats are slaughtered in unauthorized places. There is great scope for rearing goats for meat production under semi-intensive and intensive systems of management. The reproduction rates, multiple births, growth, slaughter weight, dressing percentage, bone: meat ratio and survivability can be significantly improved which in turn will provide higher returns to the goat owners. Scope for value addition in milk, meat, skin and fibre is tremendous. Opportunities for employment generation and foreign exchange earnings are very high. Increase in population, shrinkage in grazing lands, slaughter of elite breeding males, emerging diseases and shortage of trained manpower are also posing threat to development of goat industry in the country. Goat rearing for meat production needs greater attention. Strategy should aim at increasing slaughter weight produced per doe per year through improved breeding, reproduction, nutrition and health cover interventions. Attempts are also required to be made to increase the milk production in different goat breeds. The medicinal values of goat milk need to be evaluated, high lighted and exploited. Characterization of goat genetic resources, enhancing quantity and quality of meat and milk production, improving feed resources, controlling diseases, improving fertility and flock productivity, transfer of improved technologies, human resource development through organization of training programmes and extensive use of biotechnological tools are some of the identified issues for immediate attention.

Socio-Economic Gains of Goats

The socio-economic importance of goats in India is evident by the sharp increase in their numbers and contributions during the last about 30 years. Goats contribute milk, meat, fibre, skins and manure to the subsistence of small holders and landless rural poor. They play an important role in income generation, capital storage, employment generation and house hold nutrition. Importance of goats lies in the fact that human population is increasing very rapidly creating increasing demands for animal protein foods on the one hand and the feed resources for increasing large ruminants are decreasing due to shrinkage of grazing lands on the other. This demand can, therefore, be met with by increasing population of goats. It is easier to increase the population of goats than cattle and buffaloes because the capital investment is relatively low, land requirements per animal are small, reproductive rates are higher both due to shorter breeding interval and high prolificacy and they can be managed by spare family labour and do not require any serious housing facilities and management skills. There is much less risk in goat farming in drought prone areas where large mortality occurs due to frequent droughts. The goats act as an insurance against disaster under pastoral and agriculture subsistence system. Goats have religious and ritualistic importance in India. They are offered as sacrificial animals both by Muslims on Id and by Hindus especially the worshippers of Goddess Kali. They are worshipped for their creative and generative powers and sexual virility. There are no religious taboos against consumption of goat meat. Goat meat is preferred over sheep meat as it is leaner. Goat milk is easily digestible because of small sized fat globules. It has much less allergic problems than the milk of other livestock species. It also has medicinal value and can ward off many diseases as the goats browse on variety of plants including medicinal ones. Goats can be milked as often as required unlike cows and buffaloes and are, therefore, called moving refrigerator. The goatskins are highly valued and have large export potential both in the processed form and as products. The value addition in goat skin processing is much larger than other products. The bones of slaughtered and dead goats are also utilized for bone meal manufacture. A goat produces about 150 kg of dry manure per year for use in crop production and gardening. Goat browsing accelerates growth of trees, shrubs and surface vegetation. They also act as seeding machines. They are much less prone to the adverse effects of toxic plants. They have higher dry matter and fibre digestibility and can subsist on poor woody vegetation. Goats are able to obtain more nutrients from the given environment in all seasons than other livestock species and are often the last species to leave the ecology during severe and continuous drought conditions (Singh, 2007).

Goats make important economic contributions in India. The size and magnitude of the contributions have not been adequately assessed. However, a few reports available do justify their claim to equality if not superiority with other livestock. They are so vital to a very large human population that their contribution to national economy can not be over looked. Goats require relatively much lower investments and facilities in terms of housing, feed, labour and health care. There is quick pay off due to fast multiplication and early maturity. Further, the risk involved in goat farming is much lower when compared to other livestock and crop production. Goats are reported to be more economical than cattle and sheep under natural grazing/browsing on arid zone range. Estimates made at CAZRI, Jodhpur indicated that a good indigenous goat may provide Rs. 250/- per year to a small farmer besides 2 quintals of manure and benefits of clearing obnoxious weeds and thorny bushes (Ghosh and Khan, 1980). The indigenous goats were found 2.5 times more economical than indigenous sheep when maintained on a free range grazing on highly degraded land in semi arid ecology of Rajasthan at CSWRI, Avikanagar (Swain et al. 1982).

Sharma (1987) recorded significantly more meat and milk production per unit live weight per year from goats than buffalo, camel and sheep have been recorded. The cost of production of goat milk worked out to be less than half than for cow's milk while milk from buffaloes was intermediate. The results of a socio-economic survey in Rajasthan conducted by Ahuja and Rathore 1987) have revealed that the number of goats increased 3 times between 1951- 1983 and goats accounted for 28-31% of the value of the livestock assets and for 16-19% of the gross receipts from crops and livestock. Some studies have revealed that the goats contributed up to 50-55% to the total cash income of a farm family in the hot arid region of the country. There are growing scientific evidences to suggest that goat keeping is more economical in drier regions which are not suitable for water spreading and dry land grain or fodder cropping (Govt. of India, 1987). Devendra (1992) reported that the value of annual contributions of goats to the Indian economy exceeded US $ 1.6 billions and of this the value of meat accounted for 79.3%, skins 12.9% and offals 7.3% in 1990. It is thus evident that goats make very sizeable economic contributions to the economy of India. It is therefore, important that development programmes should focus on the efficient use of these renewable resources as well as explore ways and means of increasing the current level of production.

Goat and Ecology

Man, Animal and Nature are in symbiotic relationship for their survival and sustenance. The balance maintained among the three for several millennia has been disturbed by over exploitation of natural resources for meeting the demands of increasing population of men and animals. The gravity of the situation can be gauged from the fact that India, with about 2.30% of land area only 4 % of the water of the world, is maintaining nearly 17.50 of the world's human population and about 12.00% of the livestock. The livestock in general and the goats in particular have been held responsible by the foresters and environmentalists for causing ecological degradation and desertification in Asian continent. There has been a lot of controversy over the role of goats in ecological degradation and in desertification especially in India. Of the domesticated animals, no animal has suffered from so much abuse as the goat due to this rather wrong conception. Two conflicting views prevail on the goat's role in land use (i) The goat is the major cause of deforestation, rangeland destruction and soil erosion, and as such, its propagation should be checked, and (2) Goat acts as regenerator of vegetation through dispersal of seeds in its droppings and vegetative propagation through browsing. It is undoubtedly true that an uncontrolled population of goats roaming in open can be extremely harmful to the growth and development of vegetation especially where new reseeding/

plantation has been done. But it is just as true that any over grazing by any other ruminant will have similar consequence. The high pressure on and shrinking of the grazing lands, especially in view of their low carrying capacity results in their over use and depletion of natural vegetation. The man to land ratio in the arid and semi-arid zones is fast declining. The livestock population is increasing due to low farm incomes from marginalization of land holdings and increased dependence on livestock, leading to over exploitation of the shrinking grazing lands. In addition, indiscriminate felling of trees and deforestation, removal of shrubs, bushes and roots for fuel purposes, use of mechanical powers for cultivation and continuous neglect of the grazing lands are the real reasons causing damage to the ecology. Most of the deteriorated rangelands have been caused by over grazing by cattle and sheep. The scrub vegetation left behind could only be utilized by goats and goats may have stayed in such degraded ecologies when other species have already left. Sheep because of their bifid upper lip are able to graze closer to the ground than cattle and goats and in doing so in dry weather and on loose soils, frequently uproot the small grass species and thereby permanently damage the soil. The cattle with their large size hooves and heavier body weights cause more loosening of the soil rendering it susceptible to water and wind erosion. When the sheep move towards the water points or night shelters, they follow each other in a row because of the flocking instinct and generate tracks, which become free of vegetation and subject to erosion. These habits of sheep and not of the goats are distinctly destructive to the rangelands (Acharya and Singh, 1992). Mann (1977) observed that the following human interferences greatly contribute towards desertification: (a) over-and uneconomic exploitation of resources, (b) Overgrazing by domestic animals and wild ungulates, (c) deforestation and indiscriminate lopping of trees and (d) uneconomic land use practices- cultivation on sand dunes and marginal lands causing sands to shift and deposit on the adjoining fertile lands. These cause soil erosion, and creep and expansion of deserts. Agriculture operations in general and on the arid and semi arid lands in particular cause more soil erosion than by the livestock grazing (Shankarnarayan et al. 1985). Contrary to cattle and sheep, goats are mainly browsers and would seldom go for grown up trees (Lu, 1988). They dislike the leaves of timber trees. They also do not prevent the establishment and spread of grasses which are so essential for soil conservation. Acharya et al. (1980) found that goats spend more than 90% time on browsing and hardly graze for 10% time on surface vegetation. Goats have been found to defoliate smallest branches of trees without damaging the twigs in several behavioural studies conducted at CAZRI, Jodhpur. Their browsing habit tends to reclaim saline soils by consuming salt-laden leaves of the range plants and contribute fertility to soil by even distribution of essential manure on the lands they graze. Cattle exploit 3-5 times more of phytomass than goats. Prajapati et al. (1988) reported that intensity of grazing of 2-4 goats/ha had no effect on run off and soil loss in hot arid regions of Rajasthan in normal rainfall years. Stocking rate of 3 sheep or goats per hectare produced satisfactory performance especially of goats and no deterioration in the physical and chemical properties of the soil was observed for 3 years.

Goats also help in dispersal of grass, bush and tree seeds. They consume grass and bush seeds and fodder tree pods while grazing/browsing and defecate hard coat undigested seeds especially of pod-bearing and xerophytic trees after providing acid treatment while passing through digestive system and fortifying with nutrients in the form of the faecal pallets and spread more uniformly all over the grazing land. These seeds germinate in large number as soon as soil moisture conditions are favourable. Such grazing lands would require only protection from grazing for 3-4 years to develop into a high class three-tier silvi-pasture for livestock grazing. The goat saliva left on the bitten foliage adds nitrogen directly to the plant cells inducing quick regrowth. Saliva is also said to detoxify the high tannin available in many of the tree leaves found in the desert areas. The biting of tender leaves and twigs by goats also induce tillering and faster luxuriant regrowth of both branches and foliage. Sharma and Ogra (1987) reported that vegetative regeneration was found to be 27% more in goat paddocks comprising C. ciliaris, D. nutans and L. leucocephala as compared to sheep paddocks. Harsh and Shankarnarayan (1982) recorded better performance of goats than sheep in all the ecosystems in arid and semi-arid regions of Rajasthan for three successive years A task force to evaluate the impact of sheep and goat rearing in ecologically fragile zones was constituted by the Govt. of India in 1987 under the Chairmanship of Professor K. Hanumantha Rao one of the leading economists and then a Member of Planning Commission with a large number of foresters and environmentalists as members. The task force observed that there was no definite evidence to prove that goats pose a threat to the ecology as is generally believed. In any case their negative effect, if any, on ecology has been vastly exaggerated. It further observed that within the desired grazing pressure, sheep and goats are more economical and less harmful than large ruminants.

The habits and behaviours of goats are generally studied in relation to sheep to justify that there is no or very little competition in the two species for available feed resources. Goats work as leaders and the sheep act as followers while moving in a mixed flock. Goats move with their heads up and the sheep keep their heads down most of the time. Goat walks longer distances than sheep during a given time in search of their choice feed. Flocking instinct is much high in sheep as compared to goats. Sheep prefer to graze together but goats scatter all over the available grazing land. Goats prefer browsing on bushes, shrubs and lower branches of trees and sheep like grazing on surface vegetation close to the ground. Goats are much more selective than sheep. Browse and tree leaves are highly relished by goats and not so much by sheep. Goats, unlike sheep, prefer variety of feeds. Taste sensation is more discerning in goats as compared to sheep. Goats can distinguish between bitter, sweet, salty and sour tastes and show a higher tolerance for bitter substances. Goats tend to consume their diet less quickly and to take more meals as compared to sheep. Goats are able to obtain more nutrients from the environment in all seasons than sheep.

It was thus, revealed that goats should not be categorized as animals responsible for the destruction of ecology. It is unwise to put the blame on goats for land degradation and desertification caused by activities of human and other livestock. What is needed, therefore, is to maintain a balance between the number of goats and other livestock species and the quantity and type of feed resources, which are available for the combined population. In a developed three-tier silvi-pasture system, the goats should be introduced first followed by cattle and then sheep. After the sheep are out of the grazing area, the land should by given rest for regeneration/reseeding. This way the goat will go for bush, shrub and low set tree branches, cattle would utilize taller grasses and the sheep shorter and closer to surface vegetation resulting is most efficient and judicious utilization of the available feed resources.

Perspective

The goat population in India is poised to reach around 170 million and the annual meat, milk and skins production to 700, 6000 and 225 thousand MT respectively by 2025. The human population is likely to touch 1.5 billion and the annual meat requirement to 9855 thousand MT. Thus there will be a large gap between the requirement and the availability. Although buffaloes, sheep, goats, pigs, poultry and fishes contribute substantial quantities of meat, the goat meat (chevon) is the most preferred meat in India. Further, there is very high demand of live goats, goat meat and goat meat products both in developing and developed countries more so because there are no religious taboos against it and it is leaner and mostly organic. The goat milk is cheap, wholesome, easily digestible and nutritious. It is recommended for use in dyspepsia, peptic ulcer, pyloric stenosis, liver dysfunction, jaundice, biliary disorders, acidosis, insomnia and food allergies. These virtues of goat milk need to be highlighted and commercially exploited in the national and international markets. The goat skins are also a very potential source of income and a few goat breeds produce very high quality skins which are used in leather garments, shoes and fancy toys and show pieces. The goat manure and folding are very valuable source of nutrients for organic crop farming, ornamental and horticultural plantation. Studies have shown that goats are not responsible for vegetative destruction, soil erosion and desertification as blamed more than large ruminants and man made reasons.

The productivity of goats in respect of milk, meat and fibre is low in general. The major reasons for low productivity are inadequate grazing resources, poor nutrition, tropical heat, disease problems and serious lack of organized efforts for bringing genetic improvement. The studies have shown that the goat productivity can be significantly improved if the above aspects are thoroughly researched upon, required inputs are judiciously provided and the listed constraints are adequately solved. Research with goats has markedly grown in the last decade. Nonetheless, there still remain many challenges facing the goat industry, which require considerable research to adequately address there. The CIRG has an important role to play in identifying these critical research needs, seeing that it conducted, and ensuring that findings are disseminated to those who can derive benefit from them. There are a large number of aspects that will necessitate research attention in the coming years in order to maintain as well as assume greater role of goats in the mixed farming systems of our country.

Goat breeding

Genetic improvement of goats holds great promise for enhancing their productivity. Considerable genetic diversity exists in Indian goats but there has been very little effort to tap this resource. Some fragmentary efforts have been made during last 25 years to describe, evaluate, conserve and manage some of the goat breeds. These have, however, been based on small populations and generally in institutional flocks. Research programmes, therefore, need to be taken up on identification, characterization and evaluation of indigenous goat genetic resources. There is need for research on gene-based technologies to identify indigenous genetic resources possessing traits influencing adaptability to harsh environments, productivity under sub-optimal conditions and disease and parasitic resistance. Studies on improvement of goats for meat, milk and fibre through selection/grading with superior breeds have been limited to a few breeds and involving small number. Priority efforts are required on faster multiplication and production of superior germplasm of different Indian goat breeds through Open Nucleus Breeding Scheme adopting Artificial Insemination and Multiple Ovulation Embryo Transfer Techniques both under farm and field conditions. A mechanism to safeguard, conserve and increase population through reproductive biotechnology of some of the rare breeds of goats before their extinction should be a priority. Sequencing and mapping of the goat genome will allow identification of individual genes controlling specific production and economically important traits viz. growth rate, milk yield, carcass quality, feed conversion efficiency and nutrient metabolism. The potential impact of globalization on the goat gene pool should also be considered. Use of reproduction biotechnology for conservation of threatened goat breeds has to be intensified. Semen freezing technology and use of frozen semen on large scale need perfection.

Feeds and feeding

Feedstuffs represent the major cost of production of goats. The nutrition and feeding practices are of great importance and more research is needed to develop and evaluate feeding programs that enhance level, efficiency, and profitability of production of goat meat, milk and fibre. There is need for basic and applied research to address influences on nutrient requirements of factors such as genotype, diet, and environment along with physiological processes responsible for their potential impacts in order to maximize the long-term rate of progress. Goats in India are generally reared under extensive system and play only a secondary role to crop as well as other livestock production. The greatest limitation in our natural pastures is on the availability of adequate energy throughout the year and adequate protein for more than half the year. The adult body weights, reproduction rates and milk production are lower and morbidity and mortality rates are higher. The birth, weaning and slaughter weights and dressing percentage are poor and bone: meat ratio is narrow in kids maintained solely on grazing on the degraded lands. There is great prospect of improving these natural rangelands. Research is needed on development of feed and fodder resources through development of multi-tier pastures by reseeding the grazing land with nutritious perennial grasses, legumes, fodder shrubs and trees. Research is needed on harvesting and conserving the abundantly available monsoon herbage as hay, silage, feed blocks etc. for use during lean periods. Research is also needed on identification, evaluation, improvement and utilization of low-grade roughage and agro-industrial by-products. With the shrinking of community grazing lands it may not be possible to continue rearing goats under extensive system and the goat farmers will have no option but to switch over to semi-intensive and intensive systems not only to maintain but also to improve the productivity. Studies on development of strategic supplementation of goats during different physiological stages, seasons and regions for different breeds will have to be conducted to suggest suitable feeding systems for adoption by the goat keepers. Further, studies are, on priority, required on development of intensive system of goat meat production to improve the quantity and quality of meat. For this, work on development of complete feeds as mess, pallets and feed blocks using cheap and locally available ingredients needs to be undertaken. Production performance and economics of different goat breeds under different feeding management systems based on mixed farming also needs to be studied. Development of Model Goat Production Systems based on mixed farming for different regions for integrated rural development needs due emphasis. Nutrition research in regards to understanding how genes and nutrients interact and physiological mechanisms responsible for effects of nutritional manipulations will of paramount importance.

Goat health

Society is questioning the safety and acceptability of food from animals. Implications of this for the goat industry in developing countries where production is primarily extensive are unclear. Today, the public or consumer is not only interested in the wholesomeness of the product but also how, where and under what conditions the item was produced. While traditional concerns such as fat, cholesterol and drug residue levels remain at the top of the list, there is recent interest in whether animals were raised on grass or under organic system of production. Research on surveillance, monitoring and forecasting of goat diseases and systems for prophylaxis and treatment have to be continued. Development of vaccines against the emerging goat diseases has to be prioritized. Critical investigation of goat diseases and developing location specific preventive and control measures needs to be continued. Extensive use of synthetic anthelmintics is resulting in widespread resistance of internal parasites. It is uncertain if new commercial compounds will be developed before resistance to the ones currently available reaches intolerable levels. Free movement of animals from one area to another has aided the spread of resistant nematodes in the country. Therefore, there is need for internal parasite research in a number of topics, such as management strategies to avoid high nematode burdens, treatment with inherent anthelmintic activity, and use or development of goats with resistance to internal parasites and acceptable levels of performance. Studies on recurrent diarrhoea in adult goats and kids, its effective prevention and control measures also require attention.

Goat products

The living standards of a large population are rising, concomitant with an increased demand for meat and milk. Distribution of goat products is going through both horizontal and vertical integrations as well as increased globalization. Studies on evaluation of quality of meat, milk and fibre produced by various goat breeds under different management systems and development of processing technologies for new value added meat and milk products need to be undertaken. Goat researchers have not kept up with research and industry enhancements in carcass packaging, storage, grading systems, or new products, or in the development of dairy products that are beneficial for gastrointestinal and cardiovascular health. These are critical areas that deserve attention in order to add value to goat products for substantial economic returns. The advantages of consuming Indian goat meat mostly produced organically and the medicinal value of the goat milk need to be highlighted in the national and international markets to exploit the rising demands of such food products.

Studies on the socio-economic aspects of goat farming in small and commercial goat holdings have to be conducted. Studies on the marketing of live goat and goat products both in internal market and for export are very important. Participatory research projects with the farmers especially in the area of breed and feed resource development and commercialization of goat farming need priority.

There is need for greater cooperation and collaboration among research institutions and organizations in order to achieve most efficient use of limited resources. Such research should have well defined goals and realistic objectives, leading to identifiable outcomes with positive impacts. Duplication should be avoided and previous experiences built upon. Furthermore, stakeholders must be involved in research prioritization. If the above measures were adopted from planning to implementation stage, goat milk, meat and skin production would be significantly improved during the next two decades. The most important determinants of research conducted are the availability of funding and suitable scientific manpower.

Breeding Management of Goats

Goat genetic resources

Goat breeds belong to dairy, meat, fibre and pelt types or their various combinations and meat being an important component in economics of production of most of the goat breeds. In addition to the important descript breeds, there are a large number of descript but less important breeds and non-descript populations. The breeds of sheep and goats in India have been inventerized. There is serious need to describe and evaluate the existing genetic resources in terms of their population, population trends, flock size and structure, ecology, feed resources, management practices, body size and weight, confirmation and production characteristics using modern biological and statistical methods for accurate description and determining genetic differences. The availability of cytological, biochemical and molecular methods for determining gene marker characters will help in discriminating among breeds. Some breeds have been identified as improver breeds and have been utilized within the country or in the region. The important goat breeds found in different regions of the country are listed below (Singh, 2007)-

a. *Northern- Western Region:* (Punjab, Haryana, Rajasthan, Gujarat and Western U.P.) Beetal, Barbari, Jamunapari, Sirohi, Marwari, Jhakrana, Kutchi, Gohilwadi, Zalawadi, Mehsana and Surti

b. *Temperate Himalayan Region*: (Hill region of U.P., Himachal Pradesh and Jammu & Kashmir) Gaddi, Chegu and Changthangi

c. *Eastern Region*: (Eastern U.P., Bihar, Bengal, Orissa and Assam) Bengal and Ganjam

d. *Southern Peninsular Region*: (Maharashtra, Andhra Pradesh, Karnataka, Tamilnadu and Kerela) Osmanabadi, Sangamneri, Malabari and Kannaiadu

Breeding strategy

The breeding strategy involves decision on objective and approach in bringing genetic improvement. This also depends on socio-economic and ecological consideration, existing genetic resources, their productivity, possibility of their improvement through selection within a breed, grading with indigenous improver breeds, or replacing an indigenous breed within exotic breed if it is arising from similar ecological region and thus will be adaptable, or cross-breeding for evolving new breed combining the adaptation and hardiness of local breed and higher productivity of the exotic breed.

Breeding objectives

Major breeding objectives may be for milk, meat and fibre or their combinations. For milk production, major consideration will be given to lactation yield, lactation length and milk composition. In meat production, live-weight gain, efficiency of feed conversion, carcass yield and quality, reproductive performance, survival and milk yield of the dam are important characters. In fibre production the greasy fleece weight, scoured fleece yield, fibre diameter and fibre length are important characters.

Breeding approach

Breeding approach may involve (1) selection within local breeds, (ii) grading a local breed with superior indigenous breed (iii) crossbreeding with exotic breed for evolving new breeds combining adaptation and hardiness of the indigenous and higher productivity of the exotic and (iv) replacement of an indigenous breed with higher productivity exotic breed. The breeding approach will have to be decided on the basis of the performance of local breed, the size of the genetic variability in the characters of interest, physical environment and feed resources available to sustain higher production. Decision will also depend upon the demand for the products in markets close to the centre of production and distant markets, and handling, processing and marketing infrastructure available

Selection criteria

Selection criteria will depend upon the phenotypic and genetic variances in and co-variances among important characters, their relative economic values and sources of information. Attempts have been made to develop selection criteria for improving production for various economic traits. For improving milk production in goats, selection based on age at first kidding and the first lactation milk yield combined into an index will provide maximum gains in reproduction and milk production. For improving meat production, selection on 6-month body weight may be most feasible and bring reasonable genetic progress in meat production through improving reproduction, body weight gain and carcass yield.

Selection in farmer flocks

Through recording of performance of a large number of farmers' flocks preferably through objective assessment, flocks which are superior than average can be identified and used as flocks for producing genetically superior breeding males which could subsequently be distributed to the flocks which are inferior to the average of the population. This will require identification of superior males and females based on the characters of importance and mating them to produce male progeny, getting them reared in the farmers' flocks and/or in rearing centres and making them available to the flocks to be improved on male to male replacement basis.

Selection in institutional flocks

Institutional flocks may be established through selection of superior males and females from farmers' flocks. This is the usual system of genetic improvement in majority of developing countries where the government has responsibility for providing superior breeding males to the flock owners. Such flocks are properly recorded and performance data are utilized in selecting the breeding males and females. The best males are retained in the institutional flock for breeding with elite females for producing young males and the rest are made available to the flock owners. In majority of cases the flock size is small allowing a few males to be used for breeding and thus the performance, may not be better than in the private flocks and males produced on such farms are not thus acceptable to the farmers. The number of males available are also rather small than those required. In case of selection for milk the number of daughters per sire is usually small and considering the reproductive performance, the age by which the progeny performance becomes available, the male is too old to breed. The selection in such small institutional flocks is essentially based on the dam's performance or at best sister's performance. A large number of superior males can be produced by identifying superior males and females from farmers' flocks and utilizing the technique of Multiple Ovulation Embryo Transfer (MOET). This can be done either in the farmers' flocks or in institutional farms.

Selection in open nucleus flocks

It may be desirable to organize farmers' into co-operative group breeding schemes. Open nucleus flocks can be established through selection of better performing animals on the basis of objective assessment from the farmers' flock and utilized for producing breeding males through intensive selection and their distribution to the co-operating and other farmers. Keeping nucleus flock open would help in reducing in-breeding and increasing selection intensity through continuous introduction of superior animals from the farmers' flocks. After distribution of the males to the co-operating flocks, rest could be distributed to the other flocks to be improved. Subsequently, flocks of co-operating farmers could also be utilized in breeding males using the selected males from the nucleus (multiplier flocks). The use of MOET can help in improving genetic gains through selection especially in case of dairy goats where selection of males can be based on the performance of collateral relatives (full and half sisters) rather than the progeny. Such a selection scheme will bring faster genetic progress specially through reducing the generation interval, although the accuracy of selection is slightly lower than based on progeny performance. These selected breeding males could be more extensively utilized through artificial insemination with frozen semen.

Grading with superior indigenous breeds

Grading with superior breeds requires approach similar to ONBS as described earlier for producing superior males and their distribution to the flocks to be improved. This could be done with the farmers' flocks or in institutional farms. There are a number of superior breeds of goats for example Jamunapari, Beetal, and Barbari which have been utilized with success for improving body size, milk yield and prolificacy in medium and small size breeds.

Crossbreeding with exotic breeds

A number of specialized dairy breeds viz. Alpine, Toggenburg, Saanen and Anglo-Nubian from Europe and North America have been extensively utilized for improving milk production in indigenous goat The results indicate that there is substantial improvement of crossbreeds in milk production and reproductive performance specially age at first kidding although there is a slight decline in prolificacy. Half-breeds have the best performance and there is little to gain by increasing exotic inheritance beyond 50% from one or more than one exotic dairy breeds.

Utilization of MOET and AI

For bringing quicker genetic improvement for milk, it will be essential that similar approach as in dairy cattle both in selection as well as in extensive utilization of selected breeders is followed. For the purpose, utilization of embryo technology and artificial insemination with frozen semen will be necessary. Extensive research and development effort has been made in developing method for long- term preservation of semen in goats with variable success. This is an area that would need serious attention in case quicker genetic improvement in goats is desired. Similarly, for making genetic progress through selection in the indigenous breeds or their crossbreds with exotic breeds in developing countries where performance recording in farmers' flocks on continuous basis is not currently being practised and is difficult, the utilization of open nucleus breeding flocks with MOET can help in generating large superior male germ plasm which can be extensively utilized through artificial insemination with frozen semen.

Feeding Management of Goats

Management systems

India is a group of States characterized by mixed farming systems. Goats are common and essential part of existing systems and offer opportunities for high-value production. Although a large number of goat feeding management systems are in practice and vary from region to region these can essentially be described under three systems viz. extensive, semi-intensive and intensive. The ultimate aim of adopting a particular feeding system should be to bring improvement in reproduction and survival rates of the goat flock and birth, weaning and slaughter weights and dressing percentage in the male kids. Extensive studies on evaluation of community grazing lands, developed pastures, semi-intensive and intensive feeding systems vis-à-vis performance and production levels in sheep and goats have been conducted and the results have been reviewed by Singh and Patnayak (1987), Patnayak *et al.* (1995), Singh (1998), Singh *et al.* (2004) and Singh (2007). A brief account of the three systems is presented below-

Extensive system

Goats are generally reared under extensive system and play only a secondary role to crop as well as other livestock production. This system includes migratory, transhumance, free range, pasture and range grazing management of goats. It is most common system throughout the Asian region because the small size of goats has distinct advantages over other livestock species to contribute significantly to the sustainability of small farm situations which predominate in this region. The goats usually owned by small farmers and landless labourers are grazed together and tend to be herded over long distances in search of feed and water. The flock sizes are larger and goats belonging to several owners are run together. A low level of unpaid family labour represents the main input. The system is principally one of low resource use and a low level of productivity emerges from poor nutritional availability. There is marked fluctuation in feed availability and its nutritive value among different regions and among years and seasons. Extensive studies on evaluation of common grazing lands and developed pastures vis-a-vis performance and production levels in goats have been conducted.

The greatest limitation in our rangelands and natural pastures is on the availability of adequate energy throughout the year and adequate amounts of protein for more than half the year. Although the grazing on the rangelands is considered to be the cheapest method for goat production, the migration of the animals and over grazing of available lands is causing serious problems of soil erosion and degradation leading to desertification. Due to long and dry summers, erratic rains and light textured soils deficient in organic content having tendency to salination, the natural potential of these lands falls short of the nutritional requirements of goats. The adult body weights, reproduction rates and milk production are lower and morbidity and mortality rates are higher. The Beetal goat male kids reached a body weight of only 11.5 kg at weaning and 14.1 kg at 6 months age when maintained under free range grazing without any supplementary feeding with over all survivability of 87.5% up to six months age (Mishra, 1981). The birth, weaning and slaughter weights and dressing percentage are poor and bone: meat ratio is narrow in kids maintained solely on grazing on the degraded lands. The common grazing/ rangelands can hardly carry one goat per hectare per year when unprotected and two when protected from biotic factors. There is great prospect of improving these natural rangelands. Reseeding with *Cenchrus ciliaris* and *Lasiurus sindicus* perennial grasses increased the grass yield from 0.5 ton to 1.5- 2.0 ton per hectare per year. Intercropping of cowpea in *Cenchrus ciliaris* pasture increased the dry matter yield by three times. Introduction of *Dolichos lablab* legume improved the nutritive value and increased the dry matter yield from 1-1.5 ton to 2.5-3.0 ton per hectare. Such a grass-legume pasture could carry five goats per hectare for five years. Plantation of 50 fodder trees of *Prosopis cineraria, Ailanthus excelsa* and *Leucaena leucocephala* had no adverse effect on the pasture growth and provided an additional yield of 1 Ton dry matter when fully grown and lopped twice a year. Areas in arid and semi-arid regions that can't support cattle and buffaloes may be identified and utilized for small ruminant production with low investment. Rural common grazing lands should be improved through reseeding with nutritious, perennial and high yielding grasses and legumes. Fodder tree plantation should be taken up on pasturelands, wastelands, riverbanks and roadsides and bunds of ponds, canals and agriculture fields on large scale.

Semi-intensive system

A kind of compromise between extensive and intensive systems is referred to as semi-intensive system of goat management. It is a combination of limited free range grazing and feeding in stalls. Integration of goat rearing with arable cropping and tree cropping is also included wherein either the goats are tethered or cut and carry system of available fodder is employed. Animals belonging to several owners may be combined for grazing which is mostly done morning and evening for 4 to 6 hours. The animals are supplemented with kitchen wastes, concentrate mixtures, crop residues, green and dry fodders and tree leaves etc. as per the availability. Thus goats utilize all available feed resources including natural grasses, shrubs, bushes, tree leaves, crop residues, stubbles, weeds, cultivated fodders and concentrates etc. under this system. The level of nutrition is just optimum or low but surely better than that under extensive system. Production performance depends on the quantity and quality of grazing/browsing availability and supplementary feeding provided. Studies have shown that the natural range/grazing lands cannot support growth of kids for commercial meat production and additional supplementation with concentrate mixture and/or leguminous fodders and tree leaves is recommended. Supplementation of additional energy and protein through concentrates has been advantageous in increasing the growth rate and attaining the desired slaughter weights at relatively younger age. Kids raised on *Cenchrus ciliaris* pasture with concentrate supplement @ 1.5% of body weight attained body weight of 15.4 kg at 3 months and 26.0 kg at 6 months of age. It has been found that the kids fed on 50 concentrate: 50 roughage ration @ 4% of body weight attained 25 kg live weight at about 6 months of age. If the roughage portion of the diet can be made available from grazing, the remaining 50 % of the diet has to be supplemented through concentrate mixture. The Sirohi does grazing/browsing for 8 hours and supplemented with 150, 300 and 450g/h/digestibility concentrate mixture during last 45 days of pregnancy and first 150 days of lactation lost body weight when only grazed, maintained when supplemented with 150 g concentrate and gained in their live weights when supplemented with 300 and 450 g concentrate during pregnancy. The same does lost when only grazed or supplemented with 150 g concentrate and gained in their live weights when supplemented with 300 and 450 g concentrate during lactation. The milk production was improved by about 29, 31 and 67% and pre-weaning growth of the kids by 17, 20 and 35% with the three supplementary levels (Singh, 1996). Sirohi, Marwari and Kutchi does produced 84.4, 89.1 and 94.3 kg milk with no supplementation, 98.6, 96.1 and 93.2 kg with 150 g concentrate, 100.9, 115.7 and 110.0 kg with 300 g concentrate and 109.0, 106.4 and 101.1 kg with 450 g concentrate supplementation in addition to 8 hrs grazing during 150 days lactation (Singh,1992). Ad-libitum supplementation of concentrate, hay and green between 91 to 180 days of age to the Sirohi kids browsing on available range land resulted in an increase of 44.8% in pre- slaughter weights, 65.1% in carcass weights and 14.3% in dressing over the browsing alone. A growth rate of 19.4, 41.7, 111.0 and 108.2 g/h/d was obtained in Beetal weaner kids from 91 to 180 days of age on ad-libitum browsing, browsing +

green, browsing + concentrate mixture and browsing + concentrate mixture + green respectively (Parathasarthy et al.,1983). It is seen that although the goats are primarily browsing animals but a marked improvement in kidding rates, milk yield, live weight gains and quantity and quality of meat production can be achieved by grazing/browsing the goats on developed two-tier pastures and supplementing with concentrate mixtures and/or cultivated leguminous fodders and tree leaves at appropriate levels.

Intensive system

The intensive system of goat feeding management includes grazing on developed pastures and/or feeding completely in stalls on cultivated fresh or conserved fodders, crop residues and concentrates. Although goats prefer to browse as compared to grazing, they are quite capable of making efficient use of cultivated pastures for meat and milk production. Stocking rates of 16 to 60 goats per hectare are feasible depending on the type of grass, level of fertilization and the presence and absence of legumes and fodder trees. This system requires high labour and capital investment and is suitable for only intensive meat production from goats. In addition to providing better milk, growth and carcass quality it removes pressure from the grazing lands.

Several studies have been conducted on fattening of indigenous and crossbred kids under ad libitum feeding on concentrates including barley grain, maize grain, oil cakes, wheat and rice brans, molasses etc. and roughages including tree leaves, cultivated grasses and legumes in the ratio of 40:60, 50:50 and 60:40. The male kids weaned at 2-3 month age and fed under feedlot system achieved slaughter weights of 25 kg at 5 to 6 months age with a dressing % of 48 to 51 (Parathasarthy et al.,1984). The kids maintained under semi-intensive system reached the target live weight of 25 kg earlier than those under intensive system. The kids under semi-intensive required less feed for a kg of gain than those under intensive feeding. The dressing % was superior under intensive. The bone and lean % was higher under semi-intensive and fat % under intensive system (Singh and Sahu, 1997). Average daily gains under intensive system were higher in lambs than kids while they were similar under semi-intensive system (Shinde *et al.* 1995). The daily gains, milk intake, meat quantity and quality and feed efficiency were found superior in the Sirohi, Marwari and Kutchi kids maintained under semi-intensive as compared to those maintained under intensive or extensive system. The milk yield during 150 days of lactation was higher under intensive system than that under semi-intensive and extensive systems and in Sirohi does than that in Kutchi and Marwari does (Singh, 1993). The total confinement and complete feeding of goats in stalls has, however, proved to be detrimental some times. Integration of goats with tree cropping especially with reseeding area among rows of trees with shade loving nutritious perennial grasses and legumes has considerable potential. This system has the advantage of increased fertility of the soil via the return of dung and urine, control of weeds and thus reducing use of weedicides and fertilizers in addition to easier management of the crop (Devendra, 1987). The value of goats is also very significant in the wider context of massive agro-forestry programmes taken up in our country. Since liberal use of crop residues and agro- industrial by-products has to be made for compounding feedlot rations, efforts to identify and utilize them should be intensified. Production and consumption of balanced goat feed mixtures as pallets and blocks based on maximum use of by-products should be developed in a coordinated manner. Cereal straws are an important roughage resource and their nutritive value can be improved with nitrogen-supplementation, amoniation, urea solution and chemical treatments. Studies have shown that the low- grade roughages like straws and stovers added with little oil cakes, brans and mineral supplements may be very suitably compounded into Complete Feed Blocks and Pallets for goats. The technology of compounding Complete Feed Blocks is advantageous in several ways. Feed Blocks require less than one-third storage space as compared to the raw material. Transportation of Feed Blocks is easier and cheaper. Selectivity of ingredients by the animals from the Feed Blocks is avoided. Palatability and nutritive value of the ration is improved. Live weight gains and milk production are also improved.

Commercial Chevon Production

The goat meat available in the Indian markets comes either from old and culled adults or from male kids slaughtered any time between 6 months to 1 year of age. The quantity and quality of this meat is poor due to poor market weights, lower dressing percentage and narrow bone: meat ratio as these kids are maintained on scrub vegetation like their dams and hence hardly attain a body weight of 15-16 kg when they are normally marketed. The dressing percentage varies from 35 to 40 and bone: meat ratio from 1: 3 to 1: 4. The studies conducted in the country have revealed that a marked improvement can be achieved both in the finishing weights and carcass quality through intensive feeding of the male kids. Based on these findings two packages of improved practices have been developed and recommended for adoption by the goat farmers and the entrepreneurs for commercial goat meat production. A brief account of these packages is given below -

Package for progressive farmers

Sheep and goat Farmer Cooperative Societies may be formed in the goat rearing areas. These Societies in association with the village Panchyats should develop improved silvipastures on the available community grazing lands with financial assistance from the financial Institutions and subsidies being provided by the Central and State Governments. The registered goat flocks may be allowed to graze on these pastures judiciously for 6 to 8 hrs daily. In addition to grazing, the pregnant does during last 30 days of pregnancy and the lactating does during first 60 days of lactation be supplemented with 300g/h/d concentrate mixture containing 12 % DCP and 65 % TDN to ensure 2.5 to 3 kg birth weights and 14 to 15 kg weaning weights in male kids. To ensure a weaning weight of 14 to 15 kg, these male kids be provided *ad lib* suckling, creep ration and green/dry leguminous fodders during pre weaning period and completely weaned at 60 days of age. These weaners may then be fed *ad lib* on complete feeds comprised of 50 % concentrate and 50 % roughage under stalls till they attain 25 to 30 kg finishing weight at around 6 to 7 months of age. Alternatively the kids be allowed to graze on available pastures and supplemented with concentrate mixture @ 1.5 to 2% of the body weight till they reach the desired finishing weights. The finisher kids should then be sold by the farmers or their Cooperative Societies for slaughter to the consumers or the traders directly avoiding involvement of middlemen.

Package for entrepreneurs

The male kids produced and reared by the farmers in general and the progressive goat breeders following the recommended package of practices in particular may be purchased by the entrepreneurs at around 60 days of age and transferred from the villages to the Meat Production Complexes established near the cities. This will help in reducing the grazing pressure on shrinking pasturelands, early rebreeding and easy management of the flocks. These Complexes may be equipped with a Feed Compounding Plant capable of incorporating higher proportion of cheaper low grade roughages and agro industrial by-products to manufacture economic complete feeds, Feedlot Animal Houses for intensive feeding of the kids, Modern Slaughter House, Meat Processing, Product Manufacturing and By- Products Handling Machineries. The kids procured from the villages may be intensively fed on complete 50:50 or 60:40 concentrate: roughage feeds under feedlot from 2 months to 6 - 7 months age when they attain finishing weight of 25-30 kg. These intensively fed kids may then be sold as live animals in the National or International markets or slaughtered for selling fresh meat or their meat may be processed and converted into different meat products for export. Similarly the slaughterhouse by- products may be converted into commercial products for sale.

Recommendations

1. The economic and social importance of goats especially for the rural poor demands for large research and development investment.
2. The National governments and International agencies should support goat production more than other livestock species.
3. There is need for greater regional and international co-ordination in research and development programmes for goats to allow exchange of information, literature, germ plasm, diagnostics and vaccines.
4. Surveys supported by gene marker studies on existing genetic resources of goats should be carried out in the breeding tracts of different breeds.
5. Adaptation of the breed to local environment should be the key element in the choice of the breed and in determining breeding strategy.

6. Since the records available from the institutional flocks are limited, it is essential to adopt open nucleus breeding scheme using multiple ovulation embryo transfer and artificial insemination with frozen semen.
7. The extensive system of goat rearing should be replaced with semi-intensive and intensive systems for commercial meat production.
8. The two suggested Packages of Practices should be adopted by the progressive farmers and the Entrepreneurs for commercial chevon production.
9. Feed resources for goats should be developed through 3-tier system involving production of surface vegetation through reseeding with perennial and nutritious grasses and legumes creating middle level vegetation through fodder shrubs and top level through trees.
10. The feeds and fodders available during rainy season should be harvested, conserved and utilized during scarcity period. It will also be desirable to create animal feed security system utilizing chemically treated and densely baled crop residues, forest grasses and tree leaves etc.
11. There has been a large progress in disease diagnosis and control through prophylactic vaccination and dosing. It will be desirable to develop suitable disease monitoring systems and undertaking strategic control and eradication measures rather than taking up routine prophylactic measures.
12. The goat production system should be considered as a component of the agriculture production system and major research effort should involve on farm research on farming system basis.
13. A proper Animal Husbandry Extension Service similar to that for crop production should be established so that the improved technologies can reach the farmers.

REFERENCES

Acharya, R.M.; Swain, N.; Jain, P.M.; Mishra, R.K.; Mathur, P.B.; Sharma, M.M. and Mudgal, R.D. 1980. Relative productivity of sheep and goats on free-range grazing/browsing management on semi-arid rangelands of Rajasthan. Project Report, CSWRI, Avikanagar.

Acharya, R.M. and Singh N.P. 1992. The role of goat in conservation of ecology and livelihood security. Plenary Papers and Invited Lectures. V International Conference of Goats held at New Delhi.

Ahuja, K. and Rathore, M.S. 1987. Goats and Goat Keepers. Institute of Development Studies, Printwell Publishers, Jaipur.

Devendra, C. 1987. Small Ruminant Production Systems. Small Ruminants Production Systems in South and South-East Asia. International Development Research Centre, Ottawa, Canada.

Devendra, C. 1992. Goats and Rural Prosperity. Plenary Papers and Invited Lectures. V International Conference of Goats held at New Delhi on March 2-8, 1992.

F.A.O. 2005. Production Year Book 58. Food and Agriculture Organization of the United Nations, Rome.

Govt. of India. 1987. The report of the task force to evaluate the impact of sheep and goat rearing in ecological frazile zone. Ministry of Agriculture, Govt. of India, New Delhi.

Ghosh, P.K. and Khan, M.S. 1980. The Goat in the Desert Environment. Research Bulletin No.12, CAZRI, Jodhpur.

Harsh, L.N. and Shankarnarayanan, K.A. 1992. Performance of sheep and goats under different intensities of grazing on Cenchrus pasture dominated by *Zizyphus nummularia*. Seminar on Problems and Prospects of Livestock Development, CAZRI, Jodhpur, Nov. 26-30, 1982.

Lu, C.D. 1988. Grazing behaviour and diet selection of goats. *Small ruminant Research* 1, 205-216.

Mann, H.S. 1977. The desertification Process- An Overview. *Annals of Arid Zone* 16, 279-280.

Patnayak, B.C., Singh, N.P. and Karim, S.A. 1995. Transferable technologies for meat production in sheep and goats. Lead paper. III National Seminar on Sheep & Goat Production and Utilization held at CSWRI, Avikanagar on April 8-10, 1995.

Parthasarthy, M.; Singh, D. and Rawat, P.S. 1983. Effect of supplementation on the performance weaner kids. *Indian Journal of Animal Sciences* 53, 671-672.

Parthasarthy, M.; Singh, D. and Rawat, P.S. 1984. Performance of feedlot kids on different dietary regimen. *Indian Journal of Animal Sciences* 54,130-131.

Prajapati, M.C., Tiwari, A.K., Nambiar, K.T.N., Singh, J.P., Malhotra, B.N. and Sharda, B.N. 1988. Effect of different intensities of grazing goats on vegetation cover, run off and soil loss in a forest water shed Yamuna ravines via-a-vis animal production. III International Rangeland Congress, RMSI, Jhansi.

Shankarnarayanan, K.A.; Bohra, H.C. and Ghosh, P.K. 1985. The goat - an appropriate animal for arid and semi-arid regions. *Economic and Politician Weekly*. 20, 1965-1972.

Sharma, K. 1987. Goat Rearing. CIRG, Makhdoom, Mathura.

Sharma, K. and Ogra, J.L. 1987. Reaction of component plant species of synthesized pasture under three-tier system to high intensity of grazing by goats and sheep in semi-arid zone. IV International Conference on Goats, Brazilia, Brazil.

Shinde, A.K.; Karim, S.A.; Singh, N.P. and Patnayak, B.C. 1995. Growth performance of weaner lambs and kids on intensive and semi-intensive feeding management. *Indian journal of Animal Science* 65, 830-833.

Singh, N.P. and Patnayak, B.C. 1987. Feeding of sheep and goats for meat production. Lead Paper. National Seminar on Small Ruminant Production held at CSWRI, Avikanagar on January 5-7, 1987.

Singh, N.P. 1992. Effect of varied levels of supplementary concentrate on the performance of lactating does of three breeds. V International Conference on Goats held at New Delhi on March 2-8, 1992.

Singh, N.P. 1993. Milk yield and composition in three goat breeds under three systems of feeding management. VI Animal Nutrition Research Workers Conference held at OUAT, Bhubaneshwar on Sept.13-16,1993.

Singh, N.P. 1996. Development of feeding system for goat meat production. National Symposium on feeding strategies for eco-friendly animal production in India held at IVRI, Izatnagar on Feb. 14-15, 1996.

Singh, N.P. and Sahu, B.B. 1997. Chevon production in three goat breeds under two feeding systems. *Indian journal of Animal Science* 67, 87-89.

Singh, N.P. 1998. Production performance of sheep and goats under different management systems. Lead Paper. Proceedings of the Golden Jubilee Seminar on Sheep, Goat and Rabbit Production and Utilization held at CSWRI, Jaipur on April 24-26,1998.

Singh, N.P.; Sankhyan, S.K. and Shinde, A.K. 2004. *Animal Nutrition and Feed Resource Development Research.* A bulletin published by CSWRI, Avikanagar.

Singh, N.P. 2007. Conservation, improvement and management of goat genetic resources in India. Compendium on the Training Programme organized at NBAGR, Karnal on Indigenous livestock and poultry conservation in field conditions sponsored by Directorate of Extension, Ministry of Agriculture, Govt. of India on September 12-19, 2007.

Singh, N.P. 2007. Goat - Livelihood security for the rural poor. Lead Paper presented in the Seminar on Livestock linked Livelihood and its impact on Rural Economy in Orissa organized by the Orissa Livestock Development Society and Indian Society for Sheep and Goat Production and Utilization at OUAT, Bhuvneshwar on September 16, 2007.

Singh, N.P. 2007. Sustainable intensive meat production system for goats and sheep in the tropics. Lead Paper presented in the International Tropical Animal Nutrition Conference organized by ANSI at NDRI, Karnal on October 4 -7, 2007.

Swain, N.; Jain, P.M. and Acharya, R.M. 1982. Relative economics of sheep and goats. III International Conference on Goat Production and Disease, TUCSON, Arizona, USA on January 11-15, 1982.

❑❑❑

3

Contribution of Small Ruminants in Red Meat Production in India

N. KONDAIAH, Y. BABJI
National Research Centre on Meat, Hyderabad-500039, Andhra Pradesh

Introduction

Small ruminants have an unique role in Indian Meat sector having prime importance in production and consumption by large masses. Sheep and goat provide livelihood to a large section of rural population and also compliment valuable animal protein supplies to both rural and urban populations. Meat from sheep and goat has wide acceptability among Indian consumers though there are differences in the regional preferences for the two meats with higher preference for mutton in the Southern states and Jammu & Kashmir while goat meat has higher preference in majority of the states. For export lamb meat is preferred and commands higher price due to leanness (lower fat) and production by natural grazing. Clean meat production and processing sheep and goat meat for value addition would substantially increase contribution of small ruminants in Indian meat sector. Unorganized nature of sheep and goats marketing and exploitation by middle men are affecting the prospects of small ruminants. Pragmatic approaches in small ruminant production, processing and utilization are required to sustain the increased role of small ruminants in providing meat to a large population who have preference for mutton and goat meat. This would also contribute for increased income and better livelihood of millions of small, marginal and landless farmer producers. The need for exploring opportunities and face the challenges in the production of meat from sheep and goat meat has become more important in the changed land utilization pattern, socio-economic conditions and increasing consumer demand due to triple effects of increasing income, urbanization and awareness for consumption of animal protein. There is need for analyzing the opportunities and adverse impact on the prospects of sheep and goat meat in the post WTO scenario as there are countries which need Indian meat but also some countries have strategic advantage to enter the Indian consumer market in the liberalized trade scenario. Promotional efforts with appropriate R &D programs need to be initiated to sustain the unique role of small ruminants in Indian meat sector Joshi and Kondaiah, 1989).

Demand for Sheep and Goat Meat

Based on nutritional consideration of providing 20g/ per person/ day (against world average of 29 g as against Indian availability of 10g) nearly a doubling of all animal products production is desired. Sheep and goat production need to receive a better emphasis in view of their significant role in food security and livelihood. Thus the estimated consumption demand would be 687000Mt for mutton and 934000 mt for goat meat from the present production estimates of 220000 mt for mutton and 470000 mt for goat meat contributing for an increase of 196% and 100% in mutton and goat meat production respectively. How to achieve this increased production in the present social and environmental constraints affecting small ruminant production is a challenge and the opportunities for value added products need to form part of strategies for increasing per capita production and consumption. Due to low per capita animal protein consumption and wide consumer acceptability for mutton and goat meat the demand related prospects are large and production programs need to be reoriented to meet such demand driven growth prospects.

Utilization of meat animal resources largely depends on hygienic production of meat; cost efficient processing technologies; creating sustained demand for the product; innovative marketing approach; better utilization of byproducts and providing positive image to enterprise. These aspects have not received due attention partly due to the marketability of sheep and goat without much difficulties. However, with newer practices in sheep and goat production (semi-intensive and intensive production systems) and liberalized trade in post WTO period it has become important to consider appropriate strategies of utilization to ensure reasonable returns to sustain modern production system with higher production costs as compared to traditional methods due to the changes in land utilization and agriculture. Also, it is necessary to provide mutton, goat meat and products to ever growing consumer with changing life-styles demanding variety, convenience and quality. Thus appropriate strategies for better utilization of sheep and goat are important, particularly processing and utilization of meat from these ruminants through development of processed / value added products sector Kondaiah *et al.* (2006). This would play a major role in sustaining sheep and goat species production potential contributing to increased demand and marketability.

Contribution of Sheep and Goat Meat

At current prices value of meat from sheep and goat in 2004-05 was Rs 10329 crores forming 40% of total meat value at Rs 25984 crores and only next to poultry at Rs 11259 (43%). The value increased from Rs 7224 crores (38 % of total meat value) in 1999-2000 to the present value of Rs 10329 crores. At constant prices of 1999-2000 value of sheep and goat meat increased from Rs 7224 crores to Rs 8944 crores in 2004-05 registering an increase of 23.8% during the period. Whole sale prices have indicated a steady increase in case of sheep and goat meat showing the consumer demand compared to availability and registered a value of 204.6 in 2003-04 compared to base value of 100 in 1993-94 as against 91.6 for chicken in 2003-04. The shrinking land resources for livestock in general and sheep and goat in particular with increase of land under non-agricultural use by 151 per cent during the last 50 years (Table1) suggest the need for alternate production systems with appropriate R&D efforts to sustain the role of small ruminants in red meat production. Increased production and utilization efficiency need attention as alternative to traditional methods. Though the National Commission on Agriculture (1976) has recommended that the sheep population should be raised to about 45 m and goats to about 78m by 1985 and also desired that the goat population should be stabilized to about 40m by 2000 and the shortage of meat to be met by increase in population of sheep. The Commission

Table 1. Comparison of changes in land, livestock and human population

	1951	2003	% Change
Area of land utilization (m ha)	284.32	306.01	+7.6
Cultivated area (m ha)	129.4	155.90	+20.5
Net Area sown	118.8	141.10	+18.8
Area under non-agricultural uses (m ha)	9.36	23.57	+151.8
Human population (m)	360.1	1029	+185.0
Livestock population (m)	292.8	485	+65.6
Land available (ha)/1000 human	787	297	-62.3
Land available (ha)/1000 livestock	971	631	-35.0
Livestock (n)/ 1000 human	811	471	-41.9
Cattle (m)	155.3	185.2	+19.3
Buffaloes (m)	43.4	97.9	+125.6
Sheep (m)	39.1	61.5	+57.3
Goat (m)	47.2	124.4	+163.6
Pigs (m)	4.4	13.4	+204.5
Poultry (m)	73.5	489	+565.3
Total livestock and poultry (m)	366.3	974	+165.9
Milk Production (mmt)	17	87.3	+413.5
Egg Production (m)	1832	40300	+2100
Meat Production (000 mt)	631	6000	+850.9

has also suggested the slaughter of goats for meat should be increased so that the rate of growth of population may come down to a level of 67m by 1985 and go down to 40m by 2000. However the present population of sheep and goat indicate that both populations have increased and goat population has not decreased as indicated. A reconsideration is required so that prospects of both the small ruminants is sustained in order to meet the demand for red meats for the growing population. In the Report of the task force to evaluate the impact of sheep and goat rearing in ecologically frazilezone (1987) it has been recommended that sheep and goat do not pose a threat to the ecology as is generally believed and sheep and goat need to be given due priority under Central as well as State plans so as to raise per animal production and also made a number of intervention measures for increased contribution of sheep and goat.

Export of Sheep and Goat Meat

During the decade 1991-2000 sheep and goat meat exports increased by 49% registering a 5% growth rate in quantity terms and by 191% in value terms. During the period 2000-01 to 2004-05 sheep and goat meat exports decreased in quantity terms from 11903 MT to 8885 MT but, in value terms exports have increased from Rs78.16 crores to 79.36 crores. Compared to buffalo meat exports sheep and goat meat exports were of very low level (less than 5%) due to increased domestic demand at higher process. However, with modern production and processing approaches India could substantially increase meat export from the present value. Production of quality animals that can find demand in export market need consideration. The scope for export of value added processed meat products need to be studied. India need to be prepared for price competition in the export market. While disease situation in the country has improved with attaining Rinderpest free status, modern abattoir facilities for slaughter of sheep and goat need to be strengthened for any substantial increase in meat exports.

Sheep and Goat Meat Production Potential

Sheep and goat meat production potential under good management and efficient production system would be substantial from the present low input and unorganized system. Providing necessary scientific practices, technology and critical inputs would contribute for realizing production potential. Minimizing lamb and kid mortality and growing them to optimum weights would increase quality meat production. Attempts to establish rural feed industry to produce agro-byproducts based low cost complete feeds would go a long way in increasing sheep and goat meat production potential.

Complimentarity with Leather Sector

Leather sector is growing steadily and has better prospects with value addition and increasing demand both in the domestic and export market. As the skins are the valuable raw material resource for leather industry sheep and goat production could derive substantial benefit from the growth of leather sector. An integrated approach of leather, feed and meat sectors is required for the mutual benefit of all the sectors and farmers and consumers benefits as well.

Central Leather Research Institute (2005) in a survey study reported the comparative status of slaughter rate and availability of sheep and goat skins (Table 2). The data indicate that mortality rate has decreased with low proportion of skins from fallen animals and decreased number of slaughtered animals per

Table 2. Comparative status of skins availability from sheep and goat in 1987 and 2004.

Parameter	Sheep		Goat	
	1987	2004	1987	2004
Slaughter/ 1000 human population (no)	36	27	89	74
Displacement rate (%)	67	69	73	73
Net off-take rates (%)	60.4	58.5	67.9	64.4
Availability of skins (m)	31.4	29.6	75.4	87.9
Skins availability fallen: meat sources	11:89	1:99	9:91	2:98

1000 human population. Though the displacement rates and off-take rates are nearly sustained but certainly demands attention for augmenting populations of sheep and goat to meet the demand for meat and skins as well. factors of Table 1, changes a number of feed oriented constraints are affecting sustained rural livestock production. Some of these include

- Unorganized small holder livestock units.
- Diversification of traditional crop based agriculture to commercial agriculture affecting feed inputs availability.
- Decrease in grazing land
- Diverse use of crop residues for non-feed uses
- Inadequate efforts for biomass processing to feed
- Non-availability of compounded feed in easy reach and at justifiable cost
- Effect of floods and droughts on rural livestock resources
- Effect of land reforms on rural livestock
- Urban biased feed ingredients and compounded feed industry trade
- Lack of remunerative prices for rural livestock products versus feed costs
- Export of oil meals and cakes

Some Strategies for Sustaining Role of Small Ruminants

- Production and processing heavier weight lambs and kids for better deboning yields and to promote value added products.
- Hygienic meat production at rural, urban and peri-urban levels.

- Production of variety products to cover a larger number of consumers to increase demand.
- Production of convenience products for increasing per capita consumption.
- Production of products using combination of meats for better marketability.
- Incorporation of eggs, vegetables and other non-meat ingredients for a variety of benefits to processor and consumer.
- Wider use of simple technologies with Indian Culinary Practices to increase consumer demand.
- To produce a large variety of cured and smoked products to generate consumer demand.
- Evaluation of rich traditional meat cooking practices to select products and processes of better promise to promote to the present day consumers with advantage.
- Popularizing unique aspects of mutton and goat meat and developing technologies for overcoming deficiencies.
- Organizing clubs and events related to value added sheep and goat meat products.
- Processing products to sustain producer prices, when adversely affected.
- Processing meat to meet a particular demand.
- Processing technologies to promote sheep and goat into National Programs and goals.
- To induct a variety of skilled and unskilled workers into the production and processing activities.
- Technologies for promoting better utilization of tough meat and byproducts from spent sheep and goats.
- Innovative marketing technologies to facilitate availability of products to a large variety of consumers.
- Developing ready reckoner tables with growth rates, optimum slaughter weights and meat yields for different breeds of sheep and goat under different production systems to provide guidance to persons in the sheep and goat related ventures.
- Introduction of grading live animals/ carcasses to the benefit of producer/ processor/consumer
- Developing entrepreneur projects in sheep and goat meat sector for easy adoption by a variety of entrepreneurs.

REFERENCES

Central Leather Research Institute. 2005. Report of the All India Survey on Raw Hides and skins Volume-I, pp 252-254.

Joshi, H.B and Kondaiah, N.1989. Meat production and byproducts utilization from sheep and goats. Animal Productivity (P.N. Bhat, K.K.G.Menon, and .C. Srivastava, eds.) Oxford and IBH Pub. Co., National Symposium on Animal Productivity, April 17-18, Vigyan Bhavan, New Delhi, pp. 599-612

Kondaiah, N, Anjaneyulu, A.S.R. Pragati, H. and Thomas, R. 2006. Opportunities and Challenges in Value addition of Goat and Sheep Meat. In: Souvenir cum proceedings " National Workshop cum seminar on Commercial goat and sheep farming and marketing" 4-5 March, CIRG, Makhdoom, pp 46-56.

National Commission on Agriculture, 1976. Govt. of India, Min. of Agriculture, New Delhi. Pp.177-216.

Report of the Task Force to evaluate the impact of sheep & goat rearing in ecologically frazile zone. 1987. Govt. of India, Ministry of Agriculture, Department of Agriculture and Cooparation. New Delhi, pp15-18.

❑❑❑

4

Nutrition of Lambs in Pre-and Post Weaning Phase of Growth

S.A. KARIM, M.K. TRIPATHI, N.M. SOREN

Central Sheep and Wool Research Institute, Avikanagar, Rajasthan - 304 501

Introduction

Increasing animal production without increasing the number of animal population is great challenge of present animal production system. Animal production is to be enhanced in general and of human food chain in particular. Immediate measures require in increasing carcass quantity and quality with economic husbandry practices. Grass based sheep production system offer the potential of economic animal production to reduce labour and fixed costs. However, offering grass or grazing as the sole foodstuff to ewes and lambs can not support acceptable lamb birth weight, weaning weight and finishing weights thus low weight carcass yield at more days of lamb age. To achieve higher and quality carcass faster rate of growth will desire, which absolutely depend on birth weight, ewes colostrum and milk production efficiency, weaning weight and thus finishing performance, therefore necessary technological interventions have to initiate beginning from the foetal development till slaughter. The importance of nutrition for profitable livestock enterprise is well known fact. Sheep husbandry supports livelihood of fairly good number of farmers of our country. Scientific rearing of lambs has a bearing on the future sheep production which in turn adds to the income generation of the farmers. Phase of lamb nutrition or feeding can be broadly divided into pre-weaning phase which generally last up to 0-3 months, followed by weaning phase at 3 months of age and thereafter post-weaning phase. The nutrition of the dam is equally important as the nutrient available during the pregnancy is deposited/ reflected in the birth weight and overall health status of the lambs. Similarly the birth weight has effect on the pre and post weaning phase of growth.

After birth of the lambs there are two options available to the farmer: firstly to enable the lambs either to suckle their dams *ad libitum* or have to reared artificially on liquid milk substitute. Whichever choice is made it is essential to ensure first that the lamb receives adequate colostrum as soon after birth. Lambs that do not receive adequate colostrum will generally die. Generally ewes have a limited amount of colostrum that will gradually get depleted by the suckling lambs during the first day after birth.

On grass based ruminant production systems, supplementation of concentrate needed to provide adequate nutrition, supplementary feeding of concentrate is crucial for sheep in late pregnancy to achieve desired lamb birth weight, mammary development and colostrum production. Feeding of concentrate containing digestible undegradable protein 25 g/ kg DM recommended (Annett *et al.*, 2005) because fifty percent of foetal growth takes place during final 4 weeks of pregnancy with redistribution of ewe's body reserves to maintain udder development and colostrum production. Higher level of concentrate supplementation require for sheep bearing multiple foetus to meet nutrient supply for foetus and ewe's reproductive developments as ewe's failure to increase feed intake due to foetus. Colostrum contains antibodies from the ewe, which will help to protect the newborns from various diseases. Colostrum is also higher in energy than milk. As lambs are born with few body reserves, it is essential that they need energy soon after birth. If lambs are weak or the ewe has inadequate colostrum, cows' colostrum can be given at about 50 ml/kg live weight of the lambs. The important features of pre- and post weaning nutrition of lambs under scientific rearing are described in brief. Managemental practices help in enhancing performance of pre- and post weaning lambs. Castration at early age of animals improve performance and carcass value. Effect of castration on weight gain are similar in sheep and goat, and castration of kids during first week after birth increase daily gain than in intact male and female kids. Though weight gain of lamb and kid reduce after 7.5 months of age in castrated lambs (Locuca *et al.*, 1977) but for meat production small ruminants are reared up to six months of age, therefore castration would be a management tool to enhance finishing weight of lambs. Assuming a little effect of castration on weight gain Koyuncu et al. (2007) found increased dressing per cent through an increase in per cent of total dissectable muscle weight. Higher finishing weights at an early age provide heavy carcass with better quality (Karim *et al.*, 2007a). Therefore, attempts are required to enhance mutton production from existing numbers of slaughtered animals by nutritional manipulations to have higher finishing weights with better economic returns.

Nutrition of pre-weaner lambs

After ensuring proper colostrum feeding of the lambs, little attention is needed to the nutrition of sucking lambs since almost all the nutrients are supplied through dams' milk but utmost care must be taken to ensure that the ewe gets adequate quantity and quality of feed. Birth weight of lambs no doubt is an indicator of pre-weaning performance but nutritional state of lamb during pre-weaning phase of growth influence weaning weight. Weaning weight of the lambs have significant impact on post weaning performance thereby finishing weight (Fig 1) thus carcass and quantity quality. Nutritional manipulation of pre-weaner lambs enhances performance (ADG up to 165 g) and weaning weight (17.4 kg) and varies with level of nutrition (Table 1). The pre weaner lambs maintained on free suckling and *ad lib.* creep mixture with varying levels of crude protein (CP) indicated that 18 % CP in the creep mixture was adequate for the desired rate of gain of pre- weaner lambs. The lambs in pre weaning phase on an average consumed 83 g DM, 12.0 g DCP and 48 g TDN/kg $W^{0.75}$ amounting to 54 g DCP/M cal DE and had 140 g ADG. Early rumen development also known to improve weaning weights. Live microbes feeding facilitate early establishment of rumen microflora and rumen development. Three yeast strains, *Kluyveromyces marximanus, Saccharomyces cerevisiae* and *Saccharomyces uvarum,* and a mixed culture (1:1:1 ratio) were evaluated for their value as probiotics in lamb feeding. Lambs were offered *ad libitum* a creep mixture and *Zizyphus nummularia* leaves, and yeasts cultures ($1.5\text{-}2\times10^{10}$ live cells ml^{-1}) were dosed orally 1 ml / kg live weight. During 63 days of experiment lambs gained at the rate of 165 g per day with mixed yeast culture feeding and lambs consumed 50 g DCP and 5.34 MJ ME to harvest 165 g ADG. A diet containing 8.45 MJ ME and 80 g DCP in each kg of feed DM seems optimum for harvesting 165 g ADG and more than 16 kg weaning weights (Tripathi et al., 2007a). Pre-weaner lambs may consume concentrate dry matter between 3.0 to 3.6 % of live weight to have an ADG raging from 137 to 163g.

Lambs rearing on milk replacer

Alternate feeding arrangement has to be made for orphan lamb or lambs that have been confiscated from ewes with multiple births (e.g. duplets), some of the suggested methods may be that they will either have to be fostered by other ewes or be raised on milk replacer. Lambs less than four weeks of age that are being bottle fed require a replacer that is high in milk fat and good quality protein, since at this age lambs are essentially non-ruminants, where quality of protein is important than the quantity. During the first weeks of life, lambs' growth and development is dependent upon the ewe's milk production. Lambs not receiving adequate milk during the first month of life are more prone to contracting infectious diseases and will show overall a poor growth performance. Although they will start nibbling on feeds within a few days of birth, lambs less than four weeks of age are non-ruminants and will consume high levels of milk and thus very little in terms of dry matter. Although feed intake is minimal during these first weeks, it is important to introduce creep feed to the lambs at 10 days of age. Milk replacer with added prebiotic (Terre *et al.,* 2007) and probiotic (Fleige *et al.,* 2007) preparations improve pre-weaning performance, thus weaning weights. Addition of lectulose or mannan oligosaccharides 3 % in milk replacer and a probiotic (10^9 CFU of per kg) after one week of birth increased intake, body weight and affect the morphology of the small and large intestine in pre-weaning stage. Experiment with varying plane of nutrition during pre-weaning phase, were maintained under feeding regimens of grazing (C0; Control), grazing and *ad lib* creep mixture supplementation (C-AL; Concentrate) or grazing, *ad lib* creep mixture and milk replacer supplementation (C-ALMR). Accordingly, lamb during pre-weaning phage had a growth rate (ADG, g/day) of 90, 138 and 154 g respectively in C0, C-AL and C-ALMR groups (Bhatt et al., 2007 unpublished data). Lambs were allowed to suckle respective dam in morning and evening till 90 d of age, and fed *ad lib* green leaves of *Ailanthus excelsa* (AL) after grazing. This show that milk replacer feeding did not have added advantage once lambs given free access to a creep mixture.

Creep Feeding

Creep feeding is a simple way to allow lambs access to extra feed supplements or a different supplement, while excluding the ewes. This means that the lambs will still be suckling milk and grazing but will also have extra supplements to make up any shortfall in their intake. The lambs gain access to the feed through a 'creep' which is simply an opening in a fence or gate that is large enough to allow the lambs access but too small for the ewes to enter. Lambs usually commence creep feeding around 10 to 14 days of age, and the amount consumed is inversely proportional to the amount of milk consumed. Inadequate energy intake by suckling lambs is the major cause of slow weight gains. Greater efficiency and lamb weight gains occur if lambs are creep fed than if only the ewes are grain fed. The amount of creep feed consumed by lambs at 2 to 6 weeks of age is affected by the palatability of the ration (ration composition and ration form). Low milk yield of the dam tends to encourage creep consumption but in general the size of the lamb is affected by birth weight and milk consumption. Soybean meal is an important ingredient in creep diets because of its high protein content and palatability. For rapid weight gains, creep diets must be palatable and contains high energy and adequate protein (14 to 18 percent), minerals (especially calcium, since grains are low in calcium, and ideal calcium to phosphorus ratio of the diet should be 2:1) and vitamins should be supplemented.

The most important physiological factor determining successful early weaning and ability to utilize solid food is the state of rumen development. Rumen development is stimulated by the intake of solid feed, which, on fermentation, yields volatile fatty acids. Lambs suckling high-milking yielding dams are less inclined to eat solid feed. To achieve satisfactory performance and encourage rumen growth, lambs should receive a diet that ferments rapidly and does not lead to an accumulation of indigestible fibrous material within the rumen.

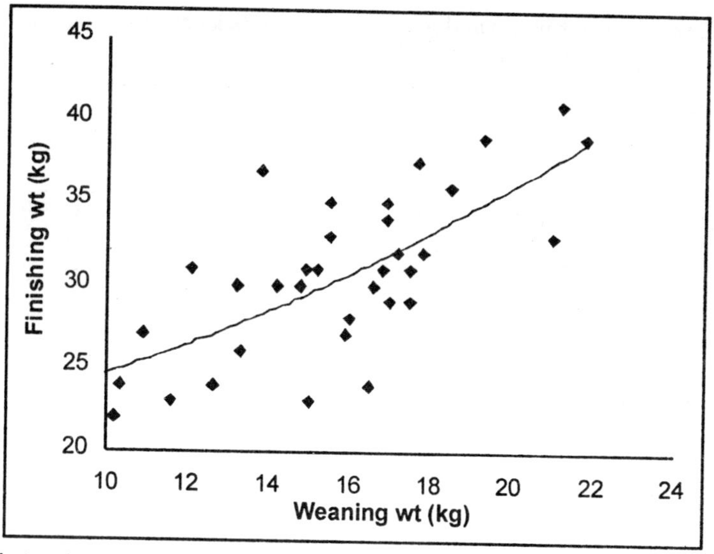

Fig 1. Relationship between weaning weight (WW) and finishing weight (FW) of mutton lambs of Malpura (Finishing live weight (FW, kg) = 19.761 + 0.1636 *(WW, kg) + 0.0325 *(WW, kg)^2)

Early weaning of lambs

As lambs approach weaning age decision must be taken whether to switch lambs to the post-weaning ration before or after weaning. Weaning age directly influences this decision. However, with younger lambs, the change needs to occur after weaning. An additional factor to consider is the cost of creep ration compared to the grower/finisher ration. Lactating ewes normally reach their peak milk production around 3 to 4 weeks postpartum and produce 75 percent of their total milk yield during the first 8 weeks of lactation. While milk production during early lactation can be stimulated through proper selection of feeds, after 6 to 8 weeks milk production declines markedly and high nutrient intake fails to stimulate production (Jordan and Hanke, 1977). There is considerable interest in early weaning of lambs because it will enable early marketing of lambs, out-of-season lambing, multiple lamb crops per year, and use of prolific breeds. Early weaning can be cost-effective because it enables higher and more-efficient gains while the lambs are young and also reduces ewe cost because the ewes can be maintained on a limited feed allowance for longer periods of time between parturitions. Lambs that are to be weaned early should receive creep feed from the time they are old enough to eat solid feed (7 to 14 days of age). At weaning, stress on the lambs should be minimized by removing the ewes and leaving the lambs in familiar surroundings. Since their source of protein from milk has been removed, the level of protein in the dry diet of a 6- to 8-week-old weaned lamb should actually be higher than that for a 3- to 5-week-old suckling lamb (Jordan and Hanke, 1970) and certainly higher than for older lambs.

Table 1. Growth performance and feed conversion efficiency of pre-weaner lambs on different feeding management

Feeding schedule	Initial wt.(kg)	Weaning wt.(kg)	ADG (g)	ECR (%)
Suckling with *ad libitum* Creep mixture and pala leaves (76 day)	4.9	14.3	123	2.5
Suckling with *ad libitum* creep mixture and cowpea hay (54 day)	3.8	11.3	140	3.4
Suckling with grazing and *ad libitum* creep mixture supplementation (77 day)	5.6	16.3	138	3.7
Suckling with grazing, *ad libitum* creep mixture and milk replacer feeding (76 day)	5.5	17.3	154	3.7
Suckling with *ad libitum* creep mixture and y east culture dosing (63 days)	6.0	16.4	165	2.3

Nutrition of post-weaning lambs

Nutrition during the post-weaning phase is very important in the sense that nutritional management will decide the marketability of lambs for different purposes. Dry matter intake for feeder lambs varies between 3.5% and 4% of their body weight. Actual intake, however, depends on several factors, namely, age, size and condition, growth rate, amount of feed offered, competition for available feed, palatability, physical form of feed, energy and fibre content of feed, ambient temperature and humidity and availability of quality water are some of the factors. The post-weaning ration should be a high-concentrate ration with a minimum of 16 percent crude protein, 0.6 percent calcium, and 0.30 percent phosphorus.

Lambs may be marketed at weaning (as new crop lambs), as light weaned lambs or as heavy finished lambs (>33kg). Marketing strategy of the farmer will determine how quickly one wants the lambs to grow (rate of gain) and how soon they will finish. But in country like India the farmers generally maintain the lambs on pasture and market straight away from the field. Rate of gain in these lambs will be slow, but the fed costs will be minimal. Many farmers may supplement grazing with concentrates to improve the rate of gain. The nutrient requirements for lambs will not only be influenced by the desired rate of gain, but also by the weaning weight of the lambs, their sex and their breed. Ewe lambs have lower voluntary feed intakes than either ram lambs or wethers and tend to deposit more fat than males. As a result they grow slower and are less efficiently converts feed to live weight gain. Marked variability has been seen in the growth rates between breeds due to genetic potential. The differences are partially due to differences in feed consumption and efficiencies in feed utilization for fat and lean deposition. Breeds also vary in their mature weights. Breeds with low adult weights will mature at a lighter body weight than large breeds. Therefore, light breeds will tend to fatten (finish) at lighter weights.

The post-weaning diet should be high energy and should contain 18 to 20 percent protein (as-fed basis) for the first 3 weeks and then 14 to 17 percent protein thereafter. It is doubtful whether higher levels of protein will result in sufficiently better lamb performance to justify the cost. Growth of post weaning lambs can be manipulated by feeding of fermented feed, using mushroom by-product, rice bran and barley bran in ratio of 50:20:25 were fermented for 30 days at 60 % moisture level. Feeding of dried fermented product upto 5 % affected the growth performance, carcass traits, meat quality and altered fatty acid content favorably (Song *et al.*, 2007).

Nutrient Requirement of Feeder Lambs

There are several nutrients and nutrient classes that need to be addressed for optimum performance of sheep and lambs. Lists of the important nutrients are summarized as follows

Water intake is usually overlooked. The majority of gain is in fact is water. Proper hydration also improves growth. It is of utmost important that availability of clean potable water is provided at all times. Waterborne minerals may interact with dietary compounds that may improve or inhibit the uptake of some nutrients.

In a typical ration, energy is usually the limiting factor. In modern farm rations, carbohydrate is the prime energy source. There are 2 major categories:

Non-structural carbohydrate (NSC)-This includes starches and sugars. In a grain based ration, starch is the primary source of NSC. NSC is the fraction of the feed that is likely to be associated with acidosis.

Structural carbohydrate (SC) : It consists of fibre and cell walls fractions. These are found in the hull of grains and in forages. SC is less digestible and lower in energy than that of NSC, but structural carbohydrates are usually less expensive. Forage maturity is the major determinant of the digestibility of this fraction.

In mature animals with fully functioning rumens, lower quality protein or non-protein nitrogen (NPN) can be converted to high quality bacterial protein. But in young lambs it is not so, since the rumen may not be fully developed. As a result, protein quality is important. Young animals are more dependent on the presence of higher quality protein in the ration.

Numerous vitamins and minerals (Ca and P) must be included in the diets. Proper ratio of Ca and P has to be maintained, most commonly recommended Ca to P ratio in the diet should be not less than 2:1. In stall fed lambs vitamin A and D_3 supplementation is essential. Feed additives like coccidiostats and ionophore can be added in the diet to improve feed conversion efficiency.

Energy is critical in meeting nutrient requirement of ruminants. Under Indian condition grazing is the major source of nutrient intake by the growing lamb. The tropical ruminants especially sheep have adapted over the years to a continuous low protein regimen and have developed the ability to recycle urea efficiently, hence can sustain themselves on relatively low protein diets. Studies have shown that these animals needs concentrate supplementation only for critical seasons and physiological stages.

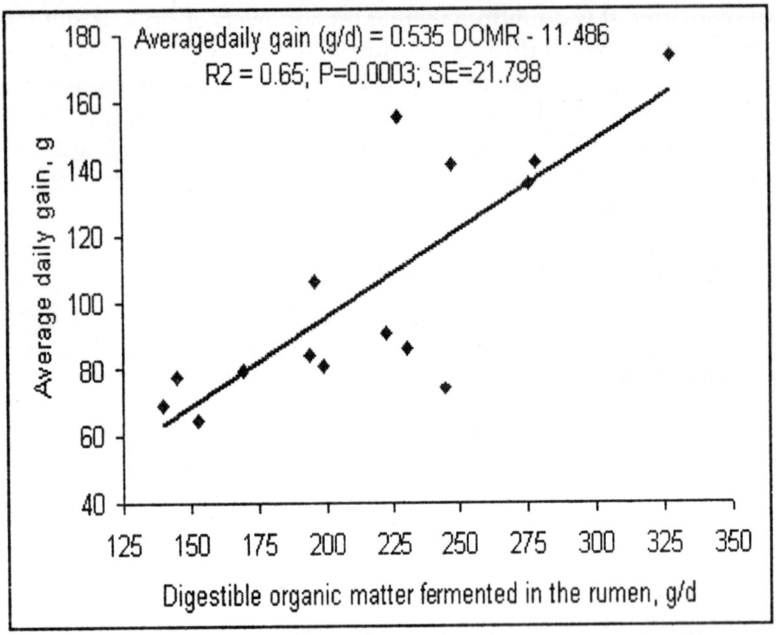

Fig. 2. Relationship between DOMR (g/d) and ADG (g/d) in lambs

Animals those maintained on grazing resources have developed efficient mechanism over the years to select plants parts rich in protein for meeting their protein needs. The sheep are able to maintain CP content of selected vegetation at steady 12-13% even on pasture having as low as 4-5% CP (Shinde *et al.*, 1996). Hence for meeting nutrient requirement, energy supplementation is critical for harvesting desired production traits in animals. For economic mutton production locally available cheaper sources of roughage, energy and protein supplements have to be used in least cost ration formulation. Studies conducted at CSWRI, Avikanagar indicated that for harvesting desired finishing weight khejri, ardu, pala leaves, conserved cultivated grasses and crop residues as roughage sources (Karim and Patnayak, 1991), maize, barley, jowar, bajra or its part replacement by livestock feed grade damaged wheat as energy (Karim and Patnayak, 1992; Tripathi *et al.*, 2007b) and groundnut cake, mustard cake (Tripathi *et al.*, 2001) , til cake, cotton seed cake, gour korma or its part replacement by urea as protein (Karim and Patnayak, 1993) supplement in 50-60 % concentrate based complete feed can be used in least cost ration formulation for intensive mutton production. 100 % concentrate feeding is not recommended in intensive feeding because of higher incidence of enterotoxaemia or over eating diseases. However, high concentrate feeding require to have enhanced growth of lamb, such diets may induce rumen acidosis, which may impair growth performance of lambs. On high concentrate diets supplementation of 1.5 % sodium bicarbonates helps to maintain rumen pH near neutrality, which enhance growth by 35 % (Tripathi *et al.*, 2004). A brief summary of performance studies conducted at CSWRI,

Avikanagar using different breed of sheep and different feeding regimen is presented in Table 2.

Table 2. Growth performance of weaner lambs under different feeding regimen

Breed	Feeding regimen	Initial wt. (kg)	Final wt. (kg)	ADG (g)	FCE ratio	References
Malpura	Grazing with supplementation	6.2	16.2	111	-	Arora et al., 1972
Malpura D×M D×S	25:75 R:C	13.8 12.9 15.1	24.6 27.7 31.1	120 165 178	7.8 5.4 5.5	Sehgal et al., 1982
Sonadi × Dorset	50:50 R:C	13.7	23.3	106	6.3	Sehgal et al., 1983
Malpura	60 days weaning and feeding on 50:50 C:R	14.1	25.4	148	6.3	Karim and Patnayak, 1991
Avibastra	50:50 R:C 90-180 days	19.2	33.5	159	8.9	Shinde et al., 1995
D×M	50:50 R:C 90-180 days	19.7	35.7	178	8.3	Karim and Rawat, 1996
Malpura	60 days weaning and 90 days feeding on 50:50 R:C ration	13.0 14.8 11.6	25.5 30.1 25.1	139 169 150	6.4 6.6 6.5	Karim and Arora, 1997
Avivastra	60 days weaning and 90 days feeding on 50:50 C:R	12.2	23.5	125.5	6.3	Santra and Karim, 1999
	40:60 C:R (3-6 month)	14.4	24.0	108	6.1	Tripathi et al., 1999
	40:60 C:R (6-9 month)	25.6	38.1	139	8.2	Tripathi et al., 1999
	40:60 C:R (0-6 month)	3.5	27.2	131		Tripathi et al., 2001
Malpura	60 days weaning and 90 days feeding on 50:50 C:R	14.8	25.1	133	6.5	Santra and Karim, 2000
Malpura	90 days weaning and 90 days feeding on 75:25 C:R	14.0	22.2	91.3	5.7	Santra et al., 2003
Malpura	75:25 C:R	14.3	25.8	128	5.6	Tripathi et al., 2004
Malpura	Concentrate 1.5% BW	13.8	20.9	77.2	9.2	Tripathi et al., 2007c
	Concentrate 2.5% BW	14.3	23.2	97.6	8.5	
	Concentrate *ad lib*	13.5	27.2	151	6.8	
Malpura	40:60 C:R Feed block (6-9 month)	22.1	31.0	101		Raghuvansi et al., 2007
Bharat Merino	75:25 C:R (3-6 month)	16.0	30.0	154	6.6	Tripathi et al., 2007b

D: Dorset; M: Malpura; S: Sonadi: R: roughage; C: Concentrate

Adequate energy and protein has to be provided in balanced amount for optimum growth of the lambs. Small ruminant production system in India and other tropical countries largely involved tree foliage based feeding system that contains condensed tannins. Maintaining a condensed tannins (CT) intake 5.8 g day^{-1} or 6.9 g kg^{-1} DMI improved microbial protein yield in lambs maintained with tree leaves and free choice of concentrate feeding (Tripathi *et al.*, 2006). Maintaining an adequate level of digestible organic matter fermented in rumen (DOMR), linear positive relationship with daily gain (Fig 2. ADG, g/d= 0.535DOMR-11.486, R^2=0.65, SE=21.8, P<0.003) and N retention (Fig 3. N retention, g/d = 0.0663DOMR-6.468, R^2=0.82, SE=1.727, P<0.001) obtained. Under field conditions feeding of lambs with concentrate supplementation at 2.5 % of their body weight with routine grazing management seems promising because of reasonable feed cost, better growth (Tripathi *et al.*, 2007b) and acceptable fat and lean meat (Karim et al., 2007b). Results of series of experiments extended over two decades at CSWRI, Avikanagar indicated that for maintenance the lambs require 120 kcal DE and 3 g DCP per kg$W^{0.75}$/d (Table 3), dietary DCP: ME ration depicts ADG of post weaner lambs (Fig 4).

Table 3. Nutrients requirements of lambs at fro growth

	ADG (g/d)	DMI % BW	FE kg (feed/ gain)	Diet ME (MJ/kg DM)	Diet DCP %	ME intake (MJ/kg gain)	Met. Protein intake (g/ kg gain)
Pre-weaner	150	3.0-3.8	2.7 – 3.0	8.2	8	-	-
Post weaner	150	3.5-4.9	5.2-7.0	10.8	8	41-60	544

Growth Manipulation of Post Weaner Lambs

The metabolic activity of the microbes determines the rate of utilization of substrate, depends on various microbial interactions and the substrate availability. The availability of easily degradable carbohydrates may impair degradation of fiber. Rumen pH affects fiber degradation and pH is influences by the concentration of lactate and VFA (Argyle and Baldwin, 1988; Tripathi *et al.*, 2007c). Thus, fermentation of starch and sugars, which may lead to large levels of VFA and lactate in the rumen fluid, decreases pH and hence affects the fermentation of fiber. The determination of rumen fluid pH, though is not an important determinant of substrate degradation in high-producing dairy cattle but important for microbial non-growth (maintenance) energy costs, remains a challenge of contribution of microbial species or microbial groups towards substrate degradation. Bacteria, protozoa and fungi all contribute to plant-cell-wall degradation in the rumen. The protozoal and bacterial metabolism, and interactions between protozoa and bacteria, was able to quantify the contributions of both groups to fiber degradation in the rumen. The results of the simulations indicate a protozoal contribution to ruminal fiber degradation on a variety of diets of 5–35 %.

The level of intake and composition of the diet causes large variation in microbial recycling due to protozoal metabolism within the rumen. Such variation has a significant impact on microbial protein yield. If 70 % of N requirements are met by ammonia and 30 % by preformed amino acids, 3·0 g OM is required to synthesize 1 g microbial protein (Dijkstra et al., 1992). However, the synthesis of 1 g microbial protein flowing from the rumen requires 3·8 g OM when recycling is 20 % of gross synthesis, and 30·0 g when recycling is 90 % of gross synthesis. This indicates the significance of protozoal activities on net yield of microbial protein. Relatively high protozoal contributions were obtained with diets high in starch and sugars fed at low levels of feed intake. With such diets, defaunation would be expected to decrease fibre degradation to a larger extent than diets that are high in fiber, and this is confirmed by *in vivo* studies on defaunation in sheep. Feed related defaunation could be achieved using 20 % cottonseed in sheep diets, by decreasing 50 % protozoa population (Dayani *et al.,* 2007). Thus, the development and application on the effect of defaunation on fibre degradation in the rumen is the interaction between protein availability and degradation of carbohydrates. This interaction is not only important in situations where the quality of the feed and levels of N are poor (mainly in the tropics and subtropics), but is also an issue in areas of intensive farming in developing countries, where concerns about high levels of N excretion to the environment by cattle constrain an efficient capture of N from the feed into end products (Tamminga, 1996). Degradation of OM in the rumen is depressed when the availability of N is low, probably because N deficiency impairs rumen microbial growth (Clark *et al.,* 1992; Tripathi *et al.,* 2006) and the resultant decrease in the amount of microbial biomass reduces substrate degradation. However, rumen bacteria are efficient scavengers of N sources, and ammonia concentrations above some 20–50 mg/l

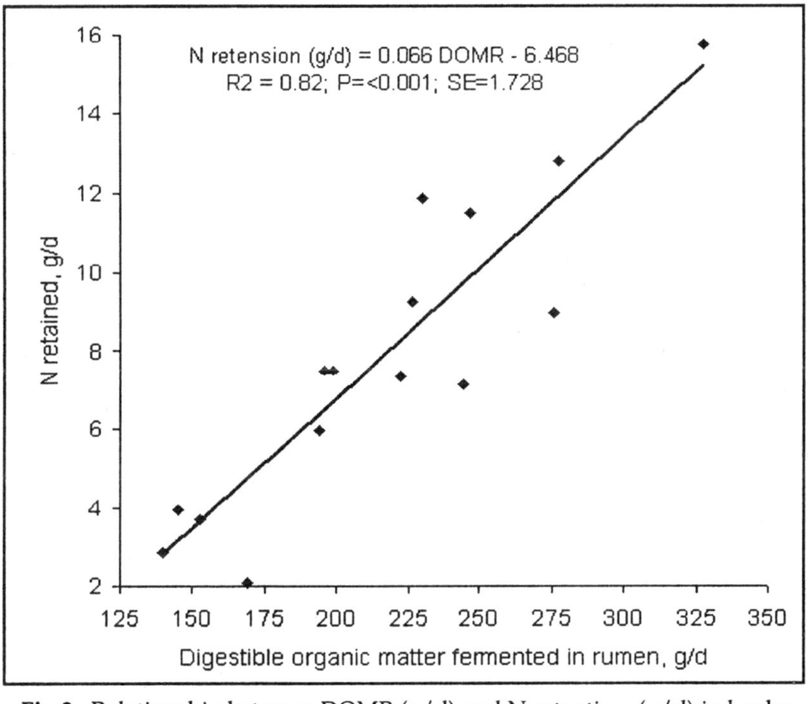

Fig 3. Relationship between DOMR (g/d) and N retention (g/d) in lambs

do not stimulate microbial growth further *in vivo* (Clark *et al.* 1992). The optimal levels of N supplementation require when the basal diet is deficient in N and this level of N supply be relative to energy supply (Tripathi *et al.*, 2007b), below which microbial synthesis is impaired. Next to substrate degradation, the growth yield is the major determinant of microbial protein synthesized, the overall use of substrates by growing microbes can be divided into substrate utilization for non-growth (maintenance) purposes and for biomass increase.

Sustainable Approach of Organic Mutton Production and Climate Change

Consumer preference toward food safety increased currently and organic food production is being propagated. Organic lamb production would improve producer income without added cost of production, at the same time organic animal products of human food chain have great market demand and at higher value. Recently developed countries European Union particular banned antibiotics use in animal feeding as growth promoters. Sustainable and alternate sources are being explored that substitute antibiotics in animal feeding. Plant secondary metabolites are being explored with great hope and a few plants have shown this ability. A mixture of three plant extract *Cynara cardunculus* subesp. *scolymus, eleutherococcus senticosus,* and *Trigonella foenum graceum. C. cardunculus* subesp. *scolymus.* (artichock) is composed of cynarin, dicaffeoylquinic acid isomer, cynaropicrin, flavonoids, chlorogenic and caffeic acid have been reported as alternate to antibiotics in high concentrate fed ruminant, supplementation of 2.8 g/ kg DM plant extract altered rumen fermentation and growth similar to monensin (32 mg/ kg DM) supplementation (Devant *et al.*, 2007). Aqueous extract of *Biophytum petersianum*, a source of saponins dosing upto 26 mg/ kg body weight as feed additive manipulate goat rumen fermentation which reduced ruminal ammonia N concentration, urinary N excretion, thereby increasing microbial N supply and retained N as proportion of N digested (Santoso *et al.*, 2007) is due to antiprotozoal activity (Baah et al., 2007). Thus saponin has positive impact on animal performance and environment. Other plant secondary metabolites such as glucosinolate (Tripathi and Mishra, 2007) and tannins (Clauss *et al.*, 2007) might contribute to overall gut health by controlling the number of potentially pathogenic bacteria.Tannins could enhance antioxidant status (Clauss *et al.*, 2003, 2006ab), and there is evidence that they reduce pathogenic organisms (Garcia *et al.*, 2007). Tannins have also been shown to be effective against helminth parasites in small ruminant and tannins are potent bacteriostatic and bacteriocidal substances for a large range of pathogenic bacteria including *Enterobacteriaceae* such as *Escherichia coli, Salmonella,* or *Proteus spp.*, or obligate anaerobes such as *Clostridium spp.* Patra et al (2007), evaluated extracts of pods of *Acacia concinna* (Shikakai), seed pulp of *Terminalia chebula* (harad), *Terminalia belerica* (bahera), *Emblica officinalis* (amla) and seed kernel of *Azadirachta indica* (neem seed) in different solvents (ethanol, methanol and water) and reported reduced methane Production. Diet fed to animal also affects health promoting properties of milk and meat. Botanical composition of grazed pasture also affects the fatty acid profile of milk and meat. Some of the changes in tissue fatty acid profile to be related to modifications of the rumen and tissue fatty acid metabolism. Grazing a botanically diverse pasture compared with a intensive mono pasture results an accumulation of biohydrogenated intermediates in rumen and in an increase of long chain poly-unsaturated fatty acids in inter muscular fat. Loureno et al. (2007) suggested that a botanically diverse pasture partially inhibit rumen biohydrogenation and stimulate desaturation and elongation of poly-unsaturated fatty acids. In general, high forage diet increase biohydrogenation of C18 unsaturated fatty acids. Low fiber diet improves level of linoleic and linolenic acid suggesting a low level of biohydrogenation by rumen microbes. Under low rumen pH conditions supplementation of 10g fish oil/ kg diet DM enhance t10 C18:1 and t10 $C:_{12}$ CLA production in rumen (AbuGhazaheh and Jacobsom, 2007). The different positioned and geometric isomers of conjugated linoleic acid (CLA) confer different health effects on mammals. The CLA and trans C18:1 fatty acids are produced during biohydrogenation of unsaturated fatty acids in rumen and are subsequently incorporated in to milk and meat of ruminant animals. Therefore, plant extracts form diverse population with reasonable concentration of secondary metabolites are good candidates for achieving rumen manipulation of rumen microbial ecosystem for organic animal production with better solutions of climate change, which enhance fibrous feed digestibility, reduce methane emission and nitrogen excretion by ruminants needs to be investigated in details, so that viable ruminant feeding strategies would be developed.

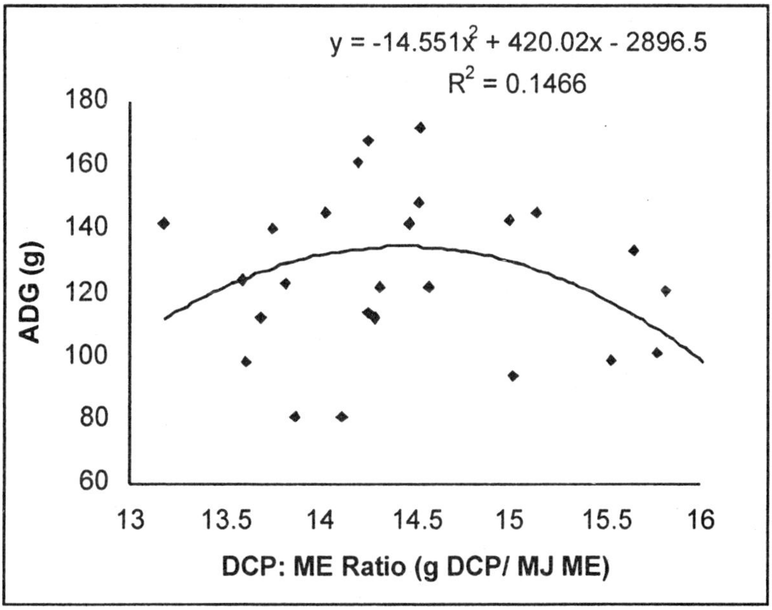

Fig 4. Dietary DCP: ME ration requirement in relation to Daily gain

Conclusion and Recommendations

The knowledge of nutritional requirement of the lambs at different phase of growth would help to formulate economic balanced ration for profitable sheep rearing. Standard husbandry practices vis-à-vis nutritional management will help to enhance the income generation of farmers for long term animal husbandry sustainability. Nutrition of lambs during pre-weaning phase of growth has significant effect on post weaning performance and finishing weight. Higher finishing weights at an early age produce good quantity and quality of carcass. Ewes' nutrition during late pregnancy is critical to meet nutrient requirement of growing foetus, its development, weight at birth and colostrums production, which indeed influence pre-weaning performance of lamb. Supplementary concentrate feeding (250 g concentrate containing 25 g RUDP/ kg DM) of ewes will provide of adequate nutrition for single foetus however multiple foetus bearing ewes would require higher concentrate supplementation. Pre-weaner lambs in addition to free suckling given *ad libitum* creep mixture and routine grazing shall be optimum to harvest enhanced weaning weight. Early rumen development techniques using yeast culture supplementation, dosing of rumen liquor from adult sheep and pre-biotic shall offer improved management of pre-weaner lambs. Post weaning phase of growth could maximize with *ad libitum* concentrate feeding under cafeteria system of feeding management, however a feeding management of free grazing with concentrate supplementation at 2.5 % of body weight would be economic under field conditions. Alternate energy and protein sources such as damaged wheat, damaged rice, mustard cake etc. can reduce cost of feeding if used judicially. Using modern feeding technologies such as microbial cultures feeding, use of plant secondary metabolites, blend of essential oil and combinations of several feed resources could manipulate growth for organic meat production with better solutions of climate change.

REFERENCES

AbuGhazaheh, A. and Jacobsom, B.N. 2007. Production of trans C18:1 and conjugated linoleic acid in continuous culture fermentor fed diets containing fish oil and sunflower oil with decreasing levels of forage. *Animal* 1, 660-665.

Annet, R.W., Carson, A.F. and Dawson, L.E.R. 2005. The effect of digestible undegradable protein (UDP) content of concentrate on colostrum production and lamb performance of triplet bearing ewes on grass-based diets during late pregnancy. *Animal Science* 80, 101-101.

Argyle, J.L. and Baldwin, R.L. 1988. Modeling of rumen water kinetics and effects of rumen pH changes. *Journal of Dairy Science* 71, 1178-1188.

Arora, C.L. Mahajan, J.S. and Nivsarkar, A.E. 1972. AICRP on sheep for mutton. Annual report, CSWRI, Avikanagar, Rajasthan, India, pp. 36-41

Baah, J., Ivan, M., Hristov, A.N., Koening, K. M., Rode, L.M. and McAllister, T.A. 2007. Effects of potential dietary antiprotozoal supplement on rumen fermentation and digestibility in heifers. *Animal Feed Science and Technology* 137, 126-137.

Clark, J.H., Klusmeyer, T.H. and Cameron, M.R. 1992. Microbial protein synthesis and flows of nitrogen fractions to the duodenum of dairy cows. *Journal of Dairy Science* 75, 2304-2323.

Clauss, M., Castell, J.C., Kienzle, E., Dierenfeld, E., Flach, E.J., Behlert, O., Ortmann, S., Streich, W.J., Hummel, J. and Hatt, J.M. 2006a. Digestion coefficients achieved by the black rhinoceros (Diceros bicornis), a large browsing hindgut fermenter. *Journal of Animal Physiology and Animal Nutrition* 90, 325-334.

Clauss, M., Lason, K., Gehrke, J., Lechner-Doll, M., Fickel, J., Grune, T. and Streich, W.J. 2003. Captive roe deer (Capreolus capreolus) select for low amounts of tannic acid but not quebracho: fluctuation of preferences and potential benefits. *Comparative Biochemistry and Physiology* B 136, 369-382.

Clauss, M., Pellegrini, N., Castell, J.C., Kienzle, E., Dierenfeld, E.S., Hummel, J., Flach, E. J., Streich, W.J. and Hatt, J. M. 2006b. Antioxidant status of feces of captive black rhinoceros (Diceros bicornis) in relation to dietary tannin supplementation. *Journal of Veterinary Medicine Series* A 53, 319-322.

Clauss, M., Wittenbrink, M.M., Castell, J.C., Kienzle, E., Dierenfeld, E.S., Flach, E.J., Macgregor, S.K., Hoppe, T., Hummel, J., Streich W.J. and Hatt, J.-M. 2007. Quantification of *enterobacteriaceae* in faeces of captive black rhinoceros (Diceros bicornis) in relation to dietary tannin supplementation. *Journal of Animal Physiology and Animal Nutrition* DOI: 10.1111/j.1439-0396.2007.00687.x

Dayani, O., ghorbani, G.R., Alikhan, M., Rahmani, H.R. and Mir, P.S. 2007. effect of dietary whole cotton seed and crude protein level on rumen protozoal population and fermentation parameters. *Small Ruminant Research* 69, 36-45.

Devant, M., Anglada, A. and Bach, A. 2007. Effect of plant extract supplementation on rumen fermentation and metabolism in young Holstein bulls consuming high level of concentrate. *Animal Feed Science and Technology* 137, 46-57.

Dijkstra, J., Neal, H.D.S.C., Beever, D.E. and France, J. 1992. Simulation of nutrient digestion, absorption and outflow in the rumen: model description. *Journal of Nutrition* 122, 2239–2256.

Fleige, S., PreBinger, W., Meyer, H.H.D. and Pfaff, M.W. 2007. Effect of lectulose on growth performance and intestinal morphology of pre-ruminant claves using milk replacer containing *Enterococcus faecium*. *Animal* 1, 367-373.

Garcia, V., Catala-Gregori, P., Mandrial, J., Hernandez, F. and Megias, A.M. 2007. Potential of carvacrol to modify in-vitro rumen fermentation as compared with monensin. *Animal* 1, 675-680.

Jordan, R.M. and Hanke, H.E. 1970 Protein requirements of young lambs. *Journal of Animal Science* 31, 593-597.

Jordan, R.M. and Hanke, H.E. 1977. Effect of level of grain fed ewes during late lactation on lamb production. *Journal of Animal Science* 45, 945-951.

Karim, S.A. and Arora, A.L. 1997. Growth performance of lambs weaned at 60 days and 90 days of age. *Indian Journal of Animal Sciences* 67, 536-537.

Karim, S.A. and Patnayak, B.C. 1991. Feeding different roughage sources in complete feed of lambs. Annual Report, CSWRI, Avikanagar, Rajasthan, India. pp-38.

Karim, S.A. and Patnayak, B.C. 1992. Growth performance of weaner lambs on rations with different energy sources. Annual Report, CSWRI, Avikanagar, Rajasthan, India. pp-25.

Karim, S.A. and Patnayak, B.C. 1993. Performance of weaner lambs on rations with different protein sources. Annual Report, CSWRI, Avikanagar, Rajasthan, India. pp-20.

Karim, S.A. and Rawat, P.S. 1996. Growth performance of native and crossbred weaner lambs under intensive feeding. *Indian Journal of Animal Sciences* 66, 830-832.

Karim, S.A., Kuldeep Porwal, Suresh Kumar and Singh. V.K. 2007a. Carcass traits of Kheri lambs maintained on different system of feeding management. *Meat Science* 76, 395- 401.

Karim, S.A., Tripathi, M.K. and Singh, V. K. 2007b. Effect of varying levels of concentrate supplementation on growth performance and carcass traits of finisher lambs. *Livestock Research for Rural Development* 19, Article No. 173.

Koyuncu, M., Duru, S., Uzun, K. and Tuncel, E. 2007. effect of castration on growth and carcass traits in goat kids under semi-intensive system in Turkey. *Small Ruminant Research* 172, 38-44.

Locuca, A., Ecconomides, S. and Handcock, J. 1977. Effects of castration on growth rate, feed conversion efficiency and carcass quality in Damascus goats. *Animal Production* 24, 387-391.

Loureno, M., SmetDe, S., Raes, K., and Fievez, V. 2007. Effect of botanical composition of silage on rumen fatty acid metabolism and fatty acid composition of longissimuus muscle and subcutaneous fat of lambs. *Animal* 1, 911-921.

Patra, A.K.,. Kamra, D.N. and Agarwal Neeta. 2006 Effect of plant extracts on *in vitro m*ethanogenesis, enzyme activities and fermentation of feed in rumen liquor of buffalo. *Animal Feed Science and Technology* 128, 276-291.

Raghuvansi, S.K.S., Prasad, R., Tripathi, M.K., Mishra, A.S., Chaturvedi, O.H., Misra, A.K., Saraswat, B.L., Jakhmola, R.C., 2007. Intake, performance, nutrient utilisation, rumen fermentation and rumen microbial enzymes of lambs maintained on complete feed blocks or grazing and supplementation. *Animal* (Cambridge) 1, 221-226.

Santoso, B., Kilmsakossu, A. and Sambodo, P. 2007. Effects of saponin from *Biophytum petersianum* Klotzsch on rumen fermentation, microbial protein synthesis and nitrogen utilization in goats. *Animal Feed Science and Technology* 137, 58-68.

Santra, A. and Karim, S.A. 2000. Growth performance of faunated and defaunated Malpura weaner lambs. *Animal Feed Science and Technology* 86, 251-260.

Santra, A. and Karim, S.A.1999. Growth performance of weaner lambs maintained on varying levels of dietary protein in pre-weaning phase. *Indian Journal of Animal Sciences* 69, 722-726.

Santra, A., Chaturvedi, O.H. Tripathi, M.K., Kumar, R. and Karim, S.A. 2003. Effect of dietary sodium bicarbonate supplementation on fermentation characteristics and ciliate protozoal population in rumen of lambs. *Small Ruminants Research* 47, 203-212.

Sehgal, J.P. Singh, M., Rawat, P.S. and Singh, R.N. 1983. Feedlot performance in growing lambs for mutton production. *Indian Journal of Animal Sciences* 53, 715-719.

Sehgal, J.P., Karim, S.A. Singh, M. and Acharya, R.M.1982. A note on cross breeding for fat lamb production in hot tropics. *Animal Production* 35, 435-438.

Shinde, A.K., Karim, S.A. Singh, N.P. and Patnayak, B.C. 1995. Growth performance of weaner lambs and kids under intensive and semi-intensive feeding management. *Indian Journal of Animal Sciences* 65, 830-833.

Shinde, A.K., Sankhyan, S.K., Karim, S.A., Singh, N.P. and Patnayak, B.C. 1996. Nutrient intake, its utilization and performance of sheep and goats on semi arid *Cenchrus ciliaris* pasture. *World Review of Animal Production* 31, 35-40.

Song, Y.M., Lee, S.D., Chowdappa, R., kim, H.Y., Jin, S.K. and Kim, I.S. 2007. Effect of fermented Oyster mushroom (*Pleurotus ostreats*) by-product supplementation on growth performance, blood composition and meat quality. *Animal* 1, 301-307.

Tamminga, S. 1996. A review on environmental impacts of nutritional strategies in ruminants. *Journal of Animal Science* 74, 3112-3124.

Terre, M., Calvo, M.A., Adelantado, C., Kocher, A. and Bach, A. 2007. Effects of mannan oligosaccharides on performance and microorganism feacal counts of calves following an enhanced growth feeding program. *Animal Feed Science and Technology* 137, 115-125.

Tripathi, M.K. and Mishra, A.S. 2007. Glucosinolate in animal nutrition, A Review. *Animal Feed Science and Technology*. 132, 1-27.

Tripathi, M.K., Chaturvedi, O.H., Karim, S.A., Singh, V.K. and Sisodiya, S.L., 2007b. Effect of different levels of concentrate allowances on rumen fluid pH, nutrient digestion, nitrogen retention and growth performance of weaner lambs. *Small Ruminant Research* 72, 178-186.

Tripathi, M.K., Karim, S.A., Chaturvedi, O.H and Verma, D.L. 2007c. Nutritional value of animal feed grade wheat as replacement of maize in lamb feeding for mutton production. *Journal of the Science of Food and Agriculture* 87, 2447-2455.

Tripathi, M.K., Karim, S.A., Chaturvedi, O.H. and Verma, D.L., 2007a. Effect of different liquid cultures of live yeast strains on performance, ruminal fermentation and microbial protein synthesis in lambs. *Journal of Animal Physiology and Animal Nutrition* DOI: 10.1111/ j. 1439. 2007.00759.x.

Tripathi, M.K., Karim, S.A., Chaturvedi, O.H. and Singh, V.K., 2006. Effect of *ad libitum* tree leaves feeding with varying levels of concentrate on intake, microbial protein yield and growth of lambs. *Livestock Research for Rural Development*. 18, 1-6.

Tripathi, M.K., Mishra, A.S., Misra, A.K., Mondal. D. and Karim, S.A. 2001. Effect of substitution of groundnut with high glucosinolate mustard (*Brassica juncea)* meal on nutrient utilization, growth, vital organ weight and blood composition of lambs. *Small Ruminant Research* 39, 261-267.

Tripathi, M.K., Santra, A., Chaturvedi, O.H. and Karim, S.A. 2004. Effect of graded levels of $NaHCO_3$ supplementation on ruminal fluid pH, feed intake, nutrient utilization, growth performance and blood biochemical constituents of lamb on high concentrate diets. *Animal Feed Science and Technology* 111, 27-39.

Tripathi, M.K., Tyági, A.K. and Karim, S.A. 1999.Performance of weaner lambs maintained on mustard and groundnut meal based diets. *Indian Journal of Animal Sciences* 69, 991-992.

5

Application of Rumen Microbiology Techniques and Biotechnology to Improve Quality and Quantity of Mutton Production

M.K. TRIPATHI, S.A. KARIM
Central Sheep and Wool Research Institute, Avikanagar (via- Jaipur), Rajasthan 304 501

Introduction

To meet increase demands for animal foods of increasing human population, increased urbanization and income growth further increased demand of food from animal origin, increased animals is to achieve without increasing animal population. Enhanced animal productivity form existing animal population is to be achieved without any adverse effect on environment. The major limitation of ruminant productivity in tropical countries is poor nutrition. The productivity is restricted by the low nitrogen and high fibre content of the feed used in ruminant feeding. Dietary agents that especially increase microbial enzymes associated with improved energetic efficiency of rumen fermentation accelerate rumen microbial propagation, increase feed digestion and animal performance will have concrete influence for improved ruminant production. Attempt of rumen manipulation which enhanced feed intake and production efficiency of ruminant would provide sustainable solution for modern animal production system. The rumen microbiology and biotechnology approach seems prosing to bring desirable modifications in rumen ecosystem in modern eco-friendly animal production systems.

Rumen is the principle site of feed fermentation by the resident microbiota to provide nutrient for the nourishment of host animal. The principal end products of the fermentation of carbohydrates in the rumen are the volatile fatty acids, carbon dioxide and methane. The rumen ecosystem comprises a complex of dense microbial communities of bacteria, archaea, protozoa, fungi and bacteriophages (Mackie *et al.* 2002). The fermentation effected by this complex microbiota is responsible for the conversion of plant feedstuffs to compounds that can be utilized by the animal. Hence, the fermentations and interactions of the microbes are central to ruminant digestion and nutrition. Cellulose, the most stable plant structural polysaccharide, is degraded by only a few species of rumen bacteria, *Fibrobacter succinogenes, Ruminococcus flavefaciens* and *Ruminococcus albus*, These species are very important, as a lack of breakdown of cellulose will impede access of other bacteria to their required substrates and reduce the efficiency of fermentation of feed material. The cellulolytic populations of the rumen are subject to lytic events that may impact on the ability of these bacteria to degrade plant fibre and on the nutrition of the animal (Klieve *et al.*, 2004).

To manipulate rumen ecosystem, several ionophores type feed additives such as monensin, salinomycin, lasaloacid and tetronacin developed for ruminant feeding, which altered rumen fermentation characteristics and metabolism favourably through modifying the rumen microflora pattern that decrease acetate and methane production, increase propionate results an improved nitrogen utilisation, dry matter digestibility and thus enhanced animal performance (Berger and Bates, 1984; Russell and Strobel, 1988; Mohammed *et al.*, 2004). Ionophores can reduce the rate of amino acid deamination (McAllister *et al.*, 1994). Low protein and high energy diet also reduce methane production. Non-ionic surfactants such as Tween 80 also improve rumen fermentation and animal performance by reducing acetate: propionate ratio through enhanced propiogenic activities (Wang *et al.*, 2004). Additives including the both surfactants and ionophores enhance the beneficial effects on rumen enegetic efficiency because of synergetic and interactive effects of increased accumulation of reducing sugars and reduced acetate: propionate ratio. The use of microbial feed additives and/ or addition of microbes in ruminant diets (microorganism, i.e., bacteria, fungi, their products or spent growth medium) are increasing because these feed additives are an alarming alternative to antibiotics. Several probiotics are known to improve animal performance and reduce methanogenesis through establishment and enrichment of beneficial microbial strains under anaerobic environments. Supplementation of probiotics (yeast cultures) or direct fed microbes and prebiotics reported to decrease methane production and improve energy and protein utlisation in ruminants (Chaucheyras *et al.*, 1995; Mwenya *et al.*, 2004; Sar *et al.*, 2005). If the methanogenesis are inhibited, hydrogen accumulation and, the hydrogenases are inhibited and the carbohydrate fermenting bacteria utilise other mechnism of reducing equivalent disposal i.e., the dehydrogenases of propionate production. Reduced ruminal pH is often regarded for the establishment of propionate producing bacteria but cellulolysis is inhibited. However, supplementation of lactic acid bacteria (*L. acidophilus*) decreased ruminal pH and increase cellulolytic bacterial numbers in the rumen (Dawson *et al.*, 1990). In the rumen methane is mainly produced from H_2, which is formed by a reductive reaction. In principal H_2 formation can be reduced by augmenting other reductive reactions through competition for electrons. Propionate formation is the most important reductive reaction in rumen, and a reciprocal relationship is generally observed between methanogenesis and propionate production. Ionophores and high concentrate diets usually decrease methane and increase propionate (Miller, 1995). Propionate is produced by rumen bacteria, *Megasphaera elsdenii* and *Provotella* utilise lactate, whereas *Selenomonas ruminantium, Succinimonas amylolytica, Propionibacterium* acnes and *Veinella parvula* used succinate pathway. Bacterium *Selenomonas ruminantium* accounts 22-51 % of the total viable bacterial counts in the rumen, therefore augmentation of propionate production by *Selenomonas ruminantium* could possibly increase propionate production and reduce methanogenesis (Asanuma and Hino, 2001). Reduced methanogenesis is coupled with higher microbial protein synthesis in the rumen. Greater diversity of rumen bacteria was reported with high grain diet with consequence of the declaimed *Streptococcus bovis* and the diversity in the high fibre diet was lower and a dominance of *Streptococcus bovis* (Krause *et al.*, 2003). Present paper provide an over view of the progress made so far to manipulate rumen fermentation using microbiology and biotechnology approaches to improve nutrient availability and efficiency in rumen for enhancing ruminant productivity.

Biotechnology in Growth Manipulation

Dietary growth enhancers

Dietary feed additives of a non-nutritional nature, based on biotechnology may have major effect on growth and performance of meat ruminant production. To separate growth from better health and welfare is difficult. An animal which in a good state of health, provided with adequate nutrition will certainly achieve good production level. Products which influence these aspects include antibiotics, non-hormonal growth promoters, microbial products, including probiotics and enzymes and oligosaccharides. Antibiotics have been used effectively for decades in a sub-therapeutic role and performance enhancers. Zinc bacitracin and virginamycin can be used without veterinary prescription and have been actively promoted as growth enhancers, making significant contribution to animal production. Rumen bacteria and other microbes are chemical processors, capable of catalyzing thousands of chemical reactions that are also useful to the host animal. Many bacteria have been used in feed industry for the production of enzymes and also as probiotic feed additives. Now days, the feed industry is actively pursuing alternatives of antibiotics with the aid of other microbiology and biotechnology approaches. Probiotics use in ruminant feeding are the current development of biotechnology and proposed as alarming alternate of antibiotics in growth promotion and gut health. Fermented feeds and use of enzymes to improve rumen energetic efficiency are the other rumen microbiology and biotechnology developments.

Pro-biotics

The term "probiotic" has been defined as "a live microbial feed supplement, which beneficially affects the host animal by improving its intestinal microbial balance" and has been used to describe viable microbial cultures, culture extracts, enzyme preparations, or various combinations of the above. Microorganisms used as probiotic for ruminants include viable cultures of fungi, yeast and bacteria. Probiotics offer potential as natural products to bridge the gap which would be left if the use of antibiotics were to be reduced. The proliferation of products described as probiotics has been enormous in recent times. Selection of correct organism, appropriate strain of that organism and the way in which the organism is cultured for the type of animal in which it is to be used is important. On the entire products host to show a response in animal production following their feeding requires an increasing microbial load on the structure and metabolic activities of the rumen. A logical approach in this direction is that the identified a particular microbial strain, when fed to ruminants, is able to survive and leading to a substantial improvement in nutrient utilization, detoxification of feed anti-nutrient and thus improve production. The Application of large amounts of nitrogen fertilizers are thought to be responsible for the accumulation of nitrates and nitrites in the stems and leaves of plants fed to ruminants. In the rumen, nitrates are converted to methaemoglobin which results in a loss of oxygen transporting capacity of the blood and anoxia. Nitrite toxicity as it should be properly known leads to reduced performance, abortion and even death. A rumen *Propionibactor* converts nitrite to nitrate thus reducing nitrite toxicity. Thus a *Propionibactor* containing probiotic would be of potential interest to tackle this problem. The probiotic supplementation is gaining popularity among fat animal production systems, because these can be fed to animal either in the form of liquid cultures or active dry powders. Dried yeast powder if supplemented as probiotic must contain not less than 15×10^9 live yeast cells/g of powder. Probiotics (*Saccharomyces cerevisiae* strains) are more potent for reducing blood base consumption than ionophores (monensin) supplementation and helps to maintain a higher rumen pH. Probiotics also exert inhibitory activity on different pathogenic bacterial strains is due to production of either bacteriostatic or bactericidal activity related to yeast metabolic compounds.

Pre-biotics

Pre-biotics also known as "Neo Sugars" or oligosaccharides are polymers of simple carbohydrates and naturally present in many raw materials used in the ruminant feeding. Many of the physiological effects of these dietary carbohydrates polymers have been recognized for gut development and health. The recognition of pre-biotics have an important role as receptors for bacterial and viral pathogens, parasites, antibodies, food lectins and in normal cell processes. These also stimulate early gut development in young animals by facilitating gut microfloral development, as they are substrate for probiotic microorganism and of natural intestinal flora. The main pre-biotics of animal feed chain are polymers of sucrose (â-Fructo-oligosaccharides), lactose (Galacto-oligosaccharides; Lactulose), starch (Maltotetraose, Isomalto-oligosaccharides), xylans (Xylo-oligosaccharides) and chitin (Chito-oligosaccharides). Pre-biotics also reduces the gut pathogenesis by reducing the colonization of pathogenic micro-organisms. A number of other advantages to the feed industry, such as their ability to withstand the high temperatures encountered in pelleting, their natural image from a product registration and consumer point of view, their stability to conditions in the intestinal tract, and their very low antigenicity. All these factors favour their more widespread use in future.

Synbiotics

A combination of pre- and pro-biotic, known as synbiotic has great potential in pre-weaner ruminant for early establishment of intestinal flora and enhances growth. Pre-biotic promotes early development of rumen and provide substrate for establishment of probiotic microbials, which facilitate early establishment of normal gut microbiota and reduce the adhesion of pathogenic microorganism with intestinal epithelium. Feeding of lectulose (pre-biotic, 3 %) and milk replacer containing a probiotic (10^9 CFU of *Enterococcus faecium* per kg) after one week of birth increased body weight and affect the morphology of the small and large intestine in pre-ruminants (Fleige *et al.,* 2007).

Enzymes

Enzymes have not generally been considered relevant for ruminant nutrition because of the role of the rumen microorganisms. However, there are now several indications that the additives of extract of *Aspergillus niger* or some strains of *Saccharomyces cerevisiae* may stimulate the number of cell wall degrading bacteria in the rumen, leading to an increase in the rate of fibre breakdown and an increased flow of microbial protein to the duodenum (Wallace and Newbold, 1992). Amylase has been shown to have small but significant benefits in some cases, where high cereal rations are used for fattening rations. Konno *et al.* (1993) found that an enzyme product obtained by culturing a basidiomycete, *Irpex lacteus,* has plant tissue digestion activity which, when the diet is also supplemented with amino acids, has substantial benefit to ruminant production both in terms of milk production and composition as well as live weight gain. Enzyme cloning and inserting genes encoding phytase enzyme form *Aspergillus niger* improved phytase utilization (Pen *et al.,* 1993), and product containing a high level of â-galactosidase activity to promote the hydrolysis of oligosaccharides found in various legumes. This spread of results has given rise to confusion, but now a better understanding is being achieved as feed and enzyme producing companies invest in research specifically studying the role of enzymes in animal feeds. The success of application of enzymes to feed will depend on following factors:

- Enzyme stability in the animal
- Enzyme stability in the feed.
- Enzyme stability during and after processing
- Interaction between the substrate and the enzyme
- Ability of the animal to utilize the reaction product of the enzyme

Pre-Weaning Animals

In ruminant production systems, the efficacy of bacterial probiotics such as species of *Lactobacillus*, *Enterococcus*, *Streptococcus*, and *Bifidobacterium*, have been studied in pre-ruminant (Newman and Jacques, 1995). Bacterial probiotics (*Lactobacillus* species) in young and/or stressed ruminant has establish and maintain "normal" intestinal microorganisms rather than as a production (gain and efficiency) stimulant. For pre-ruminant rapid adaptation to solid feed by accelerating the establishment of ruminal and intestinal microorganisms and avoiding the establishment of enteropathogens, which often results in diarrhoea, is the primary goal of probiotic feeding. In the neonate and in stressed calves, the microbial population is in transition and extremely sensitive, abrupt changes in diet or the environment can cause alterations in microbial populations in the gastrointestinal tract. Tannock (1983) reported that stress often leads to an increased incidence of diarrhoea in neonates, which is associated with decreases in the population of *Lactobacillus* in the gut. Moreover, fecal counts of lactobacilli normally are higher than coliforms in healthy animals and reversed in those suffering from diarrhoea.

Feeding calves viable cultures of species of *Lactobacillus* and *Streptococcus* has been reported to decrease the incidence of diarrhoea. Abu-Tarboush et al. (1996), fed calves with *L. acidophilus* 27SC had a significantly lower scour index, which confirmed the beneficial effect of lactobacilli in reducing the incidence of diarrhoea. The decreased incidence of diarrhoea might be associated with a consistently increased shedding of *Lactobacillus* (Abu-Tarboush et al., 1996) and an inconsistent decreased shedding of coliforms (Bruce et al., 1979) in feces in response to supplements of *Lactobacillus*. This could be related to the fact that animals were not experiencing intestinal disorders (e.g., diarrhoea). Interestingly, in experiments where there has been no advantage to feeding bacterial DFM (Jenny et al., 1991), calves were generally experiencing no health problems. Rapid adaptation to solid feed by neonatal calves also depends on the development of the ruminal epithelium and ruminal capacity. Nakanishi et al. (1993) found that added lactic acid bacteria to starter diets suggested to affect ruminal function in the young animal, supplemented calves with yogurt containing *L. acidophilus* tended to ruminate more at 30 d than untreated calves, indicating that *L. acidophilus* may promote ruminal development. Performance results for neonatal calves consuming bacterial DFM have been variable. Several reports indicate no improvement in daily gain as a result of feed ported improved (17%) rates of gain when $2.5x?10^{11}$ CFU/d of *L. acidophilus* species was added to milk or milk replacer.

Feed efficiency is generally not altered by feeding DFM (Jenny *et al.*, 1991; Abu-Tarboush *et al.*, 1996). Beeman (1985) evaluated diarrhoea and antibiotic therapy to asses the effects of feeding a culture of *Lactobacillus* on weight gain from neonatal diarrhoea. Probiotic treated animals had higher gain, benefits were hypothesized to result from improvement of intestinal conditions because of lower fecal scores in DFM fed animals. Performance response is likely not important early in the pre-ruminant's life when enteric disease is most prevalent. Improved health and reduction in the incidence or severity of diarrhoea, though difficult to measure for statistical analysis, is most likely a more important response. More experiments that include detailed information about the microbial supplement, and fecal culture data from scouring experimental animals are needed to determine the usefulness of microbial supplements in neonatal animals.

Post-Weaning Animals

The feedlot animals undergo a variety of stresses, such as recent weaning, transport, fasting, assembly, vaccination, castration, and dehorning. Such stresses can alter microorganisms in the rumen and lower gut (Williams and Mahoney, 1984), resulting in decreased performance and increased morbidity and death loss. A higher level of growth desired on high concentrate feeding, which lower the rumen pH and may induce acidosis (Tripathi *et al.*, 2004). Administration of microbial feed additives postulate that the gut might reduce these changes in the microbial population. The efficacy of single and a combination of bacterial cultures containing live cultures of *L. acidophilus, L. plantarum, L. casei*, and *S. faecium* have been demonstrated. Averaged across all trials, feeding the DFM for ?30 d resulted in a 13.2 % increase in daily gain, 2.5 % increase in feed consumption, and a 6.3 % improvement in feed:gain (Fox, 1988). The greatest performance response to the bacterial DFM generally occurred within the first 14 d of the receiving period (Crawford *et al.*, 1980; Hutcheson *et al.*, 1980). Morbidity was reduced by 27.7% in cattle receiving the bacterial DFM compared with control cattle. However, it should be pointed out that morbidity was generally low. Similarly, Gill *et al.* (1987) fed a bacterial preparation during a 28-d receiving period and reported a 9.3% increase in daily gain, 9.5% improvement in feed efficiency, and a 10.9% reduction in morbidity. High rumen microbial diversity exists on high concentrate fed post weaning animals to that occur under high fibrous ruminant feeding regimens. Therefore, microbial ecology balances are to be maintained on range of diets using rumen manipulation techniques so that better nutrient use efficiency achieve.

Contribution of Rumen Micro-Organisms to Nutrient Supply of the Host

The metabolic activity of the rumen microbial population is the major determinant of rumen fermentation and consequently of the profile of nutrients that are available for absorption and ultimately for production. With increasing knowledge of rumen metabolism which provide optimization routines to estimate parameters for the contribution of increased rumen fermentation through microbial protein supply estimate is now established. Estimation of microbial protein supply describe rumen microbial efficiency based on the amount and type of available organic matter (OM), the utilization of OM for microbial growth and non-growth processes, and, to a limited extent, microbial interactions. The differential change in microbial biomass with time is the production of microbial biomass from degradation of substrates, and is the outflow of microbial biomass from the rumen. The production of microbial biomass is calculated from a series of mostly non-linear equations, which describe the rate of degradation of various substrates and the ATP requirement per unit of growth describe the change with time of fibre, starch, sugars, lipid and protein degradation. The amount of substrate that is available for microbial metabolism is a key factor in the quantity of microbial protein formed. Rumen fermentable OM or fermentable metabolizable energy largely determines the amount of microbial protein formed. Whilst the amount of microbial biomass formed really depends on factors including the retention time of feed in the rumen, the availability of other nutrients that may stimulate or inhibit substrate degradation, etc. (Firkins, 1996).

The metabolic activity of the microbes determines the rate of utilization of substrate, which in turn depends on various microbial interactions and the substrate availability. The availability of easily degradable carbohydrates may impair degradation of fibre. Rumen pH affects fibre degradation and pH is influences by the concentration of lactate and VFA (Argyle and Baldwin, 1988; Tripathi *et al.*, 2007c). Thus, fermentation of starch and sugars, which may lead to higher levels of VFA and lactate in the rumen fluid, decreases pH and hence affects the fermentation of fibre. The determination of rumen fluid pH, though is not an important determinant of substrate degradation in high-producing dairy cattle but important for microbial non-growth (maintenance) energy costs, remains a challenge of contribution of microbial species or microbial groups towards substrate degradation. Bacteria, protozoa and fungi all contribute to plant-cell-wall degradation in the rumen, but the quantitative contributions of each group to degradation remains largely unclear. Such contributions are usually measured *in vitro* (Lee *et al.* 2000) and the results have to be interpreted and translated to quantify the contribution *in vivo*. Dijkstra (1994) used data from *in vivo* and *in vitro* studies on protozoal and bacterial metabolism, and by representing the various interactions between protozoa and bacteria, was able to quantify the contributions of both groups to fibre degradation in the rumen (Dijkstra and Tamminga, 1995). The simulations results indicated a protozoal contribution to ruminal fibre degradation on a variety of diets ranging from 5-35 %.

Relatively high protozoal contributions were obtained with diets high in starch and sugars fed at low levels of feed intake. With such diets, defaunation would be expected to decrease fibre degradation to a larger extent than diets that are high in fibre, and this is confirmed by *in vivo* studies on defaunation in sheep (Ushida and Jouany, 1990; Ushida *et al.* 1990). Thus, the development and application on the effect of defaunation on fibre degradation in the rumen is the interaction between protein availability and degradation of carbohydrates. This interaction is not only important in situations where the quality of the feed and levels of N are poor (mainly in the tropics and subtropics), but is also an issue in areas of intensive farming in developing countries, where concerns about high levels of N excretion to the environment by cattle constrain an efficient capture of N from the feed into end products (Tamminga, 1996). Degradation of OM in the rumen is depressed when the availability of N is low, probably because N deficiency impairs rumen microbial growth (Clark *et al.*, 1992; Tripathi *et al.*, 2006) and the resultant decrease in the amount of microbial biomass reduces substrate degradation. However, rumen bacteria are efficient scavengers of N sources, and ammonia concentrations above some 20–50 mg/l do not stimulate microbial growth further *in vivo* (Clark *et al.* 1992). The optimal levels of N supplementation require when the basal diet is deficient in N and this level of N supply should be relative to energy supply (Tripathi *et al.*, 2007b), below which microbial synthesis is impaired. Next to substrate degradation, the growth yield is the major determinant of microbial protein synthesized, the overall use of substrates by growing microbes can be divided into substrate utilization for non-growth (maintenance) purposes and for biomass increase.

A higher proportion of the carbohydrates taken up by the microbes than the precursors of monomers will be used for incorporation into microbial biomass. Hence, the growth efficiency per unit carbohydrate utilized and the growth rate is lower in absence of preformed monomers. The ME for rumen microbes as derived by Isaacson et al. (1975) (0·05 g glucose/g microbial DM per h) is frequently applied in rumen simulation models. However, *ME* varies with bacterial species and growth conditions. Increased utilization of energy for non-growth purposes has been demonstrated in various situations, including a lack of K, ammonia and amino acids relative to carbohydrates (Hespell and Bryant, 1979). Rooke and Armstrong (1989) showed that microbial growth yield is increased when peptides or amino acids are substituted for ammonia. However, the improvement in growth yield when the entire N required is provided by amino acids varies from 19 - 77 % (Russell *et al.*, 1992; Dijkstra *et al.*, 1992; Baldwin *et al.*, 1987). Another major factor involved in microbial protein yield is the recycling of microbial protein within the rumen. Recycling of microbial N in the rumen is mainly due to the activities of the protozoa and varies between 20–90 % of gross microbial N synthesis (Firkins, 1996). The level of intake and composition of the diet causes large variation in microbial recycling due to protozoal metabolism within the rumen (Dijkstra *et al.*, 1998b). Such a variation has a significant impact on microbial protein yield. If 70 % of N requirements are met by ammonia and 30 % by preformed amino acids, 3·0 g OM is required to synthesize 1 g microbial protein (Dijkstra *et al.*, 1992). However, the synthesis of 1 g microbial protein flowing from the rumen requires 3·8 g OM when recycling is 20 % of gross synthesis, and 30·0 g when recycling is 90 % of gross synthesis. This indicated the significance of protozoal activities on net yield of microbial protein. The fractional passage rate may be a major determinant of the composition and amino acid profile of bacteria being washed out of the rumen (Rodriguez *et al.*, 2000). The fractional degradation rate itself depends on the amount of microbes present and the microbial activity. The effect of feeding frequency (three or six meals per d or continuous feeding) and of fractional degradation rate upon microbial efficiency, indicated that both factors did not affect microbial efficiency (Dijkstra *et al.* 2000). Therefore, the effect of microbial metabolism on microbial efficiency, when animals are continuously-fed, requires further research.

Lactate Production and Utilization

The fermentation pattern in the rumen is sensitive to diurnal nutrient intake and feeding frequency. In addition, degradation rate differs between nutrients and from one nutrient source to another (Tamminga *et al.*, 1990). This variation in nutrient availability, combined with the range in substrate preferences of different microbial species, leads to fluctuations in the rumen microbial population structure. For example, ingestion of a meal rich in soluble carbohydrates encourages the proliferation of bacteria such as *S. bovis* and lactobacilli (Mackie and Gilchrist, 1979). These bacteria produce lactic acid as an end product of anaerobic fermentation, which is subsequently utilized by bacteria such as *M. elsdenii* and the protozoa. This cyclic pattern of lactate production and utilization, with the associated rise and fall in the populations of *S. bovis* and *M. elsdenii*, holds true for diets in the normal range. However, accumulation of rumen lactic acid for animals fed large quantities of degradable starch or sugars occurs where the rate of lactic acid production exceeds the capacity of the lactate utilizing microbes for uptake (Owens *et al.*, 1998). The end products of microbial species may change depending on substrate availability and pH. Moreover, large variation occurs in the sensitivity to pH of both lactate-utilizing and lactate-producing microbes (Therion *et al.*, 1982). Thus, the prediction of lactic acid dynamics is complex and current feed evaluation systems for dairy cows, because of their static and factorial nature, are unable to address the complex interrelationships in the rumen that give rise to lactic acidosis.

Biotechnological Control of Rumen Acidosis

Ruminal acidosis can be characterized by low ruminal pH (below 5.6) and high ruminal concentrations of total VFA (subacute) or lactic acid (acute). Lactate-producing bacteria (*Lactobacillus* and *Enterococcus*) might help prevent ruminal acidosis in dairy cows (Nocek *et al.*, 2002), potentially because the presence of these bacteria cause the ruminal microorganisms to adapt to the presence of lactic acid in the rumen (Yoon and Stern, 1995). Live microbes feeding shown to increase daily gain, feed efficiency, improve health and performance of feedlot animals. Bacteria used as probiotic have been defined as single or mixed cultures of live organisms, which, when fed to animals, beneficially affect the host. The potential effects of live microbes feeding were for beneficial intestinal effects, including the establishment of a desirable gut microflora and/or prevention of the establishment of pathogenic organisms. Some indication are that certain bacterial probiotics might have beneficial effects in the rumen, such as decreasing the potential for ruminal acidosis. In several experiments, supplementing feedlot cattle with lactate-utilizing and/or lactate-producing bacteria has been shown to improve feed efficiency and daily gain (approximately 2.5%), with little change in DMI (Krehbiel *et al.*, 2003). The mechanisms responsible for the beneficial effects of microbes feeding for a decrease in subacute acidosis include a decrease in the area below subacute ruminal pH, increases in ruminal propionate concentrations, increased protozoal numbers, and changes in viable bacterial counts. Effects on some blood variables (lower CO_2 and LDH) also suggest a reduced risk of metabolic acidosis. Live microbials feeding have the potential to decrease ruminal acidosis in feedlot. More research is needed to describe the mode of action, and thereby improve the efficiency of microbial use in animal nutrition.

Lactate-utilizing bacterium *Megasphaera elsdenii* has been shown to prevent lactate accumulation when a highly fermentable substrate was used (Kung and Hession, 1995). *Megasphaera elsdenii. Megasphaera elsdenii* modified ruminal fermentation and prevented the accumulation of lactate during the transition from low- to high- concentrate diets (Greening *et al.*, 1991; Kung and Hession, 1995). Kung and Hession (1995) found a higher rumen fluid pH 5.5 at 4 h with *M. elsdenii* (8.7 x10^6 CFU/ ml of culture fluid) against a 4.8 in control. Lactate concentration peaked at more than 40 m*M* in control after 8 h and remained fairly constant thereafter, but in the *M. elsdenii* treatment, it was less than 5 m*M* through incubation. Total VFA concentration of cultures treated with *M. elsdenii* was more than twice that of control (131.4 vs. 63.3 m*M*). Acetate concentration was not significantly different after 2 h. The concentration of propionate, butyrate, valerate, isobutyrate, and isovalerate for control and *M. elsdenii* inoculation at 6 h were 38:47, 2:35, 1:15, 1:11, and 1:2 (m*M*, control: *M. elsdenii*), respectively. Robinson *et al.* (1992) reported the effects of inoculation with *M. elsdenii* on feed intake, ruminal pH, smolarity, lactate, and VFA concentration in acute acidosis-induced steers fed a 90% concentrate diet. In that study, the interaction between inoculation and day of diet switch moderated pH, lactate, VFA, and feed intake significantly. Steers inoculated with *M. elsdenii* consumed 24% more DM.

The choice between *Propionibacteria* and *M. elsdenii* for ruminant feeding on readily degradable and soluble carbohydrates show that *M. elsdenii* seems to be the major ruminal lactate utilizer Furthermore, *M. elsdenii* simultaneously uses lactate, glucose, and maltose, would compete with lactate-producing organisms for substrate. Although *Propionibacterium* is a lactate utilizer, it has been focused on propionate production rather than lactate fermentation for use as a microbial feed. Propionate is quantitatively the most important single precursor of glucose synthesis among VFA, and therefore has a major impact on hormonal release and tissue distribution of nutrients (Nagaraja *et al.*, 1997). For growing ruminants, propionate has been estimated to account for 61 (Reynolds *et al.*, 1994) to 67% (Huntington, 2000) of glucose release. Propionate spares glucogenic amino acids in gluconeogenesis, and consequently reduces the maintenance cost of metabolizable protein and possible heat increment (Van Soest, 1994). Also, decreased acetate:propionate has been accompanied with a decrease in methane production according to the stoichiometric laws of chemical balance and its equation (Van Soest, 1994). When acetate:propionate decreases, CH_4 production declines, and energy retention by cattle would increase. In the rumen of sheep, *Propionibacterium* accounted for 40 to 50% of the lactate utilizers however, the population of *Propionibacterium* usually seemed very low. Therefore, the concept of daily or periodic supplementation of *Propionibacterium* may be on the basis of the increment of propionate production when cattle are fed a high concentrate diet. Feeding of *Lactobacillus* species (Huffman *et al.*, 1992; Nocek *et al.*, 2000), suggesting reduced risk of subacute ruminal acidosis. Similar to ruminal pH, blood pH was not affected by bacterial supplementation. However, feeding of *Propionibacterium* and *E. faecium* tended to have lower concentrations of blood CO_2 and had lower concentrations of LDH. The authors suggested that lower blood CO_2 and LDH indicated that feeding lactate-producing bacteria along with a lactate- utilizing bacteria reduced the risk of metabolic acidosis.

Feeding *Propionibacterium* increased protozoa (especially *Entodinium*) and decreased amylolytic bacteria in the rumen of feedlot animal (Ghorbani *et al.*, 2002). Similarly, including cultures of lactobacilli in the diet prolonged retention of protozoa. The mechanism by which bacterial feeding stimulate protozoa remains unclear. Feeding live cultures of *Kluyveromyces marximanus, Saccharomyces cerevisiae* and *Saccharomyces uvarum*, and a mixed culture of three to pre-weaner lambs, Tripathi *et al.* (2007a) also found increased rumen pH and ciliate protozoa population compared to control lambs. The yeast culture feeding to young animals involved with improved organic matter degradation (Kamel *et al.*, 2004), regulation of rumen pH (Bach *et al.*, 2007) and stimulation of total and cellulolytic bacterial numbers (Newbold *et al.*, 1995). It seems that stimulation of rumen bacterial numbers improves fiber and organic matter degradation; therefore yeast supplementation improves the availability of energy for microbial growth, and the larger organic matter degradation allows DM intake to increase. These benefits of yeast culture supplementation are only assessable once the rumen is fully developed. Ciliate protozoa are involved in fibre utilization and might therefore contribute to increased fibre utilization in yeast-culture fed ruminants (Newbold *et al.*, 1995; Kamel *et al.*, 2004; Kritas *et al.*, 2006). This effect is, however, likely to be small. Brossard *et al.* (2006) reported improved growth of protozoal population in rumen of yeast supplemented animals. The differences between yeast strains in stimulating ruminal protozoa, is partly responsible for the numerically increased fiber digestion and higher ruminal pH with yeast addition. Erasmus *et al.* (2005) and El-Ghani (2004) also reported improved ruminal pH in yeast-culture supplemented cows (0.23 % in diet) and goats (0.2 g kg^{-1} live weight).

Cell Wall Fiber Utilization

Ruminants make up a significant proportion of the domesticated animal species are the best adapted to utilization of plant cell walls. Almost half of the global carbon fixed annually by photosynthesis is incorporated into plant cell walls making it the most renewable carbon source on earth. The rumen ecosystem of ruminants comprises a complex of dense microbial communities of bacteria, archaea, protozoa, fungi and bacteriophages (Mackie *et al.*, 2002). The fermentation effected by this complex microbiota is responsible for the conversion of plant feedstuffs to compounds that can be utilized by the animal. Hence, the fermentations and interactions of the microbes are central to ruminant digestion and nutrition. Cellulose, the most stable plant structural polysaccharide, is degraded by only a few species of rumen bacteria, *Fibrobacter succinogenes, Ruminococcus flavefaciens* and *Ruminococcus albus*, These species are very important, as a lack of breakdown of cellulose will impede access of other bacteria to their required substrates and reduce the efficiency of fermentation of feed material. The cellulolytic populations of the rumen are subject to lytic events that may impact on the ability of these bacteria to degrade plant fibre and on the nutrition of the animal (Klieve *et al.*, 2004). Lignocellulose is an important constituent in ruminant diets, and even in intensive finishing systems is incorporated into the ration because it is both economical and necessary for normal healthy rumen function. Improvement in the ability of the rumen microbiota to degrade plant cell wall is generally highly desirable and usually leads to improved animal performance. In the developing world ruminants have an important role in the sustainability of village communities, and in many cases form the major source of income. Smallholders rely on subsistence farming, with few or no inputs, but forages are usually available and generally provide the sole source of nutrition to the animal. Even in the developed world, forage is the major source of nutrition in many situations. For example, in the major beef-producing regions of northern Australia, forage is virtually the only source of nutrition, and supplementation is impractical and expensive. This is also true in North and South America where grazing animals are common, and lignocellulose makes up the majority of the diet.

The symbiosis between animal and microbe in the rumen allows for a cooperative system in which both the host and animal derive a benefit. The rumen is a capacious pre-gastric fermentation chamber that sustains a rich community of microorganisms that rapidly olonize and digest feed particles. Carbohydrate polymers in plants are indigestible to most animals but can be hydrolyzed and fermented by a range of microorganisms in the rumen. The end products of this fermentation are fatty acids, which form a major metabolic fuel for the ruminant, and microbial cells that are a major source of protein and amino acids when absorbed in the lower digestive tract of the animal. One cost of this relationship is the breakdown, or sacrifice, of dietary protein by rumen microorganisms before digestion by the host enzymes. Over the last 50 years, significant improvements in our understanding of the digestion of fiber in the rumen have occurred and much of this information has been translated into practical nutritional management strategies. For example, an understanding of the importance of nitrogen to the degradation of fiber by fibrolytic microorganisms has led to the inclusion of urea supplements in ruminant diets , and mechanical and chemical treatments of forages has improved their digestibility. Probably one of the most efective means of exploiting information on rumen microbiology and digestion has been the construction of computer models that predict animal performance from the characteristics of feed ingredients. These models have the ability to make significant improvements in dietary formulation, and although they do not directly manipulate fiber digestion in the rumen, they do optimize the utilization of scarce nutrients by rumen microbiota. Models are ideal examples of how incremental advances in our knowledge of rumen microbial processes have been utilized to make significant advances in rumen function (Krause *et al.*, 2003). The supposed inability of rumen microbiota to express the appropriate suit of enzymes to maximize fiber digestion is often given as a reason for research into improving the function of rumen microbiota. The inoculation of bacteria into the rumen that involved in fiber digestion, focused attempts that have been made to improve fiber digestion with inoculants, fungal treatments, and exogenous enzymes, but future research will preface this with an update of the glycosyl hydrolases (GH), and other enzymes produced by rumen microorganisms. The applications of genomics and metagenomics to rumen microbiology will affect our knowledge of rumen cell wall degradation.

Fibrolytic Ruminal Microorganisms and Enzymes

Ruminants largely dependent upon micro-organisms of the reticulo-rumen ('rumen') for the degradation of ingested plant material, to supply the host animal with end products of fermentation, particularly volatile fatty acids (VFA) that are absorbed mainly through the rumen wall and provide energy. The micro-organisms of the rumen ecosystem also provide microbial protein is usually the most important source of amino acids that are available for the metabolism of the host. A morphologically and functionally diverse population of micro-organisms is present in the rumen. The microbial species present include eubacteria, archae, protozoa and fungi. The microbial population and activities in the rumen differ widely in substrates that they ferment and in end products that they form, whilst interactions between micro-organisms will also affect the rumen fermentation. Interspecies hydrogen transfer between hydrogen-producing bacteria, protozoa and fungi and hydrogen-utilizing methanogenic archae may increase acetate production and decrease production of intermediates such as succinate and ethanol (Wolin *et al.*, 1997). Predation on bacteria by ciliate and flagellate protozoa may reduce the outflow of microbial biomass and availability of amino acids from digested microbial protein (Williams and Coleman, 1997). Thus, this diversity gives rise to large differences in rate and extent of feed degradation, amount and type of VFA formed, and amount and composition of microbial material that is washed out from the rumen, and ultimately to differences in production level, product composition and excretion of waste material. As qualitative knowledge of microbial metabolism in the rumen increased, quantitative approaches were developed. These quantitative approaches generally aim at an improved fermentation in the rumen or an improved understanding and integration and their interactions.

A complex community of fibrolytic microorganisms catalyzes the degradation of fiber in the rumen. The major fibrolytic bacteria are the Gram-negative *Fibrobacter succinogenes*, and two species of Gram-positive bacteria, *Ruminococcus albus* and *Ruminococcus flavefaciens*. *Butyrivibrio fibrisolvens* are a group of highly *xylanolytic* Gram positive bacteria inhabiting the rumen, which have a central role in fiber digestion. *Prevotella* are not regarded as highly cellulolytic bacteria but do produce a range of xylanases. A number of less well characterized cellulolytic bacteria occur, such as *Eubacterium cellulosolvens*. In addition, the anaerobic rumen fungi are considered important in fiber digestion and one of the best-studied fungi is *Neocallimastix* sp. There is also increasing evidence that the rumen protozoa may have the capacity to digest fiber (Devillard *et al.*, 2003).

Anaerobic Rumen Fungi

The actual role of anaerobic fungi in the rumen is still an issue that has not been completely resolved. It is, however, known that anaerobic fungi colonize plant tissue and appear to degrade lignified tissue that is not degraded by other microorganisms (Lowe *et al.*, 1987), but the rates of growth and degradation of fungi are much slower than those of the bacteria, and their ability to is limited because their growth rates are much lower than the rumen dilution rate. Fungi are able to degrade up to 34% of the lignin in plant tissue, can penetrate the plant tissue as a result of their filamentous growth (McSweeney *et al.*, 1994), which have a broad range of highly active enzymes, and are the only known rumen organisms with exo-acting cellulose activity. The importance of fungi is further supported by the synergy that occurs with rumen bacteria (Dehority and Tirabasso, 2000). Cell wall fragments degraded more extensively by *Neocallimastix frontalis* and *Piromyces* isolates than by *Caecomyces* isolates, *N. frontalis* and *Piromyces* had the greatest rates of stem degradation. Interactions between *F. succinogenes* and fungi growing on ryegrass stems show that *N. frontalis* inhibited *F. succinogenes*. In contrast, a *Caecomyces* species grown with *F. succinogenes* increased stem degradation, indicating that *F. succinogenes* and Caecomyces spp. may have complementary fibrolytic activities. *Neocallimastix* spp. are the best studied of the rumen fungi and are highly active against crystalline cellulose. The cellulases a cDNA library prepared from *Neocallimastix patriciarum* CX, four cDNAs - celA (cellobiohydrolase), celB (glucanase), celC, and celD were indentified. The celA was a cellobiohydrolase that hydrolyzed crystalline cellulose, while celB and celD had endoglucanase activities. The most interesting cDNA was celD that had multiple activities including endoglucanase, licheninase, cellobiosidase, and xylanase. Other enzymes identified are glucosidase, L-glucosidase, and cellobiase from rumen fungi. Cellulase activity is also associated with anaerobic *Orpinomyces joyonii* SG4, which has significant cellulase activity. Recently identified from a cDNA library of *Orpinomyces* PC2 designated celF encoding a cellulase with a signal peptide, a carbohydrate-binding module, a linker, and a catalytic domain similar to that of celA from *N. patriciarum*. The catalytic domain was also homologous to CelA and CelC from the same fungus which contain N-terminal docking domains for a cellulose-hemicellulase complex. Xylanases have also been cloned (xynA and xynB) from *N. patriciarum*. The xynA encodes an enzyme with two catalytic domains and is highly active and this enzyme has at least a fivefold higher activity than the native. The xynB encoded both xylanase and cellobiosidase activity.

Quantification of Rumen Fungi

In the rumen, population of anaerobic fungi is generally considered to remain viable indefinitely and consist of two distinct stages, the first in which free-swimming zoospores in rumen liquid and other with particle-attached fungal thalli. Both stages must co-exist in order to achieve perpetual viability of fungal populations in the rumen, because the growth of fungal biomass in the rumen, must balance the loss of fungi by death and passage from the rumen. Thus, it follows that a non-trivial relationship must exist between the numbers of zoospores and thalli which inhabit the rumen. Several procedures have been developed for enumerating fungal populations in the rumen (Mountfort and Orpin, 1995). These are mostly limited, however, to the enumeration of zoospores, either by direct counting after dilution from rumen liquid, or indirectly from the number of colonies which develop in cellobiose-containing roll tubes. A lesser-used technique, which does not provide an absolute cell count but can be used to compare thallus populations, is to count the number of zoosporangia associated with leaf blades or impregnated agar strips after *in vitro* or *in sacco* incubation in rumen liquid. An enumeration procedure based upon most probable numbers is also available. This technique does not distinguish between stages of the life cycle, but fungal populations can be enumerated, as thallus forming units, relative to their ability to degrade plant cell walls. Recent techniques, which have to be developed further, make use of ribosomal specific oligonucleotides to measure total fungal populations (Krause *et al.*, 1999). The size of the zoospore and thallus populations is likely to be dependent on a variety of factors in the rumen, including substrate type and availability, the respective swimming and encystment times of zoospore and thallus populations, the maturation time of fungal thalli, the yield of zoospores from mature thalli and the flow properties of liquid and solids from the rumen. Review of Mountfort and Orpin (1995) described the life cycles of several monocentric rumen fungi (*Neocallimastix patriciarum, N. frontalis, N. hurleyensis, Piromonas communis, Sphaeromonas communis, Caecomyces equi*)). Nonetheless, there is insufficient quantitative information in the literature to ascribe exact values, or even accurate ranges, to all the life-cycle parameters for any individual species of rumen fungus. Feeding *ad libitum* chopped perennial ryegrass hay supplemented with concentrates to steers, applying the volume and zoospore concentration Theodorou et al. (1990) measured $7{\cdot}5\times10^{8}$ of the free-swimming zoospore population in the rumen and $4{\cdot}0$-$12{\cdot}6\times10^{8}$ of the number of particle-attached thalli, of which $0{\cdot}6$-$3{\cdot}0\times10^{8}$ are mature. These estimates are broadly consistent with the minimal ruminal enumeration of $0{\cdot}9$-$8{\cdot}4\times10^{8}$ thallus forming units (zoospores plus thalli).

Quantification of Microbial Growth and Yield

Manipulation of microbial ecosystems has often been employed in attempts to improve microbial metabolism and microbial interactions. Manipulation experiments are helpful in determining the key parameters of microbial growth and metabolism. Microbial cultures are in balanced growth (Koch, 1994) experiments indicated that, a balanced growth is obtained when extensive properties, including RNA, DNA and cell number, increase regularly with time. Considering physical analogues, the rumen ecosystem can be classified into open and closed systems. In an open system, all materials that constitute the system may enter and leave it, whereas in the closed system some essential part of the system cannot both enter and leave it. Examples of the closed and open system are the simple batch culture and the chemostat (continuous flow culture), respectively. The rumen ecosystem may be considered an open system when the ruminant is eating. During periods of rest, an essential part of the system (the feed, which provides the substrates for microbial growth) does not enter the system and the system may be considered closed for this period. In these periods of rest, secretion of substrates (for example, in saliva) and exchange and net absorption (for example, water, VFA) across the rumen wall continues. Hence, some researchers prefer to describe the rumen as semi continuous rather than closed (Wolin, 1979).

Microbial Interactions in Rumen

The manipulation of rumen ecosystem often rely on information obtained from studies in axenic cultures or co-cultures, assuming that the principles elucidated from simple systems apply to the more complex rumen ecosystem. The survival of species in the rumen can be predicted using data obtained in axenic cultures. The growth of *Selenomonas ruminantium* and *Bacteroides ruminicola* with a single growth-limiting substrate data from axenic cultures revealed that with glucose as a single growth-limiting substrate, *S. ruminantium* will eliminate *B. ruminicola* because, whatever the glucose concentration, μ of the former is always higher than that of the latter. If the substrate is maltose, the growth at a concentration of 0·54 mg/l at which both species can co-exist. Many factors control competition and allow a wider range of survival.

If the growth of micro-organisms is limited by the relative availability of substrate surface (Chesson and Forsberg, 1997) the results of competition may be predicted by replacing substrate. is given by France *et al.* (1990) described simulated effect of surface area on growth of micro-organisms in a batch culture in which wheat straw particles of different sizes were fermented by bacterial- or fungal-enriched inoculae and gas production profiles were recorded as a measure of substrate degradation rates . With bacterial-enriched inoculae, the gas production rate was lower when wheat straw particle size was increased, whereas experimental fungal fermentations did not show a relationship between gas produced and particle size. Such findings are in line with simulated profiles for degradation by surface erosion (mode of attack of bacteria) and by complete invasion (mode of attack of fungi). Since various plant tissues differ widely in ratio of exposed surface area to cell wall volume, such variation is expected to affect the abundance of microbial species that differ in their mode of attack (Lopez et al., 1993). In co-cultures, the fungal mode of attack reduces mechanical resistance of particles (Fonty et al., 1999), allowing increased bacterial attack on the damaged particles and a possible co-existence of both fungi and bacteria. Competition for the same growth-limiting substrate can be controlled by the faster-growing of the two species inhibiting its own growth rate through a product *P* (g). If growth on *S* is associated with production of *P* that competitively inhibits the uptake of *S*, the equation that describes the specific growth rate is:

$$\mu = \mu_m / (1 + (1 + [P] / J_P)K_s / [S]),$$

where [*P*] is the concentration of the inhibitor (g/l) and J_P (g/l) denotes an inhibition constant (Dijkstra et al., 2002). A co-culture of two species will become self-regulating and stable if the specific growth rate of the faster-growing species is higher than that of the slower-growing species when [*P*] is zero. Knowledge on various inhibitory factors produced by micro-organisms in the rumen is limited (Dehority and Tirabasso, 2000). Bacteriocins (antibacterial peptides and proteins characterized by their ability to inhibit strains of bacteria) may have a key role in bacterial population dynamics (Riley, 1998). A competitive advantage will be obtained by the bacteriocin-producing strains by inhibiting or killing bacteria in the same environment competing for the same substrates (Pagie and Hogeweg, 1999). Other obvious and important inhibitors of growth are the end products of fermentation, VFA. The free acid form can pass through the microbial cell membrane and affect the proton motive force gradient across microbial cell membranes. Some bacterial species however may decrease their intracellular pH to maintain a relatively low pH gradient across the cell membrane and to prevent VFA anion's accumulation in the cell (Kashket, 1987). A higher fraction of the VFA in the free acid form decreaspH in the rumen, VFA and pH of rumen fluid are related (Tamminga and Van Vuuren, 1988). The effect of pH at VFA concentration of 40 mM on the *μm* with glucose of *S. ruminantium* and *B. ruminicola* growth inhibited. This self-inhibitory type of interaction has been shown to be of crucial importance in mixed cultures to stabilize the ecosystem and prevent simulated disappearance of one of the species. Dijkstra (1994) applied this auto-inhibition principle in a rumen model that represents the metabolism of amylolytic and fibrolytic bacteria and of protozoa in the rumen. Stable co-existence under a wide range of dietary inputs was achieved by including the inhibitory effect of a relationship between substrate availability and protozoal death rate.

Another way of controlling competition is achieved when the faster-growing species produces a growth activator for the other species, a form of commensalisms. A notable area of interactions between species of the rumen microbial ecosystem is the cross-feeding of substrates. A number of bacterial and fungal species hydrolysing hemicellulose do not utilize the products of this hydrolysis, whereas others do utilize the products but are unable to hydrolyse hemicellulose (Wolin *et al.*, 1997). Cross-feeding of reduced intermediates (especially hydrogen), while cross-feeding does not necessarily change the maximum specific growth rates or saturation constants of rumen microbes, the type of fermentation end product formed may be changed drastically. Unfortunately, rumen profile of VFA formed in the rumen have not been principles to describe cross-feeding, but depend on empirical relationships between type of substrate degraded and type of VFA formed (Dijkstra and Bannink, 2000). In rumen simulation models, only limited attention has been given to explicit representation of the different microbial groups and microbial interactions. Often only one state variable, representing the microbial biomass representation of one pool of micro-organisms, but was forced to distinguish between microbes degrading starch and sugars and microbes degrading fibre to avoid incorrect interactions between fibre and starch degradation.

Fiber Digestion by Rumen Inoculation

The microbiota of the digestive tract does not always comprise an optimum balance and rumen inoculation always improves performance during transition from a forage ration to a high-grain diet. If very large amounts of rumen content transferred from grain-adapted animals to un adapted animals the transition to the high-grain diet can be made quite quickly without the problems of ruminal acidosis. Similarly, adaptation of young ruminants to a roughage ration can be accelerated if normal adult microbiota is dosed. The dosing of newborn calves with rumen fluid from adult animals causes them to gain more weight and have fewer diarrhoeas than untreated controls (Muscato et al., 2002). Forage utilization can also clearly be improved with inoculation. The forage legume *Leucaena leucocephala* produces a goitrogenic compound (3-hydroxy-4[1H]pyridone) that causes toxicity in ruminants. *Synergistes jonesii,* is able to break down this compound, and was originally found in Hawaiian goats but not in cattle in Australia. This organism when inoculated into susceptible ruminants grazing *L. leucocephala* conferred resistance to the toxin. Similarly, monofluoroacetate is found in various pasture plants at levels of up to 5 g / kg and has an LD50 in ruminants of approximately 0.3 mg / kg of body weight. Gregg *et al.* (1998) inoculated four strains of recombinant *B. fibrisolvens,* transformed with a gene encoding fluoroacetate dehalogenase into sheep. The inoculated sheep showed a significant reduction in toxicological symptoms after fluoroacetate poisoning when behavioral, physiological, and histological data were compared with those of five un-inoculated sheep. It also appeared as if the inoculated strains persisted at approximately 10^6-10^7 per ml.

Genetically Modified Rumen Bacteria

In the rumen, much effort has been expended in developing genetically modified bacteria that would have superior fiber-degrading abilities. The construction of genetically modified bacteria has proceeded under the assumption that the rumen microbiota does not produce the correct mixture of enzymes to maximize plant cell wall degradation. For example, *Ruminococcus* and *Fibrobacter* do not produce exo-cellulases that are active against crystalline cellulose, so that adding this activity would make them more potent. Transformation of *Ruminococcus* and *Fibrobacter* (the most fibrolytic rumen bacteria) has not been successful, but it has been possible with B. fibrisolvens, *S. bovis,* and *Prevotella spp. Bu. fbrisolvens* is primarily hemicellulolytic and is considered ecologically robust making it a good choice as a host for recombinant plant cell wall-degrading enzymes. The modification of *B. fibrisolvens* with GH has been successful, and in vitro digestibility of fiber can be improved, but unfortunately this still does not allow them to compete with the far more fibrolytic *Fibrobacter* and *Ruminococcus* species. Initially *B. fibrisolvens* was modified with a xylanase (family 10 GH) from N. patriciarum, a GH family different from the family 11 typically present in *B. fibrisolvens.* GH 10 and GH 11 xylanases differ in their catalytic properties and it was reasoned that the introduction of a GH 11 xylanase would increase its ability to digest fiber. Fiber digestibility could be improved *in vitro* by more than 28% in comparison with the native untransformed strain. Subsequent studies demonstrated that these recombinants were unable to compete with highly cellulolytic *Ruminococcus* strains in vitro and did not persist in the rumen beyond 10-15 days. An alternative approach is to create recombinant bacteria that can degrade fiber at low pHs. It is well known that fiber digestion declines in animals on high-grain diets because the pH of the rumen drops below 6.5 and *Ruminococcus* and *F. succinogenes* are sensitive to even mildly acidic pH. The addition of fibrolytic activity to an acid-resistant species such as *Prevotella* would create an organism which would be far more competitive because it would be filling an 'acidic niche' which autochthonous cellulolytic bacteria are unable to fill.

Non-genetically Modified Fiber-Degrading Bacteria

The establishment of non-cellulolytic bacteria in the gnotobiotic rumen is possible but persistence of fibrolytic strains is more difficult and there are no cases of *Ruminococcus* becoming. *F. succinogenes* S85 has been successfully established in gnotobiotic lambs but introduction as a member of a mixed microbial community is a prerequisite. Fonty et al. (1983) tried to define which combination of 182 rumen cultures was essential to colonization but could only conclude that increased complexity increased colonizing ability. Alternative approaches to establishing cellulolytic strains have included repeated dosing of lambs and adult sheep, and completely replacing the contents of the rumen with single strains of cellulolytic bacteria. All these attempts have been unsuccessful. Increasing the numbers of fibrolytic bacteria in the rumen by feeding a high-cellulose diet composed of purified wood cellulose increase a 10-fold in the number of cellulolytic bacteria but they could not demonstrate a significant increase in the digestion of alfalfa (lucerne) cellulose. The elevated number of *Ruminococcus* species by continuous inoculation indicated that microbial enzyme activity is not the limiting factor in ruminal fiber digestion. The dosed fibrolytic bacteria into the rumen usually disappears from the rumen very fast is one of the reasons for no effect on fiber degradation that we do not yet understand how fibrolytic bacteria exist and reproduce at the fiber surface. What we do know is that organisms do not exist in ecosystems on their own, but reproduce and persist as members of complex microbial communities. Ultimately bacteria persist, or survive, because they reproduce within the physiological and ecological limits of the ecosystem (community-level reproductive strategies). These strategies derive from the evolution of cooperative networks among microorganisms in which some members cleave specific bonds, others utilize particular substrates, and still others produce inhibitors. A ruminal example is the production of cellodextrins by cellulolytic bacteria, which are utilized by non-structural carbohydrate-fermenting bacteria (NSC). The NSC in turn produce ammonia and branched-chain volatile fatty acids that are consumed by cellulolytic bacteria. In the case of fibrolytic organisms, it is likely that our understanding of nutrient requirements and other interactions between organisms at the fiber surface is inadequate. In the continuous dosing of both lambs and adult sheep (Krause *et al.*, 1999) 16S rRNA-based observations indicated that there was a significant increase in the eukaryotic population and this appeared to be primarily the result of an increase in protozoa. Protozoa have a considerable capacity to ingest rumen bacteria and sheep rumen fluid have shown that lysis of *Methanobrevibacter*, and *Selenomonas ruminantium* decreased significantly when protozoa were absent from the rumen fluid (Newbold *et al.*, 1996), an inoculated *Lactobacillus plantarum* strain disappeared from the rumen largely because of protozoal predation. Protozoal predation of *F. succinogenes* appear to be a suppression of *F. succinogenes* compared to dosed to protozoa- free gnotobiotic lambs in the presence of a complex mixture of axenic rumen bacteria. Bacteriocin production by certain cellulolytic bacteria has only recently become a subject of research and may be an essential component in the formation of cooperative microbial networks. R. albus strains can produce bacteriocin-like substances that inhibit the growth of *R. flavefaciens* but not *F. succinogenes*. When cellulolytic bacteria are grown together in diculture, cellulose degradation is often below that of the pure culture, which is probably the result of competitive and non-competitive interactions between cellulolytic bacteria. These competitive outcomes were likely the result of the superior ability of *R. flavefaciens* to adhere to cellulose but *R. albus* survived under cellulose-limited conditions. This was probably a combination of its ability to utilize glucose (*R. flavefaciens* does not), grow at low concentrations of cellobiose, and its capacity to produce bacteriocins.

Exogenous Enzymes

The direct application of enzymes to forage before consumption can improve animal performance in terms of 6.8-24.0 % more weight and higher feed conversion efficiency of 6.0-21.2 %, when hay was treated with an enzyme cocktail containing amylolytic, proteolytic, and cellulolytic enzymes (Hristov *et al.*, 1998; Krause *et al.*, 1999; McAllister *et al.*, 1999). Four different enzyme preparations given in combination with the growth stimulant diethylstilbesterol allowed growing ruminant fed a hay diet to gain 14% more weight than controls. That enzyme supplements could improve the performance of cattle on silage diets but other studies added a fungal enzyme extract to a grain supplemented Lucerne diet fed to young calves, did not found improvement in gain or feed utilization. The potential benefits of enzymes for feedlot animal production address issues related to the inconsistency of results by investigating the effects of feed type, application levels, enzyme products, and enzyme applications. Other factors, such as whether dry forage or silage is used, the enzyme is infused directly into the rumen or not have been investigated. A mechanistic understanding of how enzymes interact with forages and how synergies between exogenous and autochthonous microbiotas in the rumen occur are still not well understood.

The enzyme inoculation into the rumen may undergo inactivation by the proteolytic activity of rumen fluid but CMCase and xylanase activities contributed from endogenous sources remain active in the rumen thus all exogenous enzyme activity not necessarily inactivated once introduced into the rumen. The survival of exogenous enzymes in the rumen may improve digestion through the direct hydrolysis of ingested feed. The exogenous enzymes may enhance fiber digestion by ruminal microorganisms (Nsereko *et al.*, 2000), release of reducing sugars by exogenous enzymes is probably an important mechanism by which exogenous enzymes operate. The degree of sugar release is dependent on the feed type as well as the type of enzyme. The enzymes most effective at releasing reducing sugars from feeds is at least partially the result of solubilization of NDF (neutral detergent fiber) and ADF (acid detergent fiber) and increases the soluble fraction and rate of digestion (Hristov et al., 2000; McAllister *et al.*, 2000) and thus added increase NDF, ADF and hemicellulose degradation (Colombatto *et al.*, 2007). The exogenous enzymes work better at rumen pH near neutrality and digest substrates that would normally be digested by enzymes produced by the autochthonous microbiota in the rumen (Colombatto *et al.*, 2007). Exogenous enzymes affect the release of soluble carbohydrates, the amount liberated represents only a minute portion of the total carbohydrate present in the diet. It would be difficult to attribute any production responses solely to the increased availability of reducing sugars given that comparable increases in yield were not seen when up to 9% of total dietary dry matter was supplied as molasses. Although adding exogenous enzymes may increase the activity of xylanases and cellulases in ruminal fluid, enzyme activity in the rumen fluid usually represents less than 30% of the total enzyme activity in the rumen, the remainder being associated with the feed particles. For example, application of fibrolytic enzymes to a grass hay diet fed to sheep prior to consumption increased endoglucanase activity and xylanase activity in ruminal fluid, but this activity accounted for only 0.5% of the total endoglucanase activity in the rumen. This show that exogenous enzymes represent only a small fraction of the ruminal enzyme activity, and that the ruminal microbiota is inherently capable of digesting fiber it is difficult to envision how exogenous enzymes would enhance ruminal fiber digestion through direct hydrolysis. The mode of action of exogenous enzymes is that they work synergistically with the rumen microbiota (Wang *et al.*, 2001). This would only be true if the exogenous enzyme inoculant contained enzymes not produced by rumen microbes and was therefore contributing unique activity. Presently, it does not seem likely that the addition of exogenous enzymes has added unique activity to the rumen as only the rate, and not the extent, of cell wall digestion has been improved. Fibrolytic enzymes normally act as members of a cellulosome complex, because there are a range of glycosidic bonds that need to be hydrolyzed synergistically. A logical extension of this reasoning would therefore be to add enzymes that are not produced by rumen microorganisms and attack structures in forage that result in an increased extent of digestion.

Microbial Genome

Sequencing is a useful tool that will allow us to extend our understanding of individual microbial species of rumen microbiota. However, we should keep these genome sequencing efforts in context. It is by no means the end game given that the majority of microbial species cannot be, or have not been, cultured. Microorganisms live in community and analyses of individual organisms do not comprise the complex ecological networks that optimize ecosystem function found in most microbial systems, including those for rumen digestion. These complex systems are not well understood and it is by studying rumen fiber digestion at a community genomic level (the metagenome) that the new clues to manipulating rumen microbial ecology and fiber digestion will emerge. The analysis of the rumen metagenome is complicated by the fact that the majority of microorganisms have not been cultured (estimates range from 85 to 95%, and probably comprise upwards of 1000 individual species of bacteria, fungi, and protozoa. This situation is not unique to the rumen and the vast majority of microbes in the biosphere, often thousands of species in a single environmental niche), cannot be grown in the laboratory. It is estimated that, on average, less than 1% have ever been identified. Traditional methods for culturing microorganisms fail to represent the scope of microbial diversity in nature and they limit analysis to those that grow under laboratory conditions.

The use of PCR in conjunction with phylogenetically stable molecular markers like 16S rDNA as species-specific identifiers has provided us with the ability to detect single cloned. An extremely useful phylogenetic spin-of is that because of the insert size many clones will contain 16S genes, which together with an analysis of associated functional genes provides a phylogenetically informed view of uncultured diversity.This technology is still in its infancy and issues such as representative cloning, quantitative lysis, and expression still need to be addressed. In particular, quantitative lysis is difficult to overcome. It is imperative that large fragments be cloned, and to do this techniques based on pulsed field gel electrophoresis must be employed. To prevent shearing of DNA in aqueous solutions cells are embedded in agarose plugs and lysed *in situ*. This immediately presents the problem of quantitative lysis of the microbiota. The major microbial species are the target of rumen activities and that the metagenomes include incredible molecular diversity. The choice of appropriate clone recipients will be important as cloning of metagenomic DNA will be in heterologous hosts, and primarily in *E. coli*, simply because high clone numbers are necessary to get anywhere near a representation of the microbial diversity present in the metagenome. However, it appears that diversity in metagenomic DNA is so great that even inherently flawed cloning strategies represent so much molecular novelty that at present optimization of these strategies seems irrelevant in comparison to the constraints in downstream processing of this 'mega-diversity'. Cells in microbial ecosystems, irrespective of their culturability, and thereby widened our view of biodiversity. The recent research in molecular microbial ecology provides compelling evidence for the existence of many novel types of microorganisms in high numbers and provides an entirely different approach for tapping into the potentially limitless resource of uncultured bacteria. The rumen is no different, and clone libraries based on 16S rRNA clearly suggest that a vast reservoir of physiologies is present in the rumen. The first coined the term 'metagenome' to describe genomic and associated functional analysis at the community level. One of the keys to metagenomic studies is the ability to clone large fragments of community DNA (s100 kb) into bacterial artificial chromosomes.

Nutritional Genomics

Over the last several decades the importance of the relationship between nutrients, metabolism, and inherent genetic potential (genomes), has evolved from a general concept into a very active field of research known as "nutritional genomics" (nutrigenomics). Nutrigenomics seeks to measure the nutrition-responsive genome-expression, thus defining how a given nutritional challenge alters the "basal" content of expressed mRNA (transcriptome). The "Genomics" refers to the study of the "genome", which can be considered to be the complete set of genes possessed by an organism. Genomic studies typically either determine the chemical identity (genome sequencing), organizational structure (structural genomics) of an organism's genetic "blueprint" (DNA), or identify the complement of gene products (mRNA transcripts) expressed by a cell at a given time (transcriptome). Although the analysis of transcriptomes by microarrays are not considered to be quantitative (yet), microarrays have become the analytical tool of choice because they can provide significant insight into which of the multitude of biochemical pathways expressed by a cell at a given time were altered. From this overview, specific hypotheses to be developed and subsequently tested by the more sensitive, mRNA-specific assays, such as Northern blots, RNAse-expression assays, and real-time reverse-transcription-PCR analysis. The genomics/ transcriptomics, proteomics and metabolomics refer to the study of the complement of proteins that were actually expressed from the transcriptome and the metabolic products resulting from their activity, respectively. From a biological perspective, these areas of omic research provide fundamentally different yet complementary information about how cells control the synthesis and function of proteins, from the turning-on (transcription) of a gene through control of the mRNA-encoded protein activity. For example, transcriptome analysis, as compared to proteomic analysis, may be more likely to reveal treatment-induced -expression of novel genes, due to its greater sensitivity.

Biotechnology in Organic Animal Production

Concern regarding the use of antibiotics and other growth stimulants in the animal feed industry has increased in recent years. There has been increasing emphasis placed on disease prevention as a means of reducing the use of antibiotics and also public concern about pathogens in meat and meat products. As result, interest in the effects of DFM on animal health and performance has increased. For ruminants, microbial cultures have been used to potentially replace or reduce the use of antibiotics in neonatal and stressed calves, to enhance milk production in dairy cows, and to improve feed efficiency and daily gain in beef cattle. Most recently, cultures of *Lactobacillus acidophilus* have been shown to reduce fecal shedding of *Escherichia coli* O157:H7 by feedlot. Although responses to DFM have been positive in many experiments, basic mechanisms are not well defined and are not clearly understood. Enhancing our understanding of the mode of action of direct fed microbials would improve our ability to select and apply appropriate microbial preparation to ruminant diets. Another interest related to probiotics is that of competitive exclusion, attempting to bring better control of pathogens entering the food chain from livestock production. Competitive exclusion has already become established in some parts of the world, as a means of reducing salmonella and campylobacter contamination. However, current products, which are a diverse mixture of organisms originating from the rumen/ caecal contents of adult animals, are likely to find widespread acceptance in modern animal husbandry systems. Further refinements of the concept of competitive exclusion together with a better knowledge of the substrates required to maintain the organism's viability in the gut environment will undoubtedly be developed in the near future. Research has shown that live microbial feeding decreased fecal shedding of *Escherichia coli* O157:H7 from infected animals (Ohya *et al.*, 2000). Therefore, a possible application for microbial feed might be to reduce shedding of this pathogen from feedlot animals. Probiotics are viable micro-organisms if administer in sufficient numbers are capable of altering the microflora of the digestive tract, improve health and production. The exact mechanisms by which they exhibit their beneficial role within an animal and against other harmful bacteria are incompletely known. Some suggestions are: (i) adhesion and colonization of the gut mucosa (competition for receptors); (ii) competition for nutrients (iii) production of anti-microbial substances and (iv) stimulation of mucosal and systemic host immunity. When they are administered to animals, they appear to improve several health and production characteristics; because now days consumers require environmentally friendly solutions in the maintenance of animal health, probiotics may represent such an alternative. The role of feeding probiotics to monogastric animals has been best studied as protection against enteric diseases. Kritas *et al* (2006) evaluated under field conditions the effect of a probiotic containing *Bacillus licheniformis* and *Bacillus subtilis* on young lamb mortality and sheep milk production when administered in the late pregnancy and lactation feed of ewes. The ewes offered a probiotic product containing at the approximate dose of 2.56 x 10^9 viable spores per ewe per day. In the non-treated control group, 13.1% mortality was observed versus 7.8% in the probiotic-treated group with mortality being mainly due to diarrhoea. Microbiological examination of diarrhoeic faeces from some of the dead lambs in both groups revealed the presence of *Escherichia coli*. Diet is the major factor regulating bacterial composition in rumen and there is positive relationship between coliform and lactic acid bacteria. Various plant secondary metabolites (saponins, tannins and gossypol) have the antimicrobial properties (Tripathi and Mishra, 2007) and can be used for inhibition of gut pathogens.

Conclusion and Recommendations

To meet increase demands for animal foods of human food chain without increasing animal population is a great change of present animal production systems. Enhanced animal productivity form existing animal population realize with less nutritious feed available for animal feeding. Alternate feed resources needs to be explored to economize production cost, with enhanced animal productivity. Microbial manipulation of rumen ecosystem require to be manipulate in a way which improve rumen energetic efficiency on our feed resources, those have low nitrogen and high fibre content. Microbial feed additives seems promising dietary agents that especially increase microbial enzymes associated with improved energetic efficiency of rumen fermentation, accelerate rumen microbial propagation, increase feed digestion and improved ruminant production. Live cultures of few mocrobials are particularly important for the control of pathogenic microorganism and early rumen developments. On high concentrate fed animals biotechnological approach of control of rumen acidosis is an approved technology to reduce production losses of feedlot animal production. Microbial feed additives of prevents rumen microbes needs to be developed to maintain optimum rumen microbial diversity on high fiber ruminant feeding system, which is generally low. Biotechnological approach would benefit ruminant production system by improving rumen microbial ecology, increasing rumen detoxification systems. Future rumen microbiology and biotechnology will focus on microbial diversity studies of rumen ecosystem under range of environment and feeding conditions. Inhibition of rumen methanogenesis will particularly important for climatic change, the use natural plant and plant secondary metabolites seems with a great hope. Attempt of rumen microbial ecology studies with modern molecular techniques, such as fluorescence *in-situ* hybridization (FISH), qPCR, RT-PCR, TTGE, DGGE and T-RFLP will facilitate the analysis of microbial diversity and activity because culture techniques provided rumen microbial diversity of less than 15 %. More knowledge of rumen microbial diversity require for the development of new feed additives using cDNA techniques to manipulation rumen ecosystem which enhanced feed intake and production efficiency of ruminant as sustainable solution for modern animal production system. The rumen microbiology and biotechnology approach seems prosing to bring desirable modifications in rumen to have eco-friendly and more healthy animal production systems.

REFERENCES

Abu-Tarboush, H.M., Al-Saiady, M.Y. and Keir El-Din, A.H. 1996. Evaluation of diet containing lactobacilli on performance, fecal coliform, and lactobacilli of young dairy calves. *Animal Feed Science and Technology* 57, 39-49.

Argyle, J.L. and Baldwin, R.L. 1988. Modeling of rumen water kinetics and effects of rumen pH changes. *Journal of Dairy Science* 71, 1178-1188.

Asanuma, N. and Hino, T. 2001. Molecular chracterisation, enzyme properties and transcriptional regulation of phosphoenolpyruvate carboxykinase and pyruvate kinase in a ruminal bacterium, *Selenomonas ruminantium. Microbiology* 147, 681-690.

Bach, A., Iglesias, C. and Devant, M. 2007: Daily rumen pH pattern of loose-housed dairy cattle as affected by feeding pattern and live yeast supplementation. *Animal Feed Science and Technology* 136, 146-153.

Baldwin, R.L., Thornley, J.H.M. and Beever, D.E. 1987. Metabolism of the lactating cow. II. Digestive elements of a mechanistic model. *Journal of Dairy Research* 54, 107-131.

Beeman, K. 1985. The effect of *Lactobacillus* spp. on convalescing calves. *Agripractice* 6, 8-10.

Berger, W.G. and Bates, D.B. 1984. Ionophores: their effect on production efficiency and mode of action. *Journal of Animal Science* 58, 1465-1483.

Bruce, B.B., Gilliland, S.E., Bush, L.J. and Staley, T.E. 1979. Influence of feeding cells of *Lactobacillus acidophilus* on the fecal flora of young dairy calves. Oklahoma Agriculture Experimentation Station MP-104, 207-209.

Chaucheyras, F.G., Fonty, G., Bertin, G. and Gouet, P. 1995. In vitro H_2 utilisation by ruminal acetogenic bacterium cultivated alone or in association with an achea methanogen is stimulated by probiotic strain of *Saccharomyces cerevisiae. Applied and Environmental Microbiology* 61, 3466-3467.

Chesson, A. and Forsberg, C.W. 1997. Polysaccharide degradation by rumen micro-organisms. In *The Rumen Microbial Eco-system*, 2nd ed., [PN Hobson and CS Stewart, editors]. London: Blackie Academic and Professional. pp. 329-381

Clark, J.H., Klusmeyer, T.H. and Cameron, M.R. 1992. Microbial protein synthesis and flows of nitrogen fractions to the duodenum of dairy cows. *Journal of Dairy Science* 75, 2304-2323.

Colombatto, D., Mould, F.L., Bhat, M.K. and Owen, E. 2007. Influence of exogenous fibrolytic enzyme level and incubation pH on the in-vitro rumen fermentation of alfalfa stem. *Animal Feed Science and Technology* 137, 150-162.

Crawford, J.S., Carver, L., Berger, J. and Dana, G. 1980. Effects of feeding a living nonfreeze-dried *Lactobacillus acidophilus* culture on performance of incoming feedlot steers. Proc. *Western Section American Society of Animal Science* 31, 210-212.

Dawson, K.A., Newman, K.E. and Boling, J.A. 1990. Effects of microbial supplements containing yeast and lactobacilli on roughase-fed ruminal microbial activities. *Journal of Animal Science* 68, 3392-3398.

Dehority, B.A. and Tirabasso, P. 2000. Antibiosis between ruminal bacteria and ruminal fungi. *Applied and Environmental Microbiology* 66, 2921-2927.

Dehority, B.A. and Tirabasso, P.A. 2000. Antibiosis between ruminal bacteria and ruminal fungi. *Applied and Environmental Microbiology* 66, 2921-2927.

Devillard, E., Bera-Maillet, C., Flint, H.J., Scott, K.P., Newbold, C.J., Wallace, R.J., Jouany, J.P. and Orano, E. 2003. Characterization of XYN10B, a modular xylanase from the ruminal protozoan Polyplastron multivesiculatum, with a family 22 carbohydrate-binding module that binds to cellulose. *Biochemistry Journal* 373, 495-503.

Dijkstra, J. 1994. Simulation of the dynamics of protozoa in the rumen. *British Journal of Nutrition* 72, 679-699.

Dijkstra, J. and Bannink, A. 2000. Analyses of modeling whole rumen function. In *Feeding Systems and Feed Evaluation Models*, chapter 13, (M.K. Theodorou and J. France, editors). Wallingford, CAB International. pp. 299-322.

Dijkstra, J., France, J. and Tamminga, S. 1998. Quantification of the recycling of microbial nitrogen in the rumen using a mechanistic model of rumen fermentation processes. *Journal of Agricultural Science Cambridge* 130, 81-94.

Dijkstra, J., France, J., Dhanoa, M.S. and Lopez, S. 2000. Simulation of substrate degradation, microbial synthesis and gas production in the gas production technique and in vivo. In *Gas Production: Fermentation Kinetics for Feed Evaluation and to Assess Microbial Activity*, Edinburgh: British Society of Animal Science. pp. 42-43.

Dijkstra, J., Neal, H.D.S.C., Beever, D.E. and France, J. 1992. Simulation of nutrient digestion, absorption and outflow in the rumen: model description. *Journal of Nutrition* 122, 2239-2256.

El-Ghani, A.A. 2004. Influence of diet supplementation with yeast culture (*Saccharomyces cerevisiae*) on performance of Zaraibi goats. *Small Ruminant Research* 52, 223-229.

Erasmus, L.J.,Robinson, P.H., Ahmadi, A., Hinders, R. and Garrett, J.E. 2005. Influence of pre-partum and postpartum supplementation of a yeast culture and monensin, or both, on ruminal fermentation and performance of multiparous dairy cows. *Animal Feed Science and Technology* 122, 219-239.

Firkins, J.L. 1996. Maximizing microbial protein synthesis in the rumen. *Journal of Nutrition* 126, 1347S-1354S.

Fleige, S., PreBinger, W., Meyer, H.H.D. and Pfaff, M.W. 2007. Effect of lectulose on growth performance and intestinal morphology of pre-ruminant claves using milk replacer containing *Enterococcus faecium*. Animal 1, 367-373.

Fonty, G., Chavarot, M., Lepetit, J., Canistro, J. and Favier, R. 1999. Mechanical resistance of wheat straw after incubation in cultures of rumen cellulolytic micro-organisms. *Animal Feed Science and Technology* 80, 297-307.

Fonty, G., Jouany, J.P., Thivend, P., Gouet, P. and Senaud, J. 1983. A descriptive study of rumen digestion in meroxenic lambs according to the nature and complexity of the microflora. *Reproduction Nutrition and Development* 23, 857-873.

Fox, S.M. 1988. Probiotics intestinal inoculants for production animals. *Veterinary Medicine* 83, 806–830.

France, J., Theodorou, M.K. and Davies, D. 1990. Use of zoospore concentrations and life cycle parameters in determining the population of anaerobic fungi in the rumen eco-system. *Journal of Theoretical Biology* 147, 413-422.

Ghorbani, G.R., Morgavi, D.P., Beauchemin, K.A. and Leedle. J.A.Z. 2002. Effects of bacterial direct-fed microbials on ruminal fermentation, blood variables, and the microbial populations of feedlot cattle. *Journal of Animal Science* 80,1977-1986.

Gill, D.R., Smith, R.A. and Ball. R.L. 1987. The effect of probiotic feeding on health and performance of newly-arrived stocker calves. *Oklahoma Agriculture Experimentation Station* MP-119, 202-204.

Greening, R.C., Smolenski, W.J., Bell, R.L., Barsuhn, K., Johnson, M.M. and Robinson, J.A. 1991. Effects of inoculation of *Megasphaera elsdenii* strain 407A(UC-12497) on ruminal pH and organic acids in beef cattle. *Journal of Animal Science* 69(Suppl. 1), 518.

Gregg, K., Hamdorf, B., Henderson, K., Kopecny, J. and Wong, C. 1998. Genetically modified ruminal bacteria protect sheep from fluoroacetate poisoning. *Applied and Environmental Microbiology* 64, 3496-3498.

Hespell, R.B. and Bryant, M.P. 1979. Efficiency of rumen microbial growth: influence of some theoretical and experimental factors on YATP. *Journal of Animal Science* 49, 1640-1659.

Hristov, A.N., McAllister, T.A. and Chéng, K.J. 1998. Stability of exogenous polysaccharide-degrading enzymes in the rumen. *Animal Feed Science and Technology* 76, 161-168.

Hristov, A.N., McAllister, T.A. and Cheng, K.J. 2000. Intraruminal supplementation with increasing levels of exogenous polysaccharide-degrading enzymes: effects on nutrient digestion in cattle fed a barley grain diet. *Journal of Animal Science* 78, 477-487.

Huffman, R.P., Karges, K.K., Klopfenstein, T.J., Stock, R.A., Britton, R.A. and Roth, L.D. 1992. The effect of Lactobacillus acidophilus on subacute ruminal acidosis. *Journal of Animal Science* 70(Suppl. 1), 87.

Huntington, G.B. 2000. High-starch rations for ruminant production discussed. *Feedstuffs* 12-13:23.

Hutcheson, D.P., Cole, N.A., Keaton, W., Graham, G., Dunlap, R. and Pittman, K. 1980. The use of a living, nonfreeze-dried *Lactobacillus acidophilus* culture for receiving feedlot calves. Proc. *Western Section American Society of Animal Science* 31, 213-215.

Isaacson, H.R., Hinds, F.C., Bryant, M.P. and Owens, F.N. 1975. Efficiency of energy utilization by mixed rumen bacteria in continuous culture. *Journal of Dairy Science* 58, 1645-1659.

Jenny, B.F., Vandijk, H.J. and Collins, J.A. 1991. Performance and fecal flora of calves fed a *Bacillus subtilis* concentrate. *Journal of Dairy Science* 74, 1968-1973.

Johnson, K.A. and Johnson, D.E.1995. Methane emission from cattle. *Journal of Animal Science* 73, 2483-2492.

Kamel, H.E.M., Sekine, J., El-Waziry, A.M. and Yacout, M.H.M. 2004, Effect of *Saccharomyces cerevisiae* on the synchronisation of organic matter and nitrogen degradation kinetics and microbial nitrogen synthesis in sheep fed Berseem hay (*Trifolium alexandrinum*). *Small Ruminant Research* 52, 211-216.

Kashket, E.R. 1987. Bioenergetics of lactic acid bacteria: cytoplasmic pH and osmotolerance. *FEMS Microbiology Reviews* 46, 233-244

Klieve, A.V., Bain, P.A., Yokoyama, M.T., Ouwerkerk, D., Forster, R.J. and Turner, A.F. 2004. Bacteriophages that infect the cellulolytic ruminal bacterium *Ruminococcus albus* AR67. *Letters in Applied Microbiology* 38, 333-338.

Klieve, A.V., Bain, P.A., Yokoyama, M.T., Ouwerkerk, D., Forster, R.J. and Turner, A.F. 2004. Bacteriophages that infect the cellulolytic ruminal bacterium *Ruminococcus albus* AR67. *Letters in Applied Microbiology* 38, 333-338.

Koch, A.L. 1994. Growth measurement. In *Methods for General and Molecular Bacteriology*, (P. Gerhardt, R.G.E. Murray, W.A. Wood and N.R. Kreig, editors). Washington: American Society for Microbiology Press. pp. 248-277.

Krause, D.O., Denman, S.E., Mackie, R.I. and Morrison, M., Rae, L.R., Attwood, G.T. and McSweeney, C.S. 2003. Opportunities to improve fiber degradation in the rumen: microbiology, ecology, and genomics. *FEMS Microbiology Reviews* 27, 663-693.

Krause, D.O., Smith, W., Dalrymple, B.P., Mackie, R.I. and McSweeney, C.S. 1999. 16S rDNA sequencing of *Ruminococcus albus* and *Ruminococcus flavefaciens*: design of a signature probe and its application in adult sheep. *Microbiology* 145, 1797-1807.

Krause, D.O., Smith, W.J.M., Conlan, L.L., Gough, J.M., Williamson, M.A. and McSweeney, C.S., 2003. Diet influences the ecology of lactic acid bacteria and Escherichia coli along the digestive tract of cattle: neural network and 16S rDNA. *Microbiology* 149, 57-65.

Krause, D.O., Smith, W.J.M., Ryan, F.M.E., Mackie, R.I. and McSweeney, C.S. 1999. Use of 16S-rRNA based techniques to investigate the ecological succession of microbial populations in the immature lamb rumen: Tracking of a specific strain of inoculated *Ruminococcus* and interactions with other microbial populations in vivo. *Microbial Ecology* 38, 365-376.

Krause, M., Beauchemin, K.A., Rode, L.M., Farr, B.I. and Norgaard, P. 1998. Fibrolytic enzyme treatment of barley grain and source of forage in high-grain diets fed to growing cattle. *Journal of Animal Science* 76, 2912-2920.

Krehbiel, C.R., Rust, S.R., Zhang, G. and Gilliland, S.E. 2003. Bacterial direct-fed microbials in ruminant diets: Performance response and mode of action. *Journal of Animal Science* 81(E. Suppl. 2), E120-E132.

Kritas, S.K.,Govaris, A., Christodoulopoulos, G. and Burriel, A.R. 2006. Effect of *Bacillus licheniformis* and *Bacillus subtilis* supplementation of ewe's feed on sheep milk production and young lamb mortality. *Journal of Veterinary Medicine A* 53, 170-173.

Kung, Jr. L. and Hession, A.O. 1995. Preventing in vitro lactate accumulation in ruminal fermentation by inoculation with *Megasphaeraelsdenii*. *Journal of Animal Science* 73, 250–256.

Lee, S.S., Ha, J.K., and Cheng, K.-J. 2000. Relative contributions of bacteria, protozoa, and fungi to in vitro degradation of orchard grass cell walls and their interactions. *Applied and Environmental Microbiology* 66, 3807-3813.

Leng, R.A.1990. Factors affecting the utilisation of poor-quality forages by ruminants particularly under tropical conditions. *Nutrition Research Review* 3, 277-303.

Lopez, S,, Murison, S.D., Travis, A.J. and Chesson, A. 1993. Degradability of parenchyma and sclerenchyma cell walls isolated at different stages from a newly extended maize internode. *Acta Botanica Neerlandica* 42, 165-174.

Lowe, S.E., Theodorou, M.K. and Trinci, A.P. 1987. Cellulases and xylanase of an anaerobic rumen fungus grown on wheat straw, wheat straw holocellulose, cellulose, and xylan. *Applied and Environmental Microbiology* 53, 1216-1223.

Mackie, R.I. and Gilchrist, F.M.C. 1979. Changes in lactate-producing and lactate-utilising bacteria in reaction to pH in the rumen of sheep during stepwise adaptation to a high-concentrate diet. *Applied and Environmental Microbiology* 38, 422-430.

Mackie, R.I., Mcsweeney, C.S. and Klieve, A.V. 2002. Microbial ecology of the ovine rumen. In Sheep Nutrition (Freer, M. and Dove, H., eds), Wallingford CAB International, pp. 71-94.

Mackie, R.I., Mcsweeney, C.S. and Klieve, A.V. 2002. Microbial ecology of the ovine rumen. In Sheep Nutrition (Freer, M. and Dove, H., eds) Wallingford CAB International, pp. 71-94.

McAllister, T.A., Moustafa, S.M.S., Cheng, K.J., Newbold, C.J., Mckain, N. and Wallace, R.J. 1994. Effect of salinomycin on fermentation and nitrogen metabolism in the artificial rumen. *Canadian Journal of Animal Science* 74, 575-578.

McAllister, T.A., Oosting, S.J., Popp, J.D., Mir, Z., Yanke, L.J., Hristov, A.N., Treacher, R.J. and Cheng, K.J. 1999. Effect of exogenous enzymes on digestibility of barley silage and growth performance of feedlot cattle. *Canadian Journal of Animal Science* 79, 353-360.

McAllister, T.A., Stanford, K., Bae, H.D., Treacher, R.J., Hristov, A.N., Baah, J., Shelford, J.A. and Cheng, K.J. 2000. Effect of a surfactant and exogenous enzymes on digestibility of feed and on growth performance and carcass traits of lambs. *Canadian Journal of Animal Science* 80, 35-44.

McSweeney, C.S., Dulieu, A., Katayama, Y. and Lowry, J.B. 1994. Solubilisation of lignin by the ruminal anaerobic fungus *Neocallimastix patriciarum*. *Applied and Environmental Microbiology* 60, 2985-2989.

Miller, T.L. 1995. Ecology of methane production and hydrogen sinks in the rumen. In Ruminant Physiology: Digestion, Metabolism, Growth and Reproduction. Eds. W.A. Engelhardt, S. leonhard-Marek, G. Breves and D. Giesecke. Stuttgart: Ferdinand Enke Verlag. pp. 317-331.

Mohammed, N., Onodera, R., Itabashi, H. and Lila, Z.A. 2004. Effects of ionohpores, vitamin V6 and distiller's grain on in vitro tryptophane biosynthesis from indolpyruvic acid, and production of other related compounds by ruminal bacteria and protozoa. *Animal Feed Science and Technology* 116, 301-311.

Moss, A.R., Jounay, J.P., Newbold, J. 2000. Methane production by ruminants: its contribution to global warming. *Annals of Zootechnology* 49, 231.

Mountfort, D.O. and Orpin, C.G. 1995. *Anaerobic Fungi*. New York: Marcel Dekker Inc.

Muscato, T.V., Tedeschi, L.O. and Russell, J.B. 2002. The effect of ruminal fluid preprations on the growth and health of newborn, milk-fed dairy calves. *Journal Dairy Science* 85, 648-656.

Mwenya, B., Santoso, B., Sar, C., Gamo, Y., Kobayashi, T., Arai, I. and Takahashi, J., 2004. Effects of including â 1-4 galacto-oligosaccharides, lactic acid bacteria or yeast culture on methanogenesis as well as energy and nitrogen metabolism in sheep. *Animal Feed Science and Technology*115, 313-326.

Nagaraja, T.G., Newbold, C.J., VanNevel, C.J. and Demeyer, D.I. 1997. Manipulation of ruminal fermentation. The Rumen Microbial Ecosystem (P.N. Hobson and C.S. Stewart eds), *Blackie Academic and Professional* London, pp. 523-632.

Nakanishi, Y., Arave, C.W. and Stewart, P.H. 1993. Effects of feeding Lactobacillus acidophilus yogurt on performance and behavior of dairy calves. *Journal of Dairy Science* 76(Suppl. 1), 244.

Newbold, C.J., Ushida, K., Morvan, B., Fonty, G. and Jouany, J.P. 1996. The role of ciliate protozoa in the lysis of methanogenic archaea in rumen fluid. *Letters in Applied Microbiology* 23, 421-425.

Newbold, C.J., Wallace, R.J., Chen, X.B. and McIntosh, F.M. 1995. Different strains of *Saccharomyces cerevisiae* differ in their effects on ruminal bacterial numbers *in-vitro* and in sheep. *Canadian Journal of Animal Science* 73, 1811-1818.

Newman, K.E. and K.A. Jacques. 1995. Microbial feed additives for pre-ruminants. Biotechnology in Animal Feeds and Animal Feeding. (R. J. Wallace and A. Chesson, ed.) VCH, Weinheim, Germany. pp. 247-258.

Nocek, J.E., Kautz, W., Leedle, J.A.Z. and Allman, J.G. 2002. Ruminal supplementation of direct-fed microbials on diurnal pH variation and in situ digestion in dairy cattle. *Journal of Dairy Science* 85, 429-433.

Nsereko, V.L., Morgavi, D.P., Rode, L.M., Beauchemin, K.A. and McAllister, T.A. 2000. Effects of fungal enzyme preparations on hydrolysis and subsequent degradation of alfalfa hay fiber by mixed rumen microorganisms in vitro. *Animal Feed Science and Technology* 88, 153-170.

Ohya, T., Marubashi, T. and Ito,H. 2000. Significance of fecal volatile fatty acids in shedding of *Escherichia coli* O157 from calves: experimental infection and preliminary use of a probiotic product. *Journal of Veterinary Medicine and Science* 62, 1151-1155.

Owens, F.N., Secrist, D.S., Hill, W.J. and Gill, D.R. 1998. Acidosis in cattle: a review. *Journal of Animal Science* 76, 275-286

Pagie, L. and Hogeweg, P. 1999. Colicin diversity: a result of eco-evolutionary dynamics. *Journal of Theoretical Biology* 60, 3688-3696.

Reynolds, C.K., Harmon, D.L. and Cecava, M.J. 1994. Absorption and delivery of nutrients for milk protein synthesis by portal-drained viscera. *Journal of Dairy Science* 77, 2787–2808.

Riley, M.A. 1998. Molecular mechanisms of bacteriocin evolution. *Annual Review of Genetics* 32, 255-278.

Rodriguez, C.A., Gonzalez, J., Alvir, M.R., Repetto, J.L., Centeno, C. and Lamrani, F. 2000. Composition of bacteria harvested from the liquid and solid fractions of the rumen of sheep as influenced by feed intake. *British Journal of Nutrition* 84, 369-376.

Rooke, J.A. and Armstrong, D.G. 1989. The importance of the form of nitrogen on microbial protein synthesis in the rumen of cattle receiving grass silage and continuous intrarumen infusions of sucrose. *British Journal of Nutrition* 61, 113-121.

Russel, J.B. and Strobel, H.J. 1988. Effect of ionophores on rumen fermentation. *Applied and Environmental Microbiology* 55, 1-6.

Russell, J.B., O'Connor. J.D., Fox, D.G., Van Soest, P.J. and Sniffen, C.J. 1992. A net carbohydrate and protein system for evaluating cattle diets: I. Ruminal fermentation. *Journal of Animal Science* 70, 3551-3561.

Sar, C., Mwenya, B., Santoso, B., Takaura, K., Morikawa, R., Isogai, N., Asakura, Y., Toride, Y. and Takahashi, J. 2005. Effect of *Eshcherichia coli W3110* on ruminal methanogenesis and nitrate/nitrite reduction in vitro. *Animal Feed Science and Technology* 118, 295-306.

Tamminga, S. 1996. A review on environmental impacts of nutritional strategies in ruminants. *Journal of Animal Science* 74, 3112-3124.

Tamminga, S. and Van Vuuren, A.M. 1988. Formation and utilization of end products of lignocellulose degradation inruminants. *Animal Feed Science and Technology* 21, 141-159.

Tamminga, S., Van Vuuren, A.M., Van der Koelen, C.J., Ketelaar, R.S. and Van der Togt, P.L. 1990. Ruminal behaviour of structural carbohydrates, non-structural carbohydrates and crude protein from concentrate ingredients in dairy cows. *Netherlands Journal of Agricultural Science* 38, 513-526.

Tannock, G.W. 1983. Effect of dietary and environmental stress on the gastrointestinal microbiota. Human Intestinal Microflora in Health and Disease, (D. Hentges, ed.) Academic Press, New York. pp 517-539.

Theodorou, M.K., Gill, M., King-Spooner, C. and Beever, D. 1990. Enumeration of anaerobic *chytridiomycetes* as thallus forming units: novel method for quantification of fibrolytic fungal populations from the digestive tract ecosystem. *Applied and Environmental Microbiology* 56, 1073-1078.

Therion, J.J., Kistner, A.A. and Kornelius, J.H. 1982. Effect of pH on growth rates of amylolytic and lactilytic bacteria. *Applied and Environmental Microbiology* 44, 428-434.

Tripathi, M.K. and Mishra, A.S. 2007. Glucosinolate in animal nutrition, A Review. *Animal Feed Science and Technology.* 132, 1-27.

Tripathi, M.K., Chaturvedi, O.H., Karim, S.A., Singh, V.K., Sisodiya, S.L., 2007c. Effect of different levels of concentrate allowances on rumen fluid pH, nutrient digestion, nitrogen retention and growth performance of weaner lambs. *Small Ruminant Research* 72, 178-186.

Tripathi, M.K., Karim, S.A., Chaturvedi, O.H and Verma, D.L. 2007b. Nutritional value of animal feed grade wheat as replacement of maize in lamb feeding for mutton production. *Journal of the Science of Food and Agriculture* 87, 2447-2455.

Tripathi, M.K., Karim, S.A., Chaturvedi, O.H. and Singh, V.K. 2006. Effect of *ad libitum* tree leaves feeding with varying levels of concentrate on intake, microbial protein yield and growth of lambs. *Livestock Research for Rural Development* 18, 1-6.

Tripathi, M.K., Karim, S.A., Chaturvedi, O.H. and Verma, D.L. 2007a. Effect of different liquid cultures of live yeast strains on performance, ruminal fermentation and microbial protein synthesis in lambs. *Journal of Animal Physiology and Animal Nutrition* DOI: 10.1111/ j. 1439. 2007.00759.x.

Tripathi, M.K., Santra, A., Chaturvedi, O.H., and Karim, S.A. 2004. Effect of graded levels of $NaHCO_3$ supplementation on ruminal fluid pH, feed intake, nutrient utilization, growth performance and blood biochemical constituents of lamb on high concentrate diets. *Animal Feed Science and Technology* 111, 27-39.

Ushida, K. and Jouany, J.P. 1990. Effect of defaunation on fibre digestion in sheep given two isonitrogenous diets. *Animal Feed Science and Technology* 29, 153-158.

Ushida, K., Kayouli, C., De Smet, S. and Jouany, J.P. 1990. Effect of defaunation on protein and fibre digestion in sheep fed on ammonia-treated straw-based diets with or without maize. *British Journal of Nutrition* 64, 765-775

Van Soest, P.J. 1994. Nutritional Ecology of the Ruminant. 2nd ed. Comstock, Ithaca, NY.

Wang, Y., Alexander, T.W. and McAllister, T.A. 2004. In Vitro effects of monensin and teween 80 on ruminal fermentation of barley grain: barley silage-based diets for beef cattle. *Animal Feed Science and Technology* 116, 197-209.

Wang, Y., McAllister, T.A., Rode, L.M., Beauchemin, K.A., Morgavi, D.P., Nsereko, V.L., Iwaasa, A.D. and Yang, W. 2001. Effects of an exogenous enzyme preparation on microbial protein synthesis, enzyme activity and attachment to feed in the Rumen Simulation Technique (Rusitec). *British Journal of Nutrition* 85, 325-332.

Williams, A.G. and Coleman, G.S. 1997. The Rumen Protozoa. In *The Rumen Microbial Eco-system* 2nd edn., (P.N. Hobson and C.S. Stewart, eds.) London, *Blackie Academic and Professional* pp. 73-139.

Williams, D.L., and J.H. Mahoney. 1984. Pre-weaning and post weaning nutrition. Proc. 17th *Annual Convention of American Association of Bovine Practice* pp. 98.

Wolin, M.J, Miller, T.L. and Stewart, C.S. 1997. Microbe-microbe interactions. In *The Rumen Microbial Eco-system*, 2nd edn., (P.N. Hobson and C.S. Stewart, eds.). London, *Blackie Academic and Professional* pp. 467-491.

Wolin, M.J. 1979. The rumen fermentation: a model for microbial interactions in anaerobic eco-systems. *Advances in Microbial Ecology 3*, (M. Alexander, eds]. New York, Plenum Press. pp. 49-77.

Yoon, I.K., and Stern, M.D. 1995. Influence of direct-fed microbials on ruminal microbial fermentation and performance of ruminants: A review. *Asian-Australasian Journal of Animal Science* 8, 533-555.

❑❑❑

6

Rumen Manipulation for Improving Small Ruminant Productivity

A. SANTRA
Eastern Regional Station, National Dairy Research Institute,
A-12 Block, Kalyani - 741 235, West Bengal, India

Introduction

Continuous increases in the human population consume an ever greater fraction of earth's food, among which proteins are extremely important. According to an estimate by FAO, the human demand for food is expected to touch 110 million metric tones by 2010 from the current level of consumption about 90 million metric tones (Makkar, 2007). Herbivores, particularly ruminants, make up large share of the domestic animals involved in production of food for human consumption. They are able to utilize the lingo-cellulosic materials and convert them to animal products having high nutritional value viz. meat, milk wool/fur and hide. Moreover ruminants do not compete with human and other livestock species for feed resource.

Ruminants are fore gut fermenters and their stomach has four distinct compartments consisting of rumen, reticulum, omasum and abomasum. The rumen, which is located at the beginning of the tract, plays a major role as at least 50 % of the total digestion occurs there. Although a myriad of microorganisms are found throughout the digestive tract of ruminant, still only the microbiota inhabiting in the rumen have true symbiotic relationship with the host. These rumen microorganisms are predominantly bacteria, protozoa and phycomycete fungi. The mammalian system is devoid of enzymes to degrade structural carbohydrate and hence the symbiotic microbes inhabiting in rumen elaborate enzymes for fermentative digestion of large amounts of fibrous feed consumed by the ruminants. By providing a suitable habitat for these microorganisms, the ruminants are able to utilize the end products of microbial fermentation to meet their own nutritional need. Rumen is an open, self-contained ecosystem in which feed consumed by the ruminant is fermented by rumen microbes to volatile fatty acids and microbial biomass those serve as source of energy and protein for the host animals. The rumen microbial ecosystem is an efficient anaerobic fermentation system that confers added advantages on ruminants over monogastric or non-ruminants animals and these are:

a. Ruminants can digest large amount of fibrous feeds (ligno-cellulosic materials) efficiently,

b. They can use non protein nitrogen sources like urea as a source of nitrogen to meet part of their protein requirement,

c. They can detoxify many toxic ingredients present in feeds of plant origin.

Why We Need Rumen Manipulation?

Anaerobic fermentation of feeds in the rumen is beneficial for the host animal. Therefore, do we really need the manipulation of the rumen ecosystem? The answer to this question is definitely yes. The rumen microbial ecosystem is not so efficient for digestion of ingested feed as evident from the presence of sizable portion of undigested feeds in the faeces and production of large amount of methane gas in the rumen which could be otherwise utilized as source of energy by the animals. Ruminant animals act as important source of animal protein in the developing countries. Ruminants of the region are mainly maintained on poor quality roughage or lignocellulosic agroindustrial by products with or without concentrate supplementation resulting in poor productivity of the animals. Therefore, considerable scope exist for manipulation of ruminal fermentation to improve the utilization of forages particularly in tropical as well as developing countries to maximize the productivity of animals by using available resources.

Manipulation of rumen fermentation can be considered as an optimization process, whereby optimal condition are sought by maximization and/or minimization of fermentation process, depending on factors such as kind and level of feeding and animal production. Some of the major objectives of rumen manipulation are:

a. *Enhance fibrolytic activity:* To increase the fibre degradation mainly through manipulation of lignocellulosic bonds in high lignocellulosic feeds as the rumen microbes are the only degraders of cellulose and hemicellulose,

b. *Increase microbial protein synthesis:* A major portion of the amino acid reaching the duodenum is of microbial protein origin. Therefore, attempts should be made to maximize microbial protein synthesis in the rumen,

c. Reduction in proteolysis: Hydrolysis of feed protein, deamination of amino acids and reutilization of ammonia for microbial protein synthesis are all energy consuming process, hence the degradation of protein and deamination of aminoacids in the rumen should be discouraged,

d. *Reduction in methanogenesis:* Methane generation in the rumen is a wasteful process as 5-10 % of GE intake of ruminants is converted in to methane. The provision of an alternate hydrogen sink in the rumen may help in increasing digestible energy (DE) availability for production,

e. *Prevention of acidosis:* In high grain fed animals, the level of lactic acid can be controlled to avoid acidosis and inhibition of feed utilization due to lowered pH of the rumen liquor,

f. Shifting acetate to propionate production: In fattening beef/lambs the production of propionate in the rumen at the expense of acetate may be helpful,

g. *Novel microbes:* The quality of protein is important in high producing ruminants. Microbes, can be tailored to synthesize the amino acids in the form of the peptides and supply to the animals in the intestine,

h. *Metabolism of plant toxins:* Rumen fermentation can be manipulated for efficient utilization of feeds which contain anti nutritional factors viz. tannin, saponin, mimosine etc,

i. Synthesis of useful secondary metabolites.

Methods of Rumen Manipulation

Several techniques of rumen manipulation have been tried in different laboratories of the world during the last two decades with varying results. Broadly the methods of rumen manipulation can be classified in two i.e., genetic manipulation and non genetic manipulation. In genetic manipulation, attempts were made to develop genetically engineered rumen microbes by gene transfer/ manipulation technique to enhance the animal productivity. However success in the field of genetic manipulation of rumen microbes is very poor/sporadic. Non genetic manipulation of the rumen can be done by physical methods (dietary manipulation) and by using suitable chemicals or feeding microbes (probiotics).

Genetic Manipulation

The potential of application of molecular techniques in achieving the goals of rumen manipulation are enormous. These techniques could allow the introduction or increase of desired activities such as cellulolysis and detoxification or reduction of undesirable activities such as proteolysis, deamination and mthanogenesis. For this purpose, one approach would be to select the desirable gene and to express them in a predominant rumen bacteria. Naturally present microorganisms in the rumen can be genetically modified to enhance their capacity of defined functions or to add new functions. Introductions of diverse genes into gut microorganisms have been extensively explored. The genetically modified microorganisms are either will be able to digest fibrous components and lignins of forage, or degrade toxins, synthesize essential amino acids, reduce ruminal methane production and tolerate acids. The second approach would be to introduce new species or strains of microorganisms into the gut. Application of the said two approaches has a great potential to increase digestibilities of feedstuffs and to improve animal production.

The physiological conditions in the rumen are not favorable for most of the non rumen microbes. The genetically engineered microbes (rumen origin or non rumen origin) mostly have low competitive ability to survive in a mixed culture. The rumen bacteria from one ruminant may not necessarily be established in the rumen of another animal. The best example of the successful introduction of a new organism (not genetically modified) in the rumen was the introduction of bacteria that was capable of degrading 3-hydroxy-4(1H)-pyridone (DHP) into Australian ruminants. These animals were unable to use *Leucaena leucocephala* which have a toxic factor mimosine which was degraded to the toxic goitrogen DHP by the rumen bacteria. Australian goats those consumed *Leucaena leucocephala* developed toxicosis, but Hawaiian goats consuming the same plant spcies did not develop toxicosis. Inoculation of enriched cultures derived from ruminal content of Hawaiian goats into Australian ruminant conferred resistance to leucaena poisoning.

The problems with the establishment of genetically engineered rumen bacteria are too many and very complex. In addition to the scientific and technical problems involved in the establishment of these bacteria in the rumen, the existing regulations about the release of genetically engineered microbes in the atmosphere is also a limitation. A more realistic approach will be to study as to whether the introduced genetic product can serve the purpose of improving rumen fermentation. The success of this approach will depend upon the stability of this gene product against its degradation in the rumen.

Non Genetic Manipulations

i) Microbial feed additive (probiotics)
ii) Defaunation

Microbial Feed Additives (Probiotics)

Probiotic defined as "A live microbial feed supplement which beneficially affects the host animals by improving its intestinal microbial balance." This definition encompasses single strain or a mixture of two or more species/strains of microbes, with or without growth medium. The commonly used probiotics for animal feeding are broadly divided into two categories i.e., bacterial origin and yeast origin. The primary micro-organisms currently used in animal feeding are:

Bacterial origin

- *Bacillus licheniformis*
- *Bacillus subtilis*
- *Bifadobacterium adolescentis*
- *Bifadobacterium animalis*
- *Bifidobacterium bifidus*
- *Bifadobacterium infantis*
- *Bifadibacterium longum*
- *Bifidobacterium pseudolongum*
- *Bifidobacterium suis*
- *Bifadobacterium thermophilum*
- *Lactobacillus acidophilus*
- *Lactobacillus bulgaricus*
- *Lactobacillus casei*
- *Lactobacillus cellobiosus*
- *Lactobacillus delbrueckii*
- *Lactobacillus fermentum*
- *Lactobacillus lactis*
- *Lactobacillus plantarum*
- *Lactobacillus salivarius*
- *Lactobacillus sporogens*
- *Lactobacillus reuteri*
- *Streptococcus intermdius*
- *Streptococcus thermophilus*

Yeast origin

- *Lactobacillus brevis*
- *Saccharomyces cerevisiae*
- *Aspergilus oryzae*

Now a days scientists are trying to isolate the superior strain of rumen fungi for better cellulytic activity and their interspecies trans-inoculation. The micro organisms which used as probiotics should possess the following properties:

- Resistance to low pH and bile salt,
- Production of lactate and other anti microbial agents,
- A normal inhabitant of the gut in the target animal species,
- Able to survive, colonize and multiply at a faster rate in the gut,
- Viable product can be formed at industrial scale for its commercialization,
- Stable and viable during long storage and field conditions,
- Must produce beneficial effect in host animals.

The effects of probiotics are greatest in the fastest growing animals and diminish with age. This age effect is consistent with the capacity of the normal gut flora to resist change as the animal grows with the most stable situation occurring in the adult animals. Administration of probiotics in livestock may be most effective under following conditions:

- After birth to encourage the early establishment of beneficial rumen microflora,
- Following antibiotic treatment,
- In the presence of enteric pathogen such as *E. coli*, *Salmonella*, Coccidia,
- During environmental or mangemental stress.

In small ruminants, administration of probiotics may be most effective under the following circumstances:

- After birth,
- Before and after transportation,
- At weaning,
- Following over eating or antibiotic administration.

Recent studies indicated that the administration of probiotics had an impact on growth performance, disease resistance, improving animal production and providing a cost effective dietary supplement. Their widespread use as manipulating agents for rumen fermentation (so called direct fed microbials) is of recent origin and most of the published research papers on the topic have been periodically reviewed. The published data indicated that microbial feed additives may benefit ruminant nutrition in terms of live weight gain and milk production of the animals in the tune of 7 - 8 % (Wallace and Newbold, 1993).

Probiotic for Neonatal Ruminant

Application of microbial preparation for newborn animals includes dosing or drenching the animals soon after birth or inclusion of direct fed microbial products (DFM) in either milk or milk replacer. The goals of microbial supplementation of the neonatal ruminant are similar to those for non ruminants like rapid adaptation to solid feed by accelerating the development of a normal adult intestinal microflora, avoiding the establishment of enteropathogens and also to stimulate the early development of rumen. Lactic acid producing bacteria are administered to lambs and kids soon after birth and/or in milk replacer. The primary goal of inoculating neonates with lactic acid bacteria is to establish a beneficial populations of bacteria in the GI tract capable of competing successfully with pathogens. There is evidence of rapid rumen development and faster growth by additions of certain Lactobacillus bacteria in the animal's feed.

Probiotics for Growing Small Ruminants

For many years, ruminant nutritionists and microbiologists have been interested in the manipulation of microbial ecosystem of the rumen to improve production efficiency of domestic ruminants. Recently the use of yeast culture and lactobacillus bacteria in ruminant diets as a probiotics has received renewed attention. Production benefits due to feeding of live yeast or lactobacillus culture to the ruminant to be caused by changes in ruminal fermentation, in particularly by increase degradibility of forage and flow of microbial protein from the rumen.

Effect of Probiotic Feeding on Rumen Function

The yeast cells were able to maintain viability throughout the digestive tract. Using yeast culture in rumen simulating continuous fermentor culture, it was observed that yeast replicated in the *in vitro* system. However, the yeast culture was unable to maintain a productive population within the rumen ecosystem. The viable cell numbers of yeast showed a decline in rumen and was due to extensive lysis and due to rumen outflow. Two type of yeast have been proposed to stabilize the rumen ecosystem: *Saccharomyces cervisiae* and *Aspergillus oryzae*. Both have been shown to stimulate VFA production *in vitro* by increasing the degradation of the dietary forage fraction in mixed diets (Jouany *et al.*, 2000).

Yeast feeding has been found to be increase the total number of rumen bacterial population along with higher proportion of cellulolytic bacteria and thus improved cellulose digestion. Different cellulolytic bacteria respond differently to yeast culture. The yeast strains that stimulate the growth of *Bacteroides succinogenes* in pure cultures did not stimulate the growth of *Ruminococcus albus* to the same extent. The viable yeast culture itself or some heat liable nutrients in the yeast may be responsible for stimulation of rumen bacterial growth. However, effect of direct fed yeast culture on rumen ciliate protozoal population is contradictory.

Yeast have a buffering effect in the rumen medium and prevent sharp drops in rumen pH and thus stabilize the pH even in the high concentrate fed animals. The higher ruminal pH in yeast culture fed animals was associated with a lower lactate concentration although, yeast itself and do not utilize lactate. Yeast culture inhibit the growth of lactic acid producing rumen bacteria by utilizing their substrate (soluble sugar) and thus decreasing lactic acid production and prevent sudden fall in pH of the rumen medium. Moreover, yeast culture stimulates the growth of lactic acid utilizing rumen bacteria as well as uptake of lactate by them.

Published reports on the effect of yeast culture feeding to the animals on ruminal VFA concentration are variable: some workers reported yeast culture had no effect while others found a stimulation in VFA production as well as proportion propionate production at the expense of acetate or even an increase in the proportion of acetate. Animals consuming yeast culture have lower ruminal ammonia nitrogen concentration and higher microbial protein synthesis (Dutta *et al.*, 2001). Rumen volume, liquid dilution rate and liquid outflow rate were not affected by the inclusion of yeast culture in animal's diet. Yeasts act as a oxygen scavenger in the rumen. During feed ingestion, some amount of oxygen enter the rumen along with feed and its adversely effect the rumen environment as well as growth of the rumen microbes.

Effect of Probiotic Feeding on Animal Performances

Effect of probiotic supplementation on dry matter intake by the animals are inconsistent. Supplementation of yeast in the animal's feed improved it palatability as glutamic acid produced by yeast is responsible for improvement in the taste of feed stuffs (Agarwal, 2002). Numerical significant improvements have been reported in digestibility of dry matter, organic matter, crude protein and fibre in yeast fed animals. Higher retention of nitrogen and energy have also been reported in yeast fed animals whereas, the effects have been variable and the response influenced by the type of diet, physiological state of the animals and microbial strain employed. Basal diet of the yeast culture fed animals influence the nutrient digestibility and productivity of the animals. Effect of yeast culture is likely to be greatest in diets containing a high proportion of readily fermentable carbohydrate, such as barley based concentrate. A significant improvement in the digestibility of crude protein and crude fibre had been also observed on supplementation of diets with *Lactobacillus acidophilus* and *Saccharomyces cervisiae* in goats (Sharma and Malik, 1992). The published reports on live weight gain and feed conversion efficiency due to yeast feeding is variable from nil to positive effect. Pandey *et al.* (2001) reported that young animals respond well to dietary supplementation of yeast in the starter diet in terms of live weight gain and feed conversion efficiency. Supplementation of lactobacillus culture to the diets also improved the growth rate of the animals (Prahalada *et al.*, 2001). Positive response on milk production have been reported in yeast fed animals. Response have been greater in early compared to mid of lactation and the response was greater with diets containing the higher proportion of concentrate.

Incidence of Diarrhea

Incidence of diarrhoea reduced due to feeding of probiotic to the young animals. There was considerable reduction in the number of total coliform bacteria in the rumen liquor as well as faeces of ruminant animals fed probiotic in their diet, irrespective of the chemical composition of the ration offered to the animals (Kamra *et al.*, 1997).

Defaunation

The process of making the rumen of animals free of rumen protozoa is called defaunation and the animal is called defaunated animal. Despite the metabolic diversity of the rumen protozoa and their large biomass in the rumen, the protozoa, unlike the bacteria are not essential for the survival and development of the host ruminants. Indeed, under certain dietary situations the elimination of the rumen protozoa (defaunation) can result in increased animal growth and productivity. Nevertheless, the reports of recent years reflect that though protozoa may be non essential for ruminant, still they have significant role to play in the rumen metabolism, specially to stabilize the rumen pH and in ruminal fibre digestion.

Methods of Defaunation

The methods of defaunation are:

i) Isolation of newborn animals (separation technique),
ii) Dietary treatment,
iii) Chemical treatment,
iv) Physical treatment,
v) Tree leaves or other bioactive agents.

Isolation of newborn animals (separation technique)

Rumen protozoa free animals can be obtained by segregating the young animal away from the dam after birth and preventing them from any contact with the adult animals. In this process the new born animals should be separated from the dam 2-3 days after birth. During this time, the new born animals get the native bacterial population but do not get the rumen ciliate protozoa. However, once the animals are separated, proper care should be taken so that the isolated animals do not come in contact with any adult animals as well as any contamination from the handlers who look after faunated and defaunated animals.

Dietary Treatment

The ciliate protozoa are very much sensitive to change in rumen pH. The activity of ciliate protozoa is adversely affected when the pH of the rumen fall below 5.8 and if the rumen pH fall below 5.0, the ciliate protozoa are be completely eliminated. Therefore, offering high energy feed (especially cereal grains like barley, maize etc) to the starved (for 24 hours) animals creates acidic condition in the rumen and rumen pH fall below 5.0. This fall in rumen pH eliminates the ciliate protozoa completely and the animals become defaunated. However one serious disadvantage of this method is that chances of developing acidosis in treated animal is more. Once rumen acidosis develops the animals will suffer form various secondary complications.

Chemical Treatment

The removal of ciliate protozoa can be achieved by dosing the rumen with an anti-protozoal compound/chemicals infused in the rumen either orally or through rumen fistula. Majority of the researchers have used this method for obtaining animals free of rumen protozoa. The chemicals which have been widely used to defaunate the animals are copper sulphate, manoxol and sodium laurel sulfate. However, these chemicals are not only toxic to the rumen protozoa but also have varying level of detrimental effect on other rumen microorganism like bacteria. These chemicals are also toxic to the host animals resulting in depressed feed intake, dehydration and some cases mortality is also reported. Ruminal infusion of sodium laurel sulfate @ 8 g/100 kg body weight of animals for consecutive two days can successfully defaunated the sheep and goats (Santra and Karim, 2000; Santra and Karim, 2002a).

Physical Treatment

The other method for defaunating the animal is the emptying rumen by removing the rumen content and treating the contents before these are replaced in to the rumen again. This is a very old method of defaunation and now a day no body follows this method. Once the rumen is emptied, the animal should be starved for 40 hours. The internal lining mucosa of rumen should be washed with water (39° C), then washed with a formaldehyde solution and immediately rinsed with water.

Tree Leaves or Other Bioactive Agents

A variety of chemicals to remove protozoa from the rumen have been tested experimentally, but none is used routinely because of toxicity problems either to the rest of the microbial population e.g., rumen bacteria or the host animals. Therefore, it is necessary to screen agro-industrial by products or tree leaves (Table 1) for elimination of ciliate protozoa from rumen so that harmful effect of defaunating agent on the host animal is minimized or eliminated. Recently, there has been an increased interest in plant secondary metabolites for use as possible defaunating agent. Dietary fats generally inhibit microbial activities, particularly rumen protozoa: oils rich in saturated medium-chain and unsaturated long chain fatty acids have been shown to inhibit ruminal protozoa. Bioactive agents like steroidal saponins, condensed tannins etc. which are present in the tree leaves inhibit the growth of ciliate protozoa. Saponin affects different microbes of the rumen differently: total bacteria and methanogens were not affected adversely by *Sapindus saponaria*, but rumen protozoa were significantly lowered (Hess *et al.*, 2003). The detergent action of saponins is believed to be responsible for killing the rumen protozoa. Results of various studies indicated that the saponins have strong antiprotozoal activity and may serve as an effective defaunating agent for ruminants.

Table 1. Tree leaves having potential defaunating activity.

Family	Genus	Species
Leguminosae	*Tephrosia*	*grandiflora* *sessiflora* *toxicara*
Piscidia	*Piscicula*	*guaricensis*
Theophrastacea	*Jacqinia*	*barbasco* *mucronulata* *rovoluta*
Sapindaceae	*Sapindus* *Cupania* *Talasia*	*saponaria* *americana* *pineceps*

Maintenance of Defaunated Sheep and Goats

Once defaunated, animals normally must be maintained isolated to avoid any contact with other ruminants to remain protozoa free which is difficult to comply under field conditions. We observed that under routine mixed herd management the rumen protozoa appeared 16 days after defaunation. *Entodinium* spp. first appear in the rumen of defaunated sheep and goats followed by spirotrichs and holotrich protozoa (Santra and Karim, 1999). Therefore, it is not possible to maintain defaunated animals with single dose of chemicals (defunating agent) and hence multiple dose of defaunating agent was required to maintain the animals rumen protozoa free for prolong period. However, chemical defaunating agents also some time produce harmful effect on the animals hence a suitable method for defaunation need to be developed to comply with the field existence of the animals.

Effect of Defaunation on Rumen Ecosystem

Defaunation causes both qualitative and quantitative change in rumen bacterial population. After defaunation the bacterial population increase as rumen protozoa feed on the rumen bacteria to meet out their nitrogen requirement. The important activity of the rumen ciliates as far as host is concerned is that they engulf, kill and digest rumen bacteria and release the digestion products which are principally amino acids back into the surrounding rumen medium. All ciliates ingest bacteria as their principal source of protein amino acids. A total of 4 to 45 g bacterial dry matter is engulfed by rumen protozoa/day/sheep (Coleman, 1975). However, on defaunation the number of free bacteria increase but not the bacteria associated with feed particles suggesting that the protozoa only engulf free floating bacteria in the rumen. The qualitative composition of the ruminal flora is also altered by the presence of ciliates. The numbers of small gram negative rods, selenomonas like bacteria and gram negative cocci were higher in defaunated animals whereas gram positive strains remained unchanged. Defaunation increased the number of amylolytic bacteria due to elimination of nutritional competition between bacteria and protozoa for using starch while the cellulolytic bacterial population decreased (Jouany et al., 1988). Fungal population in the rumen increase due to defaunaton. The concentration of viable zoospore in rumen fluid collected from defaunated sheep was 2 to 5 times higher than in rumen fluid from faunated sheep given high roughage diets.

The buffering capacity of the rumen seems to be better in presence of protozoa on a wide variety of diets. In general, the pH of the ruminal content of defaunated lambs was lower (Santra and Karim, 2002b). Rumen protozoa engulf the readily fermentable carbohydrate (starch) which is stored in their body as amylopectin thereby decreasing the rate of carbohydrate (starch) fermentation resulting in lower pH in the rumen of defaunated compared to faunated animals. Thus, protozoa have a pH stabilizing effect in the rumen. Volatile fatty acids (VFAs) are the fermentation end product of carbohydrates in the rumen, which are utilized as energy source by the host animals. Hence an increase or decrease in its concentration in the rumen will affect the energy metabolism of ruminants. It is now well established that defaunation results in higher TVAs production in the growing lambs (Santra and Karim, 2002b). The protozoa produce acetic acid and butyric acid during carbohydrate metabolism hence defaunation increase the molar proportion of propionate at the expense of butyrate and acetate. Moreover, higher propionate in defaunated lambs may also be due to shift in bacterial population. It has been observed that in defaunated animals' numbers of C_2 producing species such as *Ruminococcus* are not predominant (Kurihara et al., 1978) which may enhance propionate production in ciliate free animals. Protein and other nitrogenous compounds in the rumen are broken down to ammonia and peptide. Ammonia is then utilized by bacteria to meet their nitrogen requirement for body protein synthesis while ciliate protozoa do not use it. In defaunated animals, the number as well as activity of rumen bacteria increase resulting in more uptake/utilization of ammonia by bacteria and as a result, ruminal ammonia concentration is reduced (Pal and Kamra, 2000; Santra and Karim, 2002b). Activity of fibre degrading enzyme like carboxymethyl cellulase was decreased whereas enzyme activity of â-glucosidase, protease and urease were not affected by presence or absence of rumen ciliate protozoa in growing lambs. Defaunation results in considerable decrease (20-50 %) in ruminal methane production compared to the normal faunated state. Protozoa are responsible for hydrogen production which is used for the methane synthesis by the methanogenic bacteria. The methanogenic bacteria having ectosymbiotic attachment with ciliate protozoa loose their symbiotic partner on defaunation which results in reduced methane production.

Microbial Protein Synthesis

Maximizing microbial protein synthesis and flow to the duodenum by reducing the recycling of microbial nitrogen in the rumen offers a potential approach to improve the productive efficiency of ruminant livestock. Both, bacteria and protozoa are active in the synthesis of microbial protein whereas, outflow of microbial protein in the duodenum is primarily of rumen bacterial origin. Although half of the microbial protein in the rumen can be of protozoal origin while as a proportion of the microbial protein leaving the rumen, protozoal protein is usually under 10 %. Now it is accepted that defaunation increase rumen bacterial population density hence the rate of bacterial protein as well as microbial protein synthesis also increased.

Effect of Defaunation on Nutrient Intake and Its Digestibility

Defaunation does not have a visible adverse effect on voluntary feed intake of lambs. Daily dry matter intake in defaunated and faunated lambs is reported to be similar in sheep (Santra and Karim, 2000, 2002a). It is now well established that rumen protozoa play important role in the digestion of plant cell walls. The digestibility of fibre fractions like, NDF, ADF, hemicellulose and cellulose are decreased by defaunation. Approximately one-quarter to one-third of fibre degradation in the rumen is effected by protozoa. The protozoa also stimulate bacterial cellulolysis and their released enzymes may also be of importance in the ruminal degradation of plant constituents. Lactic acid, which has been shown to depress fibre degradation is actively taken up by the entodiniomorphid ciliates and the activities of xylanolytic enzymes are increased following refaunation.

Moreover, better digestibility of cell wall constituents in faunated animals might be due to increased retention time of feed particles in the rumen, stabilization of rumen environment favoring development of cellulolytic microbes and stimulatory effect of rumen ciliate protozoa on rumen bacteria for cellulolysis. However, defaunation had no effect on apparent protein digestibility in small ruminant.

Protein to Energy (P/E) Ratio

The priority of rumen manipulation for effective use of balance diet requires optimization of fermentative digestion. The main point to be emphasized is the need to extract maximum nutrients from the basal diet by microbial fermentation/ digestion in the rumen. At the same time, the growth efficiency of rumen microbes must be optimized to provide the animal with a high protein (amino acids) to energy ratio (P/E ratio) in the nutrients absorbed. This can be achieved by supplementation with combination of non-protein nitrogen and minerals. Once this has been achieved then the animal need small amount of extra supplementary protein that escape to the lower tract to optimize the P/E ratio in the nutrients absorbed and maximize the animal's efficient use of the diet. The major problem for most developing countries is the unavailability of the feed resources needed to provide these supplements. Tree forages seeds and pods are potential sources of supplements those provide an array of minerals and soluble nitrogen for the rumen microorganisms. The same materials can potentially provide the bypass protein either naturally, when they contain tannins or when induced by simple processing.

The P/E ratio in the nutrients absorbed by ruminants on forage-based diets depend largely on the microbial growth efficiency in the rumen and on the amount of bypass protein supplement included in the diet. It is established that the presence of rumen protozoa reduces the protein to energy ratio in the nutrients absorbed (Bird, 1991) through:

- their predation of bacteria in the rumen decreasing the flow of bacterial cells to the intestine
- their apparent retention in the rumen since only 30 % of the rumen protozoal cells actually move to the intestine
- their ability to digest particulate protein in the rumen thereby converting protein and amino acids to VFA with a net loss of dietary protein.

Blood Biochemical Changes

Blood glucose and amino acid concentration increase as a consequence of defaunation whereas the changes in the levels of blood urea are more variable. We did not observed any difference in blood urea level both in growing and adult Malpura sheep irrespective of presence or absence of rumen protozoa. However, in some studies higher blood urea level was observed in faunated in comparison to defaunated animals attributed to the elevated ammonia levels in the rumen which arise as a result of increased protein turnover in the presence of active ciliate protozoal population. Higher blood glucose levels in defaunated lambs is possibly due to higher availability of propionate in their rumen.

Animal Performance

The pot-bellied appearance of lambs without ciliate protozoa is a well established observation in many studies. The availability of metabolizable energy was higher in defaunated animals due to reduction of energy loss in ruminal methane production. Total heat production of animal was also significantly lower in absence of rumen ciliate protozoa. Therefore, defaunation has positive effect on the growth rate and feed conversion efficiency of the growing lambs. The effect of defaunation on feed conversion efficiency seems to be diet dependent. In high roughage fed defaunated animals the general function of feed degradation is taken over by the increased population of bacteria and fungi. The reduction in methanogenesis results in better feed conversion efficiency on such feed. However in animals fed on high grain diet, the ciliate protozoa have specific function to perform i.e., the pH stabilization by controlling the degradation of easily fermentable sugars by protecting them in their bodies as amylopectin. Thus in absence of rume protozoa, the pH drops below the optimum level required for various enzymes activity in the rumen. This results in reduced feed utilization and decreased feed conversion efficiency on such feeds. Therefore, defaunation protocol can be used as an important tool to improve the productivity of animals in tropical countries, where majority of livestock are maintained on sole roughage diet or low grade roughage. Defaunation is associated with a 12 % higher birth weight of single born lambs whereas the average weight of twin lambs and the body weight changes of the ewes were not affected by presence or absence of rumen protozoa.

Defaunation causes higher wool yield, fiber diameter and modulation % in lambs (Santra and Karim, 2007). In general defaunation does not have any adverse effect on carcass traits. The slaughter study conducted on defaunated, faunated and refaunated finisher lambs indicated that the carcass traits were almost similar in the three groups providing a carcass of acceptable quality (Sen *et al.*, 2000).

Area of Future Research

- Screening of non conventional animal feeds specially tree leaves for anti protozoal activity.
- Standardization of defaunation method for its implication at farmer's level.
- New species/ strains of microorganism should be screened to use as probiotic.
- Mechanism of action of probiotic should be studied thoroughly.

- Reduction in methanogenesis to improve availability of digestible energy and reduce environmental pollution.
- Production of suitable strain of recombinant microorganisms and their propagation in the rumen for efficient detoxification of plant toxins, reduction in methanogenesis, higher cellulolysis, reduced ruminal proteolysis (de-amination).

Table 2. Some effects (beneficial and detrimental) of the presence of ciliate protozoa on ruminal characteristics, ruminant performance and well-being.

Increased in faunated state		Decrease in faunated state
Beneficial effect		***Beneficial effect***
Stability of fermentation and pH		Lactic acid concentration
Physical appearance of the animals		Plasma and hepatic copper levels
Organic matter breakdown		Susceptibility to lactic acid acidosis
Fibre breakdown		Susceptibility to copper toxicity
Starch digestion		Susceptibility to scours
Adverse effect		***Adverse effect***
Ammonia concentration		Bacterial population
Proteolytic a ctivity		Microbial protein synthesis
Methanogenesis		Nitrogen flow to small intestine
Plasma saturated fatty acid levels		Plasma polyunsaturated fatty acids levels
Susceptibility to bloat		Live weight gain
Carcass fat		Wool growth
		Feed conversion efficiency

Conclusions and Recommendations

Rumen is a natural fermentative anaerobic system which should be manipulated essentially by altering the composition of rumen microflora. There is ample scope to manipulate the rumen by feeding local plants or tree leaves or agro industrial by products to defaunate the animals for improving its productivity. Introduction of naturally occurring microorganism from digestive system of one species to another species for efficient degradation of plant toxins as well as for efficient utilization of nutrients will be one of the major thrust area in near future for rumen manipulation. Genetically manipulation of rumen microorganism for efficient ruminal fermentative digestion has an enormous biotechnological potential. However in tropical countries, more emphasis should be given for manipulating the rumen to increase cellulolytic activity for efficient utilization of low grade roughage.

REFERENCES

Agarwal, Neeta. 2002. Microbial feed additives for ruminants. In: Recent Advances in Rumen Microbiology (Ed. D. N. Kamra, Neeta Agarwal, L. C. Chaudhary and D. K. Agrawal). IVRI Publication, Izatnagar, India, pp. 47-56.

Bird, S.H. 1991. The role of protozoa in relation to the nutrition of the host animal. In: Ho, Y.W., Wong, H.K., Abdullah, N. and Tajuddin, A.Z. (Eds.), *Recent Advances in the Nutrition of Herbivores*. Malaysian Society of Animal Production, pp. 171-180.

Coleman, G.S. 1975. The inter relationship between rumen ciliate protozoa and bacteria. In: Digestion and Metabolism in the Ruminant (Ed. W. Mc Donald and A. C. I. Warner). The University of New England Publishing Unit, pp. 149-164.

Dutta, T.K., Kundu S. and Sharma. D.D. 2001. Potential of probiotic supplementation on *in vitro* rumen fermentation and ^{35}S incorporation in microbial protein. *Indian Journal of Animal Nutrition* 18, 227-234.

Hess, H.D., Kreuzer, M., Diaz, T.E., Lascano, C.E., Carulla, J.E., Soliva, C.R. and Machmuller, A. 2003. Saponin rich tropical fruits affect fermentation and methanogenesis in faunated and defaunated rumen fluid. *Animal Feed Science and Technology* 109, 79-94.

Jouany, J.P., Demeyer, D.I. and Grain, J. 1988. Effect of defaunating the rumen. *Animal Feed Science and Technology* 21, 229-265.

Jouany, J.P., Michalet-Dorea, B and Doreau, M. 2000. Manipulation of rumen ecosystem to support high performance beef cattle. *Asian-Australasian Journal of Animal Sciences* 13, 96-114.

Kurihara, Y., Takechi, T. and Shibata, F. 1978. Relationship between bacteria and ciliate protozoa in the rumen of sheep fed purified diet. *Journal of Agricultural Sciences* 90, 373-381.

Makkar, H.P.S. 2007. Bioactivity of phytochemicals in some lesser known plants and their effects and potential applications in livestock and aquaculture nutrition. In: Proceeding of International Tropical Animal Nutrition Conference, October 4-7, National Dairy Research Institute, Karnal, pp. 32-48.

Pal, D.T. and Kamra, D.N. 2000. Changes in rumen fermentation pattern in faunated and defaunated cattle calves on feeding of low protein diet. *Indian Journal of Dairy Science* 53, 428-433.

Pandey, P. and Agarwal, I. S. 2001. Nutrient utilization and growth response in crossbred calves fed antibiotic and probiotic supplement diet. *Indian Journal of Animal Nutrition* 18, 15-18.

Prahalada, H.K., Kamra, D.N. and Pathak, N.N. 2001. Effect of feeding *Saccharomyces cervisiae* and *Lactobacillus acidophilus* on nutrient utilization and performance of crossbred cattle calves. *International Journal of Animal Science* 16, 103-107.

Santra, A. and Karim, S.A. 1999. Efficacy of sodium laurel sulfate as defaunating agent in sheep and goats. *International Journal of Animal Science* 14, 167-171.

Santra, A. and Karim, S.A. 2000. Growth performance of faunated and defaunated Malpura weaner lambs. *Animal Feed Science and Technology* 86, 251-260.

Santra., A and Karim, S.A. 2002a. Influence of ciliate protozoa on biochemical changes and hydrolytic enzyme profile in the rumen ecosystem. *Journal of Applied Microbiology* 92, 801-811.

Santra., A and Karim. S.A. 2002b. Nutrient utilization and growth performance of defaunated and faunated lambs maintained on complete diets containing varying proportion of roughage and concentrate. *Animal Feed Science and Technology* 101, 87-99.

Santra., A. and Karim. S A. 2007. Effect of defaunation on body conformation changes, wool yield and fibre characteristics of growing lambs in a hot semiarid environment. *Australian Journal of Experimental Agriculture* 47, 789-795.

Sen. A.R., Karim, S.A. and Santra, A. 2000. Effect of defaunation on carcass and meat characteristics of finisher lambs. *Indian journal of Animal Sciences* 70, 659-661.

Sharma, D.D. and Malik, R. 1992. Probiotic supplementation in animal feed. In: Proceeding of National Symposium on Role of Amino Acid and Feed Supplementation in Animal Feeds, Trivendum, India.

Wallace, R.J. and Newbold, C.J. 1993. Rumen fermentation and its manipulation: The development of yeast culture as feed additives. In: Biotechnology in the Feed industry (Ed. R. Fuller). Chapman and Hall, London. pp. 173.

❑❑❑

7

Nutritional Manipulation for Maximization of Pre and Post Weaning Growth

R.S. BHATT
Central Sheep & Wool Research Institute, Avikanagar
(via Jaipur) Rajasthan - 304501

Introduction

Feeding and management of the lamb crop is an important phase of the sheep production cycle. The type of feeding program will depend largely on ecological situations, production system, marketing options and available feed resources. The absolute and relative amounts of nutrients needed by lambs are dependent on body weight of lambs and the desired growth rate within the existing genetic potential. The phases of the lamb growth are often distinct by the age of lamb. Usually the phases are classified as the young nursing lamb, the weaned growing lamb and the finishing phase. Each phase has unique nutritional requirements.

Colostrum Feeding

Colostrum is essential for lamb feeding as it consists of immunoglobulins. Some of immunoglobulins line the intestinal wall while the rest are absorbed into the bloodstream to help prevent infections. Colostrum consists of nutrients which fuel heat production and help prevent hypothermia. It consists of growth factors which help to promote gut growth and differentiation especially during the first 24- 48 hrs after birth, thus helping to establish the intestinal route as the sole source of nutrients.

The amount of colostrum needed depend mainly on how much fuel it requires for heat production. Therefore, any factor which increases heat production increases the colostrums requirement. During cold, wind or rain the lamb must produce more heat to avoid hypothermia, and colostrum requirement increases. Lambs born in rainy (with wind) weather in the temperature range of 0- 10 ^{0}C need about 200- 220 ml of colostrum/ kg of body weight during the first 18 hrs. Lambs born in housed conditions will require about 160- 180 ml of colostrums/ kg of body weight (Bobb, 1999). While feeding lambs with a stomach tube a thumb rule is practiced that no more than 50 ml per kg of body weight is fed.

This quantity will normally be sufficient to protect lambs against *E.coli* infection because 200 ml of colostrum usually contain enough immuno globulins for protection. It is necessary to feed the lambs 4- 5 times in the first 24 hrs if the lamb is unable to suckle on a ewe. Colostrum production by underfed ewes (body condition score 1.5 to 2) is usually about half that of well fed ewes (body condition score 3 to 4) and in some underfed ewes no colostrums is produced at all. In case of no colostrums production by ewes frozen colostrums may be used. Goat colostrums is a good substitute, however cow colostrums has 20- 40 % less nutrients than ewes colostrums. If using cow colostrums one will need to give approximately one third more volume. If using goat colostrum it should be tested negative for caprine arthritis encephalitis and the doe should be vaccinated for *E. coli* and clostridium Type C, three weeks prior to kidding.

Young Nursing Lambs

Lambs at this stage are dependent solely on ewe milk. Ewe milk is high in protein, milk sugars, fat, calcium, phosphorus and other important nutrients. Peak milk production for ewes occurs at about 20 days of lactation, stays high for two to three weeks and then declines. Lambs grow quite well on mother milk until about 25 to 30 days of age. After about 30 to 40 days of age the lambs need for nutrients to support a reasonable growth rate are greater than most ewe ability to produce milk. Nutrient requirement during this stage are 72 to 80 % TDN, 16 to 24 % crude protein, 0.5 -0.8 % calcium and 0.25 to 0.3 % phosphorus. A targeted daily gain for lambs during this period would be from150 to 200g (Mike, 1998). If the sheep are being managed on pasture in a range based system, the lambs will begin grazing after 35- 40 days of age to compensate for increased nutrient needs. Forage quality during this period is very critical because the lambs have limited potential to consume forage as their rumen are just starting to function and they have high nutrient needs. Another management consideration in forage based production system is of internal parasites. A high parasitic load will spoil the lamb gains and even result in death of the lambs. Lamb need creep mixture as supplement in addition to ewe milk.

Creep feeding: This is the practice of providing nursing lambs the opportunity to eat supplemented feed. It becomes essential if the lambs are to be weaned early. The advantage with creep feeding program are increased weight gain, better efficiency of feed utilization, early marketing of finished lambs and moving lambs to concentrate feeding enables producers to carry more ewes on available pasture. A good creep mixture should contain 16 to 20 % crude protein, 72 to 75 % TDN content and calcium to phosphorus ratio of 2:1

Lambs will begin to nibble at a creep at about one week of age. Although lambs will not consume significant amount of feed until they are 3 to 4 weeks of age, the small amount consumed are critical for establishing rumen function and to develop the habit of eating. For making creep feeding a success it is essential that it must be started as early as possible. It must be simply formulated and palatability of ingredients must be major consideration while formulating. Once the lambs are eating the creep, ingredients composition can gradually be adjusted to reflect the relative cost of the ingredients.

The Growing Lamb

Well fed lambs have genetic potential may grow @ 200 to 250 g per day. To accomplish these gains lambs must be fed a diet consisting of 73 to 78 % TDN, 14 to 16 % protein and have around 0.5 % calcium and 0.25 % phosphorus. Lambs need to be fed daily this diet @ 2.5 to 3.5 % of their body weight or *ad libitum* to grow at these target rates. Lambs fed on high forage diet exhibit lower gain than fed on grain diet. But these gains are not undesirable if they are cost effective.

Growth is a priority function of the animals and continues even during deficient nutritional regimes by drawing from the body reserves. Karim (2000) reported that the milk/concentrate/succulent fodder fed pre weaner lambs had feed conversion efficiency ranging from 38- 46 % while it was 33 % in the findings of Santra and Karim (1999). The new born lamb (1- 3 day old) require 117 kcal ME/kg $W^{0.75}$ for maintenance and 0.41 kcal ME/g gain in weight (Kearl, 1982). Weaned lambs on an average require 32- 35 g DCP/Mcal DE consumed for ADG ranging from 130-160 g. In other study Ranjhan (1980) reported that the lambs gaining @ 115 g per day require 72.7 g dietary protein/ Mcal ME and is considered on the higher side.

Finishing Lambs

Lamb processors prefer lambs in the body weight of 45 to 55 kg with a minimum of 0.5 cm fat cover and higher dressing percent. These targets can only be achieved in high grain feeding system. This phase lasts for a period of 30 days, depending upon FCR (feed conversion ratio) and target weight at finishing. Lambs requirements for the finishing phase are 73 to 78 % TDN, 12- 14 % protein, 0.5 % calcium and 0.25 % phosphorus. The concentrate should be fed @ 3 to 4 % of body weight and well fed lambs may gain between 200 to 300 g per day (Mike, 1998). However in native breeds the well fed lambs in post weaning phase have registered peak growth rate 210 g ADG. When lambs are fed on high concentrate diets, acidosis, enterotoxaemia and urinary calculi may develop. Acidosis can be prevented by including at least 10 % roughage in the diet, by feeding rumen buffer like sodium bicarbonate and by avoiding sudden change in the diet. Urinary calculi can be avoided by maintaining calcium to phosphorus ratio of at least 2:1, with provision of salt and mineral mixture freely available.

Two feeding methods exist for feeding finishing lambs. One is hand feeding or fence-line feeding. The other is a self- feeding system in which lambs gradually are brought up to a medium to high concentrate ration and the concentrate is available round the clock. In a study it has been observed that lambs which were hand fed twice daily had low body weight and poor feed efficiency as compared to lambs given 24 hrs accesses to a self feeder.

	Method of feeding and lamb performance	
	Hand fed twice daily	**Self fed**
Average daily gain (g)	260	337
Intake kg)	14.7	15.9
Feed Efficiency	5.64	4.71

Energy level is one of the important factors which is to taken into consideration in finishing- lamb rations. For maintenance lambs require 120 kcal DE and 3 g DCP/kg $W^{0.75}$ and for growth an additional 8 kcal DE and 0.2 g DCP for each g gain in live weight. In one of the study (Stanton and LeValley, 2007) a diet having 17 percent crude protein was offered @ 100 percent, 90 percent, 75 percent and 50 percent for lamb feeding . The 75 percent concentrate level provided significantly higher average daily gains, feed efficiency was better at 100 percent concentrate level but the problem of enterotoxaemia was encountered.

A sizeable data pertaining to feeding of different ingredients in lamb diets are available. Lambs fed either whole or rolled wheat had significantly higher average daily gains than that fed pelleted wheat. Feed efficiency also favored the whole or rolled form of wheat. Lambs performance was significantly better with the feeding of whole barley. While comparing barley and wheat, wheat had a slight advantage over barley in ADG; however, barley had an advantage in feed efficiency (Stanton and LeValley, 2007). When cracked or whole shelled form of maize was tested in lambs it was found that feed efficiency and average daily gains slightly improved with the whole shelled corn feeding. While testing pelleted and mash ration feeding in lambs it was found that lambs fed pelleted ration gained significantly faster than those fed an un pelleted ration. Pelleting probably did not change the nutritive value of diet whereas it improved the palatability and forced lambs to eat grain and roughage in the proportions, thus controlling the concentrate and roughage ratio. Amongst the protein sources the soybean meal, groundnut meal had higher value as compared to cottonseed meal, sunflower meal and corn gluten meal.

In India extensive range management system with top feed supplementation is is the dominant sheep production system. However for mutton production grazing along with supplementation and intensive feeding is the system of choice. For commercial application group feeding of lambs on complete feed block having 50: 50 or 40: 60 roughage and concentrate is generally practiced. To harvest better gain under intensive feeding, cafeteria system of feeding management may be applied where both the concentrate and roughages are offered in separate containers with the lamb having free choice to consume each one of it. Under this system the roughage: concentrate ratio is is stabilized at 20:80. Individual versus group feeding and composite versus cafeteria system of feeding management had similar growth performance in many of the studies (Karim *et al.*, 2004).

Nutrition and Body Composition

When the animal is fed at maintenance there is a gain in body protein and a loss of body fat. Above maintenance (Where fat deposition commences) with the increasing intake of energy, the ratio of the gain in fat to the gain in protein is constant for each increment in energy intake hence it results in faster rate of growth and an increase in the fat content of the gain. With intake near maintenance, there is considerable loss of body fat but little change in body protein. However at intake, just half of maintenance there is severe protein catabolism. When the absorption of protein increases from deficient level the ratio of fat to protein in the body gain decreases. Hence animals growing on protein deficient diet have more fat than those receiving adequate protein (Haresign, 1983).

Nutritional Manipulations

Feeding milk replacer to lambs

Milk replacer is a complete and convenient way of delivering nutrients to the lamb. Although it is costly but it is required when lambs are born as twins and triplets or the dam has died rendering the lambs orphaned. Milk replacer is formulated in a powder form and is mixed with warm water at the time of feeding. The nutrient content of milk replacer is kept as:

Ingredients	% Range (Dry matter basis)
Crude fat	30- 32
Crude protein	22- 24
Crude fiber	0- 1
Lactose	22- 25
Ash	5- 10

Milk replacer need proper mixing which is to be done by an electric appliance such as a hand blender. Lumps will lead to abomasal bloat problems. The daily feeding rate should not be more than 10 % of a lamb's body weight and in as many as four feeding per day. Appropriate feeding frequencies at a given age are listed below, however more frequent feedings may give a better result, but are not necessary.

Age (days)	Feed source	Milk replacer frequency
0- 24 hrs	Colostrum	4 to 5 times
1- 3	Milk replacer	3 to 4 times daily
4- 7	Milk replacer +creep	2 to 3 times daily
8- 21	Milk replacer +creep	2 times daily
21- 35	Milk replacer +creep or solid feed	*Ad lib.*

Replacement of costly energy sources by economical sources

Cheaper feed ingredients may be included in lamb diet to economize feed cost and increased profitability of farmers. When maize was replaced by animal feed grade damaged wheat to economize mutton production, the weight gain and feed conversion were higher at 75 % replacement of maize (Karim *et al.*, 2001). Soybean meal and groundnut meal are the best protein sources for lamb feeding but due to their high cost it is mandatory to explore its substitute. Mustard cake can totally replace groundnut cake in harvesting desired growth, although the adaptation period to feed source is relatively longer. Animal feed grade damaged wheat can be effectively used to replace barley grains to economize the cost of lamb feeding. Once adapted to mustard cake, in long term feeding, the lambs showed better growth rate than groundnut cake fed lambs (Tripathi *et al.*, 1999; 2001). Damaged wheat can replace costly energy supplement in lamb feeding with compromising growth (Tripathi *et al.*, 2007a) and health (Tripathi *et al.*, 2007b) of growing lambs. Sehgal et al. (1983) reported a daily gain of 178 g in cross-bred weaner lambs, whereas Tripathi et al. (2007c) found 151 g ADG in native Malpura weaner lambs in intensive feeding management. In another study Karim and Rawat (1997) with a ratio of 55: 45 in concentrate: roughage reported a daily gain of 197g in weaner lambs. Weaner lambs in active phase of growth maintained on varying levels of rumen degradable protein (RDP) and undegradable protein (UDP) in the diet indicated that a ration containing 45: 55 RDP and UDP in the diet was adequate for their growth requirement.

Rumen manipulation

Many ways of rumen manipulations have been tried in different laboratories. Some of them are of significance and can be adopted in lambs during post weaning phase. In genetic manipulations attempts were made to develop genetically engineered rumen microbes which results in more propionate and less acetate as well as methane production. Non genetic manipulation of the rumen is either the dietary manipulation whereby various dietary combinations are tried to improve the rumen environment or the chemical ways by which rumen microbes like protozoa can be eliminated or by feeding of probiotic beneficial microbes can be incorporated in the rumen.

Defaunation

Cereal straws and ligno-cellulosic animal feeds can yield 10 MJ ME/ kg dry matter, whereas, only 50- 60 % of this energy is available to the animals as the digestibility of these ligno-cellulosic feed material ranges from 40- 50 %. Poor digestibility of straw based fibrous feeds lower efficiency of energy utilization of these fibrous feeds by 30- 40 per cent. Efficiency of energy utilization of these feedstuffs could be improved by defaunation. Ciliate protozoa may constitute up to 50 % of the microbial mass in the rumen and the presence of these protozoa is not essential for the ruminants. Defaunation may be beneficial in lambs receiving low protein high energy diets, as the utilization of nitrogen is improved due to lower degradation of feed nitrogen and reduced recycling of bacterial nitrogen in the rumen. Sodium lauryl sulfate @ 8 g/100 kg body weight is adequate to effectively defaunate the lambs. It yielded 20 per cent improvement in growth performance and feed conversion efficiency. Both the wool yield and staple length in lambs was improved by defaunation. Digestibility of fiber fractions (NDF, ADF and cellulose) was reduced in defaunated lambs due to lower production of cellulolytic enzymes particularly cellulose and xylanase (Demeyer.1992). Defaunation in weaner could be effectively achieved by dosing with 2 g sodium lauryl sulphate (Santra and Karim, 1999a). It was observed that with similar dry matter intake and plane of nutrition, the growth performance was better in defaunated than refaunated and faunated lambs (Santra and Karim , 2000).

Rumen Buffer

In high concentrate fed lambs there is a chronic problem of acidosis and thereby reduced feed intake resulting in depressed animal performance. It occurs due to reduced ruminal pH. It was observed that sodium bicarbonate inclusion at the rate of 1.5 % of total ration increased cellulose digestibility, ciliate protozoal number, ruminal pH, total nitrogen concentration in rumen fluid resulting in improved growth of lambs by 35 % maintained on high concentrate (R:C; 25:75) diet (Tripathi *et al.*, 2004).

Antibiotics

Antibiotic feed supplements have been widely used to combat infectious disease and their beneficial effects on the growth rate and feed efficiency has revolutionized their use. Their mode of action is either through nutrient sparing effect, by reducing competitive intestinal microflora or by preventing the thickening of gut wall to allow better nutrient absorption and inhibition of toxin producing bacteria.

Probiotics

These are defined as live microbial feed supplements which beneficially affect the host animal by improving its gastro intestinal microbial balance. One of the foremost reasons for the interest in Probiotics is that they are natural alternatives to antibiotics. The majority of probiotic products are based on *Lactobacillus acidophilus, Saccharomyces cerevisiae* although other organisms such as *Streptococcus faecium, Bacillus subtilis,* and yeasts are also used. Probiotics should be viable and capable of growing in the intestinal tract. The organism should be able to survive passage through the highly acid stomach and should be resistant to bile. Probiotics feeding could improve growth performance of pre- and post weaning lambs (Tripathi *et al.*, 2007d), thus finishing weight. Probiotic should be capable of competing with existing gut microflora and should be host specific.

Ionophore

These are a class of antibiotics those are extensively used as feed additives for ruminants. The major ionophores used are produced by various strains of *Streptomyces* fungi and include monensin, lasalocid, salinomycin, lysocellin and narasin. Their mode of action is due to their interaction with metal ions and serving as a carrier by which these ions can be transported across membranes. These inhibit the growth of gram-positive bacteria. So the gram-negative population is enriched when ionophores are fed. The feeding of ionophores to cattle consistently improved feed conversion efficiency and often improved daily gain. They have a favorable effect on carbohydrate metabolism by stimulating hepatic gluconeogenesis. Ionophores reduce lactic acidosis, aid in the control of coccidiosis, feedlot bloat and are toxic to the larvae of face and horn flies in the feces (Goodrich *et al.*, 1984). They also act as methane inhibitors by reducing the population of methane producing bacteria.

Methane Inhibitor

Methane production reduces the efficiency of rumen fermentation. Inhibition of methane production would increase the efficiency of ruminant production as well as reducing methane emissions. Various chemicals inhibit methanogenesis. These are chloroform, iodoform and other halogenated methane analogs. It has been observed that the addition of saturated and unsaturated fatty acids to the diet of sheep reduces the amount of methane produced in the rumen and this change is due to the reduction in the proportion of acetic acid and an increase in propionic acid. When trichloroacetamide was added to the diet of lambs live weight gain was unchanged but feed intake was slightly reduced, thus resulting in an improvement in feed conversion efficiency. These did not prove to be of any practical utility. The main methane inhibitors are ionophores which by altering the rumen fermentation produce more of propionate and less of acetic acid. As we know with propionate production all of the carbon and hydrogen from glucose is utilized whereas with the acetate production two of the carbon and four of the hydrogens are lost in carbon and methane production.

Beta- Adrenergic Agents (Repartitioning Agents)

These are adrenalins analogs which stimulate â- adrenergic receptors. These are called as repartitioning agents because these result in a repartitioning of nutrients from fat to protein synthesis, causing increased muscle mass and decreased body fat. Clenbuterol, cimaterol and ractopamine are examples of these compounds. There is considerable interest in reducing the fat content of meat animals and increasing the yield of lean meat. These are believed to act by increasing fat mobilization and stimulating protein synthesis.

Conclusion and Recommendations

For profitable mutton production the birth, weaning and slaughter weight are important which should be at least 4, 15 and 35 kg respectively. Efforts should be made to increase the growth rate of lambs during pre weaning stage by proper supplementation to the ewes and meeting nutritional requirement of pre weaner lambs. The sheep rearing regions of the country are usually deficient in energy and protein supplements and in other intensively cropped regions the sheep are not conventionally raised. In commercial mutton production, every effort should be made to bring feed conversion between 1: 5- 6 and the feed input cost/ kg gain in body weight between Rs 25- 28.

REFERENCES

Bobb, J.D. 1999. Colostrum Requirements of Newborn lambs. *International Sheep Letter* http :/www. pipevet. com/ articles / colostrums Requirements.htm.

Demeyer, D.I. 1992. Biotechnology and the quality of animal products in sustainable agriculture. *Journal of Applied Animal Research* 1, 66-80

Goodrich,R.D., Garrett J.E., Gast D.R., Kirick M.A., Larson D.A. and Meiske J.C. 1984. Influence of monensin on the performance of cattle. *Journal of Animal Sciences* 58,1483-1498.

Haresign, W. 1983. *Sheep Production*. Redwood Burn, Trowbridge, Wilts. Butterworths.

Karim, S.A. 2000. Feedinfg small ruminants for meat and wool production. Proceedings of *National Workshop on Animal Nutrition in Health and Disease*, 17-19 February, Veterinary College, University of Agricultural Science, Bangalore.

Karim, S.A. and Rawat, P.S. 1997. Growth performance and carcass characteristics of lambs raised on varying proportion of roughage and concentrate. *Indian Journal of Animal Sciences* 67, 902-905.

Karim, S.A., Santra, A. and Singh, V. K. 2004. *Lamb feeding for Mutton Production* Central Sheep and Wool Research Institute. Avikanagar- 304 501.

Karim, S.A., Santra, A., Chaturvedi, O.H and Tripathi, M.K. 2001. Evolving lamb feeding system for economic mutton production. *Annual Report* CSWRI, Avikanagar, Rajasthan, India, pp.21-22.

Kearl, L.C. 1982. Nutrient Requirement of Ruminants in Developing Countries. International Feedstuff Institute, Utah State University, Logan, Utah, pp.53.

Mike,N.1998. *Feeding the Lamb Crop.* Purdue University. http://ag.ansc.purdue.edu/sheep/articles/feedlamb.html.

Ranjhan, S.K. 1980. *Animal Nutrition in Tropics.* Vikas Publishing House Pvt. Ltd. Vikas House, Ghaziabad, UP.

Santra, A and Karim, S.A. 1999. Efficacy of sodium lauryl sulfate as defaunating agent in sfeep and goats. *International Journal of Animal Science* 14, 167-171.

Santra, A and Karim, S.A. 2000. Growth performance of faunated and defaunated Malpura weaner lambs. *Animal Feed Science and Technology* 86, 251-260.

Santra, A. and Karim, S.A. 1999. Growth performance of weaner lambs maintained on varying levels of dietary protein in pre weaning phase. *Indian Journal of Animal Sciences* 69 ,722-726.

Santra, A., Chaturvedi, O.H., Tripathi, M.K., Kumar, R. and Karim, S.A. 2003. Effect of dietary sodium bicarbonate supplementation on fermentation characteristics and ciliate protozoal population in rumen of lambs. *Small Ruminant Research* 47, 203-212.

Sehgal, J.P., Singh Manohar., Rawat , P.S. and Singh , R.N. 1983. Feedlot performance in growing lambs for mutton production. *Indian Journal of Animal Sciences* 53, 715-719.

Stanton, T.L. and LeValley, S.B. 2007. *Lamb Feedlot Nutrition* No. 1.613, Colorado State University Extension- Agriculture.

Tripathi, M.K., Chaturvedi, O.H., Karim, S.A., Singh, V.K.and Sisodiya, S.L. 2007c. Effect of different levels of concentrate allowances on rumen fluid pH, nutrient digestion, nitrogen retention and growth performance of weaner lambs. *Small Ruminant Research* 72, 178-186.

Tripathi, M.K., Karim, S.A., Chaturvedi, O.H and Verma, D.L. 2007a. Nutritional value of animal feed grade wheat as replacement of maize in lamb feeding for mutton production. *Journal of the Science of Food and Agriculture* 87, 2447-2455.

Tripathi, M.K., Karim, S.A., Chaturvedi, O.H. and Verma, D.L. 2007d. Effect of different liquid cultures of live yeast strains on performance, ruminal fermentation and microbial protein synthesis in lambs. *Journal of Animal Physiology and Animal Nutrition* DOI: 10.1111/ j. 1439. 2007.00759.x.

Tripathi, M.K., Mishra, A.S., Misra, A.K., Mondal. D. and Karim, S.A. 2001. Effect of substitution of groundnut with high glucosinolate mustard (*Brassica juncea)* meal on nutrient utilization, growth, vital organ weight and blood composition of lambs. *Small Ruminant Research* 39, 261-267.

Tripathi, M.K., Mondal, D. and Karim, S.A. 2007b. Growth, haematology, blood constituents and immunological status of lambs fed graded level of animal feed grade damaged wheat as substitute of maize. *Journal of Animal Physiology and Animal Nutrition.* DOI: 10.1111/j. 1439. 2007.00712.x.

Tripathi, M.K., Santra, A., Chaturvedi, O.H., and Karim, S.A. 2004. Effect of graded levels of $NaHCO_3$ supplementation on ruminal fluid pH, feed intake, nutrient utilization, growth performance and blood biochemical constituents of lamb on high concentrate diets. *Animal Feed Science and Technology* 111, 27-39.

Tripathi, M.K., Tyagi, A.K and Karim, S.A. 1999. Performance of weaner lambs maintained on mustard and groundnut meal based diets. *Indian Journal of Animal Sciences* 69, 991-992.

□□□

8

Strategic Supplementation to Meet Critical Requirement of Nutrients for Sustainable Small Ruminant Production

S.K. SANKHYAN
Central Sheep and Wool Research Institute, Avikanagar - 304 501, Rajasthan

Introduction

Sheep basically is a grazing animal and thrive only on the naturally available grazing resources accessible to them on land. Particularly in our country there is no specified well developed and protected grazing land is available. This species of livestock particularly maintained on the fallow land, degraded natural rangeland and community grazing land found in near by vicinity. The condition of these grazing area is so poor that these are not even to sustain a single ewe round the year on one acre of area. The semiarid arid vegetation is characterized by rapid growth during monsoon season followed by early maturity, flowering and setting of fruits and finally converted in to dead litter in a period of 75 to 90 days, ultimately forcing poor sheep owners either to long distance migration in search of greener pasture or adopt short distance migration in the areas of sufficiency. With onset of harsh summer the sheep flocks from critical zone of Rajasthan and Gujarat migrate to adjoining states in search of green pasture and return back to their native tracts with on set of monsoon. Similarly the animals from North Western region migrate to high ranges of Himalayan temperate pasture in summer and return back to their native tract during favourable season. These practices some times create social problem and resistance from local grazer for the sharing of common feeding resources.

Organized husbandry practices, nutritional inputs matching with critical physiological stages or desired production traits and prophylactic and curative health coverage are almost negligible. About 80 to 90 % of natural range land is under poor to very poor condition mainly due to over stocking and round the year grazing without rest to the pasture. The animals have access to adequate quantity and quality of vegetation from July to October whereas in early monsoon higher moisture content of vegetation and gut fill limitation of animal render them under fed leading to lowest weight of the year in said season (Shinde *et al.*, 1996). Although both sheep and goats are well adapted to hot environmental conditions, linear increase in goat population in semi arid and arid environment is ascribed to their wider choice of vegetation (Karim and Shinde, 1992) and better adaptability to prevailing hot climatic conditions (Karim *et al.*, 1986).

The level of production in ruminants depend on the type of the animals, their genetic potential, rearing environment and while the over riding determinants are the intake of digestible energy, protein, minerals, vitamins and absence of anti nutritional factors in the feed. Under the existing grazing management practices the lambs hardly achieve 16 kg (Kaushish *et al.*, 1990) and kids 12 to 13 kg at six months of age whereas they are marketed for slaughter around 9 to 12 months of age weighing about 20 to 22 kg with dressed carcass weight of 10 kg (Karim, 1993). Hence there is an urgent need to have a strategic supplementary feeding practices or schedules for all categories of animal so that they under minimum economic involvement maximum return could be harvested from all sheep enterprises under farmers' condition.

Systems of Small Ruminant Rearing Under Arid and Semi Arid Region

Various types of grazing management system are practiced in different parts of the world still the extensive, semi intensive and intensive system of feeding management have relevance in the Indian context. The other organized systems of grazing management viz. rotational and deferred rotational are practiced in the country either at limited scale or very rarely in some areas. These have significance but due to limitations not adopted by the farmers.

Extensive system

Sheep and goat in India are generally reared under extensive system and play very important role in the sustenance of marginal and landless livelihood. This system includes migratory, transhumance, free range, pasture and range grazing management of sheep and goats. It is the most common system in India because the small size of sheep and goats have distinct advantages over other livestock species.

Extensive range management

By and large extensive range management on the degraded public range land is the existing system of production in the country, very few reports have appeared in the published literature on this important aspect. About 80 to 90 % of the public range land is under very poor condition due to over stocking and round the year grazing. In addition growth cycle of natural vegetation in semi arid and arid region is completed between 90 to 180 day, hence the animals for most part of the year survive on sparse vegetation. Due to continuous and round the year grazing pressure on such land the optimum vegetation yield of such land is never achieved. The biomass yield of public range land in semi arid region varies from 1.47, 1.57 and 0.89 q/ha during monsoon, winter and summer season respectively (Sankhyan et al., 1999). Public rangelands are adequate to meet the nutrient requirement of sheep during monsoon and winter while grossly inadequate for summer season (Karim et al.,1997; Sankhyan and Shinde, 2005).

Extensive survey conducted to assess production performance of farmers' stock indicated that under public range management in mixed grazing, the sheep weighed 9.2, 13.1, 16.1 and 19.7 kg at 2, 4, 6 and 9 months of age, respectively and that the crossbreds in general were heavier than the natives (Kaushish et al., 1990). Another study involving sheep farmers under lab to land and ORP programme indicated that the birth, 3, 6 and 9 months, weights of indigenous lambs were 2.9, 10.9, 13.9 and 20.2 kg while corresponding weights of crossbred counterparts were 3.0, 11.6, 14.2 and 19.5 kg. Similar survey on Sirohi goats on farmers flock under arid ecology indicated that their 3, 6 and 9 months weights were 12.5, 15.6 and 20.0 kg, respectively. Regional survey with larger sample size of farmers' sheep indicated that annual lambing was 55 % with 14.5 % mortality while the wool yield ranged from 600 to 1600 g in different parts of the region (Sharma *et al.*, 1995). The studied have shown that under the existing conditions, the greatest limitation in our rangelands and natural pastures is on the availability of adequate energy throughout the year and adequate amounts of protein for more than half the year. Although the grazing on the rangelands is considered to be the cheapest method for sheep and goat production, the migration of the animals and over grazing of available lands is causing serious problems of soil erosion and degradation leading to desertification. A flock of 30 Malpura ewes and 30 local does along with one breeding male from each species was maintained for three years on natural grazing/browsing without any supplementation. A lambing rate of 82.5 % and kidding rate of 91.9 % per year was observed. The annual mortality was recorded to be 31.2 % in ewes and 7.1 % in does.

Protected range land

The protected natural range land has biomass yield ranging from 7 to 12 q/ha/year, which is sufficient to maintain 2 sheep or goats or 0.4 ACU round the year. In a long term study, 15 Malpura and 15 Malpura x Rambouillet ewes and 15 Sirohi and 15 Sirohi x Beetal does were maintained on 23 hectare protected natural range land at the rate of one animal and followers per acre in semi-arid region under free grazing management for two years. A lambing rate of 106.7 and kidding rate of 153.3 % per year was recorded. The production performance of 50 native and 50 crossbred sheep and their followers maintained on 35 hectare of natural rangeland under farmers' management was studied. A lambing rate of 92, 96, 84 and 92 in Malpura, Chokla, Avikalin and Avivastra sheep was recorded on the basis of ewes available during first year and 144, 145, 109 and 120 % at the end of the second year respectively. The adult mortality was 8, 4, 4 and 4 % during first year and 4, 3, 8 and 5% during second year in the four breeds, respectively (Sankhyan et al., 1998). Long term study (3 years) conducted on sheep and goats maintained on protected natural range land with undulating topography indicated that the lambs born to the ewes on such pasture weighed 2.6, 8.6, 12.5 and 13.7 kg at birth, 3, 6 and 12 months of age while under similar management and age the kids weighed 2.9, 9.3, 13.6 and 21.7 kg, respectively (Acharya et al., 1980). The average mortality in the 3 year was 18 % in ewes whereas during the period there was no mortality in does. Under the free mating system followed in the study, the average reproductive efficiency in ewes and does was 82.5 and 91.9 % with lambing interval of 432 and 564 days and age at first lambing 854 and 564 days respectively. Further the lambing was distributed throughout the year with maximum incidence of lambing in December and January while the kidding was uniformly distributed throughout the year. Similar study on Malpura, Chokla, Avikalin and Avivastra ewes maintained on natural range land at stocking density of 3.5 sheep/ ha indicated that the natural range land had biomass yield of 20 q/ha at the initiation of the experiment which declined to 16 q/ha after 1 year of continuous grazing whereas all the 4 breeds maintained their weight. Under the free mating system followed in the study on an average 3 lamb crops were harvested in 2 years. Their annual average wool yield was 1.35 kg (Sankhyan et al., 1998). Reproductive efficiency and wool yield declined in the second than first year while the study was discontinued in the third year due to loss of production and heavy mortality. It was evident that the natural range land even stocked as per carrying capacity required rest for rejuvenation of pasture and stabilisation of pasture composition. Further under existing system of over stocked mixed grazing on degraded range land the surplus lambs are sold for meat purpose around 9 to 12 months of age weighing about 18 to 20 kg while these lambs could achieve 20 kg at 6 months of age on protected range land stocked as per carrying capacity.

Cultivated pasture

The concept of cultivated pasture for small ruminant production is almost non- existence in the country. Although some land is used for fodder production largely to meet requirement of dairy animals, small ruminants are exclusively maintained on extensive range management. The study conducted at CSWRI Avikanagar indicated that mono-culture grass (*Cenchrus ciliaris*) pasture in semi arid region has annual bio-mass yield of 20 q/ha, which is sufficient to maintain 5 sheep/goat (1 ACU) with followers round the year. The lamb born to the ewes on such pasture had birth, weaning, 6 month and 9 months weight of 3.2, 13.9,

20.9 and 23.9 kg, respectively (Singh and Prasad, 1985). Similar studies conducted on arid sewan (*Lasiurus sindicus*) pasture at stocking density of 5 sheep indicated that pasture intake was sufficient to maintain them from January to June while the animal required supplementation during lean season due to denudation of surface vegetation. The production performance of Avivastra and Avikalin strains of sheep on a developed pasture was also studied. 20 ewes each of the two strains were maintained on *Cenchrus ciliaris* pasture at the rate of 5 ewes under rotational grazing system. No supplementation except mineral-salt bricks was provided. Routine prophylactic health cover as practiced in the Institute Farm was followed. A mortality rate of 10 % in Avivastra and 5 % in Avikalin was recorded in adult sheep. While no mortality was observed in Avikalin lambs from 0 to 3 months age, 11.5 % of the Avivastra strain lambs died up to weaning. The average weaning weight was 10.9 kg in Avivastra and 11.2 kg in Avikalin lambs. A flock of 50 mutton synthetic ewes was maintained on a *Cenchrus ciliaris* pasture at the stocking rate of 3 adults and their followers per hectare for two years. A lambing rate of 92 % per year and adult mortality rate of 3 % and lamb mortality rate of 20.5 % from 0 to 9 months of age was recorded. Birth, 3, 6 and 9 months body weights during the three lambing seasons averaged 3.2, 13.9, 20.6 and 23.9 kg respectively. 32 Avivastra sheep and 32 Marwari goats were maintained on a 16 hectare *Cenchrus ciliaris* pasture for a period of three years @ 2 sheep+2 goats and their followers per hectare. It was observed that the sheep, on an average, gained by 6.2kg and the goats by 9.4 kg in their body weights during the first year. The lambing/kidding rate was 87.5. The sheep produced 2.4 kg annual fleece and the goats 714g milk per day during first 90 days of lactation. The weaner lambs weighing 11.0 kg could attain only 16.0 kg body weight at one year of age when maintained on a *Cenchrus* pasture alone, whereas lambs grazing on *Cenchrus* + *Dolichos* pasture reached 20.5 kg. Reseeded cenchrus pasture stocked as per carrying capacity can provide 11-12 kg weaning weight in lambs and kids indicating that such pasture was satisfactory for maintaining ewes at stocking density of 4 animals with followers/ha (Shinde et al.,1997). The mixed pasture in addition to increasing biomass yield provide quality nutrient for animal in critical physiological stages. Patnayak (1980) reported that the lambs showed higher body weight gain, attaining about 27 kg at 8 months of age when grazed on *Cenchur- Dolichos* pasture.

Silvipasture

Establishment of silvipasture improves quality of available nutrients ensuring its supply round the year especially in critical summer season. In addition the system improves biomass yield of pasture adding to quality feed resources availability favouring eco-conservation. The silvipasture with various tree combinations provides biomass yield up to 20- 25 q/ ha which is adequate to maintain 6 sheep or goat (1 ACU). The male Avivastra hoggets at a stocking density of 8 animals /ha on a silvipasture attained body weight of 30 kg as against 25 kg on grazing with supplementation and their wool yield was 735-787 g/ in first clip at 6 months and 825-950 g in 2nd clip at 12 months of age

(Shinde et al., 1994). Similar studies on lambs and kids maintained on silvipasture indicated that the animals achieved 22 kg at 6 months of age (Sankhyan et al., 1996). Supplementation of tree leaves grown in silvipature at stall in addition to grazing and *ad libitum* concentrate mixture feeding appears to be most desirable combination for intensive lamb production system (Tripathi et al., 2006). In practical feeding, farmers do not maintain sheep and goats exclusively on top feed resources. The top feed resources are either fed fresh after lopping in the grazing area itself or supplemented as conserved leaves after grazing in the corals. Thus the farmers out of their wisdom based on age old experiences have developed strategies to utilise these tanniferous feed by dilution technique. It is seen that a marked improvement can be achieved in overall performance and productivity of sheep and goats by maintaining them on developed pastures. The areas in arid and semi-arid regions that cannot support cattle and buffaloes should, therefore, be identified and utilized for small ruminant production with low investment. Common grazing lands should be improved through reseeding with nutritious, perennial and high yielding grasses and legumes. Large-scale fodder tree plantations should also be taken up on pasture lands, waste lands, riverbanks and roadsides and bunds of ponds, canals and agricultural fields.

Grazing with supplementation

This system is a combination of limited free range grazing and stall-feeding. Integrations of sheep rearing with arable cropping are also included where either the sheep/goats are tethered or cut and carry system of available fodder is employed. The children and women often tend the sheep and goats in this system. Animals belonging to several owners may be combined for grazing which is mostly done in the morning and evening for 4 to 6 hours. The animals are supplemented with kitchen wastes, concentrate mixtures, crop residues, green and dry fodders and tree leaves etc. as per the availability. The level of nutrition is just optimum and surely better than that under extensive system.

Under existing small ruminant production system, the slaughter weight of lambs and kids in the country is lower and age at which it is usually achieved in much higher. The system of raising finisher lambs and kids for meat purpose under grazing with supplementation, although cost effective for the producer, has not been largely adopted by the farmers owing to their poor economic back ground and age old traditional practices. It is however realised that with progressive shrinkage of grazing land and concomitant increase in stocking density on such land, in future, majority of small ruminant will be raised under grazing with supplementation. The experiment conducted in the country on performance of sheep or goats under grazing with supplementation has been periodically reviewed (Patanayak and Singh,1972; Patanayak and Mohan, 1974; Patanayak, 1993; Karim et al., 1996). The experimental results indicated that in addition to free grazing on range land limited amount of concentrate @ 1.5 to 2.0 % of body weight would provide 25 and 30 kg finishing weight at 6 and 9 months of age, respectively.

A weaning weight of 11 kg was achieved when the Malpura and Sonadi male lambs were provided 150 g/head/day creep in addition to suckling upto 90 days age. A daily gain of 56.2 g on grazing on *Cenchrus* pasture, 91.9 g when supplemented with low energy-low protein and 112.3 g when supplemented with high energy and high protein ration fed at the rate of 300 g per day to Malpura lambs was recorded. A growth rate of 140 to 165 g per day was recorded when the mutton synthetic male lambs maintained on *ad lib.* cowpea hay meal were supplemented with 300 g maize or barley grain. The control group lambs showed a daily gain of 94 g only. Under the routine farm management, a body weight of 20-22 kg at 6 months age can be achieved in Malpura and Sonadi native lambs by supplementary feeding of 150 g creep feed in addition to suckling and grazing and 300 g concentrate mixture from weaning to six months age as against 15-16 kg obtained in the field at 8-9 months, of course without any supplementation. The dressing percentage improved from 35 to about 45 and the bone meat ratio from 1:3.5 - 1:4.5. Supplementation of 400 g concentrate mixture in addition to grazing to Malpura lambs increased the carcass yield by 30% whereas supplementation of 550 g concentrate mixture resulted in an increase of 55% in the dressed carcass yield as compared to the lambs maintained on grazing alone.

The Avivastra lambs and Marwari kids grazing on Cenchrus pasture and supplemented with concentrate @ 1.5% of body weight from 91 to 180 days of age attained 27.3 and 26.2 kg weight at six months of age. The dressing percentage on live weight basis was 44.5 in lambs and 48.9 in kids. It was observed that *ad lib* supplementation of concentrate, hay and green to kids between 91 to 180 days age, in addition to browsing resulted in an increase of 44.8 % in pre-slaughter weight, 65.1 % in carcass weight and 14.3 % in dressing over the browsing alone. The kids when fed the above ration *ad lib* in stalls showed an increase of only 21.4 % in pre-slaughter weight, 39.7 % in carcass weight and 14.8 % in dressing over the sole browsing group. In another experiment a daily gain of 37.4, 87.4 and 73.3 g was obtained from 3 to 6 months and 62.6, 139.2 and 120.0 g from 6 to 9 months of age in the Sirohi x Beetal kids maintained on browsing, browsing + 756 g/h/d concentrate and total stall feeding on feedlot ration (1038 g/h/d) respectively. The dressing % was 43.05, 43.65 and 48.50 at 6 months and 47.35, 52.10 and 53.10, at 9 months of age on the three respective feeding regimes. The kids thus showed an improvement of about 44.5 and 34.0 % between 3 to 6 months and about 66.1 and 52.2 % between 6 to 9 months respectively on browsing + supplementation and feedlot system of feeding management over the browsing alone. In another experiment, the Sirohi, Marwari and Kutchi does produced 84.4, 89.1 and 94.3 kg milk with no supplementation, 98.6, 96.1 and 93.2 kg with 150 g concentrate, 100.9, 115.7 and 110.0 kg with 300 g concentrate and 109.0, 106.4 and 101.1 kg with 450 g concentrate supplementation in addition to 8 hours grazing during 150 days lactation.

Grazing with supplementation on established pasture has potential of still higher production traits as established Cenchrus pasture stocked as per carrying capacity with concentrate supplementation @ 1.5 % of BW provided 27.3 and 26.2 kg finishing weight in lambs and kids at 6 months of age (Shinde *et al.*, 1995). However, grazing with *ad libitum* concentrate supplementation had no added advantage in the kids as they achieved 25 kg at 6 months and 30 kg at 9 months of age even under regulated level concentrate supplementation (Parthasarathy *et al.*, 1984). Porwal (2005) conducted a study using patanwadi rams under grazing plus ad libitum supplementation of concentrate mixture found that target weight of 32-35 can be achieved at the age of 6 month after weaning.

Intensive feeding

The likely changes in agricultural production system in the country, requirement of export market, emerging awareness of domestic consumers for quality meat and country's priority of converting excess cereal production in certain parts of the country to animal produce is likely to alter the existing pattern of small ruminant production favouring adoption of intensive production system in large scale. By and large intensive feeding system has not been adopted by the small ruminant farmers in the country mainly due to their poor socio economic background whereas it is envisaged that in future this system of production will be extensively used for commercial sheep and goat meat production.

The intensive system of sheep and goat production includes grazing on developed pastures and/or completely stall-feeding on cultivated fresh or conserved fodders, crop residues and concentrates. Although goats prefer to browse as compared to grazing, they are quite capable of making efficient use of cultivated pastures for meat and milk production similar to sheep. A growth rate of 92 and 100 g/head/day in Malpura and Sonadi lambs maintained on a feedlot from 91 to 180 days of age was observed. Feedlot gains in Malpura, Sonadi and their crosses with Dorset and Suffolk were studied. Malpura, Sonadi, Dorset x Sonadi, Dorset x Malpura, Suffolk x Sonadi and Suffolk x Malpura lambs reached a body weight of 26.0, 25.4, 30.5, 31.3, 32.8 and 33.0 kg at 6 months of age under feedlot from 91 to 180 days of age with a feed efficiency of 14.1, 14.5, 18.6, 18.2, 18.5 and 18.3 respectively. The dressing % was recorded to be 50.9, 52.7, 51.7, 53.3, 50.0 and 50.3, respectively. Growth rate of 150 g/head/day in Avikalin lambs during 91 to 180 days of age on 50: 50 concentrates: roughage ration fed *ad lib* was reported. Similarly, a growth rate of about 150 g in Avivastra and Avikalin male weaner lambs feed on 50: 50 concentrate: roughage ration with a feed efficiency of about 18.5 % have been recorded. Performance of half bred lambs involving Suffolk and Dorset males with Malpura and Sonadi females under individual feedlot upto 135 and 180 days of age or 22 and 30 kg body weight after weaning at 90 days on 50: 50 concentrate: roughage ration indicated that the feed efficiency at 22 kg finishing live weight was superior to that at 30 kg finishing live weight. Lambs fed on complete feed consisting of concentrate (maize, barley grains, oilcakes, wheat and rice bran and molasses) and roughage (tree leaves, cultivated grasses and legume) in ratio of 40:60, 50:50 and 60:40 attained body weights of 25-27 kg at 6 month of age. The hot carcass weight of 10.3, 14.5 and

14.3 kg in the extensive, semi intensive and intensive systems and dressing yield of 44.9% in the extensive and 48.8% in semi-intensive and 50.9 % in the intensive system has been reported. The lean, fat and bone contents of 63.40, 8.52 and 15.32 % in the extensive system, 61.85, 11.84 and 14.24% in the semi-intensive and 59.34, 16.29 and 12.34% respectively in the intensive system in lambs have been reported (Karim et al., 2007). Crossbreds were 26 and 21% superior to natives in feedlot gains and efficiency of feed conversion. ADG of 205 and 207 g were recorded in Avikalin and Avikalin x Dorset terminal cross males fed *ad lib* on 70: 30 concentrate: roughage from 91 to 180 days. The dressing percentage was 51.9 and 51.6 in the two breed crosses, respectively. The mutton synthetic lambs maintained on creep upto 67 days, on 70: 30 feedlot from 67 to 99th day and on 50: 50 feedlot ration from 99th to 130th day reached a body weight of 30 kg in a record time of only 130 days age exhibiting ADG of about 200 g through out the period. The results indicated that under intensive feeding; while the native lambs reach a body weight of 22-25 kg, the half bred reach around 30-32 kg at 150 days of age. The dressing percentage and bone: meat ratio is also improved significantly by intensive feeding in both natives and crossbreds.

The kids of Sirohi breed showed a daily live weight gain of 80 g and consumed 7.7 kg feed for every kg of live weight gain when maintained on a complete feed based on 50 % cowpea hay meal in the stalls from 91-180 days of age The male kids weaned at 2–3 months age and fed under feedlot achieved slaughter weights of 25 kg at 5 to 6 months age with a dressing percentage of 48 to 51. The kids maintained under semi-intensive system reached the target live weight of 25 kg earlier than those under intensive system. The kids under semi-intensive required less feed for a kg of gain than those under intensive feeding. The dressing % was superior under intensive system. The bone and lean % was higher under semi-intensive and fat % under intensive system. ADG was higher in lambs than kids under intensive system whereas the daily gains were similar under semi-intensive system. Dressing % in lambs and kids was found higher under semi-intensive. The daily gain, milk intake, meat quantity and quality and feed efficiency were found superior in the Sirohi, Marwari and Kutchi kids maintained under semi-intensive as compared to those maintained under intensive or extensive system. The milk yield during 150 days of lactation was higher under intensive system than that under semi-intensive and extensive systems and in Sirohi does than that in Kutchi and Marwari does. Concentrated work spread over a decade indicated that the cross bred weaner (90 days) lambs in 90 days of intensive feeding had 180 g ADG compared to 127 g ADG in native lambs. In these studies it appeared that the crossbred with different levels of exotic inheritance had similar growth performance hence they were merged together and renamed as mutton synthetic (MS). These MS, Malpura (M) and Sonadi (S) weaner lambs had 170, 150 and 118 g ADG with 1:5 to7 feed efficiency ratio under intensive feeding on 50:50 concentrate and roughage based diet (Karim and Rawat,1996). Faster gain in intensively fed lambs was associated with higher carcass fat content some time exceeding 20% hence weaning age was reduced from 90 to 60 days. The 60 days weaner lambs under intensive feeding had 160 and 151 g ADG and

16 and 12 % feed efficiency while similarly fed 90 day weaner lambs had 139 and 146 g ADG with 16 and 14 percent feed efficiency (Karim and Arora, 1995). Simultaneous slaughter studies indicated that irrespective of slaughter age and feeding regimen carcass of finisher lambs weighing around 25 kg provided desirable carcass. Subsequent studies indicated that 25 kg finishing weight could be achieved in 73, 91 and 136 days of intensive feeding with 160, 135 and 112g ADG and 18,16 and 14 % feed efficiency, respectively in MS, M selected and M control lambs (Karim and Patnayak,1988).

Confinement and stall feeding was however not successful in weaner kids (Singh and Singh,1987) due to differences in their grazing behaviour although the weaner kids were able to attain target slaughter weight of 25 kg around 5 to 6 months of age. While comparing intensive and semi intensive system of feeding management, it was evident that intensive system was effective in lambs achieving 33.5 kg finishing weight with 11 percent feed efficiency at 6 months of age, while similar feeding management was not cost effective in kids (Shinde *et al.*, 1995).

Strategic supplementation

It is very well known that feeding of sheep during critical physiological stages is very important aspect as sheep and goat owners are resource poor farmer and it is very much essential that they should get maximum benefit out of their every penny spent upon the venture. Now a days the concept of strategic supplementation has come up as there are certain farmers they can supplement good amount of feed/ concentrate to their animals but some even do not have any thing for their own consumption what we can expect for sheep. It was observed that pregnant ewes maintained on community grazing land if supplemented only 400 to 425 g ground barley or maize per animal after grazing meet the critical energy demand of growing foetus as well as of nurturing mother during winter. Similarly if same ewes required to be supplemented during summer month when very less quantity of grazing material is available on the ground and animals are collecting only dead material the quantity of supplementation will be nearly 300 g maize/barley along with 150 g mustard cake or 100 g GN cake. Similarly if lamb crop is being raised for intensive mutton production to harvest desirable weight of 25-30 kg in six month of age are to be supplemented *ad libitum* concentrate mixture containing 17 to 18 % CP and 65 % TDN after grazing on pasture gave the desirable results.

Conclusion and Recommendations

1. Supplementation of 400 to 425 g ground barley or maize per animal after grazing to pregnant ewes meet the critical energy demand of growing foetus as well as of nurturing mother during winter season.

2. Supplementation of 300 g maize/barley along with 150 g mustard cake or 100 g GN cake to pregnant ewes during summer season will meet the requirement of growing foetus as well as mother.

3. Under silvipastotral system of grazing management target weight of 22 kg can be harvested by supplementing concentrate mixture 1.5% of BW. While in case of kid same target weighs can be achieved by 2% concentrate supplementation.

4. Target weight of 25-30 kg can be harvest at six month of age by *ad libitum* concentrate mixture supplementation to lamb crop for commercial lamb production.

REFERENCES

Acharya, R.M. 1980. Relative productivity of sheep and goats on free grazing/browsing management on semi arid range land of Rajasthan. Final report, CSWRI, Avikanagar.

Karim, S.A. 1993. Nutritional manipulation for quality meat production. In: Feeding strategies for livestock and poultry in relation to available feed resources. VI Animal Nutrition Research Workers Conference, 13-16 September Bhubaneswar,India. pp.99-106,.

Karim, S.A. and Arora A.L. 1995. Growth response of intensively fed lambs. *Indian Journal of Animal Sciences* 65; 1046-1048.

Karim, S.A. and Patnayak B.C. 1988. Feed lot studies in mutton synthetic sheep to attain 25 kg. *Annual Report*, CSWRI, Avikanagar, pp.52.

Karim, S. A. and Rawat, P. S. 1996. Growth performance of native and cross bred weaner lambs under intensive feeding. *Indian Journal of Animal Sciences* 66, 79-81.

Karim, S.A. and Shinde A.K. 1992. Grazing behaviour of small ruminants on pasture. *In: Recent advances in Animal Nutrition,* CSWRI, Avikanagar.

Karim, S.A., Sankhyan S.K. and Shinde A.K. 1997. Nutritional status of sheep on public range land. *Annual Report*, CSWRI, Avikanagar.

Karim, S.A., Shinde A.K. and Sankhyan S.K. 1996. Prospects of sheep and goat for meat production in eastern region of the country. In: *Prospects and strategies for sheep and goat production and utilization in eastern region*. 9-11 October Bhubaneswar.

Karim, S.A., Singh Manohar and Rawat P.S. 1986. Growth performance of native and cross bred weaner lambs. *Indian Journal of Animal Sciences* 56, 975-977.

Karim, S.A., Porwal, Kuldeep., Kumar, Suresh and Singh, V.K. 2007. carcass traits of kheri lambs maintained on different system of feeding management. *Meat Science* 76, 395-401.

Kaushish, S.K., Rawat, P.S. and Sharma S.C.1990. Performance of native sheep (Malpura) and its crosses with Avikalin under semi-arid condition. *World Review of Animal Production* 25, 44-46.

Parthasarathy, M. Singh, D. and Rawat, P.S. 1984. Performance of feed lot kids under different dietary regimen. *Indian Journal of Animal Sciences* 54, 124-130.

Patnayak, B.C. 1980. Problem and prospects in the nutrition of arid zone livestock. In : *Arid zone research and development* Ed. H. S. Mann, CAZRI, Zodhpur

Patnayak, B.C. 1993. Nutrition of grazing ruminants In: Recent Advances in Ruminant Nutrition. CSWRI, Avikanagar.

Patnayak, B.C. and Mohan, M.1974. Performance of cross bred lambs on supplementation of concentrate and cow pea hay. *Indian Journal of Animal Sciences* 44; 376.

Patnayak, B.C. and Singh M. 1972. Sheep nutrition research in India. In: Review of research on sheep and wool production in India. CSWRI mammogram publication, pp.49.

Porwal, Kuldeep. 2005 *Status of sheep production in farmers flock and its improvement by scientific feeding practices.* Ph.D. thesis submitted to Dr B.R. Ambedkar University, Agra.

Sankhyan, S.K., Shinde A.K., Karim S.A., and patnayak B.C.1998. Production performance of sheep under farmers' management system. *World Review of Animal Production* 30, 27- 35.

Sankhyan, S.K., Shinde A.K., Karim S.A., Mann J.S., Singh N.P. and Patnayak B.C.1996. Plane of nutrition and growth response of lambs and kids under two and three tier silvipasture. *Indian Journal of Animal Sciences* 66, 1194-1197.

Sankhyan, S.K., Shinde A.K. and Karim, S.A. 1999. Seasonal changes and biomass yield, nutrient intake and utilization by sheep maintained on public rangeland. *Indian Journal of Animal Sciences* 69, 617-620.

Sankhyan, S.K., and Shinde, A.K. 2005. Utilization of different types of pastures and silvipastures for sheep production CSWRI Annual Report.

Sharma, S.C., Riyazuddin and Rawat, P.S. 1995. Evaluation of cross breeding programme in sheep in Rajasthan. Part II. Factors effecting adaptation of cross breeding programme in sheep. *Wool and Woollens of India.* January-March pp.39.

Shinde, A.K., Karim, S.A., Singh N.P. and Patnayak B.C.1995. Growth performance of weaner lambs and kids under intensive and semi intensive feeding management. *Indian Journal of Animal Sciences* 65, 830-833.

Shinde, A.K., Patnayak B.C., Karim S.A. and Mann J.S. 1994. Plane of nutrition and economics of ram raising under different system of grazing management. *Indian Journal of Animal Nutrition* 11; 85-89.

Shinde, A.K., Karim S.A. and Patnayak, B.C. 1996. Seasonal changes in circulating mineral contents of sheep and goats on different types of pastures. *Indian Journal of Small Ruminants* 2, 23-27.

Shinde, A.K., Sankhyan S.K., Karim S.A. and Patnayak B.C. 1997. Nutrient intake, its utilization and performance of sheep and goats on semi arid *Cenchrus ciliaris* pasture. *World Review of Animal Production* 31, 35-40.

Singh, N.P. and Prasad V.S.S. 1985. Effect of lambing season on growth, survivability and mutton production in Mutton Synthetic lambs. *Annual Report,* CSWRI, Avikanagar pp.43.

Singh, N.P and Singh, M. 1987. Growth performance and nutrient utilization in mutton synthetic lambs. Indian Journal of Animal Sciences 57; 986-990.

Tripathi, M.K., Karim, S.A., Chaturvedi, O.H. and Singh, V.K. 2006. Effect of *ad libitum* tree leaves feeding with varying levels of concentrate on intake, microbial protein yield and growth of lambs. *Livestock Research and Rural Development* 18,1-12.

❑❑❑

9

Nutrition of Range Managed Sheep in Semiarid and Arid Regions of India

A.K. SHINDE
Central Sheep and Wool Research Institute, Avikanagar - 304 501, Rajasthan

Introduction

Sheep play pivotal role in sustaining the livelihood of economically and socially weaker sections of population in arid and semi-arid region of the country. The raising of sheep on rangeland is cheapest system of production in the country. About 90 % of sheep flocks are maintained under extensive range management on native ranges, fallow land, wasteland and forest by them for their economic sustenance (Karim and Shinde, 2007). Over the period, lands under grazing are declining rapidly due to land reclamation for agriculture purposes on the other hand the livestock population is increasing exponentially. This is resulting in fast deterioration of grazing resources. Sheep population in 1951 was 39.1 million, increased to 58.2 million in 2003 with an overall increase of 49 percent. Total area available for grazing of animals in 1951 was 82.1 million ha, which was including uncultivable land, other uncultivated land, fellow land and permanent pastures has decreased to 43.3 million ha in 2003. Total grazing lands has declined by 47% while sheep population has increased by 49% during the last five decades. In spite of gradual deteriorating condition of grazing lands in the country still grazing lands are major source of forage supply to sheep under traditional nomadic and semi-nomadic systems. Main constraints to sheep production on rangeland in semi-arid and arid regions of the country are:

- Progressive shrinkage in grazing land over the years due to intensive land reclamation for conventional agricultural production.
- Simultaneous increase of livestock population causing overstocking on available land.
- Overgrazing gradually depleting the grazing resources leading to soil erosion and ecological degradation.
- The long dry spell of 8-9 months does not allow proper rejuvenation of vegetation as a result animal faces serous feed scarcity during lean period.
- Additionally water scarcity during summer months causes serious limitation for livestock production.
- Wide seasonal variation in vegetation availability and its nutrient contents raises valid issues as to how the animals are sustaining themselves let alone on available grazing resources providing production for the farmer.
- Poor biomass yield of grazing lands in the region is major constraint in achieving desirable production from small ruminants.

Biomass yield of grass pastures and silvipastures

Major limitation to animal production on ranges in semi arid and arid tropics is poor herbage yield. High stocking density and constant grazing pressure by different livestock species on these ranges has caused serious deterioration of vegetation (Karim et al., 1999). Herbaceous species and deciduous trees often occupy such ranges are deteriorating gradually. In arid and semiarid tropics, grasses and forbs in the ranges usually sprout immediately after rain, attain peak growth in September-October, and subsequently wither off with the onset of winter. Short life cycle and constant grazing pressure by animals does not allow vegetation to attain maximum yield. Common deciduous trees in the ranges supply forage to grazing animals during winter and summer with deterioration of ground vegetation.

Native ranges located in semiarid region under continuous grazing by all kinds of livestock yield 1.5q DM/ha and yield remains similar during all the seasons, because of poor growth of vegetation in other than monsoon seasons and continuous grazing pressure (Sankhyan et al., 1999). The studies carried out in villages adjacent to CSWRI in Rajasthan have shown that simply by protecting the ranges from grazing during active growth phase, the yield of ranges increased by 40-50% (Mann and Singh, 1982).

The native ranges in semiarid region are deteriorating at a very rapid rate; moreover, most of the palatable plant species are vanishing and gradually replacing by unpalatable one. There is an urgent need for rehabilitation of these ranges by perennial grasses to ensure adequate supply of forage to grazing animals in the years to come. Mann and Mehta (1998) reported that carrying capacity of ranges increased up to 2-3 times simply by reseeding with perennial grasses. Some of the perennial grasses suitable for rehabilitation of ranges in different regions are shown in table 1.

Table 1. Suitable grasses for different regions

Situations	Suitable grasses
Desert land	*Lasiurus sindicus, Cenchrus ciliaris, Cenchrus setigerus*
Ravine land	*Cenchrus ciliaris, Cenchrus setigerus, Pennisatum pedicellatum, Chrysopogon fulvus*
Degraded rangeland	*Cenchrus ciliaris, Cenchrus setigerus, Panicum antidotale, Sehima nervosum, Dicanthium annulatum, Chrysopogon fulvus*
Salt affected lands	*Cynodon dactylon, Paspalum notatum, Paspalum dilatum Bracheria mutica, Sporobolus marginatus*
Acidic soil	*Pennisatum polystachym, Pennisatum pedicellatum, Paspalum notatum*
Shifting cultivation affected land	*Cenchrus ciliaris, Cenchrus setigerus, Pennisatum pedicellatum, Chrysopogon fulvus*
River bed affected area	*Chrysopogon fulvus, Dicanthium annulatum, Pennisatum pedicellatum*
Temperate and sub-temperate area	*Bromus inermis, Dactylis glomerata, Phleum pratensis*

Cenchrus ciliaris, Cenchrus setigerus and *Lasiurus sindicus* are the perennial grasses recommended for rehabilitation of grazing land in the dry zones of the country. Average yield of *Cenchrus ciliaris* ranges between 27-33qDM/ha in September to 7 q/ha in May. *Cenchrus ciliaris* is found most suitable in low rainfall of <250mm and light to medium type of soil. Rai et al. (1995) reported that biomass yield of 29-47q DM/ha in *Cenchrus ciliaris* and 28-55 q/ha in *Cenchrus setigerus* under favorable rainfall and soil conditions. *Lasiurus sindicus* grows well in sandy soil of arid region and can withstand ambient temperature up to 48°C. *Lasiurus sindicus* yield up to 25-30q DM/ha under rain fed and 47q/ha under irrigation condition.

The protein contents of perennial grasses declined from 12% during August - September to 4% during May- June. The protein contents of grass pastures can be improved by mix sowing of legumes like *Clitoria ternetae,Lablab purpurium, Atylosia scarbaeoides, Stylosanthes hamata, Stylosanthes scabra* and *Rhyncosia minima* with grass pasture, such combination increased yield as well as quality.

In poor soil, water and nutrient situations where crop cultivation is not feasible, the establishment of silvipasture comprising of grasses, shrubs and trees serve the purpose of forage and wood supply with environment conservation. Simply by planting fodder trees on farm boundaries of 0.09 ha land produces 91% more dry fodder during dry season than without trees and reduce soil erosion by 57%. The association of *Prosopis cineraria, Gymnosporia spinosa, Acacia nilotica, Azardirachta indica, Albizia lebbek, Bauhinia racemosa, Morus alba* and *Leucaena leucocephala* trees assocaition with grasses *Cenchrus ciliaris, Cenchrus setigerus, Panicum antidotale* provides more fodder yield than grass alone. A plot of 1 ha having 100 *Ailanthus excelsa* trees and *Dicrostachys nutans* bushes and ground cover of *Cenchrus ciliaris* yield 5.29qDM tree leaves and 2.28qDM Dicrostachys leaves and pods and 23.5 q DM grasses, totaling to 31.00 q DM/ha (Sankhyan et al., 1996).

Assessment of dietary nutrient content of forage and dry matter intake by sheep and goat on rangeland is one of the basic necessity for formulating suitable range management strategies. Basic prerequisites for determining dry matter intake of sheep on pasture are collection of representative samples of forage consumed by the animals precisely and accurately.

Diet sampling method

The diet samples of animals on grazing lands are collected by various methods, each methods has its own merits and demerits. Oesophageal canula is widely used for collection of diet sample of grazing animal (Bhatta et al., 1997). The drying of oesophageal extrusa samples in oven influences the lignin content and thereby estimates of DM intake due to formation of artifact lignin, which gives erroneous estimates. To avoid such erroneous results freeze drying of extrusa samples is the preferred method (Bhatta et al., 2001).

Hand plucking, clipping and mouth grab sampling are also used for collection of diet samples of animals (Sankhyan et al., 1999). Similarly, accurate estimation of total faecal output is must for determining the nutrient digestibility and intakes. Faecal bag or Cr_2O_3 markers are common methods for estimation of total faecal output in grazing animals (Sankhyan et al., 1999). Some of the methods commonly used in diet sample collection in grazing animals are:

- *Clipping* : In this method, ten 1m^2 quadrate in grazing area are clipped 2-3 cm above the ground level.
- *Mouth grab* : 30-35 bites from the mouth of animals are snatched before swallowing, samples with excessive salivary contamination are discarded.
- *Plucking* : 25-30 samples of forage are collected, simulating the animal's dietary preference.

- Oesophageal extrusa : The extrusa samples are collected from fistulated animals during first 3 h of active grazing, samples are then squeezed in muslin cloth to remove excessive saliva.

Faecal sampling technique

Total faecal output in grazing animals is commonly measured either by faecal bags or chromic oxide (Cr_2O_3) as an inert tracer.

- In chromic oxide marker method, feacal samples after 5 day of initial dosing with marker is collected at 12 h intervals.
- In feacal beg method, total feacal output is daily collected

A linear equation based on studies in large number of animals on different pasture conditions has been developed for the estimation of DM intake in grazing sheep and goat grazing on semi arid ranges and pasture with > 90 % accuracy, which can be used in situation where total feacal output is known.

$$Y = -0.0107 + 2.2090\ x,$$

Where, Y = DMI in g, X = Total faecal output in g

Dry matter intake estimation techniques

Several methods for estimation of DM intakes in grazing animals has been reported in literature. Commonly following formulae is used for estimation of dry matter intake in grazing animals:

$$\text{DMI by lignin ratio} = \text{Faecal output in kg} \times \frac{\text{Lignin in faeces \%}}{\text{Lignin in feed \%}}$$

$$\text{DMI by IVDMD} = \frac{\text{Faecal output}}{1 - \text{IVDMD}}$$

Where, IVDMD = In vitro dry matter digestibility of the diet sample

Voluntary intake of sheep on pastures

Sheep in native ranges of semi arid region consumes 649 g DM/day or 49.6 g DM/kg $W^{0.75}$ (2.0% of BW) in monsoon when the vegetation of native species sprouts and abundantly available. DM intake slightly declines to 572 g/day or 43.7 g DM/kg $W^{0.75}$ in winter. In summer when the ranges are poorly occupied by the native plant species, the DM intake declined by 23 per cent from monsoon (Sankhyan et al., 1999). Native range in tropics are not meeting the DM intake of sheep for maintenance requirement, resulting in loss of body weight and production traits during winter and summer. On the contrary *Cenchrus ciliaris* pasture ensured DM intake of 665g/day or 36.9 g/kg $W^{0.75}$ in monsoon season, which increased with decrease in moisture content of forage and attained 1239g/

day or 64.0g/kg$W^{0.75}$ in winter and declined to 935 g/day in summer, although seasonal changes are wide in tropical pastures but still ensured better nutrition than native ranges (Shinde et al., 1998).

Plant and Nutrient Selectivity by Sheep

The selection of plant and nutrients in relation to others in pasture or ranges are described by preference index (PI). Preference index is defined as 'the discrimination among plants exhibited by animals when given free choice'. A preference index of one, indicates that preference for these plants by animals is similar to their prevalence in the pasture. If it is <1, it reflects an animal's avoidance of these plants.

Eleusine indica, Commelina forskalaei, Eriochloa polystachya and *Eleusine aegyptica* are some of plant species highly preferred during monsoon season in native ranges of semi arid region by sheep (Sankhyan et al., 1995). *Cenchrus ciliaris,* a perennial grass is preferred in monsoon but, its preference declined in summer due to maturity and lignification. *Crotolaria burhia,* a forb widely found in ranges preferred during February-March. The selection of protein rich vegetation progressively increases with deterioration of pasture condition (Table 2). The PI of CP increases from 1.2 in monsoon to 2.1 and 3.0 in winter and summer by sheep and avoids ADF and cellulose in their diet.

Table 2. Preference indices of nutrients in sheep on Cenchrus pasture

	Monsoon	Winter	Summer
Crude Protein	1.19	2.12	3.00
Neutral Detergent Fibre	0.99	0.77	0.85
Acid Detergent Fibre	0.99	0.95	0.76
Cellulose	0.98	0.84	0.73
Hemicellulose	0.97	0.50	1.61
Lignin	1.90	0.20	0.81

Nutrient intake

Nutrient intake of sheep on pasture is a reflection of forage intake and its utilization. In tropics the nutrient intake of animal on ranges and pastures usually declined from monsoon to winter and summer because of maturity and lignifications of forage and gradual decline of forage yield.

Protein intake

Sheep on the ranges of semi arid region consumed 3.44 g/kg $W^{0.75}$ DCP in monsoon 2.42 in winter and 1.05 in summer season. The intake of DCP in summer declines up to 70-75% compared to monsoon, with declined in forage supply from ranges and found inadequate to meet the requirements (Sankhyan et al., 1999). In Cenchrus pasture, DCP intake of sheep achieved a maximum value of 4.70 g/kg $W^{0.75}$ in winter and remained between 2.10-2.50 g/kg $W^{0.75}$ in monsoon and summer. In monsoon, higher moisture contents restricted the lower intake of DM (Shinde et al., 1998). In silvipasture animal access to tree leaves in addition to grazing registered the DCP intake of 6.76 g/kg $W^{0.75}$ (Shinde et al., 1994). Sheep during empty, pregnant and lactation stages on cenchrus pasture had DCP intake of 2.5, 2.4 and 3.1 g/kg $W^{0.75}$ (Bhatta et al., 1996) and were found short of their requirement for pregnancy and lactation stages. Sheep on sewan pasture in arid region had DCP intakes of 0.9, 0.4 and 0.5 g/kg $W^{0.75}$ during empty, pregnant and lactation stages. Sheep on pasture in both the climate be supplemented concentrate mixture or other agricultural byproducts to meet their DCP requirement of pregnancy and lactation stages.

Energy intake

Native and cultivated forages on an average contain 6-7 MJ/kg DM in tropics. Sheep consuming 1.0-1.5 kg DM/day in ranges were able to meet 6-9 MJ/day, which is found insufficient for normal maintenance and outdoor activity (Table 3). Moreover in dry season when forage matured and gets lignified, the energy intake further reduced due to poor nutrient digestibility.

Table 3. Nutrient intake of sheep on Cenchrus pasture

	DCP (g/kg $W^{0.75}$)	ME (MJ/kg $W^{0.75}$)
Monsoon	2.10	0.29
Winter	4.70	0.48
Summer	2.50	0.29

The ME intake of 0.74 MJ/kg $W^{0.75}$ in monsoon and 0.42 MJ/kg $W^{0.75}$ in summer in sheep on Cenchrus pasture in semi arid region has been recorded (Shinde et al., 1998). The ME intake in summer remained much lower than the requirement, suggesting supplementation. The dry, pregnant and lactating sheep had ME intakes of 0.37, 0.38 and 0.40 MJ/kg $W^{0.75}$ on Cenchrus pasture in late winter (Bhatta et al., 1996). A sizable reduction in the energy intake of sheep on Cenchrus pasture with progress of season from autumn to late winter has been noticed suggesting. Synchronization of critical physiological stages with autumn, when Cenchrus pasture is in the pre-flowering stage and rich in nutrients. However sheep in later season be supplemented concentrate mixture to meet their requirement of pregnancy and lactation. Similarly sheep in Sewan (*Lasirus sindicus*) pasture of arid region had ME intake of 0.40, 0.26 and 0.84 MJ/kg $W^{0.75}$ during dry, pregnant and lactation stages.

Nutrient digestibility

Semi arid range/grasslands are occupied by a wide variety of plant species: grasses, forbs, shrubs and fodder trees. The chemical/ nutrient composition of these grasslands undergoes drastic seasonal changes. This resulted in variation in nutrient intake and utilization due to changes in nutrient digestibility. The nutrient digestibility linearly declined as the season progresses from monsoon to summer and is related to maturity and age of the foliage. In general nutrient digestibility in goat in comparison to sheep is higher because of the following reasons: Goat selects plant parts of higher nutritive value, greater retention time of digesta in the rumen, higher microbial protein production in the rumen and faster fractional rate of passage of digesta through the rumen leading to higher intake.

Energy expenditure of sheep on pasture

Sheep is basically considered as grazing animals and thrive well under wide range of vegetations. Sheep are maintained on extensive system where they traveled long distance in search of food on ranges. Majority of the flocks in tropics are maintained in sedentary system where they covered about 4-5 km daily for grazing. Sheep exposed to wide range of environmental condition on ranges ranging from 10-15°C ambient temperature in winter to 42-46°C in summer besides intense solar radiation in summer in semiarid tropics. The out door activities required extra energy for walking and thermo-regulation. In many cases, the energy expenditure is more than the intake resulting in loss of production. In the country, information on energy expenditure of sheep on pastures are limited. The Scientist in CSWRI, Avikanagar has initiated studies on this aspect in 1996 and developed tracheal cannula (Karim et al., 1997) for collection of expired air and measurement of EE in sheep using indigenous skill (Shinde et al., 1998) followed by face mask for short term measurement of EE (Shinde et al., 2002).

Sheep on maintenance ration under stall feeding spend 4.16 MJ of energy per day or 0.295 MJ of energy per kg $W^{0.75}$. The energy expenditure increased to 6.71 MJ /day or 0.53 MJ/kg $W^{0.75}$ in monsoon, 4.85 MJ/day or 0.34 MJ/kg $W^{0.75}$ in winter and 5.85 MJ/day or 0.40 MJ/kg $W^{0.75}$ in summer seasons (Shinde et al., 1998) EE of sheep on pasture compared to stall feeding in monsoon, winter and summer increased by 78, 15 and 33%, respectively with an overall of 43%. Sheep consumed 286, 483 and 293 MJ ME /kg $W^{0.75}$ and spend 527, 340 and 395 MJ ME /kg $W^{0.75}$ in monsoon, winter and summer seasons respectively (Shinde et al., 1998). Sheep remains in positive energy balance in winter while in monsoon and summer in negative energy balance (Table 4). Which is reflected in loss of production and growth. In semi arid region, EE in sheep on pasture is relatively lower than monsoon and summer owing to better forage availability from land and favorable ambient conditions. In monsoon, hot and humid conditions elicited the thermo-regulatory mechanism of animals to maintain body temperature resulting in increase of EE moreover high moisture content of forage restrict the

DM intake. In summer, exposure of sheep to high ambient temperature demands greater energy for thermoregulation and low availability of forage from grazing land further increases the stress on animals and EE.

Table 4. Energy budgeting of sheep on pasture and stall-feeding

	Stall-fed		Grazing	
	Autumn	Monsoon	Winter	Summer
Energy intake (kJ/kg $W^{0.75}$)	358	286	483	293
Energy expenditure (kJ/kg $W^{0.75}$)	295	527	340	395
Energy balance (kJ/kg $W^{0.75}$)	+63	-241	+143	-103

Mineral deficiency in grazing sheep

Mineral deficiency or excess is an area specific problem reflected in soil and vegetation growing on it and animals maintained on such vegetation. One of the reasons ascribed to lower productivity of native sheep on rangeland is deficiency of phosphorus for most part of the year and other micro minerals in tropics. Large numbers of livestock are maintained in rangeland system which does not meet the exact requirement of minerals in the tropical regions (McDowell et al., 1993). This results in mineral disorder, which range from acute or severe mineral deficiency to toxicity diseases. In many cases mild deficiency of mineral occur without clinical symptoms which causes unsatisfactory growth, production and fertility in animals (Underwood and Suttle, 1999). Inadequate intake of phosphorus and excess intake of iron under prevailing feeding practices are common in animals reared in rangeland of semi arid region of Rajasthan (Shinde et al., 2006). Sheep in semiarid region of Rajasthan under prevailing feeding practices adequately received Mg, Mn, Cu and Fe minerals from grazing resources and suffered from deficiency of Ca, P and Zn (Shinde et al., 2006). The deficiency of Ca and P caused reduction in milk yield and growth and Zn deficiency caused skin keratinization in sheep. The supplementation of mineral mixture in sheep maintained on rangeland in ration is restricted as most of flocks are not supplemented concentrate mixture or grains in routine feeding practices. Oral drenching or dosing or by administrating slow releasing mineral boluses are some of option for supplementation of mineral mixture in sheep in grazing condition. The CSWRI, Avikanagar institute has developed area specific mineral mixture consisting of dicalcium phosphate, zinc sulphate, copper sulphate, cobalt chloride and common salt for sheep, goats, cattle and buffaloes of semiarid region of Rajasthan. The supplementation of mineral mixture at the rate of 5g daily brings 55-60% anoestrus sheep in estrus and improves milk yield by 15-20% in sheep.

Soil ingestion and its effect on health

Majority of the sheep flocks in the arid and semiarid regions of the country are maintained on native ranges. The grazing habits of sheep in relation to other livestock species are different, as they generally graze close to the ground and thrive well on shorter than taller grasses. The peculiar anatomical structure of their lips help them to manipulate the food that grasps between the lips and between the lower incisor and upper bare gum, sheep has split (bifid) upper lip which partly accounts for their greater ability to graze close to the ground and help them to gather food from sparse vegetation cover of the grazing lands. The close grazing habit of sheep on poor vegetation cover of arid ranges caused soil ingestions. Sheep ingest 71g to 163g of soil during different seasons in arid regions (Vaithyanathan and Singh, 1994). Soil ingestion of 144, 53 and 107 g/ day, which made 26.91, 11.97 and 41.04 % of total dry matter intake during monsoon, winter and summer respectively in sheep on community land has been reported (Shinde et al., 2005). Soil ingestion cause increase excretion of endogenous nitrogen per unit of feacal organic matter beside damage to gut mucous lining, rapid wears of incisor teeth and bruising of dental pad. The excess intakes of soil by sheep during summer and monsoon seasons adversely affect the nutrition and health. It affects dry matter intake and nutrient digestibility and body weights.

Conclusion and Recommendations

The poor availability of forage from grazing land severely influenced the nutrition resulting in overall poor production, reproduction and health of sheep in arid and semiarid region of the country. The supplementary feeding of sheep is one of the option for optimizing the production, other alternative approach is improvement of existing lands under grazing by perennial grasses and silvipastures for ensuring better nutrition to animals. Besides protein and energy mineral deficiency mainly of phosphorus in sheep on native ranges in tropical condition is common and can be prevented by supplementing area specific mineral mixture. In near future the main thrust in research on sheep production should be on development of pastures and silvipasture on existing grazing lands to sustain the production with eco-system conservation.

REFERENCES

Bhatta, R., Sankhyan, S.K., Shinde, A.K. and Karim, S.A. 1997. Nutritional evaluation using oesophageal cannulated sheep under silvipasture. *Indian Journal of Animal Sciences* 67, 234-236.

Bhatta, Raghavendra, Sankhyan, S.K, Shinde, A.K. and Karim, S.A. 2001 Composition of freeze and oven-dried esophageal extrusa samples for the estimation of dry matter intake and digestibility in range managed sheep. *Indian Journal of Animal Sciences* 71, 570-572.

Bhatta, Raghavendra, Shinde, A.K., Karim, S.A. and Patnayak, B.C. 1996. Plane of nutrition in Malpura and Mutton Synthetic ewes maintained on *Cenchrus ciliaris* pasture during different physiological stages. *Indian Journal of Animal Sciences* 66, 614-618.

Karim, S.A., Bhatta, R, Sankhyan, S.K. and Shinde, A.K. 1997 Surgical procedure of tracheal cannulation in sheep. *Indian Journal of Small Ruminants* 3, 34-38.

Karim, S.A. and Shinde, A.K. 2007. Pasture based feeding system for small ruminant production and its relevance in tropics. International Conference on Tropical Animal Nutrition, NDRI, Karnal 3-5 Oct. 2007.

Karim, S.A. Shinde A.K. and Sankhyan, S.K. 1999 Feeding and grazing management of sheep under different production systems Proceeding of Golden Jubilee Seminar on Sheep, Goat and Rabbit Production and Utilization, Jaipur 24-26 April 1999.

Mann, J. S. and Mehta, R. S. 1998. Feed and fodder resource development for small ruminants production. Golden jubilee Seminar on Sheep, Goat and Rabbit Production and Utilization. 24-26 April. pp 89-94.

Mann, J.S. and Singh, N.P. 1982. A field study on the performance of natural and reseeded grassland in the semi arid conditions of Rajasthan. *Livestock Advisor* 2, 23-29.

Mc Dowell, L.R, Cornard, J.H, Ellis, G.L. and Loosli J.K .1993. *Minerals for Grazing Ruminants in Tropical Regions*. Bulletin 2nd ed, Animal Science Department, University of Florida, Gaineville, USA.

Rai, P.K., Yadav, M.S. and Sudhakar, N. 1995. Genotype x environment interaction in buffer grass. *Annals of arid Zone* 34, 111-114.

Sankhyan S.K., Shinde A.K. and Karim S.A. 1999. Seasonal changes in biomass yield, nutrient intake and its utilization by sheep maintained on community grazing land. *Indian Journal of Animal Sciences*. 69, 617-620.

Sankhyan, S.K., Shinde, A.K., Karim, S.A., Mann J.S., Singh, N.P. and Patnayak, B.C. 1996. Plane of nutrition and growth response of lambs and kids maintained on two and three tier silvipasture. *Indian Journal of Animal Science* 66, 1194-1197.

Shinde A.K., Karim S.A., Sankhyan S.K. and Bhatta R. 1998. Seasonal changes in physiological responses and energy expenditure of sheep maintained on semi-arid pasture. *Journal of Agriculture Science, Cambridge* 131, 341-346.

Shinde A.K., Karim S.A., Sankhyan S.K. and Bhatta R. 1998. Seasonal changes in biomass growth and quality and its utilization by sheep on semiarid *Cenchrus ciliaris* pasture of India. *Small Ruminant Research* 30, 29-35.

Shinde A.K., Sankhyan, S. K. Singh N. P., Verma D.L. 2005. Soil ingestion and its effect on dry matter intake and digestibility in sheep grazing on native ranges of semiarid region of Rajasthan. *Indian Journal of Small Ruminant.* 11, 37-42.

Shinde, A.K., Patnayak, B.C., Karim, S.A. and Mann, J.S. 1994. Plane of nutrition and economics of ram raising under different systems of grazing management. *Indian Journal of Animal Nutrition* 11, 85-89.

Shinde, A.K., Sankhyan S.K., Karim, S.A., Singh, N.P. and Patnayak, B.C. 1996. Nutrient Intake, its utilization, and preference of sheep and goats on semiarid *Cenchrus ciliaris* pasture. *World Review of Animal Production* 31, 35-40.

Shinde, A.K., Sankhyan, S.K., Bhatta, R. and Verma, D.L. 2000.Seasonal changes in nutrient intake and its utilization by range goats in a semi-arid region of Indian. *Journal of Agricultural Science, Cambridge* 135, 429-436.

Shinde, A.K., Sankhyan, S.K., Karim, S.A., Kumar, Ravindra and Singh, V.K. 2006. A research bulletin. Mineral Limits of Ruminants in Rajasthan Published by Central Sheep and Wool Research Institute, Avikanagar

Underwood, E.J. and Suttle, N.F. 1999. *The Mineral Nutrition of Livestock*, 3rd Ed, CABI Publishing, CAB International Wallingford, Oxon, UK.

Vaithiyanathan, S. and Singh, M. 1994. Seasonal influence on soil ingestion by sheep in arid region. *Small Ruminant Research.* 14, 103-106.

❑❑❑

10

Synchronization of Nutrient Balance for Optimizing Sheep Production

BANDLA SRINIVAS
OM HARI CHATURVEDI & RAMAN MALIK
Central Sheep and Wool Research Institute, Avikanagar - 304 501, Rajasthan

Introduction

Sheep production systems, like any other the broader livestock systems, represent the resultant interaction between animal system, ecology (biotic and abiotic factors) and socioeconomic factors of the region. Worldwide large population of sheep are habituated in arid and semi-arid regions which are characterized by the hot weather. Temperature may go as high as 48^{0}C in summer with very scanty rainfall that seldom goes beyond 500 mm. These harsh climatic factors are profoundly influencing the animal performance by directly affecting the nutrient intake, basic metabolic processes and finally, biological output. Both input and output are economically quantifiable events while transformation process of inputs to output is a qualitative event involving the successive chemical reactions to degrade simply organized nutrients from the diets into even simple products. When events of sheep production are adjudged in terms of biological efficiency, either minimization or maximization is reliable with transformation process but not with input or output. Simply minimizing the inputs may not yield maximum biological yields and similarly, maximizing the inputs may not result in proportional increase in the yields. Optimization of production process greatly dependent on the marginal production returns such as change in unit output on an increase in every unit of input. Contrary to physical optimum, which aims for maximum yield, production optimum aims for maximum gain.

Under perfect market oriented production system (best example is meat market) production optimization is essential to enhance production efficiency (a ratio between input and output) and reducing the wastivity (a ratio between waste or futile losses and Input).

In reality, sheep farmers are constrained with limited resource base and poor knowledge of input management resulting in low income. The reason behind such underprivileged entrepreneurship is due to poor knowledge of systematic livestock production management (Srinivas *et al.*, 2004). In order to simplify the feeding practice of sheep in the arid and semi-arid regions, a concentrate supplement of 250 g/d is suggested as generalized allowance in addition to the grazing (Karim *et al.*, 2004). Nevertheless, when production becomes an intensive trade based activity, optimization of nutrient input becomes an essential requirement. The discussion presented here in three sequential events to address restraints presented in the figure 1. They are; to minimize discomfort to the animal, to optimize nutrient utilization to enhance sheep production efficiency and, to reduce wastivity by synchronizing the nutrients for production process.

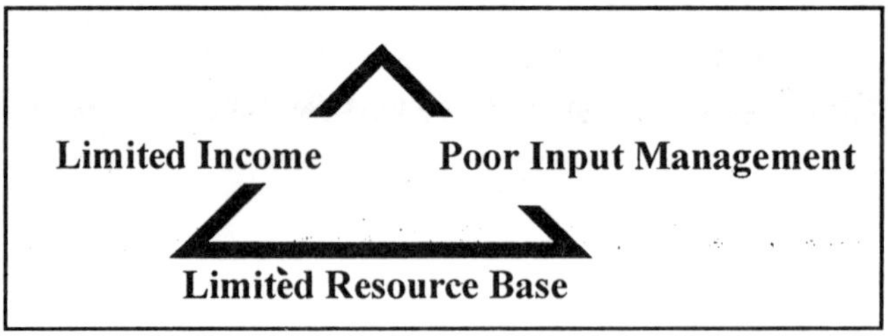

Fig 1. Farmer level sheep production system

Minimal Total Discomfort

The quantity and composition of feed affects the animal performance in many important ways. Attraction, preference, smell and texture have influence on physiological consequences. Preference in is defined as the choice of dietary component that an animal will make when there are no constrains on availability/offer of the diet components (Rutter *et al.*, 2000). Animal develops chemical senses with growing age but again starts deteriorate from the fourth quarter of its life span. Intake pattern of grazing and stall fed animals is always distinct hence, their nutrient balance and utilization pattern is also different. In the context of choice and intake, feeding system can be differentiated into closed system of feeding which includes individual or group feeding in confinement and, open system of feeding which includes the grazing in pastures, grasslands etc. In stall fed animals, intake is related to fill and break down characteristics of the forages (Demment and Vansoest, 1985). Grazing animals have to meet their nutrient requirement in finite time of grazing while moving selectively on lush green pastures or on sparse pastures. On sparse pastures, animals even have difficulty in meeting their minimum nutrient requirement. Under optimal conditions, the feed intake in ruminants appears to regulate by their capacity to dispose of energy (Weston, 1996). Inter-meal intervals would be longer with a variable meal sizes on high concentrate based feeds than roughage based diets. Inter-meal intervals are largely dependent on the size of meal ingested during the preceding periods. Ruminant animals would develop a 'nutritional wisdom' in choosing the feeds which most closely meets the animal's maintenance and production requirement when there are no constraints on feed availability (Rutter *et. al.*, 2000). When animals have choice of several foods, each meal includes a mixture of feed items rather than each meal consisting entirely of one feed. Voluntary intake of feed would approximately balances the animal requirements for nutrients when feed quality is not a limiting factor. This has been led to draw a hypothesis that ruminants learn to select a mixture of feeds that minimizes metabolic imbalance (Kyriazakis and Oldham, 1997). Contrary to nutritional wisdom, animals have 'nutritional stupeo' in choosing a wholly inappropriate mixture of feeds which may result in unhealthy or undesirable conditions (Gordon and Tribe, 1951). The word 'toxin' is usually reserved for a substance that causes obvious signs of discomfort or distress. Often animals must trade off high yield of digestible nutrients against toxins present in the forages (Hutchings et. al, 1999). All dietary components are capable of acting as toxins, if present in greater excess over requirements. Animal in a given physiological state has requirement for optimum supply rate of each nutrient. Animals may adjust their metabolism and will manage mild departures in access or deficiency from optimum levels. Only when a nutrient becomes access or deficit beyond the scope of metabolic adjustments, animals may unable to cope up from the toxicity or deficiency. Besides, the discomfort from the ingested feed, the surrounding abiotic factors to the animals such as temperature, humidity, wind velocity, rainfall, day length, evaporation, soil temperature, soil texture etc., would also create discomfort to the animal system. Integration of all factors such as health, nutrition and environment rather limiting exclusive to nutritional factors led to a concept of 'Minimal Total Discomfort' (Forbes, 1999). The 'minimal' connotes total quantitative intake of the animal to minimize total discomfort while 'total' connotes discomfort from many factors such as nutrient imbalance, hedonics, thermal environment, physiological stress, health etc., which together induce intake controlling signals mediated through central nervous system and, 'discomfort' connotes depart from the optimal supply of a nutrient. Based upon chemical composition and availability of nutrient from the diet, relative discomfort caused to the animal by the feed intake can be derived (Forbes, 2001).

Deriving Minimal Total Discomfort

The method of deriving the minimal total discomfort is explained here with the example of pregnant ewe weighing 30 Kg and kept under grazing in the pastures covered by the coarse swards. Nutrient requirements are as per the ARC (2000) with metabolizability of dietary energy (ME/GE) of 0.70. Crude protein content of the diet is 74 g/Kg DM with whole tract digestibility of 54%. Rumen degradable protein content is 48 g/Kg DM. Digestible energy content of the diet is 6.8 and metabolizable energy is 6.2 MJ/Kg DM. Total digestible protein intake is 40 g/Kg DM. Digestible protein to digestible energy ratio is 5.88 and rate of intake is 1.50 g/ min (Grazing time = 1.5 g*60 min = 90 g/h, therefore, grazing time for 720 g = 720/90 = 8 h). Limiting to the basic information, the discomfort to the animal and minimal total discomfort was derived (table 1). Total discomfort from different levels of intake is depicted as a line graph (Figure 2). Line graph showed that the total discomfort is minimal when dry matter intake was 900 g/d. At this point of feed intake, still animal is deficit of 3.5 g of rumen degradable protein and 480 KJ of metabolizable energy. The other level of dry matter intake such as 1000 g/d fulfills both rumen degradable protein and metabolizable energy but total discomfort is little higher than when the dry matter intake was 900 g/d. Such smaller deviation in minimal total discomfort may be allowed provided the animal is accessible to additional feed and able to consume 100 g/d more dry matter. Otherwise, increasing the concentration of the deficit nutrients in supplements within the limit of dry matter i.e., 900 g/d, would be the best option for the minimal total discomfort driven nutrient input manipulation. Since actual consumption of animal was only 720 g/d, it is necessary to meet the balance nutrient requirement from another 180 g of dry matter coming from either concentrate or roughages.

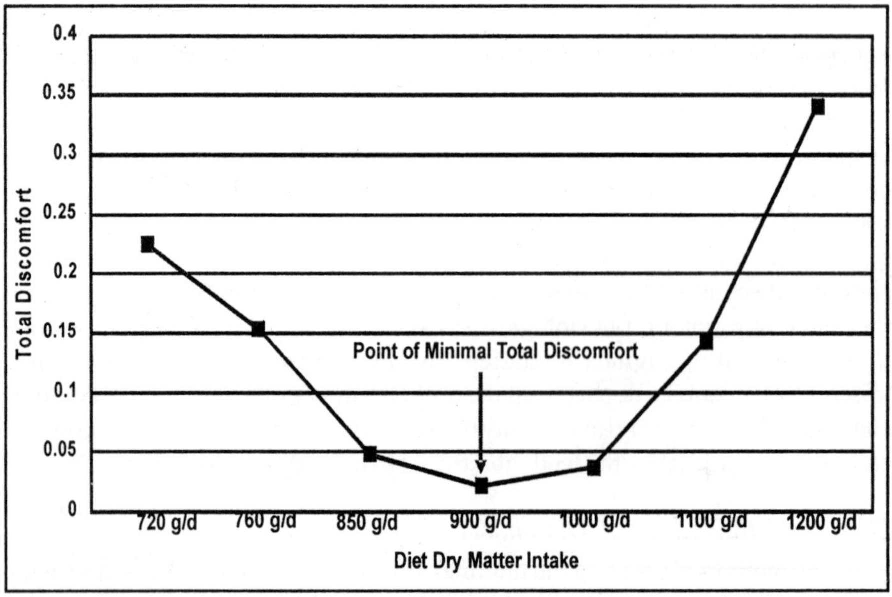

Fig 2. The Dependence of Total Discomfort on Diet Intake

Table 1. Relative discomfort to animal at different levels of feed intake.

Parameter	Requirement	Intake	Difference	Relative error	Relative Discomfort
Estimated Dry matter intake 720 g/d					
Dry matter (g/d)	865	720	-145	-145/865 = -0.17	$-0.17^2 = 0.03$
RDP (g/d)	35	25.20	-9.80	-9.8/35 = -0.28	$-0.28^2 = 0.08$
ME (MJ/d)	4.8	3.46	-1.34	-1.34/4.8 = -0.28	$-0.28^2 = 0.08$
Dig. Protein: Energy	1:6	1:5.88	-0.12	-0.12/6 = -0.02	$-0.02^2 = 0.0004$
Grazing Time (h)	10	8	-2	-2/10 = -0.20	$-0.20^2 = 0.04$
				Total Discomfort	0.2304
Simulate Dry matter intake to 760 g/d					
Dry matter (g/d)	865	760	-105	-105/865 = -0.12	$-0.12^2 = 0.01$
RDP (g/d)	35	26.60	-8.40	-8.40/35 = -0.24	$-0.24^2 = 0.06$
ME (MJ/d)	4.8	3.65	-1.15	-1.15/4.8 = -0.24	$-0.24^2 = 0.06$
Dig. Protein: Energy	1:6	1:5.88	-0.12	-0.12/6 = -0.02	$-0.02^2 = 0.0004$
Grazing Time (h)	10	8.44	-1.56	-1.56/10 = -0.16	$-0.16^2 = 0.02$
				Total Discomfort	0.1504
Simulate Dry matter intake to 850 g/d					
Dry matter (g/d)	865	850	-15	-15/865 = -0.02	$-0.02^2 = 0.0003$
RDP (g/d)	35	29.75	-5.25	-5.25/35 = -0.15	$-0.15^2 = 0.02$
ME (MJ/d)	4.8	4.08	-0.72	-0.72/4.8 = -0.15	$-0.15^2 = 0.02$
Dig. Protein: Energy	1:6	1:5.88	-0.12	-0.12/6 = -0.02	$-0.02^2 = 0.0004$
Grazing Time (h)	10	9.44	-0.56	-0.56/10 = -0.06	$-0.06^2 = 0.0031$
				Total Discomfort	0.0488
Simulate Dry matter intake to 900 g/d					
Dry matter (g/d)	865	900	+35	35/865 = 0.04	$0.04^2 = 0.0016$
RDP (g/d)	35	31.50	-3.50	-3.50/35 = -0.10	$-0.10^2 = 0.01$
ME (MJ/d)	4.8	4.32	-0.48	-0.48/4.8 = -0.10	$-0.10^2 = 0.01$
Dig. Protein: Energy	1:6	1:5.88	-0.12	-0.12/6 = -0.02	$-0.02^2 = 0.0004$
Grazing Time (h)	10	10	0	0	0
				Total Discomfort	0.022
Simulate Dry matter intake to 1000 g/d					
Dry matter (g/d)	865	1000	+135	135/865 = 0.16	$16^2 = 0.0244$
RDP (g/d)	35	35	0	0	0
ME (MJ/d)	4.8	4.8	0	0	0
Dig. Protein: Energy	1:6	1:5.88	-0.12	-0.12/6 = -0.02	$-0.02^2 = 0.0004$
Grazing Time (h)	10	11.11	+1.11	1.11/10 = 0.11	$0.11^2 = 0.0123$
				Total Discomfort	0.0371

Contd...

Table 1. Contd...

Simulate Dry matter intake to 1100 g/d					
Dry matter (g/d)	865	1100	+235	235/865 = 0.27	$0.27^2 = 0.0738$
RDP (g/d)	35	38.50	+3.50	3.50/35 = 0.10	$0.20^2 = 0.01$
ME (MJ/d)	4.8	5.28	+0.48	0.48/4.8 = 0.10	$0. 20^2 = 0.01$
Dig. Protein: Energy	1:6	1:5.88	-0.12	-0.12/6 = -0.02	$-0.02^2 = 0.0004$
Grazing Time (h)	10	12.22	+2.22	2.22/10 = 0.22	$0.22^2 = 0.0494$
				Total Discomfort	0.1436

Metabolizable energy intake has been emphasized as critical metabolic discomfort regulating the intake under the concept of minimal total discomfort. Animal would be under greater discomfort when animals metabolizable energy intake is either excess or lesser than other nutrients required (Forbes, 2001). Similarly higher or lower environmental temperature surrounding the animals is also important in terms of heat increment from the diet and heat dissipation form the body. It is considered as environmental discomfort. High dry matter content in feeds increases heat increment. Higher sward mass may also increase heat increment (Demment *et al.*, 1995). In grazing animals, the stress of walking for longer distances in search of food is considered as physical discomfort. Animals like to move in the group and each animal would try to spread within the visible vicinity of other fellow animal. Such social behaviour prevent animal to move away from the flock in search of its choice and denotes the behavioural discomforts. Therefore, animals may trade-off their intake to avoid minimal total discomfort. In practice, marginal deficiency of a nutrient may result in increase in intake but severe shortage or excess nutrient may cause intake to decline. Alleviation of deficiency is followed very quickly by a return to normal levels of intake (Forbes, 2000). Unfortunately, information regarding intake pattern of animals with relation to psychometric parameters of the weather in shorter intervals over longer periods is poorly documented. In the absence of such information, preparing complete map of bioclimatic specific fortnightly or monthly minimal total discomfort become very difficult for optimizing the sheep productivity.

Optimization of Nutrient Intake, Balance and Utilization

Current feeding systems predict that on a given diet the greater rate and efficiency of production will occur when animals are consuming the highest level of intake. Minimal total discomfort however, showed that it is not necessarily true. *Ad* libitum feeding is somewhat appropriate in low or poor quality roughages or pastures because their intake is regulated by rumen fill rather than nutrient density. Intake is often restricted by gut fill up to a breakpoint in digestibility. There after, relationship between intake and digestibility becomes negative (Conrad *et. al.*, 1964) and undigested feed residue retained in the gastrointestinal tract becomes hindrance for further feed intake (Allen, 1996). Since productivity has a functional relationship with input and output, such physical restrictions may limit the biological gains. It may also create single large total discomfort to the animal in terms of nutrient supply and affecting the animal performance. Henceforth, animal requirements should be fulfilled within the minimal total discomfort while optimizing the nutrient density through supplements. Under practical conditions there are only two choices are available for sheep producer to choose optimum method of nutrient inputs from basal roughage or supplement or combinations of both.

1. Choose the feed combinations that lead to the inputs of lowest cost of producing a given level of biological output.
2. Choose the feed resource that yield the maximum level of biological output with a given level of input and their expenditure.

The first choice is closer to the existing field practice where cost of inputs is very low and sheep are reared unto the time it attracts the buyer. Under such circumstances, principle determinant factor for profits is the availability of low cost feeds but not the out put level. As long as input constraints are not aroused in terms of availability and cost, the producer has least risk in holding the flock to maximize net returns. In contrary, the second choice represents intensive system of feeding where output is targeted than the cost of feed inputs. This system of feeding follows to achieve the targeted output in stipulated time while keeping control on the input expenditure. The major objective behind such practice is net gains. Irrespective of choice stated above, the purpose of optimization is to enhance output efficiency in following terms.

1. *Total Biological Productivity (TBP):* Increasing the nutrient intake to attain higher nutrient balance and maximum productivity. In simple terms, this is unlimited feeding without consideration to efficiency (output and input ratio).
2. *Average Biological Productivity (ABP):* Providing the feed inputs till the average output is higher. In simple terms, this is the mean value between output and input.
3. *Marginal Biological Productivity (MBP):* Feeding the animal till there is increase in additional unit of total weight gain or mutton production (or any other biological yield) from the addition of every one unit of feed input. It takes into account additional improvement in the weight gain or mutton production from every one gram of increase in feed input. In simple term, this is fractional increase in output and input.

Intricacy in three measures of optimization can be understood from the illustration given below using the concentrate supplementation to pregnant ewe weighing 30 kg (12.82 kg of metabolic body weight).

Table 2. Total, average and marginal metabolizable nitrogen (MN) and energy retention (ER) on the basis of Kg $W^{0.75}$.

Conc. Suppl. (g/d)	Total MN (g)	Total ER (MJ)	Average MN (mg/g conc.)	Average ER (KJ/g conc.)	Marginal MN (mg/g conc.)	Marginal ER (KJ/g conc.)
0	**-0.03**	-0.06	(-0.03/0)*1000 = 0.00	(-0.06/0)*1000 = 1.00	0	0
100	**0.02**	0.30	(0.02/100)*1000 = 0.20	**(0.30/100)*1000 = 3.00**	(100-0)/(-0.03-0.02)* = 0.52	**(100-0)/(-0.06-0.30)* = 3.64**
200	0.05	0.53	(0.05/200)*1000 = 0.27	(0.53/200)*1000 = 2.66	(200-100)/(0.05-0.02)* = 0.35	(200-100)/(0.53-0.30)* = 2.32
300	0.14	0.73	**(0.14/300)*1000 = 0.47**	(0.73/300)*1000 = 2.42	**(300-200)/(0.14-0.05)* = 0.87**	(300-200)/(0.73-0.53)* = 1.94
500	0.15	1.07	(0.15/500)*1000 = 0.29	(1.07/500)*1000 = 2.14	(500-300)/(0.15-0.14)* = 0.03	(500-300)/(1.07-0.73)* = 1.72
700	0.12	0.71	(0.12/700)*1000 = 0.17	(0.71/700)*1000 = 1.01	(700-500)/(0.12-0.15)* = -0.15	(700-500)/0.71-1.07)* = -1.80

* value multiplied by 1000 to express the units in either mg or KJ
Values in **bold** indicates maximum value of the respective column parameter

TBP does not necessarily increase with increasing levels of concentrate supplement. This illustrates an important principle 'the law of diminishing marginal returns' and applicable even to the biological system. The total biological products of metabolizable-N or energy retained were higher when concentrate supplement was 500 g/d. Effective weight gains in lambs have been observed with restricted level of diet compared to *ad libitum* feeding (Fluharty and McClure, 1997). The above illustration also reveals that the ABP and MBP for metabolizable-N was however, maximum when concentrate supplement was 300 g/d. Similarly ABP and MBP were maximum in case of energy retention when concentrate supplement was 100 g/d. Some trade-offs are necessary in biological system between competing parameters looking into the objective of the production requirement. For example, mutton production requires optimum yields in metabolizable nitrogen utilization for lean tissue accretion. Hence, concentrate supplement of 300 g/d could be considered as the critical level and 100 g/d could be considered as the threshold level for optimum performance. The importance of each basic approach and determining the optimum level can be visualized from the figure 3. The optimum level is the level where both MBP and ABP are at their peak. Though TBP reached its maximum only at 500 g/d of concentrate supplement, ABP is decreasing while MBP is at its lowest yields, which means that gains were decreasing per unit increase in input. Feeding *ad libitum* though meets nutrient requirements of the sheep with higher total biological product, average and marginal biological yields would not necessarily at their peak. Feeding of concentrate supplements is wastage when fed beyond the levels where ABP and MBP are highest. Under prevailing field conditions, sheep farmers are raising the sheep on free common grazing resource without any supplements. When there are no costs involved in feed inputs, the question of average or marginal biological yields does not arise. Even if feed inputs are free of cost, nutrients utilization in the system can not be free of energy requirement. In the system point of view also, feeding the animals beyond 300 g/d concentrate supplement is not viable.

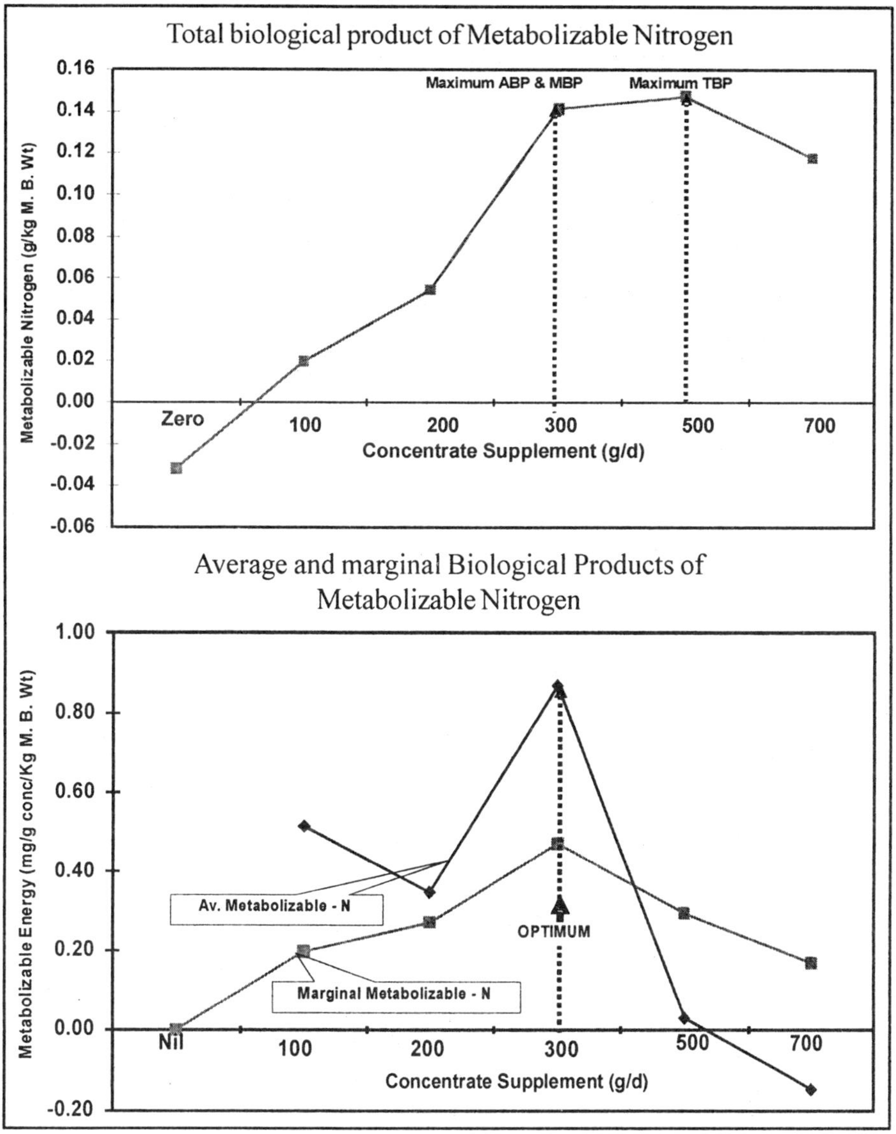

Fig 3. Relationship between TBP, ABP and MBP

Synchronization of Nutrient Utilization

From the above example, feeding animal 300 g/d would optimize production but not necessarily minimize the futile or obvious production losses. These losses can be called as obligatory process waste (OPW). Scientifically one should try to keep OPW at minimum level without loosing the production efficiency. This concept leads to genesis of synchronization or reducing asynchrony (antonym to synchrony). The major objective of the synchronization is increasing the production efficiency while reducing the wastivity. It is simply a manipulation of nutrients in the given optimum quantity of inputs as narrated below based on the above example of feeding 300 g/d of optimum quantity of concentrate supplement. Supplementation of sheep with 300 g/d concentrate supplement would provide total nitrogen of 1.52 g/kg $W^{0.75}$ and result in yielding metabolizable nitrogen of 0.14 g/kg $W^{0.75}$ that is actually available for various production processes. Efficiency of metabolizable nitrogen yield from the total nitrogen is 0.14/1.52 (a ratio between output/input) = 9.2 per cent. This means a balance of 1.38 g/kg $W^{0.75}$ is unutilized in many different ways and termed as OPW. Further improvement in nitrogen utilization may be possible if unutilized and OPW are minimized. Three options are available to execute such improvement.

1. *Option-1*: Reduce intake while maintaining same level of output so that with unit decrease in input, OPW also decreases and productivity improves per unit of input. For say, if we reduce the total nitrogen input from 1.52 to 1.25 g/kg $W^{0.75}$ while keeping the metabolizable nitrogen output to 0.14 g/kg $W^{0.75}$, then the productivity efficiency would be 11.2 Per cent. It is a way of improving the input efficiency through wastivity management.

2. *Option-2*: Keep the intake constant but enhance the output. So that, more input is diverted towards the product output and wastivity would complimentarily reduce. For say, keep the total nitrogen input to only 1.52 g/kg $W^{0.75}$ but improve the metabolizable nitrogen yield to 0.17 g/kg $W^{0.75}$, then the productivity efficiency would be 11.2 per cent. It is a way of improving the process efficiency. OPW is reduced and diverted towards the output by manipulating the system process.

3. *Option-3*: Increase out put in greater proportion than increase in input so that there will be enhancement in the output efficiency. It is a very logical option derived by combining both first and second options. However, this is only applicable in case where one is aware of the genetic potential of the animal to produce more. For example, if a producer actually feeding 250 g/d of concentrate supplement. If the level of concentrate supplement increased to 350 g/d, then total nitrogen input is 1.60 g/kg $W^{0.75}$. At any cost, the metabolizable nitrogen output should be greater than 0.14 g/kg $W^{0.75}$. Even if the metabolizable nitrogen is increased to 0.1475 g/d, still

excess feeding of concentrate supplement is of no use because production efficiency remained to be 9.2 per cent only. So, increase in output is not in a greater proportion than input (both are increased by only 5 per cent). There is only burden of excess consumption of resources and costs involved in it. One can opt for additional increase in concentrate supplement (350 g/d) only when there is any improvement in output in greater proportion (say > 9.2 per cent in this example) than additional input of 100 g/d of concentrate supplement.

Each nutrient or input has inter se relationship with other nutrient or input in synthetic processes in biological system. There is no biological product that is synthesized exclusively from single nutrient or input such as carbohydrate, fat, protein, mineral, vitamin, enzyme, reducing substrates ATP, ADP, AMP, NADP etc. Optimum level of concentrate supplementation appears to meet total nutrient requirement of animals as a whole but cannot be ascertained for synchronization in the availability of some or all nutrients in the feed. Different nutrients in a feed such as protein, soluble carbohydrates, fat, minerals, cell walls etc., would have varying rates of degradation (Smith *et. al.,* 1971). Even rate of degradation of similar nutrients, for example, energy and nitrogen, are also different among different feed stuffs (Sinclair et al., 1993). Such differences among feed stuffs and nutrients within the feedstuffs lead to uneven supply of nutrients for synthetic process at different time intervals in 24 h of biological cycle. Vast number of observations from *in sacco* nylon bag degradation and *in vitro* gas production technique lead to an opinion that earlier concept of fulfilling energy and protein requirement in the total diet has little utility unless there is synchronization among the nutrients availability at tissue level for synthesis of biological product at any given point of time in a day. This is because that synthetic processes such as hormone, enzyme, muscle, milk, conceptus production are not time specific but continuous processes in the animal system. The older concept of 'Nutritive Ratio' proposed by the Morrison (1961) only indicate digestible protein to energy ratio in the diet as a whole but does not consider in fractional terms. Synchronization of energy and protein in sheep reported to increase microbial protein production in rumen compared to diets those are fulfilling energy and protein requirements but their dietary constituents have distinct degradation rates (Sinclair *et al.,* 1995). Looking into the importance of microbial protein production in the sheep due to limitation of endogenous nitrogen recycle (Djouvinov, 1999), synchronization of nitrogen and energy in rumen is very essential in sheep as major transformation of dietary nutrients are taking place in this part of the digestive tract. Microbial flow from the rumen can meet 50 % or more of the amino acid requirement of these animals (Orskov, 1982). Such improvement will consequentially have positive impact on the production processes at tissue level too.

Table 3. Protein and carbohydrates degradation characteristics and rate during pre-gastric fermentation.

Degradation rate	Rumen available rate	Protein	Carbohydrates
Instantaneously soluble	4 to 20% $hour^{1}$	Ammonia nitrogen, non protein nitrogen, soluble protein, synthetic amino acids, amorphous urea, soya cake, groundnut cake etc.	Sugars, some starch, fructans, molasses etc.
Very rapidly degradable	1 to 30% $hour^{1}$	Salurea, urea in blocks, groundnut cake, soya cake etc.	Some starch, pectin, barley, jowar, ragi etc.
Rapidly degradable	1 to 1 5% $hour^{-1}$ and variable	Mustard cake, linseed cake, mahua cake etc.	Maize, wheat bran, starch from by-products, fruit pulp etc.
Medium degradable	1 to 8% $hour^{-1}$ and moderately variable	Cottonseed cake, coconut cake, proteins from other by products subjected to heat in their processing.	Silage, hay, any leguminous and non-leguminous green fodder, wheat bran etc.
Slowly degradable	1 to 5% $hour^{-1}$ and highly variable.	Fish meal, bone meal and other protein source from by-products.	Mature forages, low quality roughages, lengthy swards, rice, bran, tree pods etc.

Methods of synchronization of nutrients in rumen

The simplest method of synchronizing the nutrient utilization is increasing the frequency of nutrient supply by dividing the total diet into smaller parts (Chaturvedi et al., 1973). Asynchrony will reduce with increasing number of intervals with lesser time gap between any two consecutive intervals. An interval of 2 to 4 h between two meal would be more effective in creating synchrony in closed system of feeding. This method is very effective on productive animals to maximize total intake and reduce nutrient wastage (Owen, 1981). In general, synchrony in protein and energy supply from forages decreases with increasing maturity. Thickness of cell walls increases with maturity and become a hindrance for effective hydration in rumen which is prerequisite for microbial colonization and initiation of fermentative digestion (Forsberg *et al.*, 2000). Asynchrony in nutrient utilization is greater in closed type of feeding than open type. When animals are poised with feed options as in open type feeding, they avoid asynchronous diets which deliver nutrients at different rates and result in temporary imbalance. Synchronization of nutrient in diets requires information about the degradation kinetics of diet constituents. Nutrient degradation of one feed component should be complimentary to another feed component so as to maintain steady rate of nutrient supply to the production process. In general, degradation of concentrate ingredients would complete between 12-24 hrs while degradation of roughages would reach peak between 24-48 hrs and there after increase with decreasing rate till 72 h. Fermentation gap between concentrate ingredients and roughage is 2-3 folds. Nutrient release would be rapid form three classes of concentrate ingredients in the order or oil cakes followed by brans followed by grains (Srinivas and Gupta, 1994). About 80 percentile of nutrient will release from oil cakes within 10 to 12 h, while it takes 14 to 18 h in brans and 18 t0 21 h in grains. Feeds which have higher undegradable nutrients will have faster rate of nutrient release in initial stages and reach plateau sooner. Comparison among protein and carbohydrate degradation from concentrate feeds is given in table 3. Formulating the concentrate supplements with same characteristic source of protein and carbohydrates will reduce asynchrony and enhance productivity (Srinivas, 1999).

Conclusions and Recommendations

Traditional method of feeding based on the thumb rules or unlimited feeding can only workout when no costs are involved in inputs and outputs. Even inputs are available at free of cost (like in prevailing free grazing system), output is not free from biological costs. Traditional methods need to be replaced with scientific methods both in prevailing field or trade oriented feeding systems. Although unplanned feeding of animals to physical maximum appears to yield more returns, it affecting the productivity efficiency. Long term effect of such feeding to the sheep may cause adaptation plasticity to the inefficient production process. The invincible alternative lies in scientific method of synchronizing the nutrient supply at optimum quantities of input and production out put. It is recommended method because; 1. It takes care of minimal total discomfort to the animal as per the ethical strictures of animal welfare and environment, 2. It emphasizes optimization where unit increase in returns is greater than unit increase in input to sustain in competitive trade for animal products and 3. It reduces wastivity of inputs through synchronization and improves production efficiency to take care the resource shortage. Though the elaborative explanation given in this lecture for better understanding may appears to be cumbersome, in reality computer aid and application of software make it very simple.

REFERENCES

Allene, M.S. 1996. Physical constraints on voluntary intake of forages by ruminants. *Journal of Animal Science* 74, 3063-3075.

Agricultural Research Council (ARC) 2000. The nutrient requirement of ruminant livestock. Commonwealth Agricultural Bureaux. Farnham Royal, England.

Conrad, H.R., Pratt, A.D. and Hibbs, J.W. 1964. Regulation of fed ntake in dairy cows. 1. Change in importance of physical and physiological factors with increasing digestibility. *Journal of Dairy Science* 47, 54-62.

Chaturvedi, M.L., Singh, U.B. and Ranjhan, S.K. 1973. Effect of alkali treatment of wheat straw on feed consumption, digestibility and VFA production in cattle and buffalo calves. *Indian Journal of Animal Sciences* 43, 677-82.

Djouvinov, D.S. 1999. Effect of dietary protein degradability on microbial protein supply to the small intestine of sheep. *Bulgarian Journal of Agricultural Science* 5, 435-441.

Demment, M.W. and Van Soest, P.J. 1985. A nutritional explanation for body size patterns of ruminant and non-ruminant herbivores. *American Naturalist* 125, 641-672.

Demment, M.W., Peyraud, J.L. and Laca, E.A. 1995. Herbage intake at grazing: a modeling approach. In: *Recent Developments in the Nutrition of Hebivores* (Eds. M. Journet, E. Grenet, M. H. Farce, M. Theriez and C. Demarquilly). Proceedings of the IV International Symposium on *"The Nutrition of Herbivores"*, Clemont-Ferrand (France), September 11-15, 1995. Publ. Institut National De La Recherche Agronomique, 147, rue de l'Universitè - 75338 Paris Cedex 07.

Fluharty, F.L. and McClure, K.E. 1997. Effects of dietary energy intake and protein concentration on performance and visceral organ mass in lambs. *Journal of Animal Science* 75, 604-610.

Forbes, J.M. 1999. Minimal total discomfort as a concept for the control of food intake and selection. *Appetite* 33, 171.

Forbes, J.M. 2001. Consequences of feeding for future feeding. *Comparative Biotechnology and Physiology 128, 461-468.*

Forsberg, C.W., Forano, E. and Chesson, A. 2000. Microbial adherence to the plant cell wall and enzymatic hydrolysis. In: Ruminant Physiology: Digestion, Metabolsim, Growth and Reproduction (Ed. P. B. Cronje). CABI Publishing, CAB International, Wellingford, Oxon OX10 8DE, UK. pp 79-97.

Gordon, J.G. and Tribe, D.E. 1951. The self-selection of diet by pregnant ewes. *Journal of Agricultural Science* 41, 187-190.

Hutchings, M.R., Kyriazakis, I., Gordon, I.J. and Jackson, F. 1999. Trade-offs between nutrient intake and faecal avoidance in herbivore foraging decisions: the effect of animal parasitic status, level of feeding motivation and sward nitrogen content. *Journal of Animal Ecology* 68, 310-323.

Karim, S.A., Santra, A. and Singh, V.K. 2004. *Lamb feeding for mutton production*. Bulletin. Publ. Central Sheep and Wool Research Institute, Avikanagar, India.

Kyriazakis, I. and Oldham, J. D. 1997. Food intake and diet selection in sheep: The effect of manipulating the rates of digestion of carbohydrates and protein of the foods offered as a choice. *British Journal of Nutrition* 77, 243-254.

Morrison, F.B., 1961. *Feed and Feeding, Abridged*. 9th Edition, The Morrison publishing company, Clinton, IOWA. Pp 31-32.

Orskov, E.R. 1982. *Protein nutrition in ruminants*. Academic Press Inc. (London) Ltd. UK. Pp 160.

Owen, J.B. 1981. Complete-diet feeding of dairy cows. In: Recent Developments in Ruminant Nutrition (Eds. W. Haresign and D.J.A. Cole), Butterworths, London, UK. pp. 313.

Rutter, S.M., Orr, R.J. and Rook A.J. 2000. Dietary preference for grass and while clover in sheep and cattle: an overview. In: *Grazing Management* (Eds. A.J. Rook and P.D. Penning), British Grassland Society, Reading, UK. Pp 73-78

Sinclair, L.A., Garnsworthy, P.C., New Bold, J.R. and Buttery, P.J. 1993. Effect of synchronizing the rate of dietary energy and nitrogen release on rumen fermentation and microbial protein synthesis in sheep. *Journal of Agricultural Science, Cambridge* 120, 251-263.

Sinclair, L.A., Garnsworthy, P.C., New Bold, J.R. and Buttery, P.J. 1995. Effect of synchronizing the rate of dietary energy and nitrogen release in diets with a similar carbohydrate composition on rumen fermentation and microbial protein synthesis in sheep. *Journal of Agricultural Science, Cambridge* 124, 463-472.

Smith, L.W., Goering, H.K., Waldi, D.R. and Gordon, C.H. 1971. *In vitro* digestion rate of cell wall components. *Journal of Dairy Science* 54, 71-76.

Srinivas, B. 1999. Synchronizing nitrogen and energy input for optimal rumen fermentation. *Indian Journal of Dairy and Biosciences* 10, 123-131.

Srinivas, B. and Gupta, B.N. 1994. Rumen degradable protein content and gas production on rumen fermentation of some concentrate ingredients and their relationship. *Indian Journal of Animal Nutrition* 11, 171-175.

Srinivas, B., Umadevi, V. and Saran, S. 2004. Locational impact of veterinary clinics on the knowledge of livestock owners on dairying and milk production. *Journal of Dairying, Foods and Home Science* 23, 79-86.

Weston, R.H. 1996. Some aspects of constraint to forage consumption by ruminants. *Australian Journal of Agricultural Research* 24, 387-397.

❑❑❑

11

Nutrition of Small Ruminants Under Arid Environment

TRIBHUWAN SHARMA
College of Veterinary and Animal Science, Rajasthan University of Agriculture, Bikaner-334 001, Rajasthan

Introduction

Small ruminants form valuable economic and ecological niche in agricultural set up through out the arid and semiarid regions of developing and under developed countries. They are numerically important domesticated animals which contribute meat, milk, skin, fibre and manure, besides, their role as a sole or subsidiary source of livelihood for large number of rural masses which are below poverty line i.e. small/ marginal farmers or landless labourers. By and large, the importance and extent of the contributions of small ruminants to arid and semiarid zone are not given due weightage. Over and above this, rearing of these species are mostly in the hands of weaker sections of the society, as a consequence of which these precious genetic resources best suited to arid and semiarid region are continue to be backward and neglected.

Attempts to improve the relatively low level of contribution from sheep and goat, do not match their potential importance, the contribution to nutritional insecurity and livelihood of the poor of the poorest in rural areas. Infact, there is a dire need to increase the contribution to food production from components of the livestock sector, in the face of many demand driven factors such as population growth, urbanization, inability of current supplies to match requirements and changing customer preferences. This situation in tropics and more precisely in arid regions is further compounded by inefficiencies in animal production systems and natural resource management that can respond in this regard.

The arid zone of India covers about 12% of the world's hot arid region. Rajasthan contributes 61%, Punjab and Haryana 09% and Andhra pradesh and Karnataka 10%. The Indian arid zone is by far the most populated arid area in the world and even till date despite extreme degradation of grazing lands, the major stay of rural economy is grazing based livestock production system.

Of the total land of the world (13, 077 m ha) only 10.5% is arable and the rest is under pastures, forests, wastelands etc. The Asian region has 20.50% of the total and 31% of the arable land. The Asian countries with arid and semi arid areas have 15.1% of the total land and 25.1% of the arable land. This region contains 31.3% of the total livestock population, 27% of the sheep population and 49% of the goat population of the world i. e the total livestock density per unit non arable area as well as density of the small ruminants per unit of non arable area is higher in this region than the world average (FAO, 1993). It is thus clearly understood that the cause and effect of all these retrogressive changes in the grazing lands and more commonly on the ecosystem emanate from the enormous increase in human population followed by rise in livestock population far beyond the ability of land to sustain and provide for. Small ruminant population as per livestock census (2003) amounts to 137.51 million with an increase of 49% in sheep population and 155% in goat population compared to population of these species in 1951 in the country. Contrary to this, the total area available for grazing has decreased to 43.3 m ha from 82.1 m ha. In spite of aforesaid detailed adversities, small ruminant farming is still major component of arid agriculture as it makes farming system more sustainable and economically viable. Probably due to this reason the arid region is better ranked than the rest of the regions in terms of meat (720%) and fibre (725%) (Sharma, 2005). In addition, to provide quality nutrition and industrial raw material (wool, leather etc.), the contribution of small ruminants in terms of stability, round the year employment opportunity, equity distribution and being technologically simple (Choudhary *et al.*, 1999) can not be underestimated. Infact, the role of small ruminant in arid agriculture is expected to increase manifold provided inputs in terms of feed and care, an animal receive during life time are adequately provided.

Small Ruminant Production Systems

In India, grain based small ruminant rearing has not carved its place and thus these species as such do not compete with human and organized dairy sector for feed. Although a number of small ruminant production systems are in practice varying from country to country and region to region with in a country. In India, the prevailing systems are extensive, semi intensive and intensive and in arid and semi arid area small ruminants are reared following sedentary or a migratory system under extensive range management pattern. Sedentary system mainly includes stall feeding following cuts and carry system and/or extensive grazing system, whereas, migratory system follows long distance migration in dry plains. 70-80% of sheep flock in arid region are reared under migratory system which is still a effective system for small ruminant rearing in arid and semiarid zones of the country, experiences stress due to climatic variability and increasing requirements of the livestock and human population. Over exploitation of natural resources by increasing biotic pressure is though throughout the country but in this region the situation is desperate. The population pressure gets further compounded by draughts causing extreme pressure on vegetation through overstocking. The existing grazing pressure in arid area is very high i.e. 1 to 4 ACU/ha, against the normal 0.2 to 0.5 ha (Beniwal and Singh, 2005a), it has resulted in alarming scarcity of feed resource availability, posses problem in meeting nutritional needs of small ruminants under extensive system. Studies carried out indicate that in Rajasthan 0.5 million sheep are in permanent migration following established migration routes and seasons and are not brought to their homestead at any time of the year. About 1.0 million small ruminants from western districts of Rajasthan migrate for 6-9 months to the neighbouring states. Most flocks begin migration between October and February and return before the onset of the monsoon. Goat flocks are normally non migratory except in case of mixed flocks of sheep and goat.

Under the prevailing circumstances in extensive system of rearing, the small ruminants largely depend on range lands and grazing lands during the monsoon and later on stubble grazing on cropped land or forest land. Sheep and goat are exposed to barren and marginal lands of arid region where the usefulness of grazing is much less than visualized and sometimes animals have to cover long distances before getting access to any pasture. These rangelands have never been protected, reseeded or properly managed and thus are extremely inadequate for supply of nutrition round the year (Singh, 2007) with stocking rate around one small ruminant/ha or even less except during monsoon period. The dry matter yield per hectare have been recorded to be 4.92 and 1.36 q/ha from the common grazing lands of semi arid region in monsoon and winter season. Whereas, yield of stubbles after harvesting of kharif crop provide dry matter from standing biomass to be 20.4 q/ha.

Fortunately under migratory system, the small ruminant also get the opportunity to graze or browse on vegetation rich in nutrients from fallow lands of neighbouring areas of high rainfall, which help them meeting the requirements. Earlier, the migration of one to two months has been a common practice in entire arid and semiarid region. But now days, migration is very much related with the rainfall and the frequency and duration are more in famine and draught years. With the passage of time the migratory system which is the prime system for small ruminant production in arid and semiarid area is also facing problems on account of increase in area of crop cultivation, deforestation and rising incidences of violent confrontation with localities. In addition to these, there is a long list of vagaries, alarming that this viable small ruminant production system needs immediate attention for facilitating smooth functioning of migration every year on same route with provision of all basic requirements for migratory flocks as well as well as their caretakers. In cold arid regions transhumance system is followed for rearing of sheep. The flocks migrate from foot hills to high alpine ranges on typical annual migration route.

Under the compulsions of the constraints of grazing lands for raising of sheep, the desirable target production never seems to be achievable due to extremely degraded rangelands and shrinking of grazing lands. Thus, semi intensive system i.e: grazing/browsing plus supplementation seems to be best resort for sustainable small ruminant production in arid and semi arid regions. It is combination of free range grazing and stall feeding developed to take care of the limitations of extensive system. The animals are subjected to grazing in groups for about 5-6 hours and than supplemented with concentrate mixture, crop residues, green and dry fodder, top feed etc. These systems surely ensure better nutrition than that under extensive system in arid and semiarid regions. Further this system is of particular importance to goats performs well when browsed on variety vegetation and supplemented with concentration mixture. The total confinement and stall feeding of goats has been observed to be detrimental. The operational procedures in feeding and management are to be streamlined for adaptation in different systems while taking care of the cost benefit ratio into account. This system seems to be inevitable but people will adopt it extensively only when it is profitable and economical. The growth rate should be good giving higher outputs.

Ad lib supplementation of concentrate mixture, hay or green to kids between 91 to 180 days age, with browsing resulted in an increase of 44.8% in pre slaughter weight, 65.1% in carcass weight and 14.8% in dressing over the browsing alone. The kids when fed above ration *ad lib* under stall fed condition showed an increase up to 21.4% in pre slaughter weight, 39.7% in carcass weight and 14.8% in dressing over the browsing alone (Singh, 2007). The Sirohi does grazing/browsing for 8 hours and supplemented with 150, 300 and 450 g/h/d concentrate mixture during last 45 days of pregnancy and first 150 days of lactation, lost body weight when only grazed or maintained with only supplementation of 150 g concentrate mixture and gained their live weight when supplemented with 300 g and 450 g concentrate mixture during pregnancy. The same does lost weight when only grazed or supplemented with 150 g concentrate but gained live weight when supplemented with 300 and 450 g concentrate during lactation. The milk production was improved by 29.3% and 67.0% and preweaning growth of kids by 17.2 and 35.0% with the 300 g and 450 g supplementary level (Singh, 2007). Based on such findings, semi intensive system of grazing/browsing plus supplementation for different categories of sheep and goat maintained for wool, meat and milk in arid and semi arid zones was appreciated to be appropriate and packages for different categories of sheep and goat were developed.

Intensive system of sheep rearing is not common in country, except in male lambs reared with single objective to achieve maximum body weights at an early age. However, for meat and milk production from goats in urban and semi urban areas stall feeding is preferred. In intensive system, small ruminants are reared on complete stall feeding on cultivated fodder or complete feeds or crop residues or agricultural byproducts or grazing on highly developed pastures with stall feeding. This system requires high labour and capital investment and under the prevailing circumstances this system in arid region could be suitable for intensive meat production. It removes pressure from the community grazing lands besides, providing better milk, wool, growth and carcass quality. A growth rate of 92 and 100 g has been observed in Malpura and Sonadi lambs maintained on feed lot from 91 to 180 days of age (Singh, 2007). Package of practices for improving meat, wool and milk production in sheep and goat have been developed at CSWRI and CIRG.

Feed and Fodder Resources and Their Nutritive Value

According to 1997 statistics out of total 297.3 m ha land area only 3.8% area is classified under permanent pasture and grazing lands (FAO, 1993). Over and above this, what so ever small portion of community grazing lands attached to the villages, forests, marginal and wastelands, we have are in essence, private property for the community which had its own norms and established management systems among the users have been seriously eroded with the change in the political structure of the country and the laws making them the responsibility of the "panchayats" to manage. Further the inevitable continuous deforestation and removal of trees is drastically deteriorating the land fertility. About 175 million hectare of land have been classified under various kinds of wastelands and area under agricultural crop production is only 145 million hectare (Pathak, 1994). Further, in arid and semiarid region the situation regarding feed resource availability is more complex due to low erratic rain fall and the nature of soil of the region (light textured, deficient in organic matter and have a tendency to salinization). The natural potential of range and pasture land falls far short of the feeding and nutritional requirements of the animals. There are wide variations in feed availability and its nutritive value among years and seasons. The natural range/ grazing lands have been over exploited and subjected to serious problems of erosion. Even though, natural grasslands have an important place among the forage resources for the livestock population in arid and semi arid regions. The grazing lands support adapted native vegetation comprising of predominantly annual and perennial grasses (*Lasirus sindicus, Cenchrus cilliaris, Cenchrus setigerus* etc.), legumes (Indigofera spp.) forbs (*Tribulus pentantrus, Tribulus, terristris, Corchorus* spp.etc.) and woody perennials (*Prosopis cineraria, Ziziphus nummularia, Calligonum polygoinoides, Acacia* spp. etc.), used as forage by livestock as well as wild animals. In Rajasthan, vast area is available for grazing but 70% belongs to poor to very poor range condition class. The remaining 13% belongs to the good class, 14%, the fair class and only 2-3% could be assigned excellent class status. This indicates the severity of range degradation due to increasing population pressure of livestock (Shankar *et al.*, 1988). Suitable technologies for the improvement and management of these rangelands/grazing lands of arid and semi arid area through introduction of improved pasture, silvipasture and the farm forestry system increasing carrying capacity are available (Faroda, 1998). The climax vegetation of the hot rangelands in arid region is never achieved due to intensive and continuous grazing pressure. On the other hand, the climax vegetation of the protected rangelands is not effectively utilized in livestock grazing due to short spell of monsoon, fast vegetation growth, early flowering and rapid withering of vegetation due to environment related changes. The lush green monsoon herbage cannot be effectively harvested and conserved due to prevailing wet season. However the animals usually have access to adequate quantity and quality of vegetation from July to October (Shinde et al., 1996 and Singh and Karim, 1996). The growth cycle of vegetation in arid and semiarid environment is completed in less than 90 or in between 90 to 180 days and for rest of period of year the livestock survive on sparse vegetation. Forage production from different desert rangelands and effect of different fodder tree species (Karim, 1999) on forage production of rangelands have been worked out (Mann and Sharma, 2005).

The forage availability from semi arid community grazing land is recorded to be 1.47, 1.57 and 0.82 q/ha in monsoon, winter and summer season, respectively (Karim, 1999). Considering the annual biomass yield of natural rangelands to be 1.39 q/ha the carrying capacity have been estimated to be 0.3 ACU /ha round the year (Karim, 1999). The carrying capacity of the arid rangelands have also been estimated as hardly 0.2-0.5 ACU/ha, while the grazing intensity on the land far exceeds the capacity i.e. 1-4 ACU/ha (Singh and Rathore, 2005). Under this excessive grazing pressure the rangelands are under going ecological changes. The preferred plants are being eliminated and in its place the pasture is getting dominated by inedible plant species. Survival of livestock under such poor feed availability is very difficult. However, monsoon herbage is a potential resource to sustain the livestock productivity. Karim et al. (1997) reported that community grazing land in a semi arid location is adequate to sustain livestock during monsoon and winter season while it was grossly inadequate to maintain them during summer season. The forage yield of semi arid and arid rangelands is quite variable depending on soil type and agrometeriological conditions. The protected community rangelands has DM yield of 7-12 q/ha which is sufficient to maintain 2 sheep or 0.4 ACU round the year (Patnayak and Karim, 1994) while those subjected to grazing have biomass yield of 4.59 q/ha (Karim and Shinde, 2007). The average biomass yield of *Cenchrus ciliaris, Cenchrus setigerus* and *Lasirus sindicus* grass pasture range were 45, 40 and 60 q/ha in arid region (Beniwal and Singh, 2005).

It has also been estimated that 60% of the feed available for sheep, goat and camel in arid region comes from top feed only (Beniwal and Singh, 2005b). Khejri (*Prosopis cineraria*) is an important fodder tree in arid and semi arid region, which grows in rangelands and cultivated fields without any determent to the associated ground cover. The tree is lopped for its high protein (17.5%) leaves and studies

Table 1. Forage production (t ha^{-1}) from different desert rangelands

Locations	Grass cover	Forage production (t ha^{-1})
Arid		
Lawan (Jaisalmer)	*Lasiurus eleucine*	1.2
Chandan (Jaisalmer)	*Lasiurus sindicus*	2.1
Jaisalmer	*Lasiurus cymbopogon*	1.8
Beechwal (Bikaner)	*Lasiurus eleucine-Aristida*	2.4
Samadri (Barmer)	*Cenchrus eleucine-Aristida*	1.4
Semiarid		
Bhopalgarh (Jodhpur)	*Cenchrus aristida*	2.0
Borunda	*Cenchrus eleucine-Aristida*	1.4
Bisalpur (Sirohi)	*Dicanthium aristida*	3.2
Jadan (Pali)	*Cenchrus aristida-Sporobolus*	1.9
Palsana (Sikar)	*Cenchrus eleucine-Aristida*	1.6

(Shankar and Gupta, 1992)

have shown that complete lopping of Khejri tree gives an average yield of 58.7 kg/tree (Beniwal and Singh, 2005). Similar results have also been reported by Mann *et al.* (1989) from the lopping management studies. Rohida (*Tecomeea undulata*), Jal (*Salvadora oleoides*), Siris (*Albizzia lebbek*), Bordi (*Zizipus numuleria*), Mopane (*Olophaspermum mopane*), Ardu (*Ailenthus excelsa*) etc. are important top feed plants of arid and semi arid region contributing significantly to biomass availability for sustainable livestock production.

Table 2. Nutritive value of some tree leaves in goats

Top feed	DCP(%)	TDN(%)
Pala (Zizyphus numularia)	7.67	49.97
Ardu (Ailanthus excelsa)	11.72	55.42
Neem (Azadirachta indica)	9.43	48.74
Jal (Salvidora persica)	7.28	39.49
Sares (Albizia lebbek)	12.01	59.60
Tali (Dalbergia sisso)	5.15	56.62

(Sharma, 2004)

Table 3. Nutritive value (%) of fibrous crop residues in goats

Crop residues	DCP	TDN
Sorghum straw	1.8	48.1
Maize straw	3.1	50.1
Cotton straw	2.5	50.9

Contd...

Table 3. Contd...

Sunflower straw	2.4	47.7
Heteropogon contortus grass	1.8	50.5
Sehima nervosum grass	1.6	55.4
Cotton seed hulls	1.4	56.9
Groundnut hulls	3.0	23.3
Maize cobs	1.9	47.0
Sugarcane bagasse	1.7	57.6

(Reddy and Reddy, 1996)

As a result of extensive studies, silvipasture in arid and semiarid regions has been recorded to be the most effective land use system, in which trees and shrubs are grown in association with pasture grasses and legumes. This system has proved beneficial to overcome the problem of short forage supply during the lean period (Nov-June). The woody shrubs and trees in the system serve as an insurance against fodder scarcity, particularly during drought. Silvipasture system compared with other land uses in the marginal and submarginal lands provide quality forage for a larger period of the year (Shanker and Gupta, 1992).

The silvipasture with various tree combinations has biomass yield ranging from 20-25 q/ha adequate to maintain 6 sheep or 1.2 ACU/ha round the year. The carrying capacity of the semiarid *Cenchrus* dominated silvipasture indicated that annual biomass yield was 22 q/ha sufficient to maintain 1.1 ACU round the year (Shinde, 1996). Despite of most judicious utilization of all available feed resources in critical arid and semiarid area short distance and neighbouring state migration during summer month is only resort (Spare, 1976; Dave, 1977). The animals return to their native tract with the onset of monsoon.

Grazing Behaviour

Sheep and goat follow rhythmic periodicity in grazing pattern, graze actively during morning and evening hours. Forage quality particularly fibre content of forage influence the food utilization behaviour. Rumination rate (chew/bolus) in sheep increased from 62-67 with decreased mastication rate from 69-63 due to increase in ADF content of diet from 42-53%. Rumination rate increase from 45 chew /minute in monsoon, when diet contain 68% shrub and 32% grass to 70 chew/minute in summer, when diet contain shrub alone having high fibre (Karim and Shinde, 2007). Grazing hours of goat on pasture is negatively correlated with ambient temperature/relative humidity and forage supply from ranges. Goats have characteristics bipedal stance, which helps in consumption of overhead portion of shrub species (Bhatta *et al.*, 2002). This typical behaviour of goat render them to get higher protein in diet despite of appreciable decrease in protein content of ground vegetation

Herbivorous with broad and flat muzzle have less feed selectivity than species with narrow mouth and incurred incisors arcades. Sheep have a higher ability to sort preferred plant components from others. Sheep while on grazing lands travel long distances and explore wider areas to increase incidence of getting preferred species. Sheep preferred quality nutrients with deterioration of pasture to meet the requirement. The increased selection intensity of crude protein help them to increase % CP in the diet to meet the requirement. Being grazing animal diet of sheep mainly includes grasses and fibres and a little of browse species whereas, being browsing animal diet of goat includes more of browse and little of grasses. Grass is the preference of goats only in the monsoon when it is lush green. Diet of goat contained 76% shrubs and 24% grasses in monsoon, while in winter and summer, diet consisted 100% of shrubs in semiarid rangelands of India (Shinde *et al.,* 2000).

Nutritional Status of Sheep and Goat

The greatest limitation in the rangelands and the natural pasture in the arid and semiarid area is regarding availability of adequate energy throughout the year and adequate protein for more then half of the year. Sheep and goats depend on the native ranges for the main source of forage supply. The deciduous plant species and a heterogenous vegetation type of shrubs with an herbaceous understory are the main component of these ranges. Unprotected rangelands can hardly carry one sheep and by protection the carrying capacity can be increased to 2 sheep/ha and as such there is great prospect of improving rangelands. Based on assessment of seasonal changes in nutrient intake and utilization pattern in sheep maintained on protected rangelands, it was observed that the animals experienced nutritional limitation in the early monsoon and summer while it was adequate in winter season (Shinde, 1996 and Karim and Shinde, 2007). In early monsoon, in spite of access to lush green vegetation, on account of the high moisture content of the biota and gut fill limitation of the animals, the animals are underfed resulting in lower body weight of animals in this season (Sankhyan *et al.,* 1998; Shinde et al., 1998a). Whereas, the animals maintained on unprotected degraded rangelands experience nutritional limitation during winter and summer season (Sankhyan *et al.,* 1999). Goats have ability to maintain intake despite variations in forage supply across seasons due to there grazing behaviour enable them to adapt to wide variety of vegetation. Goats on native ranges consumed 64.09 g/kg $W^{0.75}$ or 2.4% of b. wt. in monsoon when ranges are sufficient with vegetation and intake decreased to 54.05 g/kg $W^{0.75}$ or 2.0 % of b. wt. in summer and winter on account of maturity and deterioration of vegetation (Shinde *et al.,* 1998a).

In general, protein intake of animals from pasture in semiarid regions is just enough for maintenance requirement during rainy season while in other seasons it is below requirement (Karim and Shinde, 2007). Studies taken up to access the plane of nutrition in sheep under different physiological stages revealed the maintenance requirement of sheep to be 27.3 g TDN or 98.5 Kcal ME and 3 g

DCP/ kg $W^{0.75}$. The requirement for wool production were 35 g TDN or 126 Kcal ME and 4.2 g DCP/ kg $W^{0.75}$, while in advance pregnancy and lactation the values were 51.0, 184.0 and 7.0 as well as 53.0, 191.0 and 8.5, respectively (Anonymous, 1998). Sheep consumed 3.44 g/kg $W^{0.75}$ DCP in monsoon, 2.42 g/ kg $W^{0.75}$ in winter and 1.05 g/kg $W^{0.75}$ in summer. Sheep on *Cenchrus* pasture consumed DM 36.5 g/kg $W^{0.75}$ in monsoon, 64 g/kg$W^{0.75}$ in winter and 53 g/ kg$W^{0.75}$ in summer (Shinde *et al.*, 1998b). On native ranges of semi arid regions, higher intake of browse species by goats let them to meet the protein requirement. Goats on native ranges in semi arid region has DCP intake of 4.81, 3.1 and 4.5g/ kg $W^{0.75}$ in monsoon, winter and summer seasons (Shinde *et al.*, 2000). Whereas, protein intake of goat on *Cenchrus* pasture declined from 4.10 g/kg$W^{0.75}$ in monsoon to 2.90 g/kg$W^{0.75}$ in summer (Shinde *et al.*, 1996). It is useful to have browse species for improving the nutrition of goat in *Cenchrus* pasture.

In arid and semiarid region the small ruminants are mainly maintained under extensive system i.e. covering long distance while foraging on ranges. In general small ruminants on ranges spend 60-70% more energy than the stall fed animals. Further, these are also exposed to wide variation in environmental temperature from 8-10°C in winter to 40-45°C in summer and require more energy for thermoregulation (Karim and Shinde, 2007). The maintenance energy requirement of sheep on pasture of semi arid Rajasthan has also been recorded to be 34% more than the stall fed. While grazing on semi arid pasture sheep spent 136.7kJ/ kg b. wt. in winter to 161.1kJ/kg b. wt in summer and 223.7 kJ/kg b. wt in monsoon (Shinde *et al.*, 1998[a]). Contrary to higher expenditure of energy for maintenance by grazing animals in arid and semi arid region, forages are usually poor in energy. Average energy content range between 6-7 MJ/kg DM. Sheep and goat on an average consumed 1.0-1.5 kg DM daily get 6-9 MJ/kg$W^{0.75}$. Goat consumed 0.90, 0.78 and 0.80 MJ ME/kg $W^{0.75}$ in monsoon, winter and summer (Shinde *et al.*, 2000). Energy intake of goat remained low during dry, pregnancy and lactation, it was estimated to be 1.2 in dry, 1.0 in pregnant and 1.04 MJ ME/ kg $W^{0.75}$ in lactating goats subjected to semi arid grazing lands.

The indigenous lambs and kids which are normally not supplemented with concentrates or legume hay attain body weight of 13-16 kg at the age of 8-9 months. The growth is only about 40 g in kids and 50 g in lambs, dressing percentage is about 35 and 40 and the bone : meat ratio is very poor (1:3.5-4.0) (Singh, 2007). The birth, weaning and six monthly body weight were 2.6, 8.6 and 12.5 kg in lambs and 2.9, 9.3 and 13.6 kg in goats, respectively. The beetal goat male kids reached a body weight of only 11.5 kg at weaning and 14.1 kg at 6 months when maintained under free range grazing with out any supplementary feeds with overall survivability of 87.5% up to six months of age (Mishra 1981). The mortality in young was 14.3, 12.1, 3.4 and 2.8 from 0-3 months and 12.5, 3.5, 5.4 and 5.7% from 3-6 months of age in crossbred sheep, native sheep, crossbred goat and native goats, respectively. In general sheep on free range grazing on natural rangelands of arid region showed decreasing trend in body weight from March to July, whereas, goats showed decreasing trend in March and April months only, there after a increasing trend in body weight was recorded. Sankhyan *et al.* (1996) studied the production performance of 50 native and 50 crossbred sheep on natural rangelands under farmers management. A lambing rate of 92, 96, 84 and 92% in Malpura, Chokla, Avikalin and Avivastra sheep was recorded on the basis of ewes available during first year and 144, 145, 109 and 120% at the end of second year respectively. The live weight of lambs harvested per year per ewe were 20, 16.65, 11.6 and 13.2 kg and per ewe per hectare were 2.28, 1.90, 1.33 and 1.50 kg in four breeds, respectively. Studies have shown that small ruminants reared on extensive system without any supplementation as a common practice in arid and semiarid area, could show a marked improvement in the overall performance including quantity and quality of mutton and chevon production by supplementary feeding (Bhatia *et al.*, 198; Krishanamohan *et al.*, 1984).

The goat is primarily a browsing animal and performs well when browsed on a variety of shrubby vegetation and supplemented with concentrate mixture in addition to browsing. Mishra et al. (1980) recorded that *ad lib* Supplementation with concentrate, hay and/or green fodder between 91 to 180 days of age, in addition to browsing resulted in an increase of 44.8% in live weight and 14.3 % in dressing percentage vs browsing alone. The kids when fed the above ration *ad lib.* in stalls performed in between these two systems. Thus, browsing *ad lib* with supplementation found to be the best method of feeding of kids for meat production. Inadequacy regarding availability of nutrients from browsing alone and need of supplementation could also be assessed from the study of Parthasarthy et al. (1984) obtained a daily gain of 37. 4,87.4 and 73.3 g from 6 months and 62.6, 139.2 and 120.0 g from 6-9 months of age in Sirohi X Beetal half bred kids maintained on browsing, browsing+750 g/head/day concentrate and total stall feeding on 1040 g/head/day feedlot ration, respectively. The dressing percentage was 43.1, 43.6 and 48.5 at six months and 47.3, 52.1 and 53.1 at 9 months of age under the three respective feeding systems. The kids thus, showed an improvement of about 44.5 and 34.0% between 3-6 months and about 66.1 and 52.2% between 6-9 months, respectively, on browsing + supplementation and feedlot system of feeding management over the browsing alone.

In present scenario of arid and semiarid zone of country, sheep and goat are not fed to a scientific plan. The availability or rather non availability of quality feed determine the nutrient input. This severely upsets the production level of small ruminants. Under the prevailing circumstances, for the provision of adequate nutrient supply, feeding strategy have to be formulated for arid and semiarid zone keeping in mind the availability of feed resources. Improvement and efficient utilization of available feed resources in combination with suitable feed technology as well as feeding practices may boost up productivity of sheep and goat economically to provide sustenance and livelihood security of a large population of rural poor in arid and semiarid regions of the country.

REFERENCES

Anonymous 1998. In: 35 years of research. Published by Central sheep and Wool Research Institute, Avikanagar, pp. 49.

Beniwal, R.K. and Singh, J.P. 2005a.Role of pasture and silvipasture development in increasing livestock productivity in arid region In: Seminar on "*Augmentation of Livestock Production Using Newer Technologies*" March 11-12, College of Veterinary and Animal Science, Bikaner.

Beniwal, R.K. and Singh, J.P. 2005b. Production systems for increasing herbage productivity in rainfed,arid and semi-arid regions. In: National Seminar on "*Conservation, Processing and Utilization of Monsoon Herbage for Augmenting Animal Production*", December 17-18, Regional Research Station, CAZRI, Bikaner, pp. 67-74.

Beniwal, R.K., Singh, J. P. and Rathore, V. S. 2005. Phog (*Calligonum polygonoides*) an important arid shrub of rain fed farming system in western Rajasthan. In: *Shrubs of Indian Arid Zone*, Arid Agro-ecosystem Directorate, CAZRI, Jodhpur, pp. 91-96.

Bhatia, D.R., Mohan, M., Patnayak, B.C. and Ratan, R. 1981. Note on growth of malpura lambs with or without supplemental concentrate. *Indian Journal of Animal Science* 51, 238-242.

Bhatta, R., Shinde, A.K., Sankhyan, S.K. and Verma, D.L. 2002.Effect of feeding tree foliage on milk yield and composition of lactating goats on seme-arid rangelands. *Indian Journal of Animal Science* 72, 84-86.

Chaudhary, A.L., Gahlot, A.K. and Beniwal, B.K. 1999. Paper presented in 4th Agricultural Science Congress, held at B. M. Birla Science and Technology Centre, February 14-21, Jaipur,

Dave, P.O. 1977. A study on migration of stationary sheep flock during scarcity. *Wool and Woolens of India* pp. 59-69.

FAO 1993. FAO Production year book., Volume 47, FAO, Rome.

Faroda, A S. 1998. Arid Zone Research-an overview. In: *Fifty years of Arid Zone Research in India*, CAZRI, Jodhpur, pp. 1-16.

Karim, S A. 1999. Current status and prospects of sheep production for wool and mutton. Lead paper In: Biannual ANA Conference "*Emerging trends for livestock and poultry feeding beyond 2000 A. D.*", December 2-4, ANG Ranga Agricultural University, Hyderabad, pp. 85-98.

Karim, S.A. and Arora, A.L. 1997. Growth performance of lambs weaned at 60 and 90 days of age. *Indian Journal of Animal Science* 67, 536-537.

Karim, S.A. and Shinde, A.K. 2007. Pasture based feeding systems for small ruminants production and its relevance in tropics. In: *International Tropical Animal Nutrition Conference, October* 4-7, NDRI, Karnal, India, pp. 80-90.

Karim, S.A., Sankhyan, S.K., Bhatta, R. and Shinde, A.K. 1997. Surgical procedure for oesophageal cannulation to assess grazing intake in sheep. *Indian Journal of Small Ruminants* 3, 72-75.

Krishnamohan, D.V.G., Reddy, K.S., Naidu, C.M., Munirathnam, D. and Reddy, K.K. 1984. Effect of different levels of energy supplementation on the performance of weaner lambs fed a basal ration of legume hay. *Indian Journal of Animal Science* 54, 1170-1172.

Mann, J. S. and Suresh, C. Sharma, 2005. Productivity of monsoon herbages in semi arid zone of India. In: National Seminar on "*Conservation, Processing and Utilization of Monsoon Herbage for Augmenting Animal Production*", December 17-18, Regional Research Station, CAZRI, Bikaner, pp. 26-29.

Mann, J.S., Hassan, B. and Beniwal, R.K. 1989. Adopt silvipasture system for more fodder in arid and semi arid regions of Rajasthan. *Indian Farming* 29, 33-35.

Mishra, R. K. 1981. Note on the performance of beetal goats under semi-arid agro climatic conditions of Rajasthan. *Indian Journal Animal Science* 51, 885-887.

Mishra, R.K., Singh, D., Gaur, D., Sharma, M.M., Rawat, P. S., Prasad, V. S. S. and Parthasarthy, M. 1980. AICRP on goat for meat production. In: Annual report, CSWRI, Avikanagar, India.

Parthasarthy, M., Singh, D. and Rawat, P.S. 1983. Effect of supplementation on the performance of weaner kids. *Indian Journal of Animal Science* 53, 671-672.

Parthasarthy, M., Singh, D. and Rawat, P.S. 1984. Performance of feedlot kids on different dietary regimen. *Indian Journal Animal Science* 54, 130-131.

Pathak, N.N. 1994. Population movement and challenges of livestock feeding in India. Summer institute on "*Recent Advances in Animal Nutrition with Special Reference to Rumen to Manipulation to Improve Nutrient Utilization and to Reduce Environmental Pollution by Decreased Methanogenesis*", May 23-June 11, IVRI, Izatnagar, 17-19.

Patnayak, B.C. and Karim, S.A. 1994. Utilization of grassland for small ruminants production. National Workshop on "*Grass land Management in India-New Opportunities and Challenges*", February 2-3, Indian Institute of Forest Management, Bhopal.

Reddy, M.R. and Reddy, G.R.V. . 1996. Urea treatment of crop residues, maize stover and sorghum straw. In: Workshop on "Small Ruminant Production and Post Production System-Current Status and Development", December 4-6, CLRI, Chennai, 7-80.

Sankhyan, S. K., Shinde, A. K. and Karim, S. A. 1999. Seasonal changes in biomass yields, nutrient intake and its utilization by sheep maintained on public rangelands. *Indian Journal Animal Science* 69, 617-620.

Sankhyan, S. K., Shinde, A. K., Karim, S. A. and Patnayak, B. C. 1996. Performance of native and crossbred sheep on natural grassland under farmer's management. *World Review Animal Production* 30, 27-35.

Sankhyan, S.K., Shinde, A.K. and Karim, S.A. 1998. Seasonal changes in biomass yield, nutrient intake and its utilization by sheep maintained on community grazing land. *Indian Journal of Animal Science* 69, 617-620.

Shanker, V. and Gupta, G. N. 1992. Restoration of degraded rangelands. In: *Restoration of Degraded Lands: Concept and Strategies (Ed.)* by Singh, J.S., Rastogi Publication, Meerut, pp. 115-155.

Shanker, V., Kumar, S. and Tyagi, A.K. 1988. Grazing resources of arid and semi arid regions. In: Pastures and Forage crops Research-A State of Knowledge Report, 3rd International Rangeland Congress, New Delhi, 63-85

Sharma, T. 2004. Feeding strategies for improving meat and milk production from sheep and goat during scarcity. In: National Seminar on "Opportunities and Challenges in Nutrition and Feeding Management of Sheep, Goat and Rabbit for Sustainable Production", February 10-12, Central Sheep and Wool Research Institute, Avikanagar, India.181-190.

Sharma, T. 2005. Prospects of fodder banking in arid zone. In: National Seminar on "*Conservation, Processing and Utilization of Monsoon Herbage for Augmenting Animal Production*", December 17-18, College of Veterinary and Animal Science, Bikaner, pp. 75-82.

Shinde,A.K. 1996. Seasonal changes in energy expenditure of sheep on pasture. Ph.D. thesis submitted to Jawaharlal Nehru Krishi Vishwa Vidyalaya,Jabalpur.

Shinde, A.K., Karim, S.A., Sankhyan, S.K. and Bhatta, R. 1998a. Seasonal changes in physiological responses and energy expenditure of sheep maintained on semi-arid pasture. *Journal of Agricultural Sciences (Camb.)* 131, 341-346.

Shinde, A.K., Karim, S.A., Sankhyan, S.K. and Bhatta, R. 1998b. Seasonal changes in biomass growth and quality and utilisation by sheep on semi-arid *Chenchrus ciliaris pasture in India. Small Ruminant Research* 30, 29-35.

Shinde, A.K., Sankhyan, S.K. and Bhatta, R. 1998. Seasonal changes in biomass yield and quality and its utilization by sheep maintained on semi arid pasture. *Journal of Agricultural Sciences (Camb.)* 131, 341-346.

Shinde, A.K., Sankhyan, S.K., Karim, S.A., Singh, N.P. and Patnayak, B.C. 1996. Nutrient intake, its utilization and performance of sheep and goat on semi-arid *Chenchrus ciliaris* pasture. *World Review of Animal Production* 31, 35-40.

Shinde, A.K., Sankhyan, S.K., Karim, S.A., Singh, N.P. and Patnayak, B.C. 1996. Nutrient intake, its utilization and performance of sheep and goats on semi arid *Cenchrus cilliaris* pasture. *World Review of Animal Production* 31,, 35-40.

Shinde, A.K., Sankhyan, S.K., Bhatta, R. and Verma, D.L. 2000. Seasonal changes in nutrient intake and its utilization by range goats in semi-arid region of India.*Journal of Agricultural Sciences* (Camb.) 135, 429-436.

Singh, J.P. and Rathore, V.S. 2005. Potential herbage resources and their availability in arid region. In: National Seminar on "*Conservation, Processing and Utilization of Monsoon Herbage for Augmenting Animal Production*", December 17-18, Regional Research Station, CAZRI, Bikaner, India. 16-25.

Singh, N.P. 2007. Sustainable intensive meat production system for goats and sheep in tropics. In: *International Tropical Animal Nutrition Conference, October 4-7*, NDRI, Karnal, India, Pp 91-104.

Singh, N.P. and Karim, S.A. 1996. Utilization of non conventional feed resources for small ruminant production. In: Nutritional Constraints and Strategies for Sustainable Milk Production. Compendium 1 (Review paper) Eds. Singhal, K.K., Thakur, S.S. and Sharma, D.D., 7th Animal Nutrition Research Workers Conference, December 12-14, TANUVAS University, Chennai, India. 91-101.

Spare, M.V. 1976. Migration of sheep in Rajasthan. *Wool and Woolens of India,* 57-61.

□□□

12

Interventions to Sustain Small Ruminant Productivity During Drought Conditions

R.C. JAKHMOLA

Arid Region Campus, Central Sheep & Wool Research Institute, Bikaner - 334 006, Rajasthan

Introduction

Indian has large variations in the temperature and rainfall pattern as such the both "*Cherapunji*" (area with world's highest rainfall) and '*Thar*' (The desert) exist here. The hot arid region of the country spread over western regions of Rajasthan and Gujrat, characterised by low precipitation, low net sown area (mainly rain fed) and low forest cover (3-5%). The livestock production remains the backbone of farming system in this area. It is generally accepted that below about 400-mm rainfall, crop cultivation is a risk-prone enterprise. With the exception of irrigation schemes, most areas with comparable precipitation have remained open rangelands. Grazing on pastures (pastoralism or agropastoralism) is the dominant form of production system adopted traditionally. Being dependent on rain, the region very often encounter drought during which there is severe loss of crop and livestock. During 2002, in Rajasthan only, approximately 40 million people, 50 million livestock were affected from drought and small ruminants were either gone to migration or sold. Due to the non-availability of water and fodder, people had started abandoning their unproductive cattle. There were reports of animals dying of hunger and thirst. Animal carcasses could be seen in the countryside. Many people have started migrating to other states with their animals. There was immense loss to livestock in the region.

Understanding of drought and its impact on livestock producers has advanced substantially in recent decades; unfortunately, the incidence and impact of such droughts also seems to have increased. This is probably because of a complex matrix of national and regional priorities coupled with forecasting, preparedness, communication and other related factors that can crosscut the subtle local appreciation of a situation that makes livestock related work less effective. The present document considers the effect of drought and technical solutions to mitigate its effect on livestock productivity with special emphasis on small ruminants.

The Drought: a Social Curse?

India has experienced 22 large scale droughts in the year 1891, 1896, 1899, 1905, 1911, 1915, 1918, 1920, 1941, 1951, 1965, 1966,1972,1974,1979, 1982, 1986, 1987, 1988, 1999, 2000 and 2002. Meteorologically a drought is a situation when the deficiency of rainfall at a meteorological sub-division level is 25 per cent or more of the long-term average (LTA) of that sub-division for a given period. The drought is considered "moderate", if the deficiency is between 26 and 50 per cent, and "severe" if it is more than 50 per cent. When meteorological drought is prolonged it becomes hydrological drought and is marked by depletion of surface water from reservoirs, lakes, streams, rivers, cessation of spring flow and fall in groundwater levels causing severe shortage of water for livestock and human needs. The agricultural drought is said to occur when soil moisture and rainfall are inadequate during the crop-growing season to support healthy crop growth to maturity. It occurs when there is severe meteorological drought with a rainfall deficiency of more than 50 per cent of the LTA within a period of four consecutive weeks or with a weekly rainfall of 5 cm or less during the period from mid-May to mid-October (the kharif season) when 80 per cent of the country's total crop is planted, or six such consecutive weeks during the rest of the year. This situation results in reduced productivity or even total loss of crop. Of the three types of drought, agricultural drought has the greatest consequences on society as it affects all sectors of life. The prediction of severity and geographic extent of drought in most parts of the world, depend on reliable weather records, or even by personal experience. The severity of drought depends by the region's ability to cope. The success of preventive or mitigating strategies depends on the strength and effectiveness of socio-political and economic institutions, as well as the relative reliability of the weather forecasting.

Drought and land degradation

The impact of grazing and browsing on land is determined by seasonality as well as intensity, and the restriction of mobility thus has more serious implications than the number of animals that are grazing on a pasture. When rangeland area is overgrazed it accelerates desertification. Reducing livestock numbers in the grazing lands and introducing range use controls on communal land is not easy. There is little likelihood that marginal dry farm lands will be returned to pastoral use unless production on the better lands is increased enough to offset the loss of cultivated land. While desertification was a long-standing problem even in the absence of droughts, the gradually increasing vulnerability of the land use made the impact of the inevitable droughts worse than ever. The factors responsible for that vulnerability are still operating, desertification continues, and future droughts will have ever-greater damaging effects on vegetation, human and animal.

Drought and Animal farmer

Primary effects of drought result from a lack of water. As a dry period progresses and water supplies dwindle, the existing water supplies are overtaxed and finally dried up. As water supplies dwindle and crops and fodder are depleted, families begin to migrate in search of better grazing lands for their livestock or move to the cities to seek jobs and alternative sources of income. If the dwindling supplies of feed are not replaced, famine can occur, and therefore further acceleration of migration out of the stricken areas to less affected zones occur. The migration may, in itself, contribute to spreading the scope of the disaster, especially if grazing animals are moved with the people. If drought is long term, it may result in permanent changes of settlement, social, and living patterns. If a drought is allowed to continue without response, the impact on development can be severe. Moreover, livestock producers in semi-arid regions are correspondingly poorly resourced. Animal health services are often of limited value, supplementary feeds are available in less quantity, and the forward planning associated with drought and emergency feed-supplies is only sporadically taken on board. The consequence is that in the face of a climatic anomaly such as drought, the impact on livestock producers in the semi-arid and arid zones is often severe.

The drought induces the long-term problem of household viability of livestock producers. Long-term rainfall deficits affect their essential capital, their herds, in a manner that can only be rectified over a period of years. Depending on severity and longevity of a drought, herd structure both in number and quality disrupt which will take several years to regain their pre-drought production/ reproduction capacity. Some farmers never rebuild their herds and simply remain engaged in the agricultural sector. The longer harsh conditions remain, the greater the herd recovery time, thereby lessening herders' capacity to cope. There is now a fairly well recognised sequence that occurs when herders are faced with extended rainfall deficits. In Indian conditions animal farmers generally affected by drought situation in following ways:

a) Animal farmers especially of small ruminants scatter in search of water and pasture. Some herders push (usually towards east or north) into higher-rainfall areas. The herd of cattle does not move to long distances and remain near fixed water-points such as ponds, where the pasture may become exhausted and in event of non-supplementation the animals die of starvation.

b) Drought causes the terms of trade to turn against producers. The price of essential items rises markedly while the sale value of their stock falls dramatically, since many other producers are also trying to sell their animals. This is accompanied by a deterioration in the quality and size of the animals on sale. Herders may also sell classes of animals, such as pregnant females, which they would normally retain. Some animals are sold for slaughter, but others that remain with farmers are maintained on already depleted resources, eventually trees become the target, which are brutally lopped creating depletion of feed resources.

c) When the rainfall levels rise again, the price of domestic stock rises as producers try to restock. Since their capital is limited, the consequence is a large number of producers with non-viable herds (i.e. herds which cannot support their household). These household herds are then even more vulnerable in the next climatic anomaly.

d) Species or breeds of animals that are thought to be unsuitable for the eco-zone are also changed. The Cattle people switch to camels and goats, or try to swap their dominant breed for one with a greater ability to digest browse or tolerate drought.

e) Several farmers leave the animal rearing occupation permanently and try to find alternate non-agriculture/ un-skilled jobs. These processes are now largely understood, by academics, aid agencies and NGOs but solutions need to be generated and applied. Long-term structural assistance is more difficult to provide than emergency aid, and it seems there is also a 'compassion fatigue'. That is why, even after providing long term assistance to the semi-arid and arid regions of the country, things never seem to improve?

Impact of Drought on livestock health and production

Unequivocally water is one of the most important nutrients for ruminants during drought. Dry environment associated with high ambient temperature induce ill effects in all categories of animals. Such conditions affect production, as increasing respiration rate necessitate reduction in feed intake and rumination. The endocrine system is also affected markedly by thermal stress. Increased sweating and evaporative water loss at the lungs are reflected in increased antidiuretic hormone concentrations in plasma. Likewise, increased electrolyte turnover is associated with altered aldosterone concentrations. Thermal stress increases turnover of Na and K. Prolectin concentrations are elevated during thermal stress. The chronic thermal stress resulted in lower metabolic rate. This is reflected in lower concentrations of growth hormone, thyroxin, and glucocorticoids in blood.

Daily water consumption (per head) by sheep and goats ranges between 2 to 8 litres, however, it increases with an increase in size of animals, lactation, dry matter intake and temperature. Extremely hot weather could increase the high values another 20 to 30 percent. Unavailability of clean drinking water for animals remain major problem in drought hit areas. Stagnant pond water that remain main source of water to animals during drought often have a high concentration of undesirable substances with high oxygen demand, which makes the water cloudy and smelly. Salinity hazard is widely known ailment of animals reared in many areas of Rajasthan. Animals, receiving water containing total soluble salt (TSS) more than 3.5g per litre shows reduction in production and reproduction traits. Further increase in concentration of TSS to 4.5 g / lit develops diarrhoea, emaciation, depression of oestrus cycle and increase incidence of abortion. Drinking water in arid region also contains high concentration of fluorine. Although, desert animals tolerate fluorine concentration up to 0.02 g/l beyond this level, the reproduction traits are affected. Continuous aeration of such water can effectively improve the quality of water. If ponds are aerated, use of water for livestock can be permitted.

Reduction in voluntary feed intake near or above the upper critical temperature of the animal is widely accepted as a major negative influence on productivity. Dry matter intake begins to decline markedly as temperatures exceed about 25-27°C. However, other environmental factors such as wind velocity, humidity, and radiation also directly affect homeothermy under natural conditions and, thus likely are interrelated with ambient temperature in affecting feed consumption. Sheep and goat has ability to decrease nitrogen losses particularly in low nitrogen intake due to renal activity and increased urea recycling to the gut. The efficiency of economizing nitrogen metabolism by recycling urea is high on low protein diet. Water deprivation is a common constraint in desert goat which decrease N losses in urine, increased urea recycled into the gut and improve N balance by the animal.

During drought, animals do graze on pastures intensively and it leads to elimination/ depletion of nutritious grasses/ Forbes, increasing concentration of weeds and toxic plants on pasture. Livestock ordinarily avoid toxic/ less nutritious plants, but during a drought, animals are forced to eat them. Such problems may be aggravated by deficiencies of phosphorus or vitamin A in livestock, which greatly alter their grazing behaviour. At times animals consume plants high in anti-nutritional/ toxic principles. Drought, followed by a rainfall, allows very high uptake of nitrate in plants. Nitrates found in forages, are converted to nitrite in rumen, and in turn the nitrite is converted to ammonia and then converted to protein by bacteria in the rumen. If animal rapidly ingest large quantities of such plants, nitrite will accumulate in the rumen. Nitrite is then absorbed into red blood cells and combines with haemoglobin to form methaemoglobin, so the animal's heart rate and respiration increases, muscle tremors can develop, staggering occurs, and the animal eventually suffocates. Nitrate toxicity may result when animals suddenly consume large amounts of forage containing 2-3 percent or more nitrate ion on a dry matter basis. Even forage with lower levels may adversely affect reproduction or become toxic if animals are nutritionally stressed or metabolically abnormal and suddenly eat a large amount of such forage. Rumen paralysis may occur. Laboured or difficult breathing may be observed. Animals may go down and die rather suddenly. Animals can be treated by intravenous injections of methylene blue. Commercial preparations intended for treatment of prussic acid poisoning only should not be used to treat nitrate poisoning. Prevention of nitrate poisoning is best achieved by controlling type and quantity of forage offered to livestock. Avoid forages with potentially toxic levels of nitrate or at least dilute them with feeds low in nitrate. When in doubt, have feeds and forages analysed for nitrate before grazing or feeding them. Forages with sub-lethal nitrate levels can be fed to livestock with appropriate precautions.

Cyanogenic glycosides occur in many *Acacia* sp. and when ingested and hydrolysed to free hydrogen cyanide (HCN) responsible for poisoning. In rumen, hydrogen cyanide is converted to thiocyanate using available sulphur. Thiocyanate is goitrogenic, inhibiting the activity of thyroid gland. Although this acid is very toxic to animals, it has little effect on anaerobic organisms, including rumen bacteria, which can use it as a nitrogen source. HCN is released through enzymatic action upon injury to the plant or through microbial action in the rumen. Soluble carbohydrates reduce cyanide toxicity. The restricted use of sorghum, sudangrass and their crosses during severely stressed condition need to be followed. Symptoms of prussic acid poisoning are very similar to those of nitrate poisoning. It is very important to differentially diagnose the form of poisoning. Blood drawn from animals suffering from prussic acid or hydrocyanic acid poisoning may turn brilliant or cherry red upon exposure to air. In contrast, blood from nitrate poisoning cases may be normal or chocolate brown in colour as drawn. Acute toxicity from prussic acid may result in lack of coordination and convulsions followed by quick death. Treatment with sodium thiosulfate-sodium nitrite solution is recommended. Repeat treatments with sodium thiosulfate if necessary.

Tree fodder, shrubs and herbaceous species, are important sources of food during drought. Several top feeds contain toxic/ anti-nutritional factors. Among these tannins and mimosine are most talked about. Much of these compounds are detoxified in the rumen and in the body tissues especially in the liver. However, problem arises when the intake exceeds the animal's capacity to detoxify it. Tannins have astringent activity by binding to taste receptor in the in the mouth. It also binds proteins, thus reduce the availability of amino acids to the animal system. Animals adapted to tannin containing diet have less adverse effect on growth than those animals not accustomed to tannin rich diet. Non-adapted animals respond more strongly to polyethylene glycol treatment than the conditioned one. Tannin affect the protein metabolism whereas mimosine is uncommon amino acid presents usually in *Leucaena* species which when ingested in large dose shows ill affects specially in sheep. The reduced nutrition of the stock during drought also increases the susceptibility of animals to the effects of diseases and parasite burden. Concentration of animals around a common and limited source of water that is generally seen during water scarcity results in accumulation of the parasitic eggs and or larvae.

How to Respond to Drought?

Livestock in arid/ semi-arid regions is major source of income and employment for livelihood of rural population. Nonetheless, the gradual depopulation of rural areas as people gravitate towards cities, and the commensurate increase in demand from the urban sector for livestock produce, means that livestock can potentially play an increasingly important economic role. However, with so many uncertainties, developing new and appropriate herd-management strategies has not always proved possible either for the pastoralists themselves or for development agencies. Research into the precise nature of controls on herd structure is hampered by the fact that pastoralists are often reluctant to develop precise production database. Solely external institutions cannot undertake development in methodology, and livestock producers are crucial for the future of the sector. The mixed farming approach, which is in vogue in the country provide a strength to household economy. However, resource shortages need to be tackled. As environmental conditions are not under the direct control of the farmers, they can restrict in the number of animals that can be reared on a given piece of land. There is need to create awareness to use monetarised market that will show importance of cash. This will raise the status of livestock holders, and the goal of market will becomes the accumulation of money rather than the accumulation of animals. Market value then will depend on its quality, and the wealth of the herder will then be dependent on the accumulated value of the individual animals, rather than on the number possessed.

Management of water resources and Rain water Harvesting

Wells/ Bawadies were traditional water points across western India. Animal power was the main method used for lifting water, and the camel was the preferred species in this region where deep wells were prevalent. Farmers/ local community owned these deep wells. The process of watering a herd at these wells was labour-intensive. However, open wells existed in the past dried away during last 20 years. The ponds are still common water source used for watering animals in this part of the country. During recent years excessive use of pond water for irrigation using pumps, started, this leads to drying up of ponds during summer. The surface water pools if not properly maintained, do get degenerated and biological oxygen demand of the water increases. Stagnant pond, particularly during hot, dry spells, provides environment for the growth of blue green algae and other toxic aquatic wild plants that can be poisonous for livestock. The continuous aeration of such water will effectively combat this cycle. Dugout rehabilitation is necessary keep water supply healthy. If ponds can be aerated, use of water for livestock can be permitted. If not, it may be necessary to fence off the pond and offer another source of water. Thus efficient use of surface water for drinking and irrigation is need of the time. There is clearly a need to regulate groundwater extraction so that an adequate share of it is used for drinking water and irrigation purposes.

Rainwater harvesting is an effective tool as conservation measure in arid regions. Apparent argument is generally given it has little role in arid (25 per cent of the country) regions and semi-arid regions (48 per cent of the country) with an annual rainfall of less than 50 cm. We should learn it from Israel, which has developed a sustainable system based on rainwater harvesting with an annual rainfall of about 23 cm. Also in the desert regions of Australia, with an annual rainfall of 7.5 cm, run-off has been brought down to 4.5 cm. Consider a place like Jaisalmer, which has an annual average rainfall of 10 cm. Over an area of 1 ha or 10,000 square metres, the total amount of water received is one million litres. There are traditional and modern improved methods for efficient rainwater harvesting, which can be adopted for location-specific environments and needs. Given the increasing scarcity of water and frequency of calamities, rainwater harvesting may prove to be the only feasible solution in the rural areas.

Effective Watershed development and maintenance

The soil and water conservation measures of may regenerate the process of vegetation development in drought prone area. The measures taken are both engineering and vegetative in nature. Providing good vegetative cover to a degraded site is the final answer for its rehabilitation. However, in highly degraded lands, establishment of vegetation is difficult due to high runoff/ debris movement, lack of moisture and absence of fertile soil. The engineering measures are therefore needed before regeneration programme to stabilize the slopes and create conditions conductive to plant growth by arresting fine soil and improve moisture status. Constructing temporary check dams or gabion check dams can do gully stabilisation. The plantation of vegetation, in catchments is simultaneously done. The choice of species should aïm not only regreening of the area but to rehabilitate the site ecology in a way that ensure sustainable utilization of eco-system components. Therefore plant species adapted to local conditions, that provide quick green cover and bind soil, have fast growth, can fix nitrogen and have social and economic value to local population and capable to provide fuel/fodder will be best choice. Plantation of grasses and legume mixtures for quick initial establishment of vegetative cover in problem land is necessary.

Once water shed is developed, compatible and suitable tree species improve soil fertility by increasing soil organic matter content, water holding capacity and have favourable influence on hydrological characteristics. The feed and fodder resources improves and so as the animal husbandry activity. However, for sustainable water shed development/management peoples participation is prerequisite. Equitable water shearing mechanism, diversification of agriculture for better utilisation of land and water resources and improving farm income, conservation of feed resources, judicious use of feed resources, and integrated approach are necessary components.

Increase nutrient availability from land

Most of the reports on the drought situation have prioritised availability of feed/ fodder as the main concern. Assuming the digestible crude protein (DCP) requirement for cattle/buffalo/camel as 2.84 g/kg $W^{0.75}$ and for sheep/goat as 2.73 g/kg $W^{0.75}$, and adding 20% allowance for meeting DCP requirement for various production functions, the annual DCP requirement would be 22.74 lakh tonne annually. Similarly, if a goat/sheep requires 424.3 kJ/kg $W^{0.75}$, buffalo/ camel requires 523 kJ/kg $W^{0.75}$ and a cattle requires 495 kJ/kg $W^{0.75}$ metabolisable energy (ME) daily, the annual ME requirement would be 477.4 X 10^9 MJ or 313.75 lakh tonne total digestible nutrients (TDN). Livestock's diet in this part of the country is roughage based, and a typical ratio of dry roughage to greens - 70:30 for cattle and buffaloes and 60:40 for sheep/ goat/ camel - in the diet is assumed. Such diets are good to support maintenance requirement, or a low rate of production. To obtain optimum production from livestock, feeding of required quantities of concentrate are essential. Thus, annual DM requirement from dry roughage, green fodder and concentrate are 488.95, 231.87 and 75.43 lakh tonne respectively.

A rapidly growing population and limited land resources make it difficult in maintaining livestock on existing eroded pastures in the country. Land pressures must somehow be relieved, and that can only be accomplished by raising productivity and yields per unit area. Weeding of pastures to keep noxious and toxic plants under control is important. During periods of drought, many of the more palatable range plants mature and dry up early. Thus, little good forage is produced during much of the grazing season. In contrast, several toxic plants have the ability to remain green during these periods. As a result the proportion of these toxic plants increases in the pasture and may have possibility to be utilized by the grazing animals.

If grazing must occur, as it usually does, livestock poisonings may be inevitable. However, there are a few general rules that help reduce or eliminate the chances for toxic plant poisonings.

- Use good grazing management practices. Grazing systems that give additional rest to plants during drought are preferred over continuous grazing, and livestock distribution within a pasture becomes even more important during drought situations. Good grazing management helps to maintain an alternative forage choice over toxic plants.
- Use good livestock management practices. Strategic supplemental feeding may help reduce losses to toxic plants. Supplementing adequate protein and phosphorus may keep animals from seeking out toxic plants that are high in protein. Vitamin A deficiency, also common on drought-stressed

rangeland, may alter the grazing habit of the animal and probable cause to consume toxic plants that would normally not be consumed. Good livestock management practices also means never releasing hungry animals into a pasture that is known to have toxic plants, especially when toxic plants are the only "green" forage available.

- Plant control measures may become necessary to avoid livestock losses to toxic plants. The mechanical, chemical, and biological control methods are valid options. Individual plant treatments may be preferred over broadcast treatments for chemical control during drought. Individual plant treatments allow the applicator to selectively control only target species leaving the other broadleaf weeds as potential feed. While some plants are toxic to one class of livestock they may not be to another class. "Cleaning up" the range by first grazing with animals not affected by the plant toxins and following later in the season with grazing animals that would be affected by the plant is an example of biological control.

Develop pastures and rangelands based production system

Where rainfall is extremely patchy, pasture resources are to be exploited opportunistically. Under these circumstances a farmer with a high level of herd mobility can maintain his animals in land that is almost unusable. Moreover, mobile graziers do not have to pay any of the fixed costs associated with protected pastures, and grazing is thus essentially a free resource. The system was in place functioning in desired manner and was sustainable for long. However, in past rangelands have been reduced in size, in part because the widespread use of irrigation technologies, both in traditional and more recently in sprinkled forms, has allowed crop cultivation to colonize much larger regions of the rangelands. Also use of rangelands for non-agricultural purpose has increased in recent past. As a result, what rangelands remain are considerably more arid than those exploited by pastoralists in sub-tropical region. Pastoralism has traditionally been oriented around camels and sheep, with sheep becoming predominant in recent times due to their greater marketability.

There is a need to protect pastures from indiscriminate grazing. Adaptation of rotational grazing along with suitable water harvesting method will create conditions for development of a pasture. Reseeding of pasture with suitable grass/ legume species is another step. The addition of some phosphoric fertilizers is also important to develop a sustainable pasture system. The legume would increase the total herbage produced directly through its own growth and indirectly through better growth of grasses responding to nitrogen fixed by them. In addition, combined grass/legume would meet the energy/protein requirement of grazing animals. The plantation of multipurpose trees -produce fodder and fuel - in the pasture, barren and uncultivable land, cultivable wasteland and fallow land would further help in increasing the productivity of pasture and at the same time would reclaim the soil. The species selected for planting should have following characteristics:

- The seedling should establish easily and after establishment should tolerate varying degree of dry spells
- The full grown tree should have wide canopy and should be able to produce fodder/ fruits for a period of about 20-30 years.
- They should produce substantial decomposable litter.
- The trees should preferably be leguminous (N-fixing capability) and should not contain toxins.
- The species should have high water and nutrient use efficiency.

On conversion of pastures into silvi-pasture and following judicious grazing/lopping management, the productivity of the pasture can increased by 2 to 3 folds and about 156.1 lakh tonne DM of high nutritive value can be produced. After introduction of agro-forestry based system in to barren and uncultivable land, 56.1 lakh tonne DM can be produced annually.

Conservation of forages

During normal years, a large proportion of seasonal/ perennial monsoon grasses/ legumes left unconsumed. Generally, these forages are harvested at a much-advanced stage of maturity when seeds are shed off and traces of green colour remain. However, for harvesting maximum nutritious fodders that are plenty in monsoon it is essential to evolve suitable and economic methods of processing or preservation. Various methods of fodder preservation/ processing are available for preserving natural grasses so that these may be used during scarcity periods.

The forages excess to that of requirement are generally preserved as hay. Fodders can be harvested at the stage when there is maximum accumulation of nutrients in the plant, for preservation. The preserved fodder can then provide nutritious food to livestock during the lean season when there is scarcity of fodders. The ration of the animals can be balanced with the help of good quality hay. The good quality legume hay may also replace certain amount of concentrate in the ration, thus reducing the cost of production. The supplementary value of hay in ration not only depends on the quality of the hay prepared but also on the method through which it has been preserved. The hay quality should, therefore, be maintained by avoiding the losses of valuable nutrients. The hay prepared from a nutritious plant species and stored properly taking necessary precautions may supply nutrients comparable with the green forages. The hay quality, a result of several factors like stage of harvest and method of processing, is generally determined by analysis of several parameters comprising of physical and chemical characteristics of forage plant.

Silage is a feed produced from crops with high moisture content through controlled, auto or induced fermentation, in a silo, achieving and maintaining anaerobic conditions, allowing lactobacilli or other similar species to proliferate and produce lactic acid and restricting the growth of spore forming anaerobes and clostridia which produce amines, ammonia and carbon dioxide. The major advantages of ensiling are: 1) Preservation of surplus green fodder abundantly available in rainy season, for feeding during lean season; 2) Silage is highly palatable and slightly laxative; 3) Weed can be utilised along with main fodder, as ensiling destroys the germination power of a majority of weeds; 4) The organic acids produced during ensiling are easily digestible by ruminants; 5) There is lesser loss of carotenes in silage making and 6) Inclusion of organic wastes (poultry litter, poultry droppings etc.) is possible in ruminants' diets through ensiling it with forages.

Efficient Utilisation of Crop Residues and Agro-Industrial by-Products

In view of the limited feed resources for livestock, livestock rearing is more an adjunct activity to crop production than an independent commercial enterprise. Future scenario relates to potential supply of crop residues. In states like Punjab and Haryana due to the extensive use of combine harvesters a large portion of the wheat and rice straw is either left in the field and burnt or becomes unfit for fodder purpose. The oilseed crop residues are used in some parts of the country for livestock feeding, still the use is more in the form of oilcakes than as Stover. Most of the crop residues have more than 60% of dry matter in the form of cellulose and hemicellulose. Cellulose and hemi-cellulose are good sources of energy for the ruminants, but these are not easily approachable by hydrolytic enzymes and microbes present in the rumen. The main hindrance in hydrolysis of cellulose is protective nature of lignin. In addition to it the presence of silica may also decrease the digestibility of certain ligno-cellulosic material. About 67 lakh tonne of unconventional straws- rape seed/ mustard straw etc.- currently available in state, remain unutilised. These crop residues are fibrous and low in digestibility, hence the intake by the animal is also low. Urea treatment of MS followed by ensiling changes its physical nature and improves digestibility as well as utilisation by the sheep. Urea treated MS in combination with tree leaves in 25:75 ratio will make it a maintenance ration. Prickly pear cactus (*Opuntia* species), a fast growing xerophytic plant well adapted to semi arid conditions, rich in energy and water and poor in protein, has been used in livestock feeding in North America since 19th century. The plant is extremely variable in nutritional value depending upon species, variety, age of the plant, sampling season and plant parts. However, in general opuntia is high in moisture (85%) and IVDMD (75%) and can be used as source of sustenance in drought feeding. Vilayati babul (*Prosopis juliflora*), a leguminous plant well adapted to semi arid and arid ecology having multifaceted utility yielding 17 to 140 kg pod having high energy value can replace 30% of concentrate in ruminant feeding. The babul (*Acacia nilotica*) pods at the rate 150 g/day with *ad libitum* cenchrus hay could maintain the pregnant ewes in positive N balance. Studies on Siris (*Albezia lebbeck*) pods indicated that the ground material can replace 20 to 40 per cent concentrate for adult sheep Similarly, cotton straw, cotton seed hulls, sunflower straw, sunflower heads, sugarcane bagasse, saw dust, wood pulp waste, maize cobs, groundnut hulls, forest dry grass and leaves can be used in ruminant feeding at appropriate level of incorporation. The ways are being developed on the use of treatment methods or supplements or combination of feeds to enhance animal performance on these basal roughage. In addition to treatments, supplements are needed that improve the pattern of rumen fermentation. The extent to which treatments/ supplements result in improved efficiency of overall feed use depends on changes in the cell wall structures, rates of fermentation of fibrous carbohydrates, synthesis of microbial protein, degradation of dietary protein and outflow of particulate matter from the rumen. The tactics in supplying supplements are all related to timing, so as to best modify these rates, which both determine and are determined by the level of intake. Alternate resources for feeding livestock have been explored for several years. Many agro by-products were identified and evaluated for inclusion in the compounded feed mixtures. They include mango seed kernel, mahowa cake, neem cake decorticated cotton seed cake, de-oiled rice bran, sal seed meal, apple pomace, soya pulp, whey powder, vegetable residues, leucaena leaf meal, karanjia cake, taramira cake etc. It is important that they find their use in the compounded feed manufacture. There is a need to develop specifications for these feed ingredients for their nutritive value as well as their nutritive limitations.

Urea-molasses mineral block licks

Urea-molasses can be fed as urea-molasses liquid feed or urea-molasses mineral block licks. For increasing the efficiency of course roughage utilization through increased rumen fermentation and increased microbial protein yield, supply of essential mineral have to be provided for sufficient rumen fermentation. Hence, based on the availability of molasses feeding systems may be adopted for improved utilization of fibrous crop residues. The lick block meets 80-90% of total mineral and vitamins requirement of animal and supplement diet with soluble sugars and performed proteins.

Similarly urea-molasses-mineral block (UMMB) are useful and these can be manufactured by various processing methods. Main constituents of the blocks are molasses, urea, mineral mixture, binding agents, vitamin supplements and fillers like GNC, mustard cake, cotton seed cake, rice polish, wheat bran etc,. Recently the cold process is in use by the manufacturers and these blocks are available in the market. These blocks will be kept in the manger and allowed to lick by the animal. It increases the intake of crop residues by 15-30 percent, reduces the requirement of concentrates, and improves daily gain and milk production by 15-20%.

Liquid licks of urea-mineral -molasses

During severe drought when the pastures dried up and animal are unable to find dry matter from such pasture that time the liquid lick of urea-minerals and molasses can be fed to them. In rehabilitation camps this liquid lick can be prepared and animals then allowed consuming it. Molasses (93%), urea (2.5%), water (2.5%), mineral mixture (1.5%), salt (0.5%) and vitamin supplement (25g/100kg) are used to prepare liquid licks.

Some precautions are necessary to avoid molasses toxicity. Prior to shift the animal to liquid licks diet adaptation is required. Animal has to provide 6-8 kg dry forage during first two days of feeding. From third day onward, dry fodder reduced up to 5 kg and provides 1kg liquid lick. During next seven days, provide animal 3kg dry fodder and 3-4 kg liquid lick. From 11th day onward, liquid lick can be provided al lib. After eating 3 kg of dry fodder. Lactating animals producing 5-10 l milk daily must provide 1kg of concentrate mixture.

Use of complete feed blocks

The low-grade cellulosic materials have low palatability; poor nutritive value; low in protein and energy. The energy content of these materials poorly utilized by microbial population in the rumen due to the presence of lignocellulose, whose components are either indigestible (lignin) or act as a barrier between the potentially digestible fraction and the digesting enzymes. Supplementation of protein, minerals and vitamins alone does not overcome problem of low intake and low metabolizable energy contents. The mechanical process of making blocks/pellets of such low-grade roughages will help to improve voluntary intake, nutritive value and their utilization. These blocks can be prepared in the surplus area and transported to the deficit regions to save heavy loss of livestock. Different types of blocks for maintenance, growth and lactation can be formulated and fed to animals under different physiological conditions. The roughages and concentrate ratio will vary according to the block type. The advantages of the complete feed blocks are:

- During storage prepared feed blocks required one-third space as compared to their mash form.
- The transportation of such blocks are easier and trouble free.
- Because of high pressure employed during block formation, the concentrate sticks on roughage part of the rations. Which ensure proper, balanced and adequate intake of concentrate and roughage in a particular proportion that helps for better fermentation in the rumen of the animal.
- The palatability, voluntary intake and nutrition utilization of low-grade roughages is increased.
- Cost of animal production on complete feed blocks will be reduced.

- Reduced loss of valuable dry matter in the orts.
- If necessary the prophylactic medicines can be added in the blocks.

Prevent Diseases and Parasitic Infestation

During drought, livestock generally die of starvation rather than of thirst; low rainfall is insufficient to produce enough grazing for ruminants, although pools usually meet their water requirements. Reduced productivity of rangeland causes reduction in feed resource availability to animals. High cost/ unavailability of feed/ water for livestock and increased cost of feed transportation adds to cost of animal maintenance. Poor nutrition causes decreased stock weights, disrupts normal reproduction cycle and reduced milk production. Hot winds and dry environment prevail during hot summer induce ill effects in all categories of animals leading to severe physical and mental stress on the animals. In general, the less digestible ration fed to a heat-stressed animal, the more will be the rate and extent of reduction in consumption. Reduced feed intake results in smaller quantities of essential nutrients and metabolisable energy being consumed by animals. This consequence may deleterious, particularly to performance of grazing ruminants, unless some alternative approach (e.g. supplementation) is employed to provide required nutrients and energy.

The large herds of poorly fed animals, often harbour sub-clinical pathogens. Most livestock diseases are now in principle treatable. The farmers are now increasingly becoming aware about medication of their herds. Timely vaccination of animals will protect them from some deadly diseases. Herd maximization during good season is justified by the argument that if there are more animals to begin with, the impact of shock events will not be so devastating. However, when resource availability becomes the single most important factor limiting herd size becomes more important.

Round worm, tapeworm, strongyles, amphistoms and liver flukes are the most common worms affecting the farm animals. Deworming with appropriate anthelmintic at regular intervals keep the animals free from parasites. Those classes of cattle which may not normally be treated (e.g. mature cows which are not normally worm drenched), may need treatment during feed shortages. Deworming with appropriate anthelmintic at regular intervals keep the animals free from parasites.

Conclusion and Recommendations

Droughts are common in arid-tropics. The effective way to fight drought is to be prepared for it. The preparedness is only solution that can be done through scientific management of community pastures, grazing management, conservation of forages, use of unconventional feeds in animal ration and observing practices for livestock health. Animal husbandry, a major agriculture activity in Rajasthan, and most areas affected from drought, can be made sustainable, if it develop as environmental friendly enterprise and able to cope with seasonal variation in nutrient availability. The rural employment and environmental development should go hand in hand and should make it economically viable, socially just and drought resistant. The areas where immediate attention needed are those related to pasture improvement, wasteland development, and watershed development for rainwater harvesting and forage conservation. All existing wasteland/ eroded terrain, if properly treated, can provide enough fodder for animals to sustain them during prolonged drought. When drought is there scientific feeding of complete feed blocks, urea molasses blocks can be useful. In addition to it all resources which are presently under utilised should be channelled for animal feeding through establishing rural agro-industries.

FURTHER READING

Blench, R.M. and Marriage, Z. 1998. Climatic Uncertainty and Natural Resource Policy: What should the role of government be?' *Natural Resource Briefing Paper* 31. Overseas Development Institute, London.

Blench, R.M. and Marriage, Z. 1999. Drought and livestock in semi-arid Africa and South west Asia. Working Paper 117. Overseas Development Institute. London.

Boda, K. 1990. *Nonconventional feed stuffs in the nutrition of farm animals*. Elsevier Science Publishers, Amsterdom, Netherlands.

Collier, R.J., Beede, D.K., Thatcher, W.W., Isreal, L.A. and Wilcox, C.J. 1982. Influences of environment and its modification on animal health production. *Journal of Dairy Science* 65, 2213-2227.

Dahl, G. and Hjort, A., 1976. What is a Herd? Primary and practical considerations'. In: *Having herds: Pastoral herd growth and household economy*, (Dahl, G. and A. Hjort.) Stockholm Studies in Social Anthropology 2. pp. 114–129. Stockholm, Sweden.

Glantz, M.H. 1997. Eradicating Famines in Theory and Practice: Thoughts on early warning systems'. *Internet Journal for African Studies* 2. 'Using science against famine: Food security, famine early warning, and El Niño.

Heffernan, C. 1995. Restocking Pastoralists: A review of the literature'. MSc dissertation, Veterinary Epidemiology and Economics Research Unit, University of Reading. December 1995.

Hiernaux, P. (1996) 'The Crisis of Sahelian Pastoralism: Ecological or economic?' *ODI Pastoral Development Network Paper* 39a. 20p. London: Overseas Development Institute.

Jakhmola, R.C. 2005. Final Report (1999-05) National Agriculture Technology Project "Evaluation of locally available feed and fodder to improve quality and formulate complete economic rations with high roughage diets". C.S.W.R.I., Avikanagar.

Jakhmola, R.C. and Jain, R.K. 2003. Sustainable Animal Production. Pointer Publishers, Jaipur .

Juyal, G.P. 2004. Bioengineering measures for rehabilitation of highly degraded lands - minespoils and landslides. In Compendium of lectures 'Winter school on

Conservation measures for reclamation of degraded lands in a watershed'. CSWCRTI, Dehradoon. (8-18, Sept., 2004). Pages 181-189.

Mitra, R., Christensen, G.I. and Johnson, H.D. 1972. Effect of prolong thermal exposer on growth hormone (GH) secretions in cattle. *Journal of Animal Science* 34, 776-779.

N.C.A. 1976. Report on Climate. National Commission on Agriculture, Govt of India New Delhi. Part IV: 1-124.

Pathak, N.N. and Jakhmola, R.C. 1983. *Forages and livestock production*. Vikas Publishing House, New Delhi.

Pathak, P.S. and Gupta, J.N. 2003. Grazing Resources in India - Problems and prospects. In *Sustainable livestock production* (Eds. R.C.Jakhmola and R.K.Jain), Pointer Publishers, Jaipur. pp 25-53.

Prasad, J.N.S. 2003. Farming system approach for drought management. In Compendium of lectures Summer school on drought Management, CRIDA, Hyderabad (6-26, May, 2003), 171-179.

Raizada, A. 2004. Silvi-pastoral practices for degraded land. In Compendium of lectures 'Winter school on Conservation measures for reclamation of degraded lands in a watershed'. CSWCRTI, Dehradoon. (8-18, Sept., 2004), 220-230.

Roy, M.M. 2003. Grassland and pasture management system in arid and semi-arid region. In *Sustainable livestock production* (Eds. R.C.Jakhmola and R.K.Jain), Pointer Publishers, Jaipur. 54-82.

Russel, C.R.1986. Cactus ecology and range management during drought. In Proc. Workshop Livestock and wild life management during drought. Kingsville, TX. pp 59-69.

Singh, D. and Swarnkar, C.P. 2003. Gastrointestinal nematodiasis in grazing ruminants. In *Sustainable livestock production* (Eds. R.C.Jakhmola and R.K.Jain), Pointer Publishers, Jaipur. 264-297.

❑❑❑

13

Nutritional Manipulation for the Optimizing Wool Production

RAMAN MALIK, BANDLA SRINIVAS
OM HARI CHATURVEDI
Central Sheep and Wool Research Institute, Avikanagar - 304 501, Rajasthan

Introduction

The effects of nutrition of sheep on the amount of wool grown, on the components of fiber growth (length and diameter), and on the protein composition and strength of the fibers have been studied extensively. Although much of this information comes from studies with Merino sheep, other breeds and crossbred sheep have also been studied. Several reviews have summarized the information available from studies with sheep (Allden, 1979; Purser, 2006; Reis, 1979, 1989a, 1992a; Russel, 2002; Sumner and Bigham, 2003). The quality of wool or mohair may be regarded as the various characteristics and defects in the fleece, both inherited and acquired, that influence its end use. Some quality factors can be influenced by the supply of nutrients available to wool follicles, and presumably also to follicles in Angora goats, during the process of fiber growth (Russel, 2002). These factors are mean fiber diameter, fiber length, and fiber strength. In both Merino sheep (Whiteley, 1984) and Angora goats (Uys *et al.*, 1985; McGregor, 1990) fiber diameter is the major determinant of the price paid for the cleaned fiber. The presence of medullated fibers is an important quality factor in the fleece (McGregor, 1990). Although medullation is influenced markedly by genetics and age, possible effects of nutrition will be considered. This lecture outlined briefly the effects of nutrition on wool production. This review is an attempt to summarize the main features of the extensive information available on sheep. Nutritional effects on both the quantity and quality of fiber produced are considered.

The Process of Fiber Growth

To understand the effects of general nutrient status and of specific nutrients on the growth and properties of fibers, the process of fiber growth needs consideration. Fiber diameter and fiber length growth determine the amount of wool produced by an animal, as well as its quality. The amount of wool grown by a sheep over a defined period of time depends on 1) the number of wool follicles growing a fiber, 2) the rate of growth of each fiber, and 3) the diameter of the fiber. Virtually all follicles in sheep are actively growing a fiber at any given time; there is little seasonal variation in the rate of fiber growth provided nutrition and environmental temperature are kept constant (Reis, 1982). The growth rate of an individual fiber depends on following parameters.

(1) The number and size of cells in the proliferative zone of the follicle bulb (bulb cells) and the proportion of newly formed cells that migrate from this zone.

(2) The rate of division of bulb cells as measured by the time taken for the total cell population in the proliferative zone to be renewed (turnover time).

(3) The proportion of migrating cells that enter the fiber.

(4) The increase in size of cells forming the fiber (Black and Reis, 1979).

Most of these events are influenced by the nutrient supply to the follicles and the amount of wool grown by a sheep depends on changes in cross-sectional area (calculated from diameter) and length growth rate of fibers.

Seasonality of Fiber Growth

Most of the breeds of sheep exhibit an annual rhythm in the rate of fiber growth that is controlled by day length unlike Merino sheep which exhibit little seasonal rhythm in wool growth (Hutchinson, 1976). Growth rates are lowest around the shortest day and may be only 20% of the maximum rate as observed in Scottish Blackface sheep (Doney, 2004). Minimum rates of fiber growth in winter in most sheep breeds are at least 50% of the maximum (Nagorcka, 1979). During winter, the rate of fiber growth is less responsive to nutritional status in breeds (Allden, 1979; Sumner and Bigham, 2003). Seasonal changes in fiber production are often confounded with seasonal changes in nutrient status. Less experimental evidence is available regarding the seasonality of fiber growth in Angora goats; the effects of season have often been confounded with nutrition (Stapleton, 1978). Fiber growth is subjected to substantial seasonal rhythm and growth rate is reduced in winter (Stapleton, 1978; Russel. 2002). During winter, fiber growth rate was lowest (approximately 60% of maximum rate), and the proportion of inactive follicles was highest. Both length growth and fiber diameter contributed to these changes.

The Relationship Between Length Growth and Diameter of Fibers

Some nutritional experiments showed fourfold increase in the rate of wool growth of individual Merino or Corriedale sheep by varying nutrition (Downes, 1971; Downes and Sharry, 1971). However, the ratio of mean length of fiber grown per day (L): mean fiber diameter (D) changed by no more than approximately 10%. Although both L and D of fibers and wool growth increased by supplying amino acids (AA) or protein to the abomasum, the ratio of L:D remained constant or varied only slightly (Reis, 1992b). Unless wool growth rate is constant, variations in emergence time of wool fibers will cause incorrect correlations of L and D when the fibers are clipped. Although the ratio of L:D may be nearly constant for individual sheep, it varies widely (from 10 to 18) between sheep (Downes, 1971). The approximate constancy of the ratio of L:D means that, whenever wool growth is increased by a nutritional treatment, D automatically increases; an increase in D reduces the quality and thereby the value of a given weight of wool. Reis (1989b) employed the constancy of the L:D ratio within sheep to calculate the relative amount of wool grown as fiber diameter increased. By accounting for changes in the output of wool, and in the price paid for wool, as fiber diameter varies, it is possible to calculate the value of wool grown by a sheep as production rate varies.

Figure 1 is produced from data for Merino wool presented by Reis (1989b). In two seasons (1979/80 and 1983/84) it proved profitable to increase wool production because the extra wool growth more than offset the drop in price due to increased fiber diameter. However, the premium paid for finer wool in 1987/88 was large, so extra wool growth actually reduced the value of the wool grown for fiber diameters up to about 25 pm. Hence, nutritional strategies to increase production of wool or mohair need to account for the current price structure in relation to fiber diameter. Changes in fiber growth in breeds of sheep that exhibit a seasonal rhythm of wool growth are due to changes in L and D of growing fibers. The relative value of wool grown as fiber diameter varies for Merino combing fleeces grown in three separate seasons. The amount of wool was obtained by regarding wool fibers as cylinders and calculating the volume grown per day from fiber diameter, assuming a value of 15 for the length: diameter ratio. A different value would alter the position, but not the shape, of the curves.

In addition, variations in the proportions of active to resting follicles can also alter fiber production appreciably. The proportion of active to resting follicles also varies seasonally in Angora goats (Stapleton, 1978; McGregor, 1990). The L:D ratio was not constant for an individual sheep; it varied inversely with the inherent seasonal pattern of wool growth (Woods and Orwin, 2000). However, monthly L and D measurements were often poorly correlated. L:D in general, ranges from 21 to 27 (Sahlu and Fernandez, 1992; Sahlu *et al.*, 1993) measured both diameter and fiber length; their data suggest that L:D ranges from 21 to 27. Data from Clarke and Smith (1975), Qi *et al.* (1992), and Sahlu *et al.* (1992), based on staple length, gave a similar range, but a few values were as high as 30. In some species, like Angora goats, fiber length growth for fibers of similar diameter may greater by at least 50% than Merino or other breeds but the change could not be in response to nutrition (Sahlu *et al.*, 1993; Badenhorst and Diedericks, 1993). Studies with sheep indicate that length growth tends to increase slightly more than diameter as wool growth is increased by nutrition. The reaction-diffision theory of hair fiber formation (Nagorcka and Mooney, 1989) proposes that changes in follicle activity alter the rate of fiber growth by influencing both length and diameter.

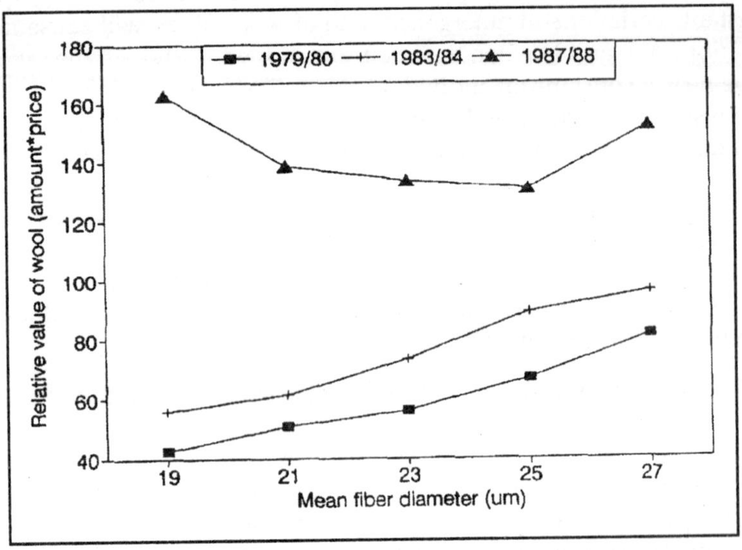

Fig. 1.

The Contribution of Length and Diameter to Changes in the Rate of Fiber Growth

Based on the constant ratio of L:D as observed in Merino sheep, calculations were drawn for the expected contribution of L and D to wool growth as the rate of growth changes. Changes in cross-sectional area of fibers (calculated from diameter) make the greatest contribution to fiber growth. The relative contributions of L and D to fiber growth are not affected by large differences in the ratio of L:D for different animals or by the magnitude of the values for L and D. However, the contributions of L and D to fiber growth change as the rate of fiber growth increases. Calculations in Table 1 show that as the rate of fiber growth increases, the contribution of L to the extra fiber growth decreases. The contribution of D to extra fiber growth increases because of a large increase in a component that we term D(L) (i.e., the increased diameter of the extra length of fiber).

The calculations in Table 1 show that the contribution of L to increased fiber growth would usually be in the range 20 to 30%: the diameter (including D(L)) accounted for the remaining 70 to 80%. These calculations kept the ratio of L:D constant. If L:D is increased by 10% from the control period to the increased growth period (from 15 to 16.5), the values for calculation 2 (Table 1) would change such that L contributed 32.6% of the increase in fiber growth. The calculations in Table 1 can be compared with observations made by Reis (1992b), when wool growth was approximately doubled by the abomasal infusion of AA into Merino sheep; the contributions of L, D, and increased diameter of the added length to increased fiber growth were 25.4, 53.6, and 21.0%, respectively. In contrast, the observations of Hynd (1989), in which wool growth of sheep was increased by 45% by altering nutrition, indicate that L may make a much greater contribution (more than 50%) to increased fiber growth. The higher length growth and diameter of fibers usually observed in Angora goats would not alter the relative contribution of these two components to the growth of mohair fibers. However, if length growth is not responsive to nutrient intake, as suggested by some published data for Angora goats, diameter changes may constitute an even larger component of increased mohair growth. This is illustrated in Figure 2.

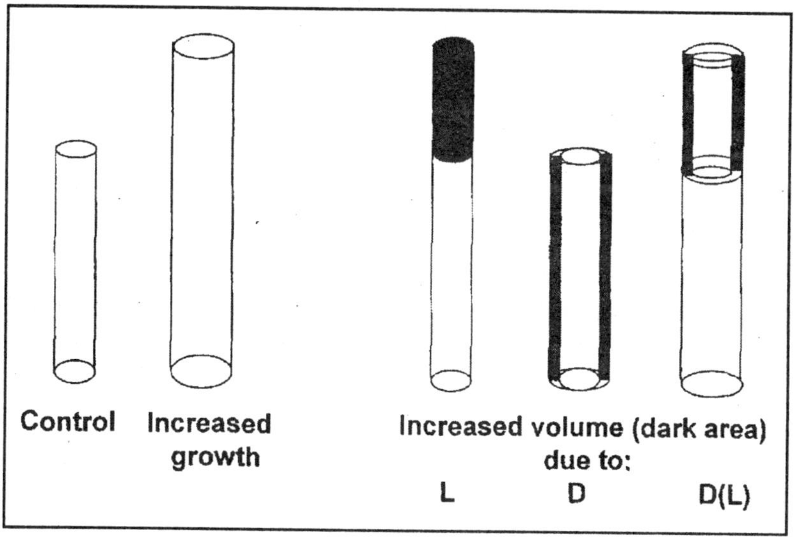

Fig. 2. Diagram of the contribution of length and diameter to increased fiber growth based on values of calculation 3, Table 1.

The Influence of Nutrient Supply on Fiber Growth

This section will consider the effects of nutrient supply on fiber growth only during adult life of nonpregnant, nonlactating animals. The effects of nutrition in pregnant and lactating animals will be discussed later. Many studies have demonstrated the large effect of feed intake on wool growth in various breeds of sheep (Allden, 1979). Likewise, fiber growth in Angora goats responds to increased feed intake (Russel, 2002). Merino sheep have been reported to respond similarly to increased feed intake at any time of the year (Allden, 1979; Reis, 1979); in contrast, most other breeds of sheep are less responsive during periods of short day length (Allden, 1979; Sumner and Bigham, 2003). The rate of growth of mohair fibers may also be less responsive to nutrition during the winter months, but definitive data are lacking. To determine which specific nutrients influence fiber growth, it is necessary to avoid the degradative activity of ruminal microbes. The major nutritional limitation to wool growth is the amount and composition of AA available to wool follicles; the supply of sulfur containing (S) AA (cyst[e]ine and methionine) often limits wool growth, and lysine supply is also important. A large amount of cyst[e]ine is required to synthesize wool proteins, and much of this can be provided by conversion of methionine. However, methionine also plays a more specific role in stimulating wool growth (Reis, 1989a). The rates of wool growth were near maximum at a maintenance energy intake when approximately 150 g of AA, given as casein via the abomasum, was provided for intestinal absorption (Reis, 1969). Protein flow of 150 g/d to the abomasum yielded the maximum rate of wool growth (Hynd and Allden,1985). Abomasul infusion of most of the nutrients showed that the rate of wool growth was markedly altered by protein supply, but not by energy supply (Reis *et al.*, 1992). However, a small, but significant, interaction between the effects of protein and energy on wool growth has been detected (Black *et al.*, 1973). The maximum wool growth rate (20 g/d of clean wool) of a very-high-producing sheep would be sustained with approximately 0.43 MJ/d (Black and Reis, 1979; Black, 2007). This value is equivalent to only 9% of the basal metabolic rate for a 40-kg sheep. If this figure is correct, the energy requirement for wool growth would usually be less than 5% of basal metabolic rate. Similar values might be expected to apply to the energy requirement for synthesis of mohair fibers. However, Herselman and Smith (1991) calculated the energy cost of fiber production to be 136 k J of ME/g of fiber synthesized which indicated that the energy requirement for fiber synthesis by Angora goats may be somewhat higher than the above estimate for sheep. Several feeding experiments with Angora goats have shown that mohair growth responds to the intake of extra nutrients in both wethers and dry does. These experiments do not define which nutrients are required by the follicles because ruminal microbes alter the nutrient content of digesta before absorption. When Angora goats were totally hand-fed, increasing dietary protein (supplied by cottonseed meal) increased clean fleece weight and fiber diameter (Stewart *et al.*, 1971). In another study, supplements providing either protein or energy increased mohair production of goats on pasture (Huston, 1980). Several studies with Angora goats have shown that both the rate of fiber growth and fiber diameter were increased as the energy (Calhoun et al., 1988) or protein (Deaville and Galbraith, 1992; Shahjalal *et al.*, 1992; Hart *et al.*, 1993; Sahlu *et al.*, 1993) in the diet was increased. When yearling Angora wethers were Angora wethers were fed isocaloric diets containing three levels of dietary protein (9, 15, and 21% CP on a DM basis), clean mohair production increased linearly with increasing protein in the diet (Sahlu *et al.*, 1993). Since rumen degradation of protein is not a very efficient process of protein conversion from true protein from the diet, proteins or AA or treated with formaldehyde (Throckmorton *et al.*, 1982) or heat-treated (Sahlu *et al.*, 1992) to reduce the extent of degradation by the ruminal microbes and thereby enhancing the availability of proteins or AA at abomasums. Both formaldehyde- treated casein and heat-treated soybean meal increased the growth rate of wool fibers. Looking into the importance of L-methionine and L-lysine in the wool growth, attempts have been made to encapsulate these amino acids to protect against ruminal degradation. The effect of encapsulated L-methionine or L-lysine or both are very inconclusive with some reports claiming small increase in wool fiber growth when 0.1 and 0.2% of a different polymer encapsulated methionine was included in the diet (Bassett *et al.*, 1981; Sahlu and Fernandez, 1992) but not with the lysine (Sahlu and Fernandez, 1992) and others claiming no increase in wool production (Craddock and Bassett, 1977). Probably, encapsulation may have been ineffective or animals might have been received a high intake of good-quality diet which it-self sustained the higher basal rate of fiber growth in sheep or Angora goat.

Table 1. Calculations of the contribution of length (LJ and diameter (D) to changes in the rate of fiber growth.

Component of fiber growth	Control Period	Increased growth period	Increased volume due to		
			L	D	D(L)[a]
Calculation-1					
L, ìm/d	300	345	---	---	---
D, ìm	20	23	---	---	---
L:D	15	15	---	---	---
Volume x 10-3, ìm 3/d	94.3	143.4	---	---	---
Increase in volume x 10-3, ìm 3/d	---	49.1	14.1	30.4	4.6
Increase in volume, %	---	52.1	---	---	---
Contribution to increased volume %	---	---	28.7	61.9	9.4
Calculation-2					
L, ìm/d	300	375	---	---	---
D, ìm	20	25	---	---	---
L:D	15	15	---	---	---
Volume x 10-3, ìm 3/d	94.3	184.1	---	---	---
Increase in volume x 10-3, ìm 3/d	---	89.8	23.5	53.0	13.2
Increase in volume, %	---	95.2	---	---	---
Contribution to increased volume %	---	---	26.2	59.0	14.8
Calculation-3					
L, ìm/d	300	450	---	---	---
D, ìm	20	30	---	---	---
L:D	15	15	---	---	---
Volume x 10-3, ìm 3/d	94.3	318.2	---	---	---
Increase in volume x 10-3, ìm 3/d	---	223.9	47.1	117.8	59.0
Increase in volume, %	---	237.4	---	---	---
Contribution to increased volume %	---	---	21.0	52.6	26.4

[a] Increased diameter of the extra length of fiber grown

Effects of Minerals and Vitamins on Fiber Growth

In addition to the major nutrients, several vitamins and trace elements are required for fiber growth (Purser, 2006; Reis, 1989a). Although vitamin supply has not been shown to restrict wool growth in adult sheep, folic acid and pyridoxine are especially important for wool growth (Reis, 1989a). Copper and Zn are required directly in the process of fiber growth (Purser, 2006), and deficiencies of these elements can substantially reduce wool growth and weaken fibers. Qi *et al.* (1992) fed Angora goats diets based on bermudagrass hay, ground peanut hulls, and ground corn, supplemented with $CaS0_4$ to provide 0.16 to 0.34% S (DM basis). Clean mohair production was increased by sulfate supplementation; for maximum mohair growth 0.267% dietary S was required. This gave N: S ratio of 7.2:1, which suggested that, the N: S ratio of 10: 1 recommended for goats (National Research Council, 1981) was inadequate for Angora goats. Wool growth requires high levels of S-amino acids (Reis, 1989a); by analogy, with sheep, the effects of *S* supplementation in the above experiment most likely were mediated by way of effects on digestion in the rumen and flow of nutrients (S-AA) from the rumen (Purser, 2006).

Effects of Nutrition on Fiber Growth During Pregnancy and Lactation

Corbett (1979) reviewed the effects of pregnancy and lactation on wool growth in sheep. Maturation of secondary follicles in lambs is retarded by a restricted nutrient supply of their dams during late pregnancy; adult wool production of these animals may be reduced if the restriction is severe. Wool growth by the ewe typically is reduced during the last 2 mo of pregnancy and during early lactation. A full cycle of reproduction in ewes reduces annual fleece production by 10 to 14% (Corbett, 1979). A reduced supply to wool follicles of one or more essential nutrients is presumably responsible for these effects (Oddy and Annison, 1979). In contrast to nonlactating, nonpregnant sheep, wool growth was not primarily limited by the availability of S-AA (cyst[e]ine and methionine) during late pregnancy and early lactation (Williams *et al.*, 2005; Oddy, 1985; Masters and Stewart, 1990). However, Stewart et al. (1993) found that methionine may limit wool growth during early lactation. On the basis of changes in plasma AA concentration, Masters et al. (1993) suggested that the supply of other AA (arginine, lysine, or threonine) may limit wool growth during pregnancy and lactation, but they failed to obtain a response to abomasal injections of these AA (Stewart *et al.*, 1993). Angora does during early lactation were offered isoenergetic diets containing 10, 13, 16, and 21% CP and observed that milk production increased as dietary protein was raised to 16% but mohair production was decreased at 16% protein (Sahlu et al 1993). Somewhat surprisingly, milk production was reduced with 21% protein in the diet, whereas mohair production was increased. This result may reflect competition for nutrients in which the requirements for milk production take priority. It is also conceivable that the 21% CP diet provided excess protein, resulting in energy expenditure for urea synthesis. This energy drain may decrease milk production (but not mohair) and may, in fact, spare AA for mohair production. Compared with three does that kidded but were unable to lactate, mohair production by lactating does was reduced by as much as 29% in the group receiving 16% protein in the diet, and an overall reduction of 14% due to lactation was observed across treatments. This study suggests that the effects of pregnancy and lactation on fiber production by Angora goats may be similar to those reported for sheep.

Fiber Strength and Composition

The strength of wool fibers is influenced markedly by the nutrient supply via effects on fiber diameter and changes in intrinsic strength (Reis, 1992a). Certain AA (methionine and lysine), trace elements (Cu and Zn), and vitamins (folic acid) are of special importance. Deficiencies or imbalances in the supply of these nutrients can severely weaken fibers and even cause shedding of the fleece. Variations in the supply of AA to wool follicles have been shown to cause marked changes in the composition of wool; in particular, the supply of cyst[eline controls the synthesis of components of the high-sulfur proteins. However, no clear association has been established between the strength of wool fibers and the proportions of the constituent proteins (Reis, 1992a); the textile properties of wool are not significantly influenced by its composition (Marshall et al., 2000). Because the high-sulfur proteins of mohair and wool are remarkably similar (Gillespie, 2005), their synthesis is probably under similar nutritional control.

Nutrition and Medullation

The presence of medullated fibers (both kemp and med) in mohair reduces the value of the fleece. Kemp fibers are produced by primary follicles. The major factors influencing the proportion of medullated fibers are genetics and age (Tiffany-Castiglioni, 2006; Lupton et al., 1991). Because many of the secondary follicles that produce mohair fibers are not mature at birth, the fleeces of young Angora goats contain a higher proportion of kemp. Data from New Zealand indicate that nutrient supply may alter the kemp content of the fleece. Thus, Nixon et al. (2005) and Badenhorst and Diedericks (1993) reported that Angora goats receiving a high level of nutrition had higher proportions of kemp fibers. This subject has been reviewed by Tiffany-Castiglioni (2006) and Lupton et al. (1991), who both concluded that the proportions of medullated and kemp fibers are not markedly affected by nutrient supply.

Conclusion and Recommendations

The quantity and quality of fiber produced by sheep is markedly influenced by nutrition. Both length growth and diameter of fibers respond to changes in nutrient supplyThe major limitation to wool growth is the amount and composition of AA available to wool follicles; the S-amino acids are especially important. Maximum rates of wool growth can be sustained on maintenance intakes of energy if the supply of essential AA is adequate. An increased protein supply to the follicles can increase mohair production, but the importance of specific AA has not been established. A protein flow of 150 g/d to the abomasums may maximize the wool growth in sheep. Protein content of diet has major affect on the wool production rather than energy but protein energy interaction may affect the wool production.For maximum wool production, 0.267% of dietary sulfur is recommended. A ratio of 7:1 is recommended between N:S to cater the needs of most of the breeds of sheep rather followed practice of 10:1 which is insufficient for some breeds of sheep. In pregnant ewes, wool growth will be reduced during the last 2 months of pregnancy term. Under nutrition during this period may affect maturation of secondary follicles in lambs. There is competition between nutrient sweep to milk production and wool production. Milk production is first priority in the ewes provided total dietary protein content limit to 16 per cent. Strength of wool fiber and composition would affected by the level of nutrition but not the medullation and kemp fiber. To maximize financial returns it is important to know the relative responses of both length growth and diameter of fibers to changes in nutrient supply.

REFERENCES

Allden, W.G. 1979. Feed intake, diet composition and wool growth. In: J. L. Black and P. J. Reis (Ed.) Physiological and Environmental Limitations to Wool Growth, p 61. Univ. of New England, Armidale, New South Wales, Australia.

Badenhorst, G. and Diedericks, J. 1993. The effect of nutrition on some mohair quality traits. In: Fibre News (April-May issue). p 6. Whangarei, New Zealand.

Bassett, J. W., Baldwin, B.C., Calhoun, M.C. and Stobart, R.H. 1981. Plasma methionine and mohair response to dietary rumen- protected methonine in Angora goats. Texas Agric. Exp. Sta., PR 3904, Texas A&M Univ., p. 68.

Black, J. L. 2007. Mechanisms controlling the rate of growth, composition and morphology of wool. In: B. J. McGuirk (Ed.) Merino Improvement Programs in Australia. p 457. Australian Wool Corporation, Melbourne, Australia.

Black, J. L, and Reis, P. J. 1979. Speculation on the control of nutrient partition between wool growth and other body functions. In: J. L. Black and P. J. Reis (Ed.) Physiological and Environmental Limitations to Wool Growth. p 269. Univ. of New England, Armidale, New South Wales, Australia.

Black, J. L., Robards, G. E. and Thomas, R. 1973. Effects of protein and energy intakes on the wool growth of Merino wethers. *Australian Journal of Agricultural Research* 24, 399-406.

Calhoun, M. C., Lupton, C. J., Kuhlmann, S. W. and Baldwin, B. C. 1988. Dietary energy intake effects on mohair growth. Texas Agric. Exp. Sta., PR 4589, Texas A&M Univ. p 53.

Clarke, W.H. and Smith, I.D. 1975. A preliminary evaluation of mohair production and the potential of Angora goats in three eastern states. *Journal of Australian Institute of Agricultural Science* 41, 220-226.

Corbett, J. L. 1979. Variation in wool growth with physiological state. In: (Eds. J.L. Black and P. J. Reis) *Physiological and Environmental Limitations to Wool Growth*, p 79. Univ. of New England, Armidale, New South Wales, Australia.

Craddock, B.F. and Bassett, J.W. 1977. Effect of feeding protected protein on mohair production in Angora goats. Texas Agric. Exp. Sta., PR 3470, Texas A&M Univ. p 98.

Deaville, E. R. and Galbraith, H. 1992. Effect of dietary protein level and yeast culture on growth, blood prolactin and mohair fibre characteristics of British Angora goats. *Animal Feed Science and Technology* 38, 123-131.

Doney, J.M. 2004. The fleece of the Scottish Blackface sheep. IV. The effect of pregnancy, lactation and nutrition on seasonal wool production. *Journal of Agricultural Science* 62, 59-67.

Downes, A.M. 1971. Variations in wool length and diameter with sheep nutrition. *Applied Polymer Symposium* No. 18. p 895.

Downes, A.M. and Sharry, L.F. 1971. Measurement of wool growth and its response to nutritional changes. *Australian Journal of Biological Science* 24, 117-124.

Gillespie, J.M. 2005. The high-sulphur proteins of normal and aberrant keratins. In: A. G. Lyne and B. F. Short (Ed.) Biology of the Skin and Hair Growth. p 377. Angus and Robertson, Sydney, Australia.

Hart, S.P., Sahlu, T. and Fernandez, J.M. 1993. Efficiency of utilization of high and low quality forage by three goat breeds. *Small Ruminant Research* 10, 293-301.

Herselman, M.J. and Smith, W.A. 1991. The energy requirements of Angora goats. *Karoo Agriculture* 4,18-25.

Huston, J. E. 1980. Supplemental energy and protein effects on growth rate and mohair production in weaned Angora female kids. Texas Agric. Exp. Sta., PR 3706, Texas A&M Univ., p 58.

Hutchinson, J.C.D. 1976. Photoperiodic effects on hair and wool growth of domestic animals. In: *Progress in Biometeorology*, Division B, Vol. I, Part 2, 196311973. p 47. Swets and Zeitlinger, Amsterdam, The Netherlands.

Hynd, P.I. 1989. Effects of nutrition on wool follicle cell kinetics in sheep differing in efficiency of wool production. *Australian Journal of Agricultural Research* 40, 409-417.

Lupton, C. ., Pfeiffer, F.A. and Blakeman, N.E. 1991. Medullation in mohair. *Small Ruminant Research.* 5, 357-363.

Marshall, R.C., de Boos, A., Gillespie, J.M., Marler, J.W., Reis, P.J., Rogan, I.M., White, M.A. and Whiteley, K.J. 2000. *An investigation of the relationship of wool textile properties to fibre protein composition.* Proc. 8th Int. Wool Text. Res. Conf. 1, 266-274.

Masters, D.G. and Stewart, C.A. 1990. Wool growth and reproduction. In: C. M. Oldham, G. B. Martin, and I. W. Purvis (Ed.) Reproductive Physiology of Merino Sheep. Concepts and Consequences. p 265. Univ. of Western Australia, Perth, Australia.

Masters, D.G., Stewart, C.A. and Connell, P.J. 1993. Changes in plasma amino acid patterns and wool growth during late pregnancy and early lactation in the ewe. *Australian Journal of Agricultural Research* 44, 945-952.

McGregor, B.A. 1990. Nutritional, seasonal, management and processing factors affecting mohair quality and quantity. In: K. Thompson (Co-ordinator) Goat Health and Production, Refresher Course for Veterinarians, Proc. 134. p 347. Univ. of Sydney, Sydney, Australia.

Nagorcka, B.N. 1979. The effect of photoperiod on wool growth. In: *Physiological and Environmental Limitations to Wool Growth* (Eds. J.L. Black and P.J. Reis). p 127. Univ. of New England, Armidale, New South Wales, Australia.

Nagorcka, B.N. and Mooney, J.R. 1989. The reaction-diffusion system as a spatial organizer during initiation and development of hair follicles and formation of the fibre. In: *The Biology of Wool and Hair* (Eds. G. E. Rogers, P. J. Reis, K.A. Ward, and R.C. Marshall). p 365. Chapman and Hall, London.

Nixon, A.J., Saywell, D.P. and Bown, M.D. 2005. Nutritional effects on fibre growth cycles and medullated fibre production in Angora goats. Proc. N.Z. SOC. *Animal Production* 51, 359-367.

NRC. 1981. *Nutrient Requirements of Goats: Angora, Dairy and Meat Goats in Temperate and Tropical Countries*. National Academy Press, Washington, DC.

Oddy, V.H. 1985. Wool growth of pregnant and lactating Merino ewes. *Journal of Agricultural Science* 105, 613-620.

Oddy, V.H. and Annison, E.F. 1979. Possible mechanisms by which physiological state influences the rate of wool growth. In: *Physiological and Environmental Limitations to Wool Growth* (Eds. J. L. Black and P. J. Reis). p 295. Univ. of New England, Armidale, New South Wales, Australia.

Purser, D.B. 2006. Effects of minerals upon wool growth. In: *Physiological and Environmental Limitations to Wool Growth* (Eds. J. L. Black and P. J. Reis). p 243. Univ. of New England, Armidale, New South Wales, Australia.

Qi, K., Lu, C.D., Owens, F.N. and Lupton, C.J. 1992. Sulfate supplementation of Angora goats: Metabolic and mohair responses. *Journal of Animal Science* 70, 2828-2836.

Reis, P.J. 1969. The growth and composition of wool V. Stimulation of wool growth by the abomasal administration of varying amounts of casein. *Australian Journal of Biological Science* 22, 745-752.

Reis, P. . 1979. Effects of amino acids on the growth and properties of wool. In: *Physiological and Environmental Limitations to Wool Growth* (Ed. J. L. Black and P. J. Reis). p 223. University of New England, Armidale, New South Wales, Australia.

Reis, P.J. 1982. Growth and characteristics of wool and hair. In: *Sheep and Goat Production: World Animal Science* (Ed. I. E. Coop), C1. p 205. Elsevier, Amsterdam, The Netherlands.

Reis, P. J. 1989a. The influence of absorbed nutrients on wool growth. In: *The Biology of Wool and Hair* (Eds. G.E. Rogers, P.J. Reis, K.A. Ward, and R. . Marshall). p 185.Chapman and Hall, London.

Reis, P.J. 1989b. Pastures and wool quality. *Proceeding of Annual Conference of Grassland Society of New South Wales* 4, 63-72.

Reis, P. J. 1992a. Variations in the strength of wool fibres - a review. *Australian Journal of Agricultural Research* 43, 1337-1344.

Reis, P.J. 1992b. Length growth and diameter relationships of Merino wool fibres. *Wool Technology and Sheep Breed* 40, 52-59.

Reis, P.J., Tunks, D. A. and Munro, S. G. 1992. Effects of abomasal protein and energy supply on wool growth in Merino sheep. *Australian Journal Agricultural Research* 43, 1353-1361.

Russel, A.J.F. 2002. Fibre production from sheep and goats. In: *Progress in Sheep and Goat Research* (Ed. A. W. Speedy). Chap. 11. p 235. CAB International, Wallingford, U.K.

Sahlu, T. and Fernandez, J.M. 1992. Effect of intraperitoneal administration of lysine and methionine on mohair yield and quality in Angora goats. *Journal of Animal Science* 70, 3188-3194.

Sahlu, T., Fernandez, J.M., Lu, C.D. and Manning, R. 1992. Dietary protein level and ruminal degradability for mohair production in Angora goats. *Journal of Animal Science* 70, 1526-1535.

Sahlu, T., Hart, S.P. and Fernandez, J.M. 1993. Nitrogen metabolism and blood metabolites in three goat breeds fed increasing amounts of protein. *Small Ruminant Research* 10, 281-289.

Shahjalal, Md., Galbraith, H. and Topps, J.H. 1992. The effect of changes in dietary protein and energy on growth, body composition and mohair fibre characteristics of British Angora goats. *Animal Production* 54, 405-412.

Stapleton, D.L. 1978. Mohair production and seasonal variability in the fleece of the Australian Angora goat. Ph.D. Thesis. Univ. of New England, Armidale, New South Wales, Australia.

Stewart, C.A., Masters, D.G.., Williams, I.H. and Connell, P.J. 1993. Changes in plasma amino acid patterns and wool growth in response to abomasal injections of amino acids during late pregnancy and early lactation. Australian *Journal of Agricultural Research* 44, 959-966.

Stewart, J.R., Shelton, M. and Haby, H.G. 1971. Nutritional investigations with Angora goats. Texas Agric. Exp. Sta., PR 2933, Texas A&M Univ. p 64.

Sumner, R.M.W. and Bigham, M.L. 2003. Biology of fibre growth and possible genetic and non-genetic means of influencing fibre growth in sheep and goats - a review. *Livestock Production Science* 33, 1-8.

Throckmorton, J.C., Ffoulkes, D., Leng, R.A. and Evans, J.V. 1982. Response to bypass protein and starch in Merino sheep and Angora goats. *Proceedings of Australian Society of Animal Production* 14, 661-667.

Tiffany-Castiglioni, E. 2006. Review: Genetics and management of kemp in mohair. Texas Agric. Exp. Sta., PR 4403, Texas A&M Univ. p 60.

Uys, D.S., van der Westhuysen, J.M. and Wentzel, D. 1985. Biologiese faktore wat die gehalte van sybokhaar be'invloed. *South African Journal of Animal Science* 15, 128-134.

Whiteley, K.J. 1984. Quantitative importance of wool characteristics in processing. In: *Proceedings of a Seminar on Wool Production in Western Australia* (Eds. S. K. Baker, D.G. Masters, and I.H. Williams). p 1. *Australian Society of Animal Production,* Perth, Australia.

Williams, A.J., Tyrrell, R.N. and Gilmour, A.R. 2005. Effects on wool production during pregnancy and lactation of twice daily abomasal supplementation with casein or methionine and cystine. *Australian Journal of Experimental Agriculture and Animal Husbandry* 18, 52-58.

Woods, J.L. and Orwin, D.F.G. 2000. Seasonal variations in the dimensions of individual Romney wool fibres determined by a rapid auto radiographic technique. *New Zealand Journal of Agricultural Research* 31, 311-317.

□□□

14

Recent Advances in Reducing Enteric Methane Emission in Ruminants Through Feeding Strategies

R.C. JAKHMOLA

Arid Region Campus, Central Sheep and Wool Research Institute
Bikaner, Rajasthan - 334 006

Introduction

The ruminants thrive on roughages rich in plant cell walls - cellulose and other plant fibre polymers that are fermented in the rumen by a complex mixture of microbial species that have cellulolytic capabilities. Owing to these capabilities ruminants have traditionally been seen as "mobile fermentors" (Hungate, 1966). The rumen microbial ecosystem comprises consortium of different microbial groups living in symbiotic relationship with the host, where they convert lignocellulosic feeds into volatile fatty acids that in turn serve as a source of energy for the animals. The rumen microbial ecosystem consists of bacteria (10^{10}–10^{11} cells/ml), ciliate protozoa (10^4–10^6/ml), anaerobic fungi (10^3–10^5 zoospores/ml) and bacteriophages (10^8–10^9/ml) (Kamra, 2005). Gases - carbon dioxide, methane and traces of hydrogen that are generated during fermentation processes help in maintaining anaerobic conditions in the rumen. Some of the oxygen entrapped in the feed consumed by the animal is utilized by the facultative anaerobes present in the rumen and thus a perfect anaerobic condition is generated and maintained. Therefore, only the microbes that are able to tolerate a low redox potential (-350 mV) are able to survive and the rest are eliminated from the rumen system. Further, a narrow range of pH is maintained in the rumen which is not generally disturbed due to its high buffering capacity.

In the recent time there is increasing awareness on global warming and contribution there in by methane is considered important. Methane is a potent greenhouse gas and its release into the atmosphere is directly linked with animal agriculture, particularly ruminant production. Methane emissions result during microbial fermentation of feed within the rumen and represent a loss in productive energy for the animal. Methanogenesis in the ruminant gut is calculated to account for 15-25% of the global methane emission of around 7.5×10^{12} g yr-1 (Kirchgessner et al, 1995). Energy lost as enteric methane from mature ruminants (cattle) ranges from 2 to nearly 12% of GE intake (Johnson et al., 2000). Mitigating methane losses from ruminants will have long-term environmental benefits in terms of decreasing agriculture's contribution to greenhouse gas emissions. There is a growing interest now a day in decreasing the potential for global warming by reducing emissions of greenhouse gases into our atmosphere (Moss et al., 2002). Thus there is a need to understand the factors affecting enteric methane production to decrease the greenhouse gas emission and develop reduction strategies.

The Methanogens

The methanogens play a vital role in the rumen by scavenging molecular hydrogen generated during rumen fermentation while converting hydrogen oin in methane. Major methanogenic archea isolated from rumen are *Methanobrevibacter ruminantium, M. smithii, Methanobacterium formicicum Methanosarcina barkeri and Methanomicrobium mobile* (Kandler and Hippe, 1977; Patterson and Hespell, 1979; Lovely, et al., 1984; Miller, et a.l, 1986). In rumen methanogens live in close association with entodiniomorphid protozoa and bacteria possibly to get a constant supply of hydrogen because on pumping hydrogen in the rumen, the methanogens get detached from the protozoa (Stumm et al., 1982).

Rumen Fermentation

Methane is the major electron sink product of the fibre-degrading cellulolytic-methanogen consortium; by comparison the amylolytic-methanogen consortium produces relatively more propionate and less methane (Kirchgessner et al., 1995). Methanogenesis is high because bicarbonate is by far the most important extracellular electron acceptor in the ruminant gut; alternative electron acceptors including sulphate and nitrate are present in only small amounts in the feed. Unsaturated fatty acids arising from lipid hydrolysis are hydrogenated (Hungate, 1966). Rapid hydrogenation of the phenolic compounds ferulic, coumaric and caffeic acids also occurs, but unlike bicarbonate, these compounds are not renewed as a result of fermentation so their capacity to act as alternative sinks is limited (Prins and Stewart, 1997). Formation of acetate in the rumen promotes methane production, whereas propionate production and methanogenesis are competing processes (Moss et al., 2000). Fermentation acids produced in the rumen are toxic to methanogenic bacteria at pH less than 6 (Van Kessel and Russell, 1996).

Interactions in cellulolytic microbes

The synergism and antagonism among the different groups of microbes and even among different genera of the same group is so diverse and complicated that it is difficult to quantify the role played by any particular group of microbes present in the rumen. The net result of these reactions undergoing in the rumen is responsible for the bioconversion of feed into such form that is utilizable by the animal. These microbes survive in the rumen in spite of presence of anti-nutritional factors, which might limit the growth of some of these natural microbial inhabitants.

Cellulolytic microbes in the rumen have both synergetic and competitive relationships. Nutritional interactions are based primarily upon interspecies competition for and transfer of nutrients and/or protons. Many saccharolytic bacteria and methanogens stimulate microbial growth and cellulolysis by removing products or supplying nutrients such as branched-chain VFA (Wolin et al., 1997). For example the propionate-producing bacterium *Selenomonas ruminantium* uses sugars and H_2 produced by the anaerobic cellulolytic fungus *Neocallimastix frontalis* (Marvin-Sikkema et al., 1990) and the presence of *S. ruminantium* enhances cellulolysis. In contrast, addition of *S. ruminantium* to co-cultures of *N. frontalis* and *Methanobrevibacter smithii* reduced methanogenensis, presumably because the bacterium, which competes successfully for sugars released upon cellulose hydrolysis, produces less H_2 than the fungus (Table 1). Co-culture of cellulolytic species reveals the existence of competitive and inhibitory interactions, co-cultures often demonstrating reduced fibre degradation compared to that achieved by the more active organisms grown in pure culture. This may be based on competition for cellulose, perhaps through competition for attachment sites as the bacteria spread from their initial point of attachment, or competition for soluble nutrients. Thus, cultures of *Ruminococcus flavefaciens* strains were found to be more active in degrading fibrous substrates such as barley straw and clover than were cultures of *Fibrobacter succinogenes* strains or mixed cultures containing both species (Saluzzi et al., 1993). Some cellulolytic bacteria produce anticompetitor compounds (Bernalier et al, 1992, Odenyo et al., 1994, Stewart et al., 1992, Kopecny et al., 1996). The chitinolytic rumen bacterium *Clostridium tertium* inhibits fungal cellulolysis by an as yet unknown mechanism (Kopecny et al., 1996), presumably involving chitinase. Also, inhibitor production *in vivo* is presumably regulated to ensure the survival of producers, rather than the complete elimination of competitors. In the rumen, *Ruminococcus* species and anaerobic fungi co-exist, despite the almost total inhibition of fungal growth on cellulose by ruminococci *in vitro* (Stewart et al, 1992). Although inhibitor-susceptible strains may be eliminated in laboratory co-culture, they may survive in high numbers in the gut. This may reflect the wide variety of micro- habitats present *in vivo* (Stewart, 1999).

Dietary factors and enteric methane emission

Feed intake and composition

Methane emission increases with feed intake, but the high degree of variation between individual animals exists. The level of feed intake and the composition of the diet (Johnson and Johnson, 1995; Moss et al., 2000; Benchaar et al., 2001) affect the methane production. On intensive feedlot systems less methane is produced per unit of meat produced compared with extensive grazing systems due to faster growth rate (Clemens and Ahlgrimm, 2001) and higher feed efficiency. On diets containing mainly maize grain it was estimated that feedlot cattle lose 3.5% of GE intake as methane (Houghton et al., 1996). Johnson et al. (2000) speculated that high-grain barley diets result in greater methane emissions than would be expected from high-grain corn diets, based on methane losses of 6.5 to 12% of GE reported previously for barley diets (Hashizume et al., 1968; Whitelaw et al., 1984).

Table 1. Fate of hexose C in mono, di- and tri-cultures grown on cellulose (Stewart, 1999)

Inoculuma	CO_2 [c]	CH_4	Hfor[b]	Hac	Hpr	Hlac	Eth	Hsucc
N. frontalis strain (RE 1)	10.0	-	23.0	32.0	-	20.0	11.0	4.0
RE1 + *M. smithii* strain (PS)	10.1	16.6	-	60.8	-	3.1	5.8	3.5
RE1 + PS + *S.ruminatium* strain JW13	7.5	6.0	-	40.0	36.2	2.1	6.0	-

[a]Species N = Neocallimastix, M = Methanobrevibacter, S = Selenomonas.

[b]Hfor, Hac, Hpr, Eth, Hsucc = formate, acetate, propionate, lactate, ethanol and succinate respectively; - = not detected.

[c]Data shows the C in each product expressed as a percentage of the total C recovered .

Methane emissions decrease as the proportion of concentrate in the diet increases (Johnson and Johnson, 1995). It is also evident that the source of grain can also have dramatic consequences on enteric methane production from feedlot cattle. The cattle had greater enteric methane production when fed a corn diet compared with those fed a barley diet (Beauchemin and McGinn, 2005). Lower methane emissions can also be due to a shift in the site of digestion from the rumen to the intestines, and rolled corn is typically less extensively digested in the rumen than rolled barley (Yang et al., 1997). The cows consuming 5.2 to 8.6 kg/d diets that contained 26 to 41% ground or steam-rolled barley lost 7.4 to 11.6% of GE in the form of methane (Hashizume et al., 1968). On a pelleted diet containing 85% barley and 15% protein supplement loss of energy through methane ranged from 10.6 to 11.5% of GE (Whitelaw et al. 1984). But they were fed restricted amounts (70% of ad libitum intake, approximately 3.8 kg/d) of feed. It seems that methane emissions from feedlot cattle fed high-grain barley diets are slightly greater than would be expected from cattle fed corn diets (Beauchemin and McGinn, 2005), and slightly greater than the value of 3.5% of GE (Houghton et al., 1996).

Table 2. Methane and CO_2 emissions from feedlot cattle fed barley or maize diets (Beauchemin and McGinn 2005)

Item	Maize	Barley	SEM
DMI kg/d	6.94	5.34	0.55
Methane			
g/(heifer, d)	170.6	129.7	7.9
g/kg of DMI	24.8	24.3	0.1
% of GE intake	7.55	7.28	0.29
CO_2 kg/d	3.50	2.78	0.62

Feed additives

A wide range of feed additives was proposed to reduce methane. These include alternative hydrogen acceptors (e.g. malate, fumarate), halogenated methane analogues (e.g. chloroform, bromoethanesulphonic acid), antibiotics (e.g. monensin, mevastatin), defaunating agents (e.g. manoxol, teric), and probiotics, bacteriocins and naturally occurring plant compounds (e.g. condensed tannins). Davies et al. (1982) successfully inhibited methanogenesis with a series of 2,4-bis (trichloromethyl)benzo[1,3] dioxins in mixed ruminal cultures, in acute intraruminal models, and in chronic (4-week) experiments with sheep and cattle. Concurrent with significantly less methanogenesis, VFA production was in some cases unchanged, and in other cases it was increased. Feed efficiency during 14 weeks of treatment was improved in cattle fed the methanogenesis inhibitors. Monensin, an ionophore and gram-positive antibiotic decrease methanogenesis (McGuffey et al., 2001, Thornton and Owens, 1981). An increase in VFA production was observed when methanogenesis was inhibited by the ionophore monensin (Thornton and Owens, 1981). The increased feed efficiency achieved by feeding monensin to cattle was interpreted as a result, at least in part, of decreased methanogenesis (Thornton and Owens, 1981). Bromoethanesulfonate, a specific inhibitor of a key enzyme in methane formation (methyl-coenzyme M reductase)(Gunsalus et al., 1978), has been tested in an infused sheep rumen; however, resistance develops within a few days (Immig et al., 1996). The studies of Razwan et al. (2003) are based on the hypothesis that treating animals with a specific inhibitor of methanogenesis will have a beneficial effect on the animal by increasing the levels of VFAs in the rumen. However, there are problems with many of these compounds, such as toxicity to the microbes and the animal, short-lived effects due to microbial adaptation and failure to meet consumer acceptance. Although, wide range of feed additives for reducing methane has been tested over the last 40 years, not even one has been incorporated into farm practice.

Fat and oils

There is possibility of using naturally occurring plant products to reduce methane production in rumen. For example, addition of lipids to the diet can reduce methane emission. Three factors - the quantity, the degree of unsaturation and the chain length of the lipid - can have an effect (Johnson and Johnson, 2002). McGinn et al. (2004) compared various methane inhibitors in cattle. Approximately 6.5% of the GE consumed was lost in the form of methane emissions from animals fed the control diet. Sunflower oil decreased methane emissions by 22% compared with the control, but also reduced energy digestibility. It appears that the effect of degree of unsaturation is relatively small and that the effect of lipid is mainly in depressing digestion (Johnson and Johnson, 1995; Mathison et al., 1998). Coconut oil, reduce methane, possibly by suppressing protozoa (Machmuller et al., 1998; Dohme et al., 1999). Fish oil was also found potent rumen methane inhibitor (Fievez et al., 2003).

Plant secondary metabolites

Rearing systems of small ruminants in India and most Asian countries are nomadic in nature where animals graze on an array of plant species. There are naturally occurring compounds in some plants that appear to have ability to modify rumen metabolism. At times the livestock are fed on agro-industrial by-products, which contain a lot of factors which may act as inhibitor for some of the rumen microbes. The anti-nutritional compounds like tannins, lignins, saponins, mimosine, etc. are synthesized in the plants to protect them against the invading microbes. Therefore when consumed by the animals, these compounds limit the growth of different types of microbes (useful and undesirable) in the rumen.

Tannins

Tannins are phenolic compounds with molecular weight ranging from 500 to 3000 Da, which are water soluble, able to precipitate proteins, alkaloids and gelatin. Based on the differences in their chemical structure and relative ratios of different phenolic monomers, tannins are classified into two groups, e.g. hydrolysable and condensed tannins. The hydrolysable tannins are mostly the esters of gallic acid (gallotannins) or ellagic acid (ellagitannins) with a sugar core (usually glucose), which are readily hydrolysed by acids or enzymes into monomeric products, on the other hand the condensed tannins are polymeric proanthocyanidins. The fibre-degrading bacteria are susceptible to tannins. In the animals fed tannin rich *Calliandra calothyrsus*, the population of *Ruminococcus* spp. and *Fibrobacter* spp. reduced considerably, but fungi, protozoa and proteolytic bacteria were less affected (McSweeney et al., 1995). The number of total bacteria in the rumen of goats decreased significantly when the animals were fed tannin-rich plant (*Acacia nilotica*) and the decrease in the numbers was directly proportional to the level of this feed in the diet (Sotohy et al., 1997). Condensed tannins were known to have the properties to prevent bloat in cattle (Jones et al, 1973) and have the potential to improve the efficiency of protein utilisation by the animal (Reid et al., 1974). Condensed tannins also decrease the protozoal number (Waghorn et al., 1994). A depression of methane emission was found while feeding condensed tannin-containing legume *Lotus corniculatus* and sulla (*Hedysarum coronarium*) to sheep and dairy cows (Woodward et al., 2001, Waghorn et al., 2002, 2006, Woodward et al., 2002) possibly by increasing microbial protein synthesis and decreasing protein degradability in rumen (Makkar, 2003). Similarly, condensed tannins from the leaves of sainfoin (*Onobrychis viciifolia*) inhibited growth and protease activity in *Butyrivibrio fibrisolvens* A38 and *Streptococcus bovis* 45S1, the growth of *Prevotella ruminicola* B14 and *Ruminobacter amylophilus* WP225 was not much affected (Jones et al., 1994). Although tannins do have harmful effect on rumen microbes, but on prolonged feeding some of the tannin-degrading bacteria proliferate and increase in numbers and make the animals more tolerant to higher levels of tannins in their diet. Supplementation of the tannin extract decreased methane release (kJ/MJ gross energy intake) by 13% on average without affecting body energy retention, thus can be useful in mitigating methane emission without major losses in feeding value of the diet. Although, tannin-rich shrub legumes such as *C. calothyrsus*, was also effective in limiting methanogenesis, their use in animal feeding may be restricted due to the simultaneous depression of the feeding value of the diet (Hess et al., 2006). Similarly, methane emission expressed as both quantity per day or relative to DMI was lower for *Sericea lespedeza* than for crabgrass/tall fescue (7.4 vs. 10.6 g/d and 6.9 vs. 16.2 g/kg DMI). Substantial differences between the forages in condensed tannins concentration and methane emission by Angora goats suggest that condensed tannins decreased methane emission (Puchala et al., 2005).

Saponins

Saponins are steroidal sapogenin covalently linked to oligosaccharide moiety. The saponins of lucerne origin have detrimental effect on rumen fermentation and caused a reduction in total volatile fatty acid production and acetate to propionate ratio from 1.93 to 1.37 in the presence of 1% saponin in the medium. The rumen microbes are able to deglycosylate the saponins to release the steroid moiety, which affects rumen fermentation. Gupta et al (1993) claimed that the leaves of the tropical plant *Enterolobium timbouva* defaunated the rumen of buffalo. *Yucca schidigera* extract has been found to alter rumen fermentation favourably, increase animal growth and milk production (Kil et al., 1994,) The yucca extract is also inhibitory for some of the rumen ciliate protozoa and bacteria (Wallace et al, 1994). Yucca extract also had anti-protozoal activity but it helped in increasing the protease activity of the rumen microbes (Wang et al. 1998). Saponins of *Sapindus rarak* fruit (methanol extract) caused a 57% reduction in the number of protozoa and 69% increase in bacterial population which resulted in improved feed conversion efficiency and better gain in body weight of the animals (Thalib et al.1996). Broudiscou et al. (2003) found methane inhibiting properties in the extracts of *Equisetum opense* and *Saliva officianalis*. Plant saponins also known to inhibit rumen methanogenesis by way of decreasing protozoal count (Hess et al., 2003). Saponins from tea seeds have the potential to modulate the rumen fermentation pattern (Liu et al., 2003) and have anti-protozoa effect. Inclusion of 2, 4, 6 and 8 mg of tea saponins in per 200 mg DM decreased methane concentration by 13, 22, 25 and 26%, respectively in vitro (Hu et al, 2005). Since the depression of methane release due to tea saponins addition was found only in faunated rumen fluid, it could be assumed that this effect was mediated through associated effects on protozoa.

Strategies for reducing enteric methane production

Strategies for methane mitigation require a robust prediction of methane emissions from rumen function. The relationship with digestibility and gross chemical composition of diets with methanogenesis exists. The plant, animal, microbial and perhaps adaptation variables are also important. Basic concepts of digestibility, rumen residence time, physical aspects of the diet including rates of degradation may have impact on methanogenesis. In addition to the impact of archaea on methanogenesis, hydrogen sinks such as pectin and ammonia are likely to affect net methane production. There are possibilities for reducing methane and/or increasing fibre digestibility via improvements in efficiency of animal's rumen. However, there is a stronger negative relationship between methane emitted per unit of feed intake and feed intake. So there is an advantage, in terms of methane emission, to feed animals on as high an intake as possible. It is generally accepted that digestion of cell wall carbohydrates produces more methane than the digestion of soluble carbohydrates. Protein and lipids appear to have a negative effect on methane production, but the effects are variable, and in the case of lipids toxicity to the rumen microbes can be a problem.

The strategy to increase feed energy utilisation by ruminant may be achieved through increasing feed intake, manipulation of dietary composition to increase feed quality (e.g. decrease cell wall carbohydrate), increasing metabolic efficiency and genetic improvement in rumen microbes. The judicious use of plant secondary metabolites that affect rumen fermentation in a way that less methane is produced is also important. Any attempt to reduce methane emissions from livestock is unlikely to be adopted unless production efficiency is enhanced. The relationship with digestibility and gross chemical composition of diets demands an integrated evaluation of methanogenesis to include plant, animal, microbial and perhaps adaptation variables. Basic concepts of digestibility, rumen residence time, physical aspects of the diet including rates of degradation may affect methanogenesis. There are alternate approaches like dietary approaches (e.g. use of polyunsaturated fatty acids or ingredients containing these acids), vaccination against methanogens, supplementation strategies, etc. that have the potential to reduce methane. The clear differences between individual animals in their microflora, digestion and performance have been masked when groups are used to evaluate treatments. Animal variation also provides a key to methane mitigation and must be given greater priority in future research. With grazing animals, a delivery system would be required to ensure regular delivery of metabolic modifiers/ methane inhibiting substances into the rumen. The complete feed blocks could be one of these. As there is immense loss of energy by animals in terms of methane particularly on those fed near maintenance, inclusion of methane inhibiting plants in animal diet is necessary. Thus, there is a need to explore plants that are grazed by animals or available locally for their properties to decrease methanogenesis.

REFERENCES

Beauchemin, K.A. and McGinn, S.M. 2005. Methane emissions from feedlot cattle fed barley or corn diets. *Journal of Animal Science* 83,653-661.

Benchaar, C., Pomar, C. and Chiquette, J. 2001. Evaluation of dietary strategies to reduce methane production in ruminants: A modeling approach. *Canadian Journal of Animal Science* 81,563-574.

Bernalier, A., Fonty, G., Bonnemoy, F. and Gouet, P. 1992. Inhibition of the cellulolytic activity of *Neocallimastix frontalis* by *Ruminococcus flavefaciens*. *Journal of General Microbiology* 139, 873- 880.

Broudiscou, L.P., Papon, Y. and Broudiscou, A.F. 2000. Effects of dry plant extracts on fermentation and methanogenesis in continous culture of rumen microbes. *Animal Feed Science and Technology*. 87, 263-277.

Clemens, J., and Ahlgrimm, H.-J. 2001. Greenhouse gases from animal husbandry: Mitigation options. *Nutrition Cyclone and Agro-ecosystem* 60, 287-300.

Davies, A., Nwaonu, H.N., Stanier, G. and Boyle, F.T.. 1982. Properties of a novel series of inhibitors of rumen methanogenesis: in vitro and in vivo experiments including growth trials on 2, 4-bis (trichloromethyl)-benzo [1, 3]dioxin-6-carboxylic acid. *British Journal of Nutrition* 47, 565-576.

Dohme, F., Machmuller, A., Estermann, B.L., Pfister, P., Wasserfallen, A. and Kreuzer, M. 1999. The role of the rumen ciliate protozoa for methane suppression caused by coconut oil. *Letters in Applied Microbiology* 29, 187-192.

Fievez, V., Dohne, F., Daneels, M., Raes, K. and Demeyer, D. 2003. Fish oils as potent rumen methane inhibitors and associated effects on rumen fermentation in vitro and in vivo. *Animal Feed Science and Technology* 104, 41-58.

Gunsalus, R.P., Romesser, J.A. and Wolfe, R.S. 1978. Preparation of coenzyme M analogues and their activity in the methylcoenzyme M reductase system of *Methanobacterium thermoautotrophicum*. *Biochemistry* 17, 2374-2377.

Hashizume, T., Morimoto, H., Haryu, T., Itch, M. and Tanabe, S., 1968. Utilization of the energy of fattening rations containing ground or steam-rolled barley by Japanese Black Breed cattle. *Energy Metabolism of Farm Animals*, EAAP Publ. No. 12. K. L. Blaxter, J. Kielanowski, and G. Thorbek, ed. Oriel Press, Newcastle upon Tyne, U.K. pp 261-271.

Hess, H.D., Kreuzer, M., Diaz, T.E., Lascano, C.E., Carulla, J.e., Soliva, C.R. and Machmuller, A. 2003. Saponin rich tropical fruits effects fermentation and methanogenesis in faunated and defaunated rumen fluid. *Animal Feed Science and Technology*. 109, 79-94.

Hess H.D., Tiemann, T.T., Noto, F., Carulla J.E. and Kreuzer M. 2006 Strategic use of tannins as means to limit methane emission from ruminant. *livestock International Congress Series* 1293, 164-167.

Houghton, J.T., Meira Foilho, L.G., Lim, B., Tre´anton, K., Mamaty, I., Bonduki, Y., Griggs, D.J. and Callander, B. A. 1996. Greenhouse Gas Inventory Reference Manual. Revised 1996 IPCC Guidelines for National Greenhouse Gas Inventories. Vol. 3. Agriculture, Intergovernmental Panel on Climate Change, Geneva, Switzerland.

Hu, W., Liu, J., Wu, Y. and Guo, Y. 2005. Effect of tea saponins on rumen fermentation in vitro. *Animal Feed Science and technology* 120: 333-339.

Hungate, R.E. 1966. The rumen and its microbes. Academic Press, London & NY

Immig, I., Demeyer, D., Fiedler, D., Van Nevel, C. and Mbanzamihigo, L. 1996. Attempts to induce reductive acetogenesis into a sheep rumen. *Archives of Tierernahr*. 49, 363-370.

Johnson, K.A., and D.E. Johnson. 1995. Methane emissions from cattle. Journal of Animal Science 73, 2483-2492.

Johnson, D.E., Johnson, K.A., Ward, G.M. and Branine, M.E. 2000. Ruminants and other animals, Chapter 8. *Atmospheric Methane: Its Role in the Global Environment*. M.A.K. Khalil, ed. Springer-Verlag, Berlin Heidelberg, Germany. pp 112133.

Jones, W.T., Anderson, L.B. and Ross, M.D. 1973. Bloat in cattle XXXIX. Detection of protein precipitants (flavolans) in legumes. *New Zealand Journal of Agricultural Research* 16, 441-446.

Jones, J. ., McAllister, T.A., Muir, A.D. and Cheng, K. -J. 1994. Effects of sanfoin (*Onobrychis vicifolia* Scop.) condensed tannins on growth and proteolysis by four strains of ruminal bacteria. *Applied and Environmental Microbiology* 60, 1374-1378.

Kamra D.N. 2005. Rumen microbial ecosystem. Special Section: Microbial Diversity. *Current Science* 89, 124-135.

Kandler, O. and Hippe, H. 1977. Lack of peptidoglycan in the cell walls of *Methanosarcina barkeri*. *Achieve of Microbiology* 118, 57–60.

Kil, J.Y., Cho, N.K., Kim, B.S., Lee, S.R. and Maeng, W.J. 1994, Effects of yucca extract addition on *in vitro* fermentation characteristics of feed and faeces and on the milk yields in lactating cows. *Korean Journal of Animal Science* 36, 698-709.

Kirchgessner, M., Windisch, W., Muller, H.L. 1995. Nutritional factors for the quantification of methane production. In: v. Engelhardt W, Leonhardt-Marek S, Breves G, Gieseke D (Eds) Ruminant physiology, digestion, metabolism, growth and reproduction. Proc 8th International Symposium on Ruminant Physiology. Ferdinande Enke Verlag, Stuttgart, pp. 333-348.

Kopecny, J., Hodrova, B. and Stewart, C.S. 1996. The effect of rumen chitinolytic bacteria on cellulolytic anaerobic fungi. *Letters in Applied Microbiology* 23,199-202

Liu, J.X., Yuan, W.Z., Ye, J.A. and Wu, Y.M., 2003. Effect of tea (*Camellia sineis*) saponin addition on rumen fermentation in vitro. *Tropical Subtropical Agroecosystem*, 3, 561-564.

Lovely, D.R., Greening, R. and Ferry, J.G. 1984, Rapidly growing rumen methanogenic organism that synthesizes coenzyme M and has a high affinity for formate. *Applied and Environmental Microbiology* 48, 81–87.

Macmuller, A., Ossowski, D.A., Wanner, M. and Kruezer, M. 1998. Potential of various fatty feeds to reduce methane release from rumen fermentation *in vitro* (RUSITEC). *Animal Feed Science and Technology* 71, 117-130.

Makkar, H.P.S. 2003. Effects and fate of tannins in ruminant animals, adaptation to tannins, and strategies to overcome detrimental effects of feeding tannin-rich feeds. *Small Ruminant Research* 49, 241-256.

Marvin-Sikkema, F.D., Richardson, A.J., Stewart, C.S., Gottschal, J.C. and Prins, R.A. 1990. Influence of hydrogen-consuming bacteria on cellulose degradation by anaerobic fungi. *Applied and Environmental Microbiology* 56, 3793-3797.

Mathison, G.W., Okine, E.K., McAllister, T.A., Dong, Y., Galbraith, J., and Dmytruk, O.I.N. 1998. Reducing methane emission from ruminant animals. *Journal of Applied Animal Research* 14, 1-28.

McGinn, S.M., Beauchemin, K.A., Coates, T. and Colombatto. D. 2004. Methane emissions from beef cattle: Effects of monensin, sunflower oil, enzymes, yeast, and fumaric acid. *Journal of Animal Science* 82, 3346–3356.

McGuffey, R.K., Richardson, L.F. and Wilkinson. J I.D. 2001. Ionophores for dairycattle: current status and future outlook. *Journal of Dairy Science* 84(Suppl. E), E194-E203.

McSweeney, C.S., Palmer, B., Krause, D.O. and Brooker, J.D., 1999. Rumen microbial ecology and physiology in sheep and goats fed a tannin containing diet. In Tannins in Livestock and Human Nutrition. Proc. International Workshop, Adelaide, Australia, 31 May 2 June, 1999, pp. 140-145.

Moss, A.R. 2002. Environmental control of methane production by ruminants. Pages 67–76 in Greenhouse gases and animal agriculture. J. Takahashi and B. A. Young, ed. Elsevier, Amsterdam, The Netherlands.

Moss, A.R., Jouany, J.-P. and Newbold, J. 2000. Methane production by ruminants: Its contribution to global warming. *Annals of Zootechnol* 49, 231–253.

Miller, T.L., Wolin, M.J., Zhao, H. and Bryant, M.P. 1986, Characteristics of methanogens isolated from bovine rumen. *Applied and Environmental Microbiology* 51, 201–202.

Odenyo, A., Mackie, R.A., Stahl, D.A. and Whyte, B.A. 1994. The use of 16sRNA probes to study competition between rumen fibrolytic bacteria: development of probes for *Ruminococcus* species and evidence for bacteriocin production. *Applied and Environmental Microbiology* 60, 3688-3699.

Patterson, J.A. and Hespell, R.B. 1979, Trimethylamine and methylamine as growth substrates for rumen bacteria and *Methanosarcina barkeri*. *Current Microbiology* 3, 79–83.

Prins, R.A. and Stewart, C.S. 1997. Microbial interactions in the rumen. *Reproduction Nutrition and Development* 5(Suppl.), 1-4.

Puchala, R. Min, B. R. Goetsch, A. L. and Sahlu, T. 2005. The effect of condensed tannin-containing forage on methane emission by goats. Journal of Animal Science 83,182-186.

Razvan, D., Palencia, H., Schroeder, S.D., DeMontigny, B.A., Takacs, J.M., Rasche, M.E., Miner, J.L. and Ragsdale, S.W. 2003. Targeting methanoprotein biosynthesis to Inhibit methanogenesis. *Applied and Environmental Microbiology* 69, 7236-7241.

Reid, C.S.W., Ulyatt, M.J and Wilson, J.M. 1974. Plant tannins and nutritive value. *Proceedings of the New Zealand Society of Animal Production* 34, 82-92.

Saluzzi, L., Smith, A. and Stewart, C.S. 1993. Analysis of bacterial phospholipid markers and plant monosaccharides during forage degradation by *Ruminococcus flavefaciens* and *Fibrobacter succinogenes* in co-culture. *Journal of General Microbiology* 139, 2865-2873.

Sotohy, S.A., Sayed, A.N. and Ahmed, M.M. 1997. Effect of tanninrich plant (Acacia nilotica) on some nutritional and bacteriological parameters in goats. *Deutsch-Tierarztl Wochenschr* 104, 432-435.

Stewart C.S. 1999 Microbial Interactions in the Rumen and Their Potential Impact on the Survival of *Escherichia coli* O157 *Microbial Biosystems: New Frontiers* Proceedings of the 8 th International Symposium on Microbial EcologyBell, C.R., Brylinsky, M., Johnson-Green, P. (eds) Atlantic Canada Society for Microbial Ecology, Halifax, Canada.

Stewart, C.S., Duncan, S.H., Richardson, A.J., Backwell, C. and Begbie, R. 1992. The inhibition of fungal cellulolysis by cell-free preparation from ruminococci. *FEMS Microbiology Letters* 9, 83-88.

Stumm, C.K., Gijzen, H.J. and Vogels, G.D. 1982, Association of methanogenic bacteria with ovine rumen ciliates. *British Journal of Nutrition* 47, 95–99.

Thalib, A., Widiawati, Y., Hamid, H., Suharman, D. and Sabrani, M. 1996. The effects of saponin from *Sapindus rarak* fruit on rumen microbes and performance of sheep. *Jurnal Ilmu Ternak dan Veteriner* 2, 17–21.

Thornton, J.H., and Owens. F.N. 1981. Monensin supplementation and in vivo methane production by steers. Journal of Animal Science 52, 628-634.

Van Kessel, J.A.S. and Russell, J.B. 1996. The effect of pH on ruminal methanogenesis. *FEMS Microbial Ecology* 20, 205–210.

Waghorn, G.C., Tavendale, M.H. and Woodfield, D.R. 2002. Methanogenesis from forages fed to sheep. *Proceedings of the New Zealand Grassland Association* 64,167-171.

Waghorn, G.C., Woodward, S.L., Tavendale M. and Clark D.A. 2006 Inconsistencies in rumen methane production – effects of forage composition and animal genotype. *International Congress Series* 1293, 115-118.

Wallace, R.J., Arthauo, L. and Newbold, C.J. 1994, Influence of *Yucca schidigera* extract on ruminal ammonia concentrations and ruminal microorganisms. *Applied and Environmental Microbiology* 60, 1762-1767.

Wang, Y., McAllister, T. A., Newbold, C. J., Rode, L. M., Cheeke, P. R. and Cheng, K. J. 1998, Effects of *Yucca schidegera* extract on fermentation and degradation of steroidal saponins in the rumen simulation technique (RUSITEC). *Animal Feed Science and Technol*ogy 74, 143-153.

Whitelaw, F., Eadie, J.M., Bruce, L.A. and Shand, W.J. 1984. Methane formation in faunated and ciliate-free cattle and its relationship with rumen volatile fatty acid proportions. *British Journal of Nutrition* 52, 261-275.

Wolin M.J., Miller, T.L. and Stewart, C.S. 1997. Microbe-microbe interactions. In: Hobson PN, Stewart CS (Eds) The Rumen microbial ecosystem. Chapman & Hall, London, p 467-491.

Woodward, S.L., Waghorn, G.C., Lassey, K.R. and Laboyrie, P.G. 2002. Does feeding sulla (*Hedysarum coronarium*) reduce methane emission from dairy cows? *Proceeding of New Zealand Society of Animal Science* 62, 227-230.

Woodward, S.L., Waghorn, G., Ulyatt, M.J. and Lassey, K.R. 2001. Early indications that feeding Lotus will reduce methane emission from ruminants. *Proceeding of New Zealand Animal Production* 61, 23-26.

Yang, W.Z., K.A. Beauchemin, K.M. Koenig, and L.M. Rode. 1997. Comparison of hull-less barley, barley, or corn for lactating cows: Effects on extent of digestion and milk production. *Journal of Dairy Science* 80, 2475-2486.

□□□

Wang, Y., McAllister, T. A., Newbold, C. J., Rode, L. M., Cheeke, P. R. and Cheng, K. J. 1998. Effects of Yucca schidigera extract on fermentation and degradation of steroidal saponins in the rumen simulation [illegible]. Animal Feed Science and Technology, 74: 143-153.

[illegible]

[illegible] pp. [illegible]

[illegible]

15

Bio-Conversion of Agro-Wastes into Microbial Feed for Feed Security to Small Ruminants

A.K. MISRA, M.M. DAS, K.K. SINGH
Indian Grassland & Fodder Research Institute, Jhansi-284003, Uttar Pradesh

Introduction

The livestock sector plays a significant economic role in most developing countries and is essential for the food security of their rural population. It is contributing to poverty alleviation and provides elements that are essential to the national economy, such as: traction power, transport, manure as fertilizer and fuel, food, fibre, leather, savings bank, and by generating significant household cash income through sales of live animals or livestock products. However, among the major constraints limiting the development of livestock production in many developing countries, inadequacy of animal feed resources is most often the crucial factor. Feed shortages, both quantitatively and qualitatively, limiting livestock productivity. In many tropical countries of Asia the bulk of livestock feed resources comes from grazing mainly poor quality annual and perennial grasses from natural pastures often at a late stage of maturity and the crop residues. The most abundant residues are cereal straws, sugarcane tops, sugarcane bagasse, pulse straws, millet straws, etc. These feeds are unable to meet the maintenance requirement of animals because of the low digestibility, influenced by high fiber, lignin and silica content.

Structural Components of Crop Residues

Plant cell wall material of commonly fed straws and stovers is composed of three important constituents: cellulose, lignin, and hemicellulose. Lignin is particularly difficult to biodegrade, and reduces the bio-availability of the other cell wall constituents. Cellulose is a long chain of glucose molecules, linked to one another primarily with glycosidic bonds. The simplicity of the cellulosic structure, using repeated identical bonds, means that only a small number of enzymes are required to degrade this material. Ruminants create an environment in their rumen, which encourages this microbial degradation, converting cellulose to volatile fatty acids, and microbial biomass, which the ruminant can then digest, and use. Hemicelluloses are branched polymers of xylose, arabinose, galactose, mannose, and glucose. Hemicelluloses bind bundles of cellulose fibrils to form microfibrils, which enhance the stability of the cell wall. They also cross-link with lignin, creating a complex web of bonds, which provide structural strength, but also challenge microbial degradation (Lynch, 1992). Lignin is a complex polymer of phenyl propane units, which are cross-linked to each other with a variety of different chemical bonds. This complexity has thus far proven as resistant to detailed biochemical characterization as it is to microbial degradation, which greatly impedes our understanding of its effects. Nonetheless, some organisms, particularly fungi, have developed the necessary enzymes to break lignin apart. The initial reactions are mediated by extra cellular lignin and manganese peroxidases, primarily produced by white-rot fungi. Actinomycetes can also decompose lignin, but typically degrade less than 20 percent of the total lignin present (Crawford, 1986). Lignin degradation is primarily an aerobic process, and in an anaerobic environment lignin can persist for very long periods (Van Soest, 1994). Lignin is the most intractable component of the plant cell wall, and a direct negative relationship between the proportions of lignin and the bioavailability of the substrate exists. The effect of lignin on the bioavailability of other cell wall components is thought to be largely a physical restriction, with lignin molecules reducing the surface area available to enzymatic penetration and activity (Haug, 1993).

Microbiological Processes that Upgrade Straw Quality

Considerable efforts have been made to upgrade the nutritive value of straw (Jackson, 1977). Several physical treatments, such as pelleting; grinding; steaming and irradiation, as well as chemical treatments, particularly alkali addition (Sundstol and Owen 1984), have been tested in attempts to improve dry matter digestibility and voluntary intake of straw. The practical use of these treatments is limited by safety concerns, costs and potentially negative environmental consequences.

Single cell protein production from lingo-cellulosic wastes

Han and co-workers (Han *et al.*, 1975; Han, 1978) developed a fermentation process essentially involving growth of microorganisms other than fungi on straw in solid state fermentation (SSF). Since straws normally resistance to microbial growth, pre-treatment was done before yeast cultivation. There are two processes: In the first process the straw is hydrolyzed with three parts of 0.5 N H_2SO_4 and 1 part of the straw which is steam treated in a pressure cooker under 10-15 *psi* of steam pressure for 30 minutes. The acid hydrolyzed straw is then treated with NH_3 to raise the pH to 4.5 and it has been estimated that at this stage it contains 20% fermentable sugars and 2.3% N. This is then incubated with suitable organisms like *Candida ntilis* or *Aspergllus pullulans*. In the second process, straw is treated with 4% (dry weight) NaOH and after neutrilisation of excess alkali, ammonium sulphate is added. This pretreated straw is then inoculated with the cellulolytic organisms like mixed culture of *Cellulomonas* sp. or *Alcalegenes faecalis*.

Waterloo Process

The waterloo single celled protein process as developed by Moo-Young et al. (1981) is based on initial treatment of straw with alkali with high temperature and subsequent inoculation with the soft rot fungus *Chaetomium cellulolyticum*. The dispersion of mycelium with cellulose facilitates the organisms to utilize cellulose and sugar like xylose and mannose derived from the hydrolysis of hemicellulose. Because of environmental pollution and problem of toxin production on cellulosic wastes, either glucose or molasses later replaced the straw substrate. However, time of harvest and nature of substrate had a large influence on the microbial biomass protein composition and nutritive evaluation (Chavez et al., 1988). The amino acid profiles of the two fungal samples produced glucose and molasses compared relatively well with that of dried skim milk although the total amino acid analyzed were lower than the dried skim milk. Usually the amount of nucleic acid reportedly present in this kind of Microbial Biomass protein (MBP) was relatively small (Moo-Young *et al.*, 1981) which seemed to be due contaminant NPN. MBP sample were freeze dried at significant responses in body weight gain, feed consumption and feed to gain ratios among rats (Chavez *et al.*, 1988). The absolute nutritive value and acceptability of cellulolytic fungus as MBP varied not only with the substrate provided for fungal growth, but also with the production process. In a practical diet for a weaning pigs and chicks, the fungal MBP grown on molasses replaced 50% of the dietary protein supplied by soybean meal with excellent results (Chavez *et al.*, 1988).

The 'palo podrido'

In South Chile, fungal delignification of wood has been observed under natural conditions. The product after delignification viz. 'palo podrido' is used as animal feed which is a white decomposed (Zadrazil, 1984. The fungal responsible for degradation of wood belonged either to *Ganoderma applanatum* or *Arnillareilla sp*. However some of the fungi like *Phanerochaete chrysosporium* degrade hemicellulose, cellulose and lignin non-selectively, which adversely affects the digestibility of cell walls of the product in the rumen (Agosine *et al.*, 1987). Zadrazil (1980) steam treated wheat straw, inoculated it with white rot fungus like decomitus squalens and incubated for 6 weeks. No nitrogen source was used in this process. There was a marginal increase in N content of the treated straw, which was mainly due to carbon loss during the incubation period. However the lignin contents decreased and the digestibility of the wheat straw improved. The optimum pH during fermentation was 3.0.

Mushroom production and the utilization of spent straws as animal feed

Many Asian families grow edible mushroom, often using straw as a partial or complete growth medium. In a study of Zadrazil, the use of large used fertilizer bag was suggested as they could be stuffed with the straw and then steamed as such. After cooling, the bags could also be inocubated with the white rot fungus (Pleurotus spp.) and incubated for several weeks. When the bags had been filled with mycelium they were slit open at the top whch allowed fruiting, from often could be harvested. The residual straw after the harvest of musroom could be fed to ruminant animals. Zadrazil and Brunnert (1981) and Gupta (1987) confirmed that at lower temperatures *Ganoderma applanatum* preferentially degrades lignin in wheat straw and helps to increase the digestibility of the fermented product.

The main problem of biological upgrading of low quality roughages into feed is to find suitable microorganism with metabolic patterns different from those of the rumen flora and fauna for use in the process of delignification and upgrading. "Ideal microorganisms" for upgrading lignocellulose crop residues in to animal feed should have a strong lignin metabolism with poor degradation of cellulose and hemicellulose so that in the treated material, carbohydrates are preserved for their utilization by rumen microbes as an energy source in ruminants (Zadrazil,1986).

How White Rot Fungi Improve the Nutritional Value of Crop Residues

Fungal life cycle we divided into two distinct phases primary and secondary phase. In primary phase active mycelia growth takes place and the available nutrients are utilized. The secondary phase start when there is scarcity of any of the essential nutrients like carbon, nitrogen, and sulphur, and during this phase distinct changes in the behavior of fungus occur- sporulation or a decline in the growth rate of fungus perhaps to compensate the scarcity of that nutrient. It has been observed that lignin degradation triggered on only in the secondary phase of growth and that to in response to nitrogen limitation (Villas-Boas et al., 2002). When the filamentous white rot fungi are grown on lignocellulosic materials, the tubular filamentous grow along side of solid particle utilizing liquid film covering the solid particle as moisture and nutrients. The growing tip of filaments (hypha) secretes powerful extra cellular lignin degrading enzymes, which degrade the lignin into in its metabolic products. The presence of lignin in the culture medium is not essential for the synthesis of lignolytic enzyme system, but increasing the partial pressure of oxygen in the gas phase significantly increasing its synthesis. The white rot fungi attack and cause oxidative changes in the phenylpropane subunits and cause ring cleavage while the lignin polymer is intact i.e. no de-polymerization has occurred.

Characterization of Lignolytic Fungi

Lignin degrading microorganisms are mainly wood inhabiting fungi which are also capable to colonize different plant residues. These fungi can be classified into different groups: Sweet-rot, brown-rot, and white-rot. Sweet-rot fungi only metabolize soluble sugars or other easily digestible cell constituents of crop residues. Brown -rot fungi primarily decompose cellulose and hemicellulose at an extremely low pH (2.0-3.0) and decomposed material is characterized by lignin accumulation. White-rot fungi are able to decompose and mineralize all plant cell constituent-cellulose, hemicellulose and lignin. The use of white -rot fungi for conversion of lingo cellulose in to feed has been studied and the positive effect to the nutritional value determined (Hartley *et al.*1974; Ibrahim and Pierce 1980; Zadrazil and Brunnert,1981). The results indicated that the in vitro digestibility of fungal substrates decreases at the initial stage of colonization by white rot fungi and increases afterwards (Zadrazil and Brunnert 1981). During the incubation period, the content of soluble substances are increased. The increase in digestibility depends on the fungal species, cultivation time, temperature, composition of substrate (Zadrazil and Brunnert 1981). Zadrazil (1986) tested 115 species and a total of 235 different strains of white rot fungi (belonging to the group *Basidiomycetes*) for growth on wheat straw: 85 species (115 strains) showed intensive growth and were tested at 22,25 and 30°C for decomposition rate of organic matter, lignin, liberation of water soluble substance from plant polymers and in vitro organic matter digestibility. Fungi may be divided in to four groups according to the different physiological behavior shown by them:

- Fungi which degrade little or no lignin and decompose cellulose and hemicellulose selectively. The in vitro dry matter digestibility (IVDMD) of untreated substrate (typical fungi of this group are *Agrocybe aegerita, Flamulina velutipes, and Volvariella volvacea*).
- Fungi which decompose lignin selectively and other substrate components (cellulose, hemicellulose) only partially IVDMD of these substrates is increased. This groups includes *Dichomitus squalens, Abortiporus biennis, Pleurotus spp.,and Stropharia rugosaannulata.*
- Fungi which decompose lignin, cellulose and hemicellulose as group B, but IVDMD is decreased or remains unchanged. Strains of this group are *Polyporus, Ganoderma and Pleurotus spp.*
- Fungi which decompose lignin and other substrates (cellulose and hemicellulose.) rapidly and change digestibility of the substrate only partly. The main fungus of this group is *Phanerochaete chrysosporium*.

White-rot Basidiomycetes fungi degrade the polymers cellulose, hemicellulose and lignin at different rates and extents. In general, strains of white-rot fungi are the most effective lignin-modifying enzyme, of which lignin-Peroxidases (liP), manganese dependent Peroxidase (mnP), laccase and H_2O_2-producing, oxidase (aryl alcohol oxidase and glyoxal oxidase) are the most important (Buswell 1992).The oxidative enzyme lignase, manganese peroxidase and laccase may be defined as phenol oxidase belong to the class of peroxidases. According to production patterns of extracellular lignolytic enzymes, white rot fungi may be divided in to four groups: (1) theLiP-MnP group, (2) the MnP-Laccas group, (3) the laccase-AAO group.AAO is aryl alcohol oxidase (Hatakka 1994). Studies on lignin biodegradation have been carried out mostly using *Phanerochaete chrysosporium* is a typical representative of the fungi belonging to the LiP-MnP group. *P.chrysosporium* is a very efficient degrader of lignin. Typical fungi producing (1) LiP-MnP are *Phlebia radiata, Coriolus versicolor and Trametes gibbosa,* (2) fungi producing MnP-Laccase are *Panus tigrinus and Dichomitus squalens.* (3) LiP-Laccase producing fungi are *Junghuhnia separabilima and Phlebia ochraceofulva,* and (4) fungi producing AAO are *Pleurotus ostreatus* and *Pleuotus Sajor-caju.*

Types of fermentation process for bioconversion

Following fermentation process was developed for bio-processing

- Solid-state fermentation
- Surface fermentation
- Submerged fermentation
- Solid/gas fluidized bed fermentation

Solid-state fermentation can be defined as a process in which solid substrates are decomposed by fungi, which are able to grow on and through the substance (lingo-cellulose crop residue) under controlled environmental conditions with the aim of producing a high quality product. It is the oldest fermentation process for the bioconversions, being used by man since centuries, when man was not knowing about the microbial process involved in the process. After become knowledgeable about the microbial process undergoing during bioconversions, some improvements were made and surface fermentation were developed. In this process, the fungi were allowed to grow as a mycelia mat on the surface of liquid medium. In submerged fermentation process, the submerged culture was allowed to grow on high-diluted substrate. The process was more efficient with higher performance and easy operations. But due to high degree of dilution, the process is costly as compared to solid-state fermentation where there is no dilution, also per unit processing of substrate a high reactor volume is required. To overcome all these problems an advanced fermentation system was developed where gas/ solid fluidized bed was used. This process has number of advantages in term of heat and mass transfer and substrate is in continuous operation. But again these processes are very expensive especially for the delignification of the ligno-cellulosic substrate because delignification process is very slow. Therefore, sold fermentation is widely accepted method for bioconversions of lingo-cellulosic feed to make it better utilized by the animal. We also followed the solid state fermentation to improve the feeding value of mustard straw. In solid state fermentation, filamentous fungi are grown because this only can grow without free water for their growth, and this process is very close to nature. Secondly because of no dilution, for the bioconversions of per unit of substrate, small reactor volume is required and there is no need to recover the feed, since it can be fed to animal directly. Solid-state fermentation with white-rot fungi is influenced by such factors as the species of the fungus, substrates, temperature (Zadrazil 1980), moisture content (Zadrazil and Brunnert 1981) and composition of the gas phase (Chiavari *et al.* 1989). All these factors determine the sequence of degradation of the cell wall polymers and final digestibility after solid-state fermentation. Detailed studies carried out with wheat straw as substrate indicated that:

1. *Pre-selection of fungi:* Various species of fungi, mainly basidiomycetes (115 species, 235 different strains), were tested for their ability to grow on sterile straw substrate. Some species (40) grew very slowly or were not able to grow on wheat straw. This group includes the species of *Agaricus, Amanita, Armillariella, Boletus, Cantharellus, Lactarius, Serpula* and other species (Zadrazil, 1986). Fungi with on good growth performance on wheat straw (80 species, 115 strains) were tested the lignin decomposition and in vitro dry matter digestibility during 60 days of incubation,

2. *Temperature and pH:* Temperature of 22,25 and 30°C influenced the decomposition of wheat straw by white-rot fungi. A small number of the fungi reached their growth and decomposition of lignin at 22°C (*Hymenochaeta tabacina*) or at 25 °C (*Flammulina velutipes, Collybia radiacata, Pleurotus serotinus*), but the majority of fungi showed best growth and substrate decomposition at 30°C (Zadrazil, 1986). Higher temperature have an adverse effect on the pocess of fermentation and lead to less lignin degradation as well as to decreased digestibility of wheat straw. Temperatures of 35,37,&32-40°C were optimal for the SSF of wheat straw by *P. chrysosporium, Chaetomium cellulolyticum* (Abdullah *et al.*, 1985) and by *Coprinus fimetarius* (Flegel and Meevootisom, 1986) respectively. Lignin degradation by most fungi occurs at an acidic pH (Reid, 1985),

3. *Moisture content:* The water content (65-75%) of the substrates should be optimal for the growth of various fungi (Zadrazil and Brunnert 1981; Babitskaja et al. 1986). A water content of more than 75% resulted in the reduction of the gas phase while in the reduction of the gas phase while a lower water content reduce fungal growth. The fungi showed specific growth optimal on wheat straw at a solid/liquid ratio of 1:3 (Zadrazil and Brunnert, 1981),

4. *Nitrogen Content*: Some fungi-*Stropharia rugusoannulata, lentinus edodes* and *Pleurotus* species Florida-were stimulated during decomposition of wheat straw by the addition of lower values of NH_4NO_3 while *Pleurotus eryngii*

was inhibited by this nitrogen source (Zadrazil and Brunnert 1981). Generally, lignin removal from lingocellulosic crop residues was suppressed by high nitrogen levels (Misra *et al.*, 2007),

5. Composition of the Gas phase: Solid-state fermentation is affected by the penetration and diffusion of gas through the substrate.

Conclusion and Recommendations

The use of micro organism to improve straw quality has been considered an alternative to chemical or physical treatments with promising results and a large variety of biological treatments were tried on lingo cellulosic materials. However, the methods differ in substrate (type of straw), type of fermentation and purpose of treatment. Not many methods are actually applied in practice and many require industrial approaches. The major problems of microbial methods are difficulties in process control (SSF), organic matter losses, isolation of substrate specific microorganism.

Given the theoretical and practical problems, there are issues related to microbiological treatment of straw and stovers to be addressed:

- Better microbiological and bio chemical understanding of the basic process
- Better definition of the goals of treatment in terms of protein quality, digestibility, organic matter losses and intake.
- Consideration of possibilities of mixed cultures
- Costs and gain in terms of improved nutritive value.

REFERENCES

Abdullah, A.L., Tangeraly, R.P. and Murphy, V.G. 1985. Optimization of solid substrate fermentation of wheat straw. *Biotechnology Bioengineering*. 27: 20-27.

Agosin, E., Tollier, M. T., Heckmann, E., Brilliuet, J. M., Thivend, P. and Monties, B. 1987. Effect of fungal treatment of lignocellulosics on biodegradability, In: Van der meer J M, Rijkens BA ferranti M P (eds) Degradation of lignocellulosic in ruminants and in industrial processes, Commission of the European communities, Elsevier. New York, pp 35-43

Babitskaja, V. G., Stakhev, I.V., Shcherba, V.V. and Vadetsky, B. Y. 1986. Bioconversion of lignocellulosic substrates by mycelial fungi in solid state culture. *Prikl Biokhim Mikrobiol* 22, 531-539.

Buswell, J. A. 1992. Fungal degradation of lignin. In: Arora D K, Rai Mukeiji KA, Kundsen G (eds) *Hand book of applied mycology*, Vol.1 Marcel Dekker' New York, pp 425-480.

Chavez, E.R., Touchburn, S.P. and Moo-Young, M. 1988. Microbial biomass protein from the fungus *Chaetomium cellulolyticum*. II. Effect of method of drying and response to amino acid supplementation. *Animal Feed Science and Technology*. 22: 13.

Chiavari, G., Francioso, O., Galletti, G. C., Piccaglia, R. and Zadrazil, F.1989. Characterization by pyrolysis- gas chromatography of wheat straw fermented with white -rot fungus stropharia rugosannulata. *Biological Wastes* 22, 303-309.

Crawford, D.L. 1986. The role of actinomycetes in the decomposition of lignocellulose. FEMS Symposium 34,715-728.

Flegel, T.W. and Meetvootisom, V. 1986. Biological treatment of straw for animal feed. In: Ibrahim MNM, Schiere JB (eds) Rice straw and related feeds in ruminants rations, Straw utilization project publication 2 kandy, Sri Lanka, pp 181-191.

Gupta, B. N. 1987. Fungal treatment of cereals straws for ruminant feeding. In: Sing T,. Flegel, T.W ,Schiere J.B. (eds) Biological, chemical and physical treatment of crop residues for use as animal feeds. ICAR, New Delhi, India 26-36.

Han, Y.W., Lee, J.S. and Anderson, A.W. 1975. Chemical composition and digestibility of ryegrass straw. J Agric. Food Chem. 23: 929.

Han, Y.W. 1978. Microbial utilization of straw (a review). *Advances in Applied Microbiology* 23, 119.

Hartley, R.D., Jones, E. C., King, N.J. and Smith, G.A. 1974. Modified wood waste and straws as potential components of animal feeds. *Journal of The Science Food and Agriculture* 25,433-437.

Hatakka, A.I. 1994. Lignin modifying enzyme from selected white–rot fungi: production and role in lignin degradation. *FEMS Microbial Reviews* 13, 125-135.

Haug, R.T. 1993. *The Practical Handbook of Compost Engineering*. Lewis Publishers, Boca Raton, Fl. 717 pages.

Ibrahim, M.N. M. and Pierce, G.R. 1980. Effect of white-rot Fungi on the composition and in vitro digestibility of crop by products. *Agricultural Wastes* 2, 199-205.

Jackson, M.G. 1977. Review article: the alkali treatment of straws. *Animal Feed Science and Technology* 2, 105-130.

Lynch, J.M. 1992. *Substrate availability in the production of composts*. Proceedings of the International Composting Research Symposium. H.A.J. Hoitink and H. Keener (Editors). pp 24-35.

Misra, A.K., Mishra, A.S., Tripathi, M.K., Prasad, R., Vaithiyanathan, S. and Jakhmola, R.C. 2007. Optimization of Solid State Fermentation of Mustard (*Brassica Campestris*) Straw For Production of Animal Feed By White Rot Fungi (*Ganoderma lucidum*). *Asian-Australasian Journal of Animal sciences* 20, 208-213.

Moo Young, M., Dougalis, A.S., Chahal, D.S., and McDonald, P. 1981. Bioconversion of lingo-cellulosics in animal feed with Chaetomium cellulolyticum, Dev. Ind. Microbiol. 23: 143-159.

Reid, I.D. 1985. Biological deligniification of aspen wood by solid state fermentation with Phlebia tremollosus. Lignin Biodegradation workshop. Univ. of British Columbia, Vancouver, Canada.

Sundstol, F., and Owen, E. 1984. Straw and other Fibrous by-products as feed. Elsevier Science Publishers B.V., Amsterdam, the Netherlands. Developments n Animals and Veterinary Sciences 14, ISBN 0-444-42302-8, 604 pp.

Van Soest, P.J. 1994. The Nutritional Ecology of the Ruminant, 2nd edition. Cornell University Press. Ithaca, NY. 476 pp.

Villas Boas, S.G., Esposito, E. and Mitchel, D.A. 2002. Microbial composition of lignocellulosic residues for production of animal feeds. *Animal Feed Science and Technology* 98, 1-12.

Zadrazil, F. 1980. Conversion of different plant wastes into feed by Basidiomycetes. *European Journal of Applied Microbiology and Biotechnology* 9, 243.

Zadrazil, F. 1984. Micrbial conversion of lignicellulose in to feed. Pp.276-292, In : F. Sundstol and E. Owen (eds): Straw and other fibrous byproducts as feed. Elsvier Sci. Publ. B.V., Amsterdum. The Netherlands.

Zadrazil, F. 1986. White-rot fungi and mushrooms grown on cereal straws: aim of the process, final products, scope of the future, pp 55. Proc. of workshop on degradation of lingo-cellulosic in ruminants and in industrial process, 17-20 March, Lelystad, The Netherlands.

Zadrazil, F. and Brunnert, H. 1981. Investigation of physical parameters important for the solid-state fermentation of straw by white-rot fungi. *European Journal of Applied Microbiology and Biotechnology* 11, 183.

□□□

16

Application of In Vitro Gas Production Technique for Feed Resources Evaluation

NARAYAN DUTTA

Centre of Advanced Studies in Animal Nutrition, Indian Veterinary Research Institute, Izatnagar, Bareilly - 243122

Introduction

The *in-sacco* and *in-vitro* techniques for feed and fodder evaluation are laborious, expensive and not always accurate enough. Moreover, during the course of fermentation soluble substrate components can not be quantified. Alternatively, the kinetics of feedstuff degradation can be determined from fermentative gas production, which measures the amount of gas released directly as a product of the fermentation and indirectly from buffered rumen fluid. The soluble components of feed are rapidly fermentable after incubation. A gradual shift subsequently occurs towards the fermentation of the insoluble parts, which need to be hydrated and colonized by rumen microbes before they can be fermented. Substrates and their components can be resistant to these processes to different extents, resulting in substantially different gas production profiles. The *in-sacco* and *in-vitro* techniques are based on the gravimetric determination of residues and lead to the solubilization of anti-nutritional factors thus not contributing to energy production in the system but being measured as dry matter digestibility. This could lead to misleading information, whereas in the *in-vitro* gas method the effects of anti-nutritional factors on rumen fermentation are reflected in gas production.

Origin of Gas

When feedstuff is incubated with buffered rumen fluid *in vitro*, the carbohydrates are fermented to short chain fatty acids (SCFA), gases (mainly Co_2 and CH_4) and microbial cells. Gas production is basically the result of fermentation of carbohydrates to acetate, propionate and butyrate. Gas production from protein fermentation is relatively small as compared to carbohydrate fermentation. The contribution of fat to gas production is negligible. The gas produced in this technique is the direct gas produced as a result of fermentation (CO_2 and CH_4) and the indirect gas produced from the buffering of SCFA (CO_2 released from the bicarbonate buffer). For roughages, when bicarbonate buffer is used, about 50% of the total gas is generated from buffering of the SCFAs and the rest is evolved directly from fermentation. At very high molar proportion of propionate (concentrate diets) the amount of CO_2 generated from buffering of SCFA is about 60% of total gas production. Each mmol of SCFA produced from fermentation releases 0.8 - 1.0 mmol of CO_2 from the buffered rumen fluid solution, depending on the amount of phosphate buffer present. Gas is produced mainly when substrate is fermented to acetate and butyrate. Substrate fermentation to propionate yields gas only from buffering of the acid and, therefore relatively lower gas production is associated with the propionate production.

Stoichiometry of Gas Production

An important aspect of gas production of the anaerobic system is that stoichiometric laws of fermentation balance can be applied since fermentation products must be derived from substrate incubated. High correlation between stoichiometrically calculated gas and actually recorded values have been reported. The stoichiometric balance allows the theoretical calculation of equilibrated amounts of the products (SCFA and gases). If the molar proportion and amount of SCFA are known, the theoretical amounts of CH_4 and CO_2 expected from the rumen fermentation can be calculated.

Implications of the Relationship between Gas and SCFA Production

Short chain fatty acids can provide most or all of the ruminants' energy requirements. *In vitro* gas production can, therefore, be considered as an estimate of the energy potential of a feed. SCFA's are only one of the two major useful products of rumen degradation, the other being microbial biomass (MB). SCFAs and MB are linked in that adenosine triphosphate (ATP) is generated along the SCFA and ATP is a major source of energy for maintenance and growth of rumen microorganism. It is generally assumed that each mmol of acetate, propionate and butyrate results in the production of 2, 3 and 3 mmol of ATP, respectively. Additionally, 1 mmol of ATP is produced with each mmol of CH_4. Thus, 1 mmol of SCFA with 0.65 A, 0.25 P and 0.1 B would yield 2.66 mmol of ATP. It is important to realize that all products, i.e. SCFA, gases and MB must be derived from the degraded feed, which they must balance.

$$OMTDR = SCFA + Gas + MBP$$

A general inverse relationship between SCFA and MBP production per unit substrate degraded has been observed. However, SCFA and gas production is closely related.

Applicability of gas production techniques

1. *In vivo* organic matter digestibility, production of metabolizable energy and protein degradability.
2. Kinetics of fermentation and mathematical description of gas production profiles.
3. Prediction of voluntary intake.
4. Evaluation of anti-nutritive factors.

I. Measurement of in vitro gas production kinetics

The preliminary screening of feed and fodder samples can be made by *in vitro* gas production technique. The fermentation may be conducted in the glass syringes containing the substrate and buffered rumen fluid at various times of incubation. The rate and extent of gas production can be estimated by fitting modified Mitscherlich equation in model of single exponential equation with and without lag phase to study the gas production kinetics of substrate.

Weighing of samples and greasing of piston

Weigh accurately about 200 mg of air-dried (1.0 mm particle size) sample and transfer it into the syringes. After weighing, grease the piston with paraffin soft white (M.P. 39-56^{0}C) and push into the barrel of the syringe. Lubrication with paraffin soft makes syringes gas-water tight and prevents the piston from getting stuck in the barrel. For each set, 30 syringes can be used (4 blank and 2 standard having known gas production) and two sets of incubations are to be conducted.

Preparation of buffer media

The composition of different solutions is given below.

1. Micro mineral solution

$CaCl_2 . 2H_2O$	13.2 g
$MnCl_2 . 4H_2O$	10.0 g
$CoCl_2 . 6H_2O$	1.0 g
$FeCl_3 . 6H_2O$	8.0 g

Dissolved in 100 ml of water

2. *Buffer solution*

NH_4HCO_3	4.0 g
$NaHCO_3$	35.0 g

Dissolved in 1000 ml of water

3. *Macro mineral solution*

Na_2HPO_4 anhydrous	5.70 g
KH_2PO_4 anhydrous	6.20 g
$MgSO_4$. 7 H2O	0.60 g

Dissolved in 1000 ml of water

4. *Resazurine solution* 0.10 % w/v

5. *Reduction solution*

1N NaOH	2.0 ml
Na_2S. 9 H_2O	285 mg
Water	47.5 ml *aqua dest*

All the solutions may be prepared well before the start of incubation except the reduction solution, which should be prepared freshly each time shortly before the rumen fluid was collected. The solutions will pour into a Woulff flask, mixed with a magnetic stirrer and warmed to 39^{0}C in a PVC water bath with digital thermostat in the following order:

Solutions	Quantity (ml)
Distilled water	365
Micro mineral solution	0.1
Buffer solution	183
Macro mineral solution	183
Reduction solution	38.8
Rumen liquor	330
Total volume	**1099.9**

CO_2 gas should be passed through the submerged tube in the Woulff flask continuously during buffer media preparation.

On the day of incubation, the mixture of rumen liquor and particulate matter (*approximately 60:40*) should be collected from the fistulated donor animal (before feeding) into pre-warmed CO_2 filled thermos and carried to the laboratory. The rumen fluid must be bubbled with CO_2 gas for few minutes and then add it to the buffer media when it will become colorless. All handling of rumen liquor should be done under continuous flushing with CO_2.

Filling of syringes and incubation

Dispense 30 ml buffered fluid to each syringe by an automatic dispenser. After recording initial volume (±0.5ml), place the syringes in the incubator maintained at 39°C. The gas production may be recorded at 0, 3, 6, 9, 12, 18, 24, 36, 48, 72 and 96 hrs of incubation and syringes should be shaken by hand intermittently. If the gas volume exceeded 80 ml mark gas should be released. The fermentation should be terminated after 96 hrs of incubation. The fermentation kinetics can be estimated by means of nonlinear regression as follows.

(a) $Y= b \times (1-e^{-ct})$

(b) $Y= b \times (1-e^{-c} \times {}^{(t-L)})$

Where y = gas production (ml) at time t
b = potential gas production (ml)
c =fractional rate of fermentation
L = discrete lag time (h)

The average fermentation rate (ml gas h^{-1}) may be defined as the average gas production rate between the start of incubation and the time at which the cumulative gas production was half of its asymptotic value, and may be calculated

$$\text{Rate} = b\,(c/2\,(\text{In}2+c\text{x}L))$$

Substrate specific times are defined by the half time ($T_{1/2}$) of asymptotic gas production and calculated as

$$T_{1/2} = \text{In}2/c$$

Voluntary feed intake potential (g DM kg^{-1} $BW^{0.75}$) can be calculated from *in vitro* gas production kinetics parameters using the equations

$$\text{Intake (g DM kg}^{-1}\text{ BW}^{0.75}) = 18.90 + 0.23\,(a+b) + 687\text{x}c + 0.11\text{x CP}$$

II. Truly degradability of dry matter and organic matter and efficiency of microbial bio-mass production

After the preliminary screening of the feed samples on the basis of gas production at half time ($T_{1/2}$), true degradability of dry matter (TDDM) and organic matter (TDOM) and microbial bio-mass production may be determined by incubating the substrate up to half time ($T_{1/2}$) and/ or 24 h. The contents of the syringes are to be transferred to 500ml spoutless beakers by repeated washings with neutral detergent solution without sodium sulphate. The contents are to be refluxed for 1 hour and the residue may be recovered in pre-tarred filter crucibles.

After drying the crucibles (with residue) to constant weight, ash at 450°C for 30 minutes. TDDM, TDOM, OMD and MBP may be calculated as follows:

TDDM = Feed (DM) incubated - residue (DM)

TDOM = Feed (OM) incubated - residue (OM)

OMD% =14.88+0.889 x net gas production (24 h) +0.45 x CP (%DM)

MBP = TDOM - (2.2´net gas volume)

Partitioning factor (PF) and Metabolisable energy

PF may be calculated as the ratio of mg of TDOM to ml of gas thereby produced. Metabolisable energy (ME, MJ kg^{-1} DM) may be estimated from the gas volume using the equation

ME (MJ kg^{-1} DM) = 2.20+0.136´gas (ml 200 mg^{-1} DM)+0.0057´CP (%DM)

Limitation of *in vitro* gas technique for tannin rich feeds

The adverse effects of tannins are associated with their ability to bind with dietary proteins, carbohydrates and minerals as well as inhibition of rumen microorganism. As gas production is mainly from carbohydrate fermentation, the binding of tannins with fibre fractions reduces the gas production and thereby SCFA production, which ultimately reduces the energy value of tannin rich, feeds. Binding of tannins with macromolecules also creates difficulties in the analysis of fibre using the detergent method. For conventional feeds the ratio of substrate truly degraded: gas volume, termed as the partition factor, ranges from 2.74-4.65 mg/ml. The high partitioning factor for tannin rich forages due to various artifacts suggests limitation of residue determination using NDS. The developed a bioassay for tannins, which is based on incubation of a feed in the presence of polyethylene glycol 6000 (PEG 6000), the most effective tannin-complexing agent in the *in-vitro* gas method. The PEG 6000 (400mg/200mg feed) binds to tannins forming inert PEG-tannin complexes, which result in an increase in gas production. Using gas volume and equations o, the metabolizable energy (ME) and organic matter digestibility (OMD) were quantified, which allowed expression of tannin activity in term of their potential to decrease ME and OMD in tannin rich feeds. The use of PEG, to bind specifically with tannins increases the gas production in the presence of PEG is significantly correlated with total phenolics (TP), total tannins and protein precipitation capacity (PPC) of browse species, indicating their depression effect on gas production.

Conclusion and Recommendations

The application of In-vitro gas test techniques is quick and can be evaluate large number of feed resources in reasonable period. Data obtained on feed quality through gas test are very near to in-vivo experimental results. Hence, this technique is easy and economic for feed evaluation and feed quality assessment.

❑❑❑

17

Sheep Genetic Resource and Strategic Breeding for Improving Mutton Production: National Perspective

A. L. ARORA, L. L. L. PRINCE
Central Sheep and Wool Research Institute, Avikanagar, Rajasthan - 304 501

Introduction

The sheep population in India is about 61.46 million (2003 census) ranking third in the world and accounting for about 5.50 % of the total sheep population. The population of sheep has increased at an annual rate of 6.91 % from 1997 to 2003. About 36 % of the total sheep population is slaughtered every year for meat purposes. Sheep represents biodiversity in the form of breeds and strains/types. The Indigenous sheep breeds/types are rich in variability and are endowed with many positive traits like disease resistance, better tolerance to heat stress and humidity etc. It has also been noted that indigenous breeds are more efficient in feed conversion. They generally perform better than exotic breeds under low input conditions, climatic stresses and especially during times of draught. The different breeds of sheep have been developed in response to economic requirement of man and agro ecological conditions. A few breeds have been evolved from the base populations created by crossing native and exotic breeds.

Sheep Genetic Resources

The sheep breeds can be classified on the basis of Agro-ecological regions (Bhat *et al*, 1980; Acharya, 1982; Acharya and Bhat 1984) *viz*. a) North temperate region, b) North Western arid and semi arid region c) Southern penninsular region and d) Eastern region. The breeds of sheep of these regions are depicted in Table 1.

Table 1. Breeds of Sheep in different agro-ecological regions in India and their major products.

North Temperate	North-Western Arid and Semi Arid	Southern Peninsular	Eastern
Bhakarwal (CW) Changthangi (CW) Gaddi (CW) Gurez (CW) Karnah (AW) KashmirMerino(AW) Poonchi (CW) Rampur Bushair(CW)	Chokla (CW) Hissardale (AW) Jaisalmeri (MCW) Jalauni (MCW) Kheri (MCW) Magra (CW) Malpura (MCW) Marwari (MCW) Muzaffarinagri(MCW) Nali (CW) Pattanwadi (CW) Pugal (MCW) Sonadi (MCW) Munjal(M)	Bellary (MCW) Coimbatore (MCW) Daccani (M) Hassan (M) Kanguri (M) Kilakarsal (M) Madras Red (M) Mandya (M) Mecheri (M) Nellore (M) Nilgiri (AW) Rammand White (M) Tiruchy Black (M) Vembur (M)	Balangir (MCW) Bonpala (MCW) Chottanagpuri (MCW) Ganjam (MCW) Garole (M) Tibetan (CW)

Within parenthesis is the major product from the breed. Apparel wool (AW); Carpet wool (CW); Meat and Carpet wool (MCW) & Meat (M)

North Temperate Region: comprises of Jammu and Kashmir (J&K), Himachal Pradesh and Uttaranchal having medium wool type breeds viz. Gaddi, Rampur Bushair, Bhakarwal, Poonchi, Karnah, Gurez, Kashmir Valley and Changthangi. Large scale. crossbreeding during last 25 years has endangered number of indigenous breeds especially from Jammu and Kashmir, where almost all the native breeds of sheep have been involved in crossbreeding with exotic fine wool breeds and about 80% of sheep are covered under this programme and about 43% of sheep population has been converted into crossbred. Changthangi breed of Ladakh which is hardy and yields good carpet/medium apparel wool has relatively longer staple length than other breeds of this region. In Sikkim and Arunachal Pradesh, Tibetan sheep are available which have good carpet quality.

Northwestern arid and semi-arid region: includes the States of Rajasthan, Punjab, Haryana, the plains of Uttar Pradesh, Gujarat, Madhya Pradesh and Chattisgarh having carpet wool type sheep breeds of Chokla, Nali, Marwari, Magra, Jaisalmeri, Pugal, Malpura, Sonadi, Pattanwadi, Muzaffarnagri, Jalauni, Hissardale and Kheri. Marwari sheep has been numerically the most important and largest contributor to carpet wool production in the country. Inter mixing of breeds especially on the migratory tract of Marwari has a resulted in development of Kheri sheep in the field conditions. This region has the second largest population of sheep of the four regions and wool produced is mostly suited for manufacture of carpets, felts and blankets.

Southern peninsular region: covers the states of Maharashtra, Andhra Pradesh, Karnataka and Tamil Nadu having meat type breeds of Deccani, Nellore, Bellary, Hassan, Mandya, Mecheri, Kilakarsal, Vembur, Coimbatore, Nilgiri, Ramnad white, Madras Red, Trichi Black and Kenguri. Deccani sheep is numerically most important and is largest contributor to the meat production in the country . Most sheep in this region, except the Nilgiri, are maintained primarily for meat.

Eastern region: Comprises of the states of Bihar, Jharkhand, West Bengal, Orissa, Assam and North Eastern states having hairy breeds of Chottanagpuri, Bolangir, Ganjam, Tibetan, Bonpala and Garole. Garole sheep in Sunderban areas of West Bengal are reputed for multiple births.

Sheep assets

India ranks fourth in sheep population and account for 4.57 % of world population. In India, the sheep population had been almost static (around 40 million) from post independence till late 1980's. It increased to 48.76 million in 1982 followed by a decline to 45.71 million in 1987 and again raised to 50.80 million in 1992 and at reached 61.46 million during 2003. At present the estimated sheep population is 62.85 million (FAOSTAT, 2006). It is estimated that about 238 million kg of mutton are produced annually from 19.77milloin sheep slaughtered in the country. The sheep population figures and the annual production of mutton as per the FAO estimates (2006) are given in table 2 along with the figures for Asia and World.

Diversity in Indian Sheep Breeds

According to the FAO World Watch List (1995), there are 57 breeds of sheep in India. This list includes both well-recognized and lesser-known breeds along with some wild species. Although there are about 42 breeds of sheep, majority of sheep population (around 75%) do not belong to any of the defined breeds. India's share to total number of breeds of world is 6.41 per cent.

India is endowed with biologically diverse sheep population comprising of sheep producing fleeces having fibre diameter below 30 μ to totally hairy sheep. Some of the breeds from the adjoining areas have been identified as separate breeds although they may be similar to each other genetically. The phenotypic differences may be due to selection for adaptation to specific agro-climatic conditions. The Madras Red in Tamil Nadu, Red Nellore in Andhra Pradesh and Ganjam sheep in Orissa although identified as separate breeds appear to be similar but their actual genetic differences are not known. Similarly in Rajasthan Marwari, Jaisalmeri and Kheri are black face sheep. The so-called Bikaneri sheep include brown face breeds, viz. Chokla, Magra and Nali and the phenotypic differences might have been due to assortative mating. Real genetic differences, if any, among similar groups need to be established.

Breeds Exhibiting More than One Strain

There are more than one strains in some of the breeds viz. Deccani, Patanwadi, Nellore etc. The sheep of Deccani breed are not uniform and their coat colour varies from black to white and the mixture of the two. The varieties of Deccani are Lonad, Sangamneri, Solhapuri (Sangola) and Kolhapuri. Sangamneri type sheep has a stronger body constitution and higher wool yield. Sangola type sheep have higher proportion of black colour sheep and produce finer fleece.

Patanwadi sheep has three types (a) non-migratory; small bodied and fine fleeced Patanwadi, (b) Migratory; leggy with long tubular ears and relatively coarse fleeced-Kathiawadi and (c) low stuttered, meat type, very coarse fleeced-Charotari also called Sonadi in Rajasthan However, looking to the phenotypic appearance of Sonadi and Patanwadi their genetic similarity needs to be verified.

Nellore sheep in Andhra Pradesh has three varieties distinguished primarily on the basis of colour viz., Jodipi, Palla and Dora (a) Jodpi (also called Jodimpu) with black spots particularly around the lips, eyes and lower jaw and also on belly and legs (b) Pala completely white or white with light brown spots on head, neck, back and legs and (c) Dora, completely brown.

Table 2. Mutton production status

	India	Asia	World
Sheep population (m)	62.85	471.23	1101.64
No. of sheep slaughtered/annum (m)	19.77	292.13	539.02
Mutton production (m. kg)	238	4652	8632
Average yield/ animal (kg)	12	15	16

Source: FAOSTAT (2006) http://faostat.fao.org

Breeds with Unique Characteristics

All the indigenous sheep breeds are known for adaptability under varied climate conditions, which is obvious from their long struggle against the natural forces. Some of the unique features have been identified in few indigenous breeds, which are summarized in Table 3.

Garole sheep are reputed for multiple births and are found in the Sunderban area of West Bengal. These sheep are reported to have contributed prolificacy to the Booroola Merino sheep. Litter size at first lambing is two and at subsequent lambing is 2 to 3. Prolificacy reported is singles 25-30%, twin's 55-60%, triplet's 15-20% and quadruplets 1-2%.

Kendrapada sheep is identified as another prolific sheep of India after Garole of West Bengal. It is 2nd sheep breed of India and 6th in world which carry *FecB* mutation. Kendrapada is distributed in Bhadrak, Konark and Puri district of coastal area of Orissa. A survey was made to study the prolificacy of Kendrapada sheep of Orissa around Kendrapada district and Nimapara (Konark) during April 2007. The survey reveals that Kendrapada is a prolific and excellent medium stature meat type sheep. In the flocks surveyed more than 75 % ewes produces multiple births and the adult body weight of sheep was about 23 kg.

Table 3. Breeds with unique characteristics

Breed	**Unique characteristics**
Magra	Lustrous wool, excellent for carpet manufacture
Changthangi	Alpine sheep of high altitude for fine wool
Chokla	Fine carpet quality fleece
Garole	High fecundity - twins and triplets common. Survival under saline conditions.
Mecheri, Madras Red	High quality skin and mutton
Mandya	Excellent meaty conformation, high quality and meat palatability.
Patanwadi	Carpet wool and good milk producer
Nellore	Tallest sheep breed of India.
Muzaffarnagari	Best mutton producing breeds
Marwari, Deccani Jaisalmeri	Hardy and capable of walking long distances during migration.
Kendrapada	High fecundity- twins and triplets common.

Kheri sheep evolved in the farmers flock under the field conditions especially in the areas that are important for sheep migration in Rajasthan appears to be a need of migratory sheep breeders. The animals of this breed have established owing to their ability to walk long distances and ability to survive on limited amount of coarse feed. The animals of this breed can sustain stress and on return of favorable condition regain faster resulting in better reproduction rate and growth of lambs as compared to prevalent breeds in the area.

Munjal sheep is the least discussed among Indian sheep breeds. Information on population and production performance of this breed are scanty. Munjal is one of the important mutton type sheep. Animals of this breed are quite big and tall. Average adult body weight is around 45 kg. In recently conducted survey it was observed that the population of this breed is 90 thousand. Animals of this breed are found around Muktsar, Bhatinda, Firozpur and Faridkot district of Punjab.

Special Characteristics of Indigenous Breeds

All the indigenous sheep breeds are known for adaptability under varied climatic conditions, which is obvious from their long struggle against the natural forces. Indigenous animals during its evolution have developed superior heat tolerance as well as disease resistance. Most of the breeds are very well adapted to harsh climate, long migration, tropical diseases and ability to thrive and produce with poor nutrition and shortage of drinking water and its poor quality. These qualities are found lacking in the exotic sheep breeds and crosses developed from them.

Heat stress

Indigenous sheep breeds are hardy and are relatively better adapted to local climatic conditions and can sustain extremely high temperature during summer season. Crossbred sheep can't be sent for migration and start panting during hot summer months. Hence crossbred and exotic sheep have to be grazed during cooler hours i.e. morning, evening and night grazing.

Disease Resistance

There are substantial evidences for genetic variation between indigenous sheep and exotic crosses in resistance to internal parasites such as *Haemonchus contortus, Ostertagia circumcincta* and *Trichostrongylus colubriformis*. Studies conducted on some of the indigenous sheep and exotic crosses with respect to infection with gastrointestinal nematodes revealed that there is significant differences in FEC (Fecal egg count), hematological and biochemical parameters between the breeds. Indigenous sheep excreted fever worm eggs in faeces and had lower morbidity and mortality rates compared to exotic breed and their crosses. Garole sheep are naturally resistance to haemonchosis.

Breeding Technologies

Important traits for mutton production

Meat production traits mainly include age and body weight at slaughter, efficiency of feed conversion, dressed carcass weight and dressing percentage. Optimum weight of about 30 kg at about six month of age is extremely important from international market standard point of view. Further it has direct bearing on meat traits and taste of consumers. Milk yield and prolificacy also contribute significantly towards mutton productivity per animal. There is a need to evaluate the mutton production traits of native breeds of the region before developing breeding strategies to usher in further improvement. Recording of meat production data is hardly done under field condition. Information on carcass characteristics is available only on the breeds maintained in the research farms. Here again the performance of the breeds are not strictly comparable due to variation in age at slaughter, feeding regimen, managemental practices and locational effect of the flocks. Still, an overall picture of the mutton production traits of native and crossbred sheep has been presented (Table 4) depicting the extent of variability in these traits. Most of these breeds (Malpura, Sonadi, Avimaans, and Avikalin) were reared under uniform managemental practices in the same location. Almost all growth and feedlot traits showed wide variability. This clearly implies that there is enough scope for further improvement in mutton production traits through selection.

Table 4. Comparative growth performance of native and crossbred mutton type sheep

Breed/Genetic group	Birth wt. (kg)	3-month wt. (kg)	6-month wt. (kg)	9-month wt. (kg)	12-month wt. (kg)	Source
Malpura	2.91 (2167)	12.86 (1734)	19.24 (1247)	22.26 (767)	25.62 (629)	1
Sonadi	2.79 (972)	12.53 (740)	18.50 (515)	22.06 (323)	25.38 (274)	1
Avimaans/ Mutton Synth.	3.21 (4570)	14.11 (3429)	22.08 (2729)	23.75 (1903)	26.71 (1579)	1
Mandya	2.15 (358)	13.45 (291)	17.28 (204)	20.11 (108)	22.91 (94)	2
Mandya Synth.	2.60 (801)	14.89 (709)	22.17 (558)	22.17 (392)	25.08 (326)	2
Nellore	3.00 (603)	13.45 (571)	17.81 (435)	21.23 (338)	24.54 (270)	2
Nellore Synth.	3.17 (757)	14.78 (711)	21.00 (611)	23.32 (477)	26.41 (385)	2
Muzaffarnagri	3.26 (1813)	14.55 (1704)	19.68 (1627)	23.88 (1177)	27.98 (1057)	2
Dorset x Muzaffarnagari	3.54 (122)	15.21 (120)	21.26 (116)	25.67 (95)	30.26 (76)	2
Suffolk x Muzaffarnagari	3.89 (26)	16.25 (26)	23.56 (146)	27.74 (46)	31.99 (44)	2
Deccani	2.88 (1668)	12.59 (1412)	16.73 (973)	18.83 (429)	22.24 (423)	2
Dorset x Deccani	3.52 (1043)	13.83 (893)	20.28 (784)	22.08 (533)	25.05 (450)	2
Madras Red	2.23 (1137)	9.03 (1025)	12.50 (843)	15.25 (616)	16.15 (519)	2
Dorset x Madras Red	2.98 (92)	12.01 (92)	16.49 (81)	19.40 (70)	21.16 (60)	2
Awassi x Malpura	3.59 (126)	16.53 (96)	24.39 (111)	27.08 (104)	30.40 (94)	3
Avikalin	2.95 (120)	14.92 (113)	19.92 (109)	23.03 (106)	30.15 (82)	3
Avikalin x Avimaans	3.04 (75)	15.81 (68)	20.79 (66)	25.17 (62)	30.62 (54)	3

Source : 1.Final Report (Mutton unit), CSWRI, Avikanagar;
2. Final Report, 1993 Project Coordinator, Sheep Breeding (AICRP);
3. Report 1996 - Annual Report CSWRI, Avikanagar

Selective breeding and crossbreeding

Several experiments for increasing body weight in sheep through selection within indigenous and crossbred population have been attempted. In mutton production programs, more weightage has been given to six month body weight in comparison to greasy fleece. Since the characteristics connected with mutton production viz. body weight gains, efficiency of feed conversion and dressing percentage are moderate to highly heritable, the selection within the indigenous breeds can bring about considerable improvement. Countries that have developed sheep industries, maintain special breeding flocks to supply superior germ plasm to commercial flocks. Performance with respect to growth, feed efficiency and carcass traits in indigenous and crossbred sheep is presented in Table 5.

Table 5. Feed lot performance of native and crossbred Mutton type sheep.

Breed/Gen. Group	Initial wt. Dressing (kg)	Final wt. (kg)	Av. daily gain (g)	Feed eff. %	Hot carcass weight(kg)	Dressing (%) EWB
Malpura	15.99	29.14	146	15.62	-	59.13
Avimaans	18.93	32.54	151	15.71	-	59.12
Mandya	10.23	18.05	87	13.46	8.48	54.5
Mandya Syn.	13.89	23.79	110	14.48	10.56	55.2
Nellore	13.45	20.01	91	14.60	10.63	58.22
Nellore Syn.	14.78	26.53	134	14.74	12.96	59.34
Muzaffarnagri	16.71	26.67	110	14.08	-	57.26
Dorset x Muzaffarnagri	18.57	32.29	152	19.46	-	58.82
Suffolk x Muzaffarnagri	13.33	26.41	145	23.62	-	56.94
Deccani	11.18	25.29	119	13.02	8.33	52.65
Dorset x Deccani	12.17	27.41	143	13.87	10.84	54.97
Madras Red	9.29	17.72	87	12.34	8.16	51.43
Dorset x Madras Red	12.02	26.04	123	14.07	11.07	57.39

Results on growth performance reveals that Malpura and Muzaffarnagri out of North-Western breeds and Nellore and Mandya out of Southern breeds have great potential for their use as improver breeds for mutton productions. New Synthetics have generally attained live weight of about 30 kg at 6 months of age under intensive feeding conditions. The results of cross breeding on the whole, revealed that percent improvement in body weight up to 6 months of age is conspicuous in different crossbreds over contemporary natives but only marginal improvement is observed at 12 months of age. This is indicative of the fact that crossbreds/ new synthetics required high plane of nutrition and if they are managed on the same feeding regimen as for natives, the differences which were conspicuous up to 6 months age, narrow down gradually and become marginal at the age of one year. Due to non-availability of required plane of nutrition, mutton type strains developed by crossing Suffolk and Dorset with the indigenous breeds could not outdo the natives under village managemental conditions despite the improvement in body size as well as in wool quality and quantity. Through selective breeding and intensive selection some of the important indigenous sheep breeds like Malpura, Sonadi, Muzzaffarnagari, Madras Red, Mandya, Nellore, Deccani have been improved and rams of these breeds are available for enhancing mutton production in farmers flocks.

Crossbreeding for heavy live weight

A need was felt to enhance milk and mutton production in Malpura sheep by crossing it with Awassi rams (well known for heavy body weight and milk production). A pilot project on crossbreeding of Awassi rams with Malpura ewes was started in 1994. Awassi x Malpura halfbreds have demonstrated their production potential for increased growth, milk and wool yield in harsh climatic conditions of semi-arid region of Rajasthan. These half-bred lambs exhibited about 20% and 15% improvement over Malpura lambs at 6 and 12 month of age with 37% more wool at first clip. Wool produced by these crossbred was of carpet quality having lustrous property. Crossbred ewes produced 29% more milk than that of contemporaries Malpura ewes. The survivability of this genotype was almost at par with native Malpura. Awassi rams were also used to cross Malpura ewes in farmers' flocks and local farmers are interested in crossbreeding of their animals with Awassi rams. Therefore, this program may be strengthened by procuring Awassi sheep from foreign countries.

Crossbreeding for multiple birth

The lambing percentage in indigenous breeds of sheep varied from 60 to 95 % mostly with single born lambs. Considering the importance of multiple births in sheep breeding in special reference to mutton production, Garole a prolific breed of sheep was introduced in 1997 in Mutton Project at CSWRI, Avikanagar. Crossbreeding of Malpura ewes with Garole rams was undertaken considering the milk availability for twins/triplets in crosses. Body weights at different ages revealed that Garole crosses were little bit on lower side in body weights compared to their counterparts Malpura lambs. Garole x Malpura half-bred ewes yielded encouraging results in terms of twin lambing percentage about 51% twining was observed with overall litter size of 1.61. Survivability of Garole x Malpura halfbreds was at par with the Malpura sheep in semi-arid conditions. To ensure that the prolificacy will be introduced into Malpura without compromising body weight, backcrossing programme was started. In backcrossing programme GM rams were used as sire and Malpura ewes as dam breed to produce the GM x Malpura (GMM ie. 75 % Malpura and 25 % Garole) and reciprocal crosses were also made to produce MGM (Malpura x GM). There was 53.76, 49.13 and 33.52 percent increase in body weight of GMM lambs over GM half-breds at birth, weaning and 6-month of age, respectively. The body weight of GMM was non-significantly different from Malpura at birth and weaning stage. These results indicate that the breeding programme is in right direction.

Results obtained in respect of prolificacy and reproduction indicated that crossing of Garole with Malpura sheep increased prolificacy and fecundity gene have been incorporated into the GM genotype. This genetic group is adaptable under semi-arid region of the country and it might prove a valuable germplasm towards evolving a new prolific strain of sheep, having optimum body weight for enhancing mutton production in the country. However, there is a need to further strengthen the population of GM half-breds and backcross (GMM) population. Simultaneously it has to be ensured that incorporated prolific genes do not get lost. Scope does exist to harvest better returns through enhancing Ewe Efficiency by improving managemental practices for this newly evolved (GM) genetic group.

For efficient and profitable mutton production programs following points may be kept in view

- It would be worthwhile to plan in advance the annual management programme aimed at supply of lambs to targeted markets based on their requirements for carcass weight and leanness.
- It has been found that lambs should preferably be weaned at two month age for putting them under intensive feeding to harvest economic market weight around 5 - 5½ month of age.
- Buy sires from reputed breeders who are making an active attempt to breed faster growing, leaner animals and whose sheep, regardless of breed produce the best lambs.
- While selecting sires within a breed, favour rams that have genetically superior growth rates and are leaner than average.
- Dorset and Suffolk are the breeds of choice in crossbreeding programmes for producing meatier lambs in high plane of nutrition. Trials have shown little difference, on average, between Dorset and Suffolk sires in growth rate, feed efficiency and carcass traits. Differences between sires of the same breed are much greater. In general, growth rate and leanness of breeds tend to be related to their mature live weight. Therefore greater emphasis should be placed on selection of right type of sire(s) rather than a change of breed.
- Use live weight scales and fat scoring techniques to market meat and meat products of acecptable quality keeping in mind the consumers' choice. Live weight should in no case exceed 30 kg and should preferably be around 25 kg. At the same time, fat percent in carcass should not exceed 18-20.
- Ability to produce fast growing lean lambs can be introduced through use of right type of sire breed. In crossbreeding programme, choice of ewe breed should be made on the basis of other factors (viz. lambing rate, mothering ability, wool production, wool quality, availability and cost). Growth rate and leanness vary greatly within a breed. From literature it is evident that these traits are medium to highly heritable. This means that if lean, fast growing rams are selected for breeding, selection will be effective and these characteristics will be passed on to their lamb progeny.
- Ewes in advance stage of pregnancy should be given due care to obtain healthy and heavier lambs.
- Extensive utilization of proven rams through use of frozen semen technology and elite females using embryo transfer technology must be carried out for getting rapid genetic progress.
- Prolific breeds like Garole may be introduced for multiple births. Before we make attempt to improve prolificacy, we may first improve mothering ability

by introducing breeds like Awassi which is well known for high milk production. It will definitely help to sustain twin/triplet born lambs.

Conclusions and Recommendations

- For profitable meat production, we need to evolve a breed, which must have fast and efficient growing lambs/kinds. The ewes/does should be prolific and must have good mothering ability. Fast growth and early maturity are the most important characteristics for meat production. Selection programs were undertaken in indigenous stock to increase their productivity.
- Potential mutton type breeds such as Malpura, Sonadi, Muzzaffarnagari, Madras Red, Mandya, Nellore, Deccani, Mecheri, Ganjam etc., may be restricted to selective breeding by using superior rams of the respective breeds for enhancing mutton production. These are promising sheep breeds for increasing mutton production and they can also be used as improver breed for upgradation of native stock
- Crossbreds may be tried in the area where feed and fodder are available adequately and they are not required to be sent on long distance in search of food.
- Elite male germplasm centres should be developed for providing good breeding rams. Frozen semen banks need to be established.
- National Ram Production Programme being implemented since long back may be suitably modified to take up production of genetically superior rams preferably through selection in open nucleus breeding flocks.
- ONBS can effectively overcome the problem of non-availability of genetically superior breeding males required for bringing improvement in production through breeding.
- Cross breeding of prolific Garole sheep with other indigenous sheep subject to availability of sufficient milk with the ewes to sustain twin/triplet should be attempted at different locations of the country so that their reproductive efficiency could be increased. The fecundity gene should be detected at an early age in the ram lambs born out from such crosses.

References

Acharya, R.M. 1982. Sheep and goat breeds of India. Rome : F.A.O. 1982. 190p (FAO Animal Production and Health paper No. 30).

Acharya, R.M and Bhat, P.N. 1984. Livestock and Poultry genetic Resources in India. IVRI, Piblication.

Arora, A.L. 2004. "Strategies for Improvement of Sheep for Mutton Production". Chapter contributed in training manual on "Advances in Sheep Goat and Rabbit Production and their product Utilization". pp 18-26. Published by Director, C. S.W.R.I., Avikanagar.

Arora, A.L. 2005 Potential of Indian Sheep Genotype and Strategy for Improvement. In Compendium of Lectures, Training Programme on " Advances in semi - intensive sheep management" sponsored by Directorate of Animal Husbandry, Govt. of Andra Pradesh organized by HRD Section of CSWRI, Avikanagar. pp 11-18.

Arora, A.L. and Kushwaha, B.P. 2001. Sheep Genetic Resources, Production Potential and Improvement Strategies. Invited Paper Presented in 3rd Indian Agricultural Scientists and Farmers' Congress Organised by Bioved Research and Communication Centre Allahabad from 3-5 Feb., 2001.

Arora, A.L. and Prince L.L.L 2004. Contribution and significance of Indigenous sheep in India. In Compendium of Lectures, Training Programme on " Chacterization and conservation of Animal Genetic Resources" conducted at NBAGR, Karnal during 2-11th August 2004 pp.95-103.

Arora, A.L., Sharma, R.C. and Kushwaha, B.P. 2001. Breeding technologies for increasing mutton production in sheep breeds of Rajasthan. Invited paper In: 4th National Seminar of Indian Society for sheep & goat production & utilization on scope and strategies for increasing meat production from small ruminants and rabbit in new millenium, held at Bharthiar University, Coimbatore (Tamilnadu) during November 9-11, 2001.

Arora, AL, Miishra AK and Prince, L L L 2007. Sheep genetic resources in India: Their conservation and improvement. Lead paper in Proceedings of National symposium on role of animal genetic resources in rural livelihood security at Ranchi during Feb, 8-9,2007.

Bhat, P.N.; Bhat, P.P.; Khan, B.U; Goswami, O.B. and Singh, B. 1980 . Animal Genetic resources in India. Publication No. 192, NDRI, Karnal.

World Watch List for Domestic Animal Diversity, second Edition (Eds Beate D.Sherf), F.A.O. Rome, September, 1995.

□□□

18

Significance of Prolificacy Trait in Intensive Mutton Production

A.K. MISHRA

Central Sheep & Wool Research Institute, Avikanagar - 304 501, Rajasthan

Introduction

India is endowed with wide diversity of sheep genetic resources, which forms the backbone of its rural livelihood security systems. Sheep rearing now faces a dilemma to produce more mutton and wool for the growing human population against the realization of shrinking grazing resources and saturation in the further growth of sheep population. In the present scenario the demand for meat in India has increased rapidly an d the emphasis has shifted from wool towards mutton as the main produce from sheep rearing. There is acute shortage of meat for domestic needs besides huge demand in the international market. The gap between the demand and production of mutton is to be bridged by augmenting the reproductive rate of low producing Indian sheep breeds. The low prolificacy of all indigenous sheep breeds, barring Garole, is a stumbling block for converting the available bio resources into more remunerative economic wealth on sustainable basis to cope up with the rising demand of end users. In order to improve the fecundity of sheep, incorporation of genetic material of prolific sheep is an ideal approach to evolve a large size breed capable of multiple births for economic and remunerating mutton production.

Prolificacy is one of important aspects in sheep breeding in special reference to mutton production because it has direct relevance on ewe productivity as well as net economic return from sheep rearing profession. Since level of production is also influenced by rate of reproduction, the reproductive efficiency of sheep becomes an important aspect in determining the profitability of sheep farming. The importance of fecundity is well understood from the fact that an increase in number of lambs weaned per ewe offers the greatest single opportunity for increasing the efficiency of sheep farming (Shaoqui, 1997). The number of ovulated follicle depends on many factors in sheep, e.g. nutrition, season, age and mainly on breed. Use of prolific Garole sheep is one of the possibilities to improve the reproductive efficiency of the sheep flocks. Reproductive traits such as fertility, prolificacy and lamb survival have been recognized as major contributing factors influencing profitability of sheep rearing. Among reproductive traits, litter size is of main interest because an increase in the number of lamb marketed per ewe per year offers the greatest single opportunity for net economic gain (Shelton, 1971). Turner (1969) concluded that litter size seemed to be the most useful selection criterion for genetic improvement of reproduction. Davis *et al.* (2002) have identified the *FecB* mutation in Garole and Indonesian Javanese sheep and hypothesized that *FecB* gene is the same as that reported in Australian Booroola Merinos. There has been increasing interest worldwide in the identification and utilization of major genes affecting litter size in sheep since last two decades. The *FecB* gene mutation has been discovered in only five breeds in the world i.e. Garole, Hu, Han, Javanese and Booroola Merino excluding the breeds into which gene has been introgressed from the Booroola Merino. Recent research has confirmed that *FecB* mutation is fixed in populations of Garole and Hu sheep but segregating in the Javanese, Booroola and Han breeds. Ewe productivity is dependent upon the component traits of fertility, litter size, lamb survival and growth (Fogarty *et al.* 1985) and is also a major concern for the sheep industry. Improving female reproductive performance is an important objective for increasing the prolificacy of sheep (Abdulkhaliq *et al.* 1989). There is great potential for increasing both biological and economic efficiency of lamb production through genetic improvement in reproductive rate than through improvement in growth rate or body composition (Dickerson, 1978). Table 1 presents the comparative performance of different breeds for prolificacy. Ewes of Romanov and Finnsheep have the highest prolificacy.

The crossing of prolific Garole sheep with non-prolific Malpura sheep was started in 1997, aiming to augment reproductive efficiency of non-prolific sheep at Central Sheep and Wool Research Institute, Avikanagar located in semi-arid region of the country. Malpura, non-prolific mutton type sheep breed of semi arid region of Rajasthan, was used as dam breed and Garole, a small sized prolific sheep breed of hot humid region of West Bengal, was used as sire breed in developing Garole X Malpura (GM) half-breds that supposed to have fecundity genes. Malpura, a famous mutton type sheep breed of semi arid tropical environment of Rajasthan (India), usually produce single lamb. The Malpura is known for its hardiness and adaptability to the local environment. As lamb production is an important source of income in all sheep flocks, increasing prolificacy of the Malpura has always been an important breeding goal. Most of the sheep breeds of India produce single lambs barring Garole; a small sized breed of sheep found in hot humid coastal belt of Sunderban in Southern part of West Bengal and is a highly prolific breed of sheep (Ghalsasi and Nimbkar, 1993; Sharma *et al.* 1999; Davis *et al.* 2002; Nimbkar *et al.* 2003a, b, Mishra *et al.* 2005). The average adult body weight of this breed ranges between 10-14 kg with mean litter size 2.27 in the native tract (Bose *et al.* 1999; Ghalsasi and Nimbkar, 1993; Fahmy and Mason, 1996; Nimbkar *et al.* 1998; Sharma *et al.*1999; Sharma *et al.*2001).

Table 1. Average Litter size of prolific sheep breeds

S N	Breed	Litter size	Reference
1	Garole	2.3	Nimbkar, 2002
2	Hu	2.1	Feng *et al.*, 1996 Davis *et al*,2005
3	Han	2.4	Feng *et al.*, 1996
4	Javanese (Thin-tailed)	1.8	Fletcher *et al.*, 1985, Davis *et al*,2005
5	Booroola Merino	2.6	Piper and Bindon, 1996
6	Afec-Awassi	2.1	Gootwine *et al.*, 1995
7	Rambouillet x Kazakh	1.55	Dobson *et al.*, 2000
8	Rambouillet X Booroola Merino	2.51	Southey, *et al*, 1996
9	Rambouillet X Booroola Merino	2.33	Crooks *et al*, 1998
10	Garole x Malpura	1.60	Mishra *et al.*, 2006
11	Garole x Deccani	2.12	Nimbkar *et al.*, 2002
12	Finnsheep	2.56	Thomas 2003
13	Barbados	1.92	Thomas 2003
14	Targhee	1.56	Thomas 2003
15	Romanov	3.40	Thomas 2003
16	Barbados	1.92	Thomas 2003

Thomas 2003.

Prolificacy and Weaning Rate

Reproductive efficiency of sheep is an important component of flock productivity and its augmentation is essential for improving the low reproductive performance of native sheep breeds. The prolificacy (litter size at birth) and weaning rate was 1.05 ± 0.01 and 1.00 ± 0.01 in Malpura ewes which increases up to 1.60 ± 0.04 and 1.41 ± 0.04 in GM ewes respectively (Table 2). Nimbkar *et al.* (2005) reported that B+ Garole x Deccani creosses gave birth to 0.5 more lambs per ewe lambing and weaned 0.3 more lambs per ewe lambing. Bunge *et al.* (1995)

also reported that prolificacy was higher in Booroola Merino- and Finnsheep-sired ewes. Rosati *et al.* (2001) reported number of lambs born alive and number of lambs alive at weaning as 1.68 and 1.48. The prolificacy in Garole that was used as sire breed in evolving GM was 1.80 (Mishra *et al.* 2005) under similar environmental conditions. Iman and Slyter (1996) also reported that Finn-Dorset-Targhee had higher prolificacy and weaned more lambs per ewe exposed (1.41 vs 1.18) than Targhee and gain in prolificacy was 48 %. In breeding experiment conducted in Egypt, using Finn sheep to increase prolificacy of Ossimi and Rohmani breeds resulted in an increase in prolificacy from 1.15 in local breeds to 1.65 in ¼ Finn crosses (Aboul-Naga, 2000). The year and parity did not affect significantly prolificacy in Malpura as well as in GM. The overall percent gain in prolificacy was 52.38, which increases up to 75.73 in third parity indicates that GM ewes were more prolific at the 2nd and 3rd parity. Bathei (1994) also reported that prolificacy increased with parity to the fourth parity whereas, Young and Dickerson (1998) reported that BM sired ewes were more prolific at 2nd and 3rd parity.

Table 2. Litter size at birth and weaning in Garole, Malpura and GM ewes

Particulars	Garole			Malpura			GM		
	Birth	Weaning	N	Birth	Weaning	N	Birth	Weaning	N
Overall	1.73	1.04	227	1.05	1.00	561	1.60 4	1.41	209
2000	1.56	1.00	9	1.00	0.96	24	1.27	0.82	11
2001	1.65	1.00	54	1.01	0.95	75	1.50	1.29	24
2002	1.87	1.29	45	1.02	0.95	100	1.51	1.31	39
2003	1.79	1.22	63	1.08	1.04	166	1.64	1.51	67
2004	1.64	0.70	56	1.	1.02	196	1.71	1.51	68
Parity									
1	1.57	0.87	98	1.04	0.97	269	1.53	1.31	93
2	1.82	1.17	65	1.07	1.04	146	1.62	1.43	63
3	1.86	1.24	37	1.03	0.98	91	1.81	1.71	31
4	1.72	1.17	18	1.05	1.05	43	1.59	1.41	17
5	2.22	1.00	9	1.17	1.17	12	1.60	1.20	05

Growth Performance

It was obvious that on crossing the small sized Garole rams with large sized Malpura ewes the half-bred progenies would weigh less than the dam parent and higher than the sire parent. The main purpose was to attain multiple births from the FecB introgressed GM sheep. Analyzing the GM lambs in terms of attainment of body weight vis-à-vis their counterpart Malpura lambs was essential for formulating further breeding and selection strategies. The least squares mean of body weight attained at birth, 3 months, 6 months and 12-months of age was 1.90±0.04, 8.53 ± 0.25, 14.72 ± 0.38 and 19.74 ±0.65 kg, respectively (Table 3). The body weights at different ages and wool yield of GM half-breds were lower, side compared to contemporary Malpura. However, the least squares means of litter weight on the basis of ewes lambed of GM were significantly higher than Malpura and Garole (Table 4). Two reasons contributed to these differences, one; the fact that the cross-contained genes from a significantly lighter breed, the second; many of the crossbred lambs were born twins or triplets which are mostly lighter than single-born lambs. Iman and Slyter (1996) reported that birth type significantly affected (P<0.01) birth weight and weaning weight of lambs. Significantly lower body weights and growth rate of lambs sired by Garole rams with Deccani/Bannur as dam breed compared to the lambs sired by Deccani and Bannur rams with Deccani/Bannur as dam breed was also reported by Nimbkar *et al.* (2000) and Nimbkar *et al.* (2003a).

Table 3. Least squares means for body weights (kg) in GM lambs.

Particulars	Body weights of lambs			
	Birth	3- month	6- month	12- month
Overall	1.90 (617)	8.53 (564)	14.72 (517)	19.74 (363)
Generation	NS	NS	NS	NS
F1	1.96 (247)	8.29 (238)	14.31 (230)	20.57 (166)
F2	1.86 (310)	9.01 (276)	14.80 (248)	19.04 (170)
F3	1.90 (60)	8.28 (50)	15.04 (39)	19.60 (27)
Parity				
1	1.74 (205)	8.25 (183)	14.52 (166)	19.23 (106)
2	1.88 (172)	8.84 (154)	15.31 (138)	19.94 (108)
3	1.93 (106)	8.65 (101)	14.71 (91)	19.64 (62)
4	1.92 (74)	8.58 (69)	15.07 (68)	19.66 (56)
5	1.93 (39)	8.39 (37)	14.39 (35)	19.78 (20)
6	2.03 (21)	8.44 (20)	14.31 (19)	20.17 (11)
Year				
1998	1.87 (45)	7.15 (43)	12.05 (42)	15.88 (27)
1999	1.91 (48)	8.47 (44)	14.59 (43)	18.22 (28)

Reproductive Performance

Reproduction is a good indicator of adaptation and contributes to the overall economics of the flock. The GM ewes had higher lambing % (90.44) on ewes' available basis than Malpura ewes (87.70 %). Lambing rate on ewes available basis which determine the flock reproductive efficiency as a whole and also covering unrecognised embryo/foetus losses i.e. it is measured in terms of live lambs born out of available ewes in the flock in a specified period. It averaged 91.75 and 148.21 % in Malpura and GM ewes, respectively (Mishra *et al.* 2006). These are in agreement with other studies (Bathei, 1994; Iman and Slyter, 1996; Mishra *et al.* 2005). The twin lambing percentage was only 4.71 % in Malpura flock. It had reached up to 51.10 % in GM flock within 5 years of inter mating between GM crosses. The 5.72 % of GM ewes produced triplets also (Mishra *et al*

2006). There was tremendous increase in multiple births and in other reproductive efficiency traits in GM ewes compared to contemporary Malpura. These results provided first evidence for increased lambing rate and litter size for their further exploitation through appropriate breeding and selection strategies. Other studies (Nimbkar *et al.* 1998; 2000; 2003b) also found Garole crosses to have the genetic potential for high reproductive efficiency. The prolificacy of Garole was 1.80 in similar environmental conditions (Mishra *et al.*, 2005). The over all distribution of single, twin, triplet and quadruplets lambing was 33.51, 47.94, 17.01 and 1.55 % respectively in Garole (Mishra *et al* 2005). It shows that twining is more common in Garole than single birth. Sharma *et al.* (2004) reported prolificacy of GM 1.19. The percent increase in observed fecundity over expected was 19.51 and maximum (1.66 vs 1.23) in year 2004. The crossing of Finnsheep also increases prolificacy (Aboul- Naga, 2000) and thus fecundity.

Table 4. Least square means for litter weight (kg) in Garole, Malpura and GM lambs

Particulars	N	Birth	3-Month	6- Month
Overall	998	2.78	11.96	17.57
Genetic group				
Garole	226	2.46	10.17	12.82
Malpura	561	2.89	12.17	18.62
G x M	211	3.00	13.52	21.26
Year				
2000	44	2.64	11.80	16.51
2001	152	2.79	11.62	16.63
2002	184	2.85	11.40	17.59
2003	296	2.91	13.88	20.44
2004	322	2.72	11.07	16.66
Parity				
1	461	2.53	11.32	16.85
2	274	2.83	12.25	17.80
3	159	2.75	12.06	17.46
4	78	2.79	12.61	18.34
5	26	3.02	11.54	17.38

Ewe Productivity Efficiency (EPE)

The ewe productivity efficiency in terms of kg weaned per ewe lambed or weaning efficiency (W.E.) of the ewe is an extremely important characteristic of economic mutton production programme because it covers two important factors: (i) Mortality of the lambs up to weaning stage (90 days) and (ii) Weaning weight of available lambs. Therefore, to improve the W.E. of the ewe due care has to be given to both the factors because if the contribution of the either factor or both is low then automatically weaning efficiency will be affected adversely.

The ewes productivity efficiency of GM sheep in terms of kg of lamb harvested per kg of ewes body weight was significantly higher than Garole and Malpura at all the ages except at birth and at birth, 3, 6 and 12 month of age it produced 0.118, 0.540, 0.855 and 1.177 kg, respectively on the basis of ewes lambed (Table 5). The results indicated that GM females are more efficient than Garole and Malpura with respect to kg lamb harvested per kg body weight of ewes. The EPE in terms of Total lambs weight produced at birth (TLWB), at weaning (TLWW) and at 6-month of age (TLWM) of G, M and GM are given in table 3. The GM sheep weaned 32.94 % more total lamb weight ($P<0.01$) over their sire breed Garole and 11.09% more over dam breed Malpura. At marketing age (6month) the increase in total weight was 65.83 and 14.18% respectively. The present finding of TLWW of GM was higher than (11.07 kg) than that reported by Sharma *et al* (2004). Nimbkar *et al.* (2005) reported that B+ Garole x Deccani ewes produced 1.1 kg greater lamb weight at 105 days, and this was a significant effect. Other studies have found that Finn cross ewes weaned more total lamb weight than other breeds (Thomas and Whiteman 1979; Fogarty *et al.* 1984; Iman and Slyter 1996). In contrast, Ramdas *et al.* (1993) reported non-significant difference between genetic groups for total lamb weight weaned per ewe exposed. The genetic group significantly ($P<0.01$) affects TLWB, TLWW and TLWM. The year significantly affects TLWW and TLWM. The difference due to year of birth was attributed to variations in environmental conditions as well as availability of other inputs.

Significance of Multiple Births in Garole Crosses

Increasing multiple births is vital for augmenting mutton production and economic returns from sheep rearing profession. The benefits of multiple births in GM ewes vis a vis single birth in Malpura ewes are reflected in terms of high (i) per kg of lamb production per ewe (ii) per kg of lamb production per kg of ewe's body weight and (iii) litter size at birth, weaning and 6-month of age. Comparative evaluation of twin and triplet-bearing ewes of GM with the monotocus Malpura ewes reveals the significant potential of prolific GM germplasm for enhancing the mutton production. The total lamb production from GM ewes bearing twin and triplet was more than Malpura ewes at birth, weaning, 6 and 12-month of age (Table 6). The difference in kilogram of lamb harvested between GM ewes bearing multiple birth and Malpura ewes constantly increased with age. The ewe productivity of GM ewes bearing twin and triplet lambs in terms of per kg body weight of ewes was 0.63 and 0.65 kg at weaning, 1.04 and 1.11 kg at 6-month of age and 1.41 and 1.35 kg at 12-month of age, respectively. The corresponding values of Malpura ewes very much low and were 0.44, 0.67 and 0.86 kg at weaning, 6-month and 12-month of age, respectively. The litter size of GM ewes bearing twin and triplet lambs were 1.98 and 2.71 at birth, 1.72 and 2.14 at weaning and 1.64 and 2.00 at 6-month of age, respectively. Nimbkar *et al.* (2005) reported that in Garole X Deccani crosses compared with single bearing

Table 5. Comparative Ewes Productivity Efficiency of Garole, Malpura and GM ewes

Particulars	N	Birth	3-Month	6- Month	12- Month
Overall	998	0.112	0.459	0.664	0.884
Genetic group		NS			
Garole	226	0.104	0.378	0.434	0.578
Malpura	561	0.115	0.460	0.702	0.897
G x M	211	0.118	0.540	0.855	1.177
Year	998				
2000	44	0.106	0.456	0.630	0.872
2001	152	0.111	0.459	0.638	0.917
2002	184	0.120	0.450	0.680	1.028
2003	296	0.117	0.524	0.762	0.880
2004	322	0.108	0.409	0.609	0.723
Parity					
1	461	0.099	0.431	0.633	0.867
2	274	0.116	0.465	0.666	0.864
3	159	0.111	0.467	0.661	0.868
4	78	0.113	0.494	0.700	0.934
5	26	0.123	0.440	0.658	0.888

ewes, twin-bearing ewes weaned 0.7 more lambs and produced 2.4 kg more weight of lambs at 105 days and triplet bearing ewes weaned 0.5 lambs and 4.7 kg more weight of lamb than twin-bearing ewes.

The results indicate that there is a high incidence of multiple births in GM ewes. The majority of GM ewes possessing *FecB* gene were found capable of producing twin lambs and they showed good survival under the harsh climatic conditions prevailing in the semi-arid region of Rajasthan. There is need to increase the number of GM ewes that produces twins or triplets by *inter-se*-mating among *FecB* carriers and discard non-carrier GM ewes for formulating intensive breeding programme through Marker Assisted Selection.

Survivability

Survivability is a very crucial factor in deciding the economic viability of any breed improvement programme. It is the indicator of adaptability of a genetic group under prevailing climatic and management conditions of a particular region. Lamb survival percentage during pre-weaning, weaning to 6 months and 6 to12 months were 97.3, 94.1 and 96.8 % for M and 94.2, 92.2 and 95.9 % for GM, respectively. The survivability of single, twin and triplet lambs of GM up to 12-month of age was 91.34, 82.22 and 62.16 %, respectively. The lower survival in GM may be due to higher twining (Mishra *et al.* 2006). Similar trend of survivability was reported in Garole sheep (Mishra *et al.* 2005). Moreover, the survivability of Garole being the native of hot humid coastal belt of west Bengal (India) was poor under semi-arid tropical region (Sharma *et al.* 2001). It appears that crossbred progenies are having genes of adaptation from Malpura sheep and the prolific genes from Garole. The results obtained revealed that GM genotype has almost similar survivability as observed in case of Malpura sheep.

Table 6. Mean and no of observations of kg lambs harvested from ewes bearing single and multiple lambs in Malpura and GM

Particulars	Live weight of lamb (kg)			
	birth	weaning	6-month	12-month
Malpura	3.02 (561)	14.03 (561)	21.19 (561)	26.88 (561)
GM				
Single	2.26 (93)	10.76 (93)	16.33 (93)	20.54 (93)
Twins	3.23 (104)	14.41 (104)	23.71 (104)	32.21 (104)
Triplets	3.95 (14)	15.82 (14)	26.95 (14)	32.51 (14)

Conclusion and Recommendations

The success achieved in evolving GM germplasm through crossing G with M has provided an ideal opportunity to substantially enhance market-oriented lamb production without necessarily increasing the size of the breeding flock balancing with the natural resource management. The viable benefits accrued by crossing Garole with Malpura are reflected by 52.3 and 41% increase in the litter size of GM lambs at birth and weaning stage, respectively compared to Malpura lambs. Likewise, the litter weight of GM lambs showed upward trend during the growth phase and was 14.2 % higher at 6 month of age as compared to Malpura lambs. The ewe productivity efficiency of GM ewes showed marginal hike at birth but increased tremendously during the growth stage and at the yearling stage was 31.2% higher, compared to Malpura ewes. Inspite of very low twinning rate Malpura ewes are capable to bear two lambs. The impact of crossing Garole with Malpura was significant as the twinning rate of GM ewes was 46.4 % higher than Malpura ewes. The litter weight of twin and triplet born GM lambs at weaning was 2.71 and 12.76% higher over Malpura, respectively. At 6-month of age the gain in the litter twin and triplet lambs was 11.89 and 27.18%, respectively. At the yearling stage the litter weight gain of twin and triplet born lambs over contemporary Malpura lambs was approximately 20%. The crossing of Garole with Deccani/ Bannur/ Patanwadi also shows similar results. Hence, there is a tremendous scope to increase the number of Garole crosses ewes that will produce twins or triplets. The encouraging productivity trends achieved after backcrossing Garole half-breds with Malpura/Deccani will be useful to further boost lamb production and make it more economically remunerative to the farmers. Garole sheep is habituated to hot and humid climatic conditions. It has shown remarkable adaptation and reproductive performance in the hot and semi-arid tropical environment. The adaptability and reproductive efficiency of Garole breed should also be tested in varying agro-climatic zones of the country for adjudging its suitability for *FecB* gene introgression in the native tract of important mutton sheep breeds of India.

The success achieved in evolving prolific GM and Garole x Deccani sheep in a semi-arid tropical environment through crossing of dwarf size Garole with large size mutton breed Malpura/Deccani should be further intensified through intensive interbreeding programme and reproductive technologies for producing a large reservoir of homozygous animals. There is an ample scope to exploit the *FecB* gene introgressed Garole crossses rams and ewes of sound breeding efficiency for backcrossing with Malpura/Deccani breed to achieve higher weaning weight and growth rate for propagation of homozygous fecundity gene carrier progeny. Under the harsh agroclimatic conditions the sheep owners' interest will be best served in increasing fecundity and lamb growth to compliment with the ability of ewes to raise two lambs under the prevailing management system. Emphasis should be given to identify *FecB* gene carrier ewes that regularly give birth to twins. The origin of Booroola is linked to Garole sheep and the success achieved in introgression of *FecB* gene through this evolved breed in other countries highlights the need to conserve and utilize Garole sheep for evolving new prolific germplasm for increasing prolificacy. Few attempts have now been made in India also to introgress *FecB* gene in non-prolific mutton type sheep breeds through Garole inheritance.

REFERENCES

Abdulkhaliq, A.M., Harwey, W.R and Parker. C.F. 1989. Genetic parameters for ewe productivity traits in the Columbia, Sufolk and Targhee breeds. *Journal of Animal Science* 67, 3250-3257.

Aboul-Naga, A.M. 2000. Improving subtropical Egyptian fat –tailed sheep through cross-breeding with the prolific Finnsheep. In: developing breeding strategies for lower input animal production environments (ed. Galal, S., Boyazoghu, J., Hammond, K), ICAR technical series no. 3. p. 293-300. ICAR, Rome.

Arora, A.L., Sharma, R.C. and Narula, H.K., 2004. Evaluation of Awassi X Malpura half-bred sheep in semi-arid region of Rajasthan. *Indian Journal of Animal Science.* 74, 1219-1222.

Arora, A.L., Sharma, R.C., Narula, H.K., Ravindra, Kumar. 1999. Comparative performance of Avikalin and Malpura sheep. *Indian Journal of Small Ruminants* 5, 4-8.

Bathei, S.S. 1994. Influence of ewe age on reproductive performance of Iranian fat-tailed Mehraban breed of sheep. *World Review of Animal Production* 29, 55-60.

Bose, S., Dutta Gupta, R., Moitra, D.N. 1999. Reproductive performance of Bengal sheep in Sunderbans. *Indian Journal of Animal Production and Management* 15, 157-160.

Bunge, R., Thomas, D.L., Nash, T.G., 1995. Performance of hair breeds and prolific wool breeds of sheep in southern Illinois: Lamb production of F_1 adult ewes. *Journal of Animal Science* 73, 1602-1608.

Crooks, A.E., Thomas, D.L., Zelinsky, R.D., Gottfredson, R.G. and McKusick, B.C. (1998). Introgression of the *Fec*^{B} allele of the Booroola Merino into a Rambouillet flock- A Progress Report. In: Proceedings of the 46th Annual Spooner Sheep Day, University of Wisconsin-Madison, U.S.A., pp 53-54.

Davis, G.H., Balakrishnan, L., Ross, I.K., Wilson,T., Galloway, S. M.,Lumsden, B.M., Hanrahan, J. P., Mullen.,M., Mao, X.Z., Wang, G.L., Zhao, Z.S., Zeng, Y.Q., Robinson, J.J., Mavrogenis, A.P., Papachristoforou, C., Peter,C., Baumung., R., Cardyn., P.,Boujenane, I., Cockett, N.E., Eythorsdottir, Arranz, J.J. & Notter, D.R. 2005. Investigation of the Boroola (*FecB*) and Inverdale (*FecX*^{I}) mutations in 21 prolific breeds and strains of sheep sampled in 13 countries. *Animal Reproduction Science* 92, 87-96.

Davis, G.H., Galloway S.M., Ross I.K., Gregan S.M., Ward J., Nimbkar B.V., Ghalsasi P.M., Nimbkar C., Gray, G.D., Subandriyo, Inounu I., Tiesnamurti B., Martyniuk E., Eythorsdottir E., Mulsant P., Lecerf F., Hanrahan J.P., Bradford G.C and Wilson T., 2002. DNA test in prolific sheep from eight countries provide new evidence on origin of the Booroola (*FecB)* mutation. *Biology of Reproduction* 66, 1869-1874.

Dickerson, G. E. 1978. Animal size and efficiency: Basic concepts. *Animal Production* 27, 367-379.

Dobson, W., Khazanov, A., Shapiro, K., Thomas, D.L., Gottfredson, R., Zanca, R., Brent, L., Abuseitova, M., Esenova, S., Kalyshev., K.A., Kasymov, M., Klyashtornyi, S., Malmakov., N.I., Masanov, N., Medeubekov, K.U., Naumova, O. and Sagnaeva. 2000. Impacts of economic reform on the livestock sector of Central Asia, Ministry of Agriculture, Republic of Kazakhstan, pp 1-22.

F.A.O. 2004. Production Year Book. Food and Agriculture Organization of the United Nations, Rome.

Fahmy, M.H., Mason, I.L., 1996. Less known and rare breed. In Prolific Sheep (Eds. Fahmy, M.H.). CAB international U.K. p. 178-186.

Feng,W., Ma, Y., Zhang, Z. and Zhou, D. 1996. Prolific breeds in China. In: Prolific sheep. Ed. M.H. Fahmy, CAB International, Wallingford, U.K., pp.146-151.

Fletcher, I.C., Gunawan, B., Hetzel, D.J.S., Bakrie, B., Yates, N.G. and Chaniago, T.D. (1985). Comparison of lamb production from indigenous and exotic x indigenous ewes in Indonesia. Tropical Animal Health, 17: 127-134.

Fogarty, N. M., Dickerson, G. E and Young, L. B. 1984. Lamb production and its components in purebreds and composite lines II. Breed effects and heterosis. *Journal of Animal Science* 58, 301-311.

Ghalsasi, P.M and Nimbkar, B.F. 1993. The "Garole" Micro sheep of Bengal, India. *Animal Genetic Resource Information* 12, 73-79.

Gootwine, E. 1995. Increasing prolificacy of the fat tail Awassi sheep using the Booroola *FecB* gene. CIHEAM-Options Mediterraneennes, 79-87.

Iman, N. Y., Slyter, A L., 1996. Lifetime and wool production of Targhee or Finn-Dorset-Targhee ewes managed as farm or range flock: 1. Average annual ewe performance. *Journal of Animal Science* 74, 1757-1765.

Kishore, K., Gour, D., Rawat, P.S. and Malik, R.C. 1982. Note on some reproductive traits in three- quarter-bred strain of sheep. *Indian Journal of Animal Science* 52, 1108-1110.

Kiyanzad, M.R., Panadam, J.M., Emamjomek Kashjan, N, Jelan, Z. A and Dahlan, I. 2003. Reproductive performance of three Iranian sheep breeds. *Asian-Australasian Journal of Animal Science* 16, 11-14.

Kumar, S., Kolte, A.P, Mishra, A.K., Arora, A.L and Singh, V.K. 2006. Identification of *Fec B* mutation in Garole x Malpura sheep and its effect on litter size. *Small Ruminant Research* 64, 305-310.

Mishra A K, Arora A L, Kumar Sushil and Singh V K. 2005a. Prolificacy of Garole in semi - arid tropics of Rajasthan. *Indian Journal of Small Ruminants* 11, 1-5.

Mishra, A.K., Arora, A.L., Kumar, Sushil and Singh, V.K., 2005a. Prolificacy of Garole in semi – arid tropics of Rajasthan. . *Indian Journal of Small Ruminants* 11, 1-5.

Mishra, A.K., Arora, A.L., Kumar, Sushil., Sharma, R.C. and Singh, V. K., 2005b. Malpura: A Mutton type sheep breed. Res. Bulletin, CSWRI, Avikanagar, via. Jaipur (Rajasthan)-304 501.

Mishra A K, Arora A L, Kumar Sushil, Prince L L L and Singh V K. 2006. Introgression of *FecB* gene in non-prolific Malpura Indian sheep for increasing litter size. In: 8th World congress on Genetics Applied to Livestock Production, August 13-18, 2006, belo Horizonte, MG, Brasil.

Mishra, A.K., Arora, A.L., Kumar, S. and Singh, V.K. 2007. Improving productivity of Malpura breed by crossbreeding with prolific Garole sheep in India. *Small Ruminant Research* 70, 159-164.

Mishra, A.K., Arora, A.L. Kumar Sushil, Gupta D C and Singh, V.K. 2005. Studies on ewes' efficiency of single and multiple bearing lambs of non-prolific Malpura, prolific Garole and Garole x Malpura sheep. *Indian Journal of Animal Sciences* 77, 759-762

Nimbkar C., Ghalsasi P.M., Walkden-Brown S.W. and Kahn L.P.2002. Breeding program for the genetic improvement of Deccani sheep of Maharashtra, India. In: Proceedings of the Seventh World Congress on Genetics Applied to Livestock Production, Montpellier, France, CD-ROM Communication, 2002, 25-11, 4pp.

Nimbkar, C., Ghalsasi, P.M., Swan, A., Walkkden-Brown, S.W. and Kahn, L.P. 2003a. Evaluation of growth rates and worm resistance of Deccani and Bannur lambs and their crosses with Garole. *Animal Science* 76, 503-515.

Nimbkar, C., Ghalsasi P.M., Maddox, J.F., Pardeshi, V.C., Sainani, M.N., Gupta, V. and Walkden-Brown, S.W. 2003b. Expression of *FecB* gene in Garole and Crossbred ewes in Maharastra, India. In the Proceedings of the fifteenth conference of AAABG, Melbourne, Australia. p.111-114.

Nimbkar, C., Ghalsasi, P.M., Ghatge, R.R. and Gray, G.D., 1998. Establishment of prolific Garole sheep from West Bengal in the semi-arid Deccan plateau of Maharastra. Proc. 6th World Cong. on Genetics App. to Livestock Prod., Armidale. pp. 257-260.

Nimbkar, C., Ghalsasi, P.M., Walkden-Brown, S. W., Kahn, L.P. and Gray, G.D., 2000. A comparison of the growth performance and worm resistance of lambs produced by diallel crossing of three Indian sheep breeds. *Asian-Australasian Journal of Animal Science* 13, 72-75.

Nimbkar, C., Pardeshi V C and Ghalsasi, P.M. 2005. Evaluation of the utility of the FecB gene to improve the productivity of Deccani sheep in Maharashtra, India, Applications of gene based techynologoies for improving Animal Production and Health in Developing countries, (Eds) Makkar HP S and Vilijoen G J, IAEA Netherland.

Piper, L.R. and Bindon, B.M. 1996. The Booroola Merino. In: Prolific sheep. Ed. M.H. Fahmy, CAB International, Wallingford, U.K., pp. 152-160.

Ramdas, S., Dally, Bradford, E and Sakul, H. 1993. Lamb, wool production of Targhee and prolific breed crossbred ewes. *Sheep Research Journal* 2, 62-70.

Rosati, A., Mousa, E., Van, Vlesck, L. D., Young. L. D., 2002. Genetic parameters of reproductive traits in sheep. Small Rumin. Res. 43, 65-74.

Shaoqui, Rao. 1997. Genetic analysis of sheep discrete reproductive traits using simulation and field data, Ph. D thesis submitted to Virginia Polytechnic Institute and State University, Virginia.

Sharma, R.C., Arora, A.L. and Khan, B.U., 2001. Garole: A Prolific sheep of India. Central Sheep and Wool Research Institute, Avikanagar. Res. Bulletin, pp. 17.

Sharma, R.C., Arora, A.L., Mishra, A. K., Kumar, S. and Singh, V. K. 2004. . Breeding prolific Garole with Malpura sheep for increased reproductive efficiency in semi-arid tropics of India. *Asian-Australasian Journal of Animal Science* 17, 737-742.

Sharma, R.C., Arora, A.L., Narula, H.K and Singh, R.N. 1999. Characteristics of Garole sheep in India. *Animal Genetic Resource Information* 26, 57-64.

Shelton, M.,1971. Some factors affecting efficiency of lamb production. Texas Agric. Exp. Sta. Tech. Rep. No. 26.

Singh, V.K., Mishra, A K and Kumar Sushil.2005. Genetic Improvement of Sheep for wool and Mutton Production in India: A Review. *Indian Journal of Animal Sciences.* 75, 356-364.

Southey, B.R., Thomas, D.L., Gottfredson, R.G. and Zelinsky, R. 1996. Introgression of the *Fec*[B] allele of the Booroola Merino into a Rambouillet flock- 1996 Progress Report. In: Proceedings of the 44th Annual Spooner Sheep Day, University of Wisconsin-Madison, U.S.A., pp 54-56.

Steinhagen, O. and Wet, P.J.De. 1986. The influence of age and generation number of the Dohne Merino on different wool production traits. *South African Journal of Animal Science* 16, 101-102.

Thomas, D. L and Whiteman, J.V. 1979. Effects of substituting Finnsheep and Dorset breeding for Rambouillet breeding. I. Productivity of young, spring-lambing ewes. *Journal of Animal Science* 48, 256-264.

Turner, H. N., 1969. Genetic improvement of reproduction in sheep. *Animal Breeding* 37, 545 (Abstr.).

Young, L.D and Dickerson, G.E., 1988. Performance of Booroola Merino and Finn sheep crossbred lambs and ewes. *Journal of Animal Science* Finland 60, 492-499.

□□□

19

Sheep for Mutton Production in Southern India

L. LESLIE LEO PRINCE
Central Sheep and Wool Research Institute, Avikanagar, Rajasthan - 304501

Introduction

Sheep husbandry plays an important role in livelihood security and economic sustenance of rural people in southern states of the country. Sheep is important livestock species in the southern part of India and mainly contribute meat, skin and wool. States and union territory that constitute southern India are Andhra Pradesh, Karnataka, Maharashtra, Kerala, Tamil Nadu and Pondicherry. The sheep in these southern states form 60.7% of the sheep population in India and Andhra Pradesh ranks first in sheep population (34.8 % of the country).

Sheep Population

The sheep population is increasing exponentially in the southern region of the country. As per livestock census, sheep population was 9.74 million in Andhra Pradesh, 5.26million in Tamil Nadu and 8.00 million in Karnataka in 1997, which modified to 21.37, 5.59 and 7.25 million respectively by 2003. There is a tremendous increase in the sheep population in AP i.e. 119.40 %. As per 2003 census, sheep population and annual growth rate during 1997-2003 in southern states are given in table 1.

Table 1. Sheep population in southern states

State	Population (thousands)		Annual Growth rate (%)
	1997	2003	
Andhra Pradesh	9743	21376	13.99
Karnataka	8004	7256	-1.62
Kerala	03	04	3.23
Maharashtra	3368	3094	-1.40
Tamil Nadu	5259	5593	1.03
Pondicherry	02	03	4.40
South India	28376	37326	
India	57494	61469	
% share of South	49.4	60.7	

Sheep Genetic Resources

In southern region meat type breeds viz. Deccani, Nellore, Bellary, Hassan, Mandya, Mecheri, Kilakarsal, Vembur, Coimbatore, Nilgiri, Ramnad white, Madras Red, Trichy Black and Kenguri were predominant. Majority of these breeds have evolved by natural selection due to centuries of adaptation to the agro climatic conditions and have coarse, open and hairy fleece cover hence primarily reared for mutton production. Nilgiri is the only apparel wool breed available developed in the region. Wool produced from Mandya and Coimbatore is of coarse carpet type. Distribution of sheep breeds with their major utility is given in table 2. All the indigenous sheep breeds are known for adaptability under varied climate conditions, which is obvious from their long struggle against the natural forces. Some of the unique features have been identified in few south Indian sheep breeds which are summarized in table 3 (Arora and Prince, 2004), Salient features of south Indian sheep breeds are described bellow:

Deccani

The Deccani breed of sheep is widely distributed in the Deccan plateau across the three states of Maharashtra, Andhra Pradesh and Karnataka. The breed has a thin neck, narrow chest, prominent spinal processes. It has Roman nose and dropping ears. The colour is dominantly black, with some grey and roan. Different strains (or within breed types) are observed in the breed tract. Four types have been noticed. Local people term them as Viz. Lonand, Sangamneri, Solapuri (Sangola) and Kolhapuri. Means of body weights at birth, three, six, nine and twelve months of age were 3.13, 14.30, 18.20, 20.10 and 22.57 Kg, respectively for Deccani sheep maintained under NWPSI (Terminal Report, 2007). Sheep were sheared two times a year and the overall means for lambs first six monthly clip and second six monthly clip and adult annual clips were 467, 499 and 488 g, respectively. Deccani lambs were put on feedlot trails (90 days) after weaning with an initial average weaning weight of 13.23 kg. Final weight at six months of age was 22.65. Average gain of 9.13 kg was obtained with a with a feed efficiency ratio of 1: 5.91 (Terminal Report, 1986). Results shows that the Deccani has the great potential for mutton production under intensive system of management.

Table 2. Breeds of Sheep in southern states of India.

Breed	Utility	Native state
Deccani	MCW	Maharashtra, AP & Karnataka
Hassan	M	Karnataka
Mandya	M	Karnataka
Bellary	MCW	Karnataka
Kenguri	M	Karnataka
Coimbatore	MCW	Tamil Nadu
Kilakarsal	M	Tamil Nadu
Madras Red	M	Tamil Nadu
Mecheri	M	Tamil Nadu
Nilgiri	AW	Tamil Nadu
Ramnad White	M	Tamil Nadu
Tiruchy Black	M	Tamil Nadu
Vembur	M	Tamil Nadu
Sandyno	AW	Tamil Nadu
Nellore	M	Andhra Pradesh (AP)

Apparel wool (AW); Meat and Carpet wool (MCW) & Meat (M)
Source : Arora et al. (2007)

Table 3. Breeds with unique characteristics

Breed	Unique characteristics
Mecheri, Madras Red	High quality skin and mutton
Mandya	Excellent meaty conformation, high quality meat and palatability.
Nellore	Tallest sheep breed of India.

Bellary

Distributed in Bellary district of Karnataka. Sheep found to the north of the Tungabhadra River are called "Deccani" and those to the south of it are called as, "Bellary". Medium-sized animals, with body colour ranging from white through various combinations of white and black to black. Fleece is extremely coarse, hairy and open; belly and legs are devoid of wool. About 89% males and 9.5% females were horned (DARE report, 2005). Means of body weights at birth, three, six and twelve months of age were 2.60, 11.09, 16.28 and 18.68 Kg, respectively. Average 6-monthly greasy fleece weight was 300g with average fibre diameter of 59.03 micron and Medullation percent of 43.43. Lambing rate is about 80% with an average lambing interval of about 12– 15 months. Litter size is single with rare case of twinning.

Nellore

Distributed in Nellore district and neighbouring areas of Prakasham and Ongole districts of Andhra Pradesh. Three varieties are distinguished, primarily on the basis of colour: "Palla", completely white or white with light brown spots on head, neck, back and legs; "Jodipi", white with black spots, particularly around the lips, eyes and lower jaw, but also on belly and legs and "Dora", completely brown. Relatively tall animals with little hair except at brisket, withers and breech. The rams are horned and ewes are almost always polled. The ears are long and drooping and most of the animals carry wattles. The overall averages of body weight of Nellore sheep maintained under NWPSI were 2.95, 13.91, 17.38, 22.39 and 26.61 kg respectively at birth, 3, 6, 9 and 12 month age (Viroji Rao *et al.*, 2002).

Mandya

It is also known as Bannur and Bandur. Distributed in Mandya district and bordering Mysore district of Karnataka. Relatively small animals. Colour is white, but in some cases face is light brown, and this colour may extend to the neck. Compact body with a typical reversed U-shape conformation from the rear. Ears are long, leafy and drooping. Tail is short and thin. A large percentage of animals carry wattles. Slightly Roman nose. Both sexes are polled. Coat is extremely coarse and hairy. Evenly placed short and stumpy legs and wide apart hipbones indicated a square type meaty conformation of the breed. Sheep are mainly kept on grazing, however, a few farmers keep the animals on stall-feeding. Some of the farmers practice sheep tethering in their cropland, shifting from place to place and bringing back to their homestead in the evening. Narayanaswami, *et al.*, 1975 reported average body weights at birth, three, six and twelve months of age as 2.02, 9.44, 14.61 and 21.34 kg, respectively. Nimbkar *et al.*, 2003 observed means of body weights at birth, three and six months of age were 2.10, 9.0 and 13.2 Kg, respectively. Average 6-monthly greasy fleece weight was 298g.

Hassan

Distributed in Hassan district of Karnataka. Small sized animals. Animals are having white body with light brown or black spots. Ears are medium-long and drooping; ear length. Majority of males are horned and females are usually polled. Fleece is white, extremely coarse and open; legs and belly are generally devoid of wool. This breed is very hardy and capable to travel long distances.

Kenguri

Distribution is hilly tracts of Raichur district (particularly Lingasagar, Sethanaur and Gangarati taluks) of Karnataka. Medium-sized animals. The body coat is dark brown or of coconut colour. The breed is also known as Tenguri. Some Kenguri sheep have black belly and are known as Jodka. A few sheep, known as Masaka, are a mixture of brown and black colour. Ears are medium long and drooping. Males are horned; females are generally polled. Kenguri sheep are maintained for mutton. Lactation length varies from 3 to 5 months, and the peak milk production is 300 to 500 ml/day (DARE report, 2005).

Mecheri

Distributed in Mecheri, Kolathoor, Nangavalli, Omalur and Tarmangalam Panchayat Union areas of Salem district and Bhavani taluk of Coimbatore district of Tamil Nadu. Medium-sized animals, and light brown in colour. Body is covered with very short hairs which are not shorn. The skin is of the highest quality of sheep breeds in India and is highly prized. Average body weights at birth, three, six, nine and twelve months of age were 2.23, 10.07, 13.73, 18.62 and 21.12 for male lambs and corresponding value for female lambs were 2.28, 9.70, 13.53, 15.84 and 18.02 kg, respectively under farm conditions (Richard Jagatheesan *et al.*, 2003b). Corresponding growth performances recorded in farmers' flocks were 2.21, 9.64, 13.25, 16.54 and 18.72kg, respectively (Richard Jagatheesan *et al.*, 2003a).

Kilakarsal

Distributed in Ramnathpuram, Madurai, Thanjavur and Ramnad districts of Tamil Nadu. Medium-sized animals. Coat is dark tan, with black spots on head (particularly the eyelids and lower jaw), belly and legs. Ears are medium sized. Tail is small and thin. Males have thick twisted horns. Most animals have wattles. Average body weights at birth, three, six and twelve months of age were 2.29, 8.53, 14.15 and 27.26 Kg, respectively (Acharya, 1982).

Vembur

Distributed in Vembur, Melakkarandhai, Keezhakarandhai, Nagalpuram, Kavundhanapatty, Achangulam and some other villages of Pudur Panchayat Union and Vilathikulam Panchayat areas of Tirunelveli district of Tamil Nadu. Tall animals. Colour is white, with irregular red and fawn patches all over the body. Ears are medium-sized and drooping. Tail is thin and short. Males are horned and ewes are polled. The body is covered with short hairs which are not shorn. Average body weights at birth, three, six and twelve months of age were 1.97, 8.42, 10.50 and 16.50 Kg, respectively (Acharya, 1982).

Coimbatore

Distributed in Coimbatore district and adjoining Dindigul district of Tamil Nadu. Medium-sized animals, white with black or brown spots. Ears are medium-sized and directed outward and backward. Tail is small and thin Fleece is white, coarse, hairy and open. Most flocks are migratory and animals were allowed to graze extensively as a flock for a distance of 5 to 8 km per day (Rajapandi *et al.*, 2006). The animals are maintained on grazing without any supplementation. Sheep are penned in harvested fields. Rams are both horned and polled while ewes are hornless. Average body weights at birth, three, six and twelve months of age were 2.16, 7.50, 10.83 and 14.77 Kg, respectively. Average 6-monthly greasy fleece weight was 365g with average fibre diameter of 41 micron and Medullation percent of 58.

Nilgiri

Nilgiri, evolved during the 19th century, originated from a cross-breed base and contains an unknown level of inheritance of Coimbatore, the local hairy breed, Tasmanian Merino, Cheviot and South Down. Distributed in Nilgiri hills in Tamil Nadu. The breed is well adapted to the conditions of the Nilgiri hills and produces fine fleeces. Medium-sized animals. Body colour is white; exceptionally there are brown patches on face and body. Face line is convex, giving a typical Roman nose. Ears are broad, flat and drooping. The tail is medium and thin. The fleece is fine and dense. Average body weights at birth, three, six and twelve months of age were 2.96, 11.84, 15.04 and 19.77 Kg, respectively. Average 6-monthly greasy fleece weight was 615g with average fibre diameter of 27 micron and Medullation percent of 11 (Acharya, 1982).

Ramnad White

Distributed in Ramnad district and adjoining areas of Tirunalveli district of Tamil Nadu. Medium-sized animals, predominantly white; some animals have fawn or black markings over the body. The ears are medium-sized and directed outward and downward. Males have twisted horns and females are polled. Tail is short and thin. Average body weights at birth, three, six and twelve months of age were 1.68, 7.31, 8.45 and 16.30 Kg, respectively.

Madras Red

Distributed in Chennai, Kancheepuram, Tiruvellore, Villupuram and adjoining areas of Vellore, Cuddalore and Thiruvannamalai districts of Tamil Nadu. Medium-sized animals. Body colour is predominantly brown, the intensity varying from light tan to dark brown; some animals have white markings on the forehead, inside the thighs and on the lower abdomen. Ears are medium long and drooping; ear length. Tail is short and thin. Rams have strong corrugated and twisted horns; the ewes are polled. The body is covered with short hairs which are not shorn. Average body weights at birth, three, six and twelve months of age were 2.71, 9.97, 14.84 and 21.03 kg, respectively (Terminal Report, 2007). Dressing percentage was around 49 (Raman *et al.*, 2003).

Tiruchy Black

Distributed in Perambalur, Ariyalur and Tiruchy, Kallakurichy taluk of South- Arcot district, Viraganur area of Attur taluk of Salem district, Tirupathur and Tiruvannamalai taluks of North Arcot district of Tamil Nadu. Males are horned and ewes are polled. Ears are short and directed downward and forward. Tail is short and thin. The fleece is extremely coarse, hairy and open. Average body weights at birth, three, six and twelve months of age were 2.13, 9.46, 10.73 and 16.80 Kg, respectively.

Sandyno

Sandyno breed was developed by crossing Nilagiri sheep with Merino and Rambouillet. The breed has ¾ inheritance from Nilagiri and 5/8 inheritance from Merino / Rambouillet breeds. The breed has been field tested and found to adapt well to the local climatic and environmental conditions. The Sandyno sheep grow to a body weight of 24 kg in males and 20 kg in females at yearling stage. The average wool yield is 2.2 kg. A flock of 320 sheep of Sandyno breed is maintained at Sheep Breeding Research Station, Sandynallah, TamilNadu.

Mutton Production Status

Mutton production from sheep was 97.75 thousand tones in Andhra Pradesh, 17.17 thousand tones in Tamil Nadu and 30.03 thousand tones in Karnataka and 26.36 thousand tones in Maharastra (DAHD report, 2006). A total of 178 thousand tones of mutton were produced in the southern India, which accounts for 73% share of India's mutton production. State wise mutton production, number of sheep slaughtered and average yield per sheep are given in table 4. Average yield per sheep ranged between 10.93 kg and 14.42 kg in different states.

Selective Breeding and Crossbreeding

The East India Company, which imported exotic breeds of sheep for crossbreeding with the indigenous sheep breeds, started sheep development activity in 19th century and isolated crossbreeding experiments using exotic breeds were undertaken. Most of these studies were carried out on small populations and except the Nilgiri, a fine wool breed of sheep evolved in the Nilgiri hills of Tamil Nadu. Some crossbreeding trials were also undertaken in Mysore and Madras using Southdown and Cape Merino exotic breeds.

With the establishment of Imperial (now Indian) Council of Agricultural Research (ICAR), research programmes were taken up on sheep development on regional basis, which included selective breeding within native stock and crossbreeding them with exotic fine wool breeds, and covered almost all the important sheep rearing states. Major emphasis was however placed on sheep development after the country attained independence and initiated its five-year development plans. During the First Five Year-Plan regional centre were set up in Deccan plateau where planned crossbreeding trials were undertaken.

In 1965, realizing the importance of sheep in the agrarian economy, the Southern Regional Research Station of CSWRI, Avikanagar was established at Mannavanur (TamilNadu) for undertaking the fundamental and applied research in sheep production and wool utilization and to provide post-graduate training in sheep and wool sciences.

The Government of India constituted the ad-hoc Sheep Breeding Policy Committee in 1970, which suggested that in order to meet the shortage of stud rams, 20,000 exotic sheep of fine wool breeds for producing breeding rams should be imported. Central Sheep Breeding Farms were established in Andhra Pradesh (Mamidiapally- rangareddy, Cintaldevi- Nellore, Penkonda- Anantpur, Manhar-warangal, Palamner- Chittor) and Karnataka (Challakere- Chitradurga, Kurikuppi- Bellary, Dhanagur- Mandyal, Angawadi- Bijapur, guttal- Dharwad, suttatti- Belgaum) to produce exotic purebred or crossbred rams.

All India Co-ordinate Research Programme (AICRP) on sheep breeding initiated by the ICAR in 1971. Crossbreeding experiments for improving mutton involved exotic mutton breeds viz. Dorset and Sufflok with indigenous breeds, viz. Nellore, Mandya, Deccani and Madras Red were done and crossbreds shown improvement in meat production as reflected by improvement in body weight gains, efficiency of feed conversion and dressing percentage (Khan *et al.*, 2001) (Table 5). Sandyno breed was developed by crossing Nilagiri sheep with Merino and Rambouillet at Sheep Breeding Research Station, Sandynallah, Tamil Nadu. Crossbreds of different exotic breeds like Rambouillet, Dorset, Suffolk, and Corriedale were generated in number of states. Cross-breeding work carried out in the country in general indicated the superiority in body weight, greasy fleece weight and fleece quality of half-breds over the indigenous pure-breds involved in crosses. The survival of half-breds was similar to that of indigenous breeds involved in the crosses, but crosses containing higher exotic inheritance did show problems of survival. Crossbreds performed well in the temperate part of south India such as Nilgiri and Kodai Hills.

Table 4. Mutton production status

State	No. of sheep slaughtered (in 000 No.s)	Average yield/ animal (kg)	Mutton production (in 000 Tonnes)
Andhra Pradesh	7205	13.57	97.75
Karnataka	2083	14.42	30.03
Kerala	731*	9.17*	6.70*
Maharashtra	2249	11.72	26.36
Tamil Nadu	1571	10.93	17.17
Pondicherry	22	12.68	0.27
South India	13861	12.86	178.30
India	19110	12.59	241
% share of South	72.5		73.0

* Sheep and Goat combined data; Source: Basic Animal Husbandry Statistics (2006)

Table 5. Feed lot performance of native and crossbred Mutton type sheep India.

Genetic Group	Initial wt. Dressing (kg)	Final wt. (kg)	Av. daily gain (g)	FE %	Hot carcass weight (kg)	Dressing (%) EWB
Mandya	10.23	18.05	87	13.46	8.48	54.5
Mandya Syn.	13.89	23.79	110	14.48	10.56	55.2
Nellore	13.45	20.01	91	14.60	10.63	58.22
Nellore Syn.	14.78	26.53	134	14.74	12.96	59.34
Deccani	11.18	25.29	119	13.02	8.33	52.65
Dorset x Deccani	12.17	27.41	143	13.87	10.84	54.97
Madras Red	9.29	17.72	87	12.34	8.16	51.43
Dorset x Madras Red	12.02	26.04	123	14.07	11.07	57.39

Source: Khan *et al.* (2001)

The Network Project on Sheep Improvement (NWPSI) was initiated on 1.4.1990 for survey, evaluation and improvement of indigenous sheep breeds under native environment. Different breeds of sheep are being improved through selection and inter-se mating for wool and mutton production. Presently, there are eight ongoing centres of NWPSI, out of which three are operating in south India, including farm based units at MPKV, Rahuri (Deccani, Dual type) and SVVU, Palamner (Nellore, Mutton) and field-based unit at TNAUVAS, Kattupakkam (Madras Red, Mutton).

Bharat Merino, a fine wool breed, developed at CSWRI, Avikanagar was introduced to SRRC, CSWRI, Mannavanur, Kodai Hills, T.N and performing well (Swain et al., 2007). This breed is well received by farmers for their good growth, higher wool yield and quality. A total of 1700 Bharat Merino sheep have been sold among interested farmers of Tamilnadu, Karnataka, Kerala and Andhra Pradesh for breeding and its multiplication. Majority of Bharat Merino sheep were distributed to Kolar District of Karnataka and in Sathyamangalam and Thalavadi areas of Erode District in Tamilnadu. The average body weight of Bharat Merino sheep at Kolar District was found to be 45 kg and 32 kg, ranged from 38-70 kg and 28-40 kg for adult male and female respectively. Birth weight of lamb was 4.0 kg with range of 3.5-4.5 kg. The weight at 3, 6, and 12 months were 18, 25 and 30 kg, respectively (Shakyawar *et al.*, 2007).

Sheep Development Status in Different States

Andhra Pradesh

State of Andhra Pradesh has the largest sheep population in the country (213 lakhs as of 2003). The local recognised breeds of sheep are Nellore, Deccani and Bellary. The first is a hairy type while the latter two are coarse wool types. Mostly pure breeding was adopted in all the three breeds with emphasis on body size, as the income from the wool was negligible. Cross breeding of Deccani and

Bellary with corriedale improved body weight, but cross-breds are not popular with the farmers due to problem of adaptability and survivability. There have been problems in marketing of wool. Sheep rearing is very popular in Telangana, Rayalaseema and south coastal districts. 95 % of the Sheep growers belong to the weaker sections of the society and any support to the sheep sector directly benefits the weaker sections.

Karnataka

There are four breeds in the state. Deccani produces coarse wool of black, white and grey colour. They were mostly purebred. In some areas crossbreeding has been undertaken with Merino and Corriedale for improving wool production and quality and for mutton production. Hassan has white body with light, brown or black spots and produces coarse wool. Bellary produces coarse wool. Mandya is a best breed of the country as far as conformation for Mutton production is concerned, although body weights, weight gains, feed conversion efficiency and carcass yield are not very superior to other breeds. Large movement of Mandya sheep out of the Mandya district because of an increase in irrigated cultivated area, leaving little land with natural vegetation for sheep grazing was reported.

Tamil Nadu

Sheep husbandry is an important subsidiary animal husbandry activity in the State. It acts as a primary source of income for majority of landless agricultural labourers and supplements the income of small and marginal farmers. Besides skin and meat, income is also derived through penning which is a very common practice followed to fertilise the fields. In Tamil Nadu closed nucleus flocks are being maintained in government/university farms for production and supply of purebreds and synthetics viz., Madras Red, Mecheri, Mandya, Ramanathapuram white, Vembur, Nilagiri and Nilagiri synthetic. Ram multiplication units have been established for Mecheri and madras Red. Some cross breeding of hairy breeds with exotic was carried out in university farms. The results were not encouraging. Cross breeding in hills were successful. Nilgiri synthetic has been developed in ICAR collaborated research programme by crossing Nilagiri ewes with Merino and stabilizing the exotic inheritance level at about 75 per cent. They are performing well under field conditions. Conservation and propagatation of Nilagiri and Sandyno breeds and to improve their performance by intensive selection methods is under progress at Sheep Breeding Research Station, Sandynallah. Improved breeding of rams and ewes of Nilagiri and Sandyno are being supplied to the sheep breeders of the Nilgiris and adjacent districts. The breeds also have wide acceptance in hilly areas of Karnataka state. Quality rams and ewes produced in these farms are sold to local farmers to improve the progeny of local sheep and sheep are also sold for breeding purpose. Seven sheep breeding farms are maintained by state animal husbandry departmental for breed improvement and distribution of superior rams.

Maharashtra

The sheep population are found concentrated mainly in dry regions or western Maharashtra i.e. Pune and Nasik region. The breed of sheep reared in Maharashtra is Deccani breed has some prominent types such as Lonad, Sangamneri, Solapur (Sangola) and Kolhapuri. The Kolhapuri type is akin to Bellary type of Karnataka. The average annual wool yield per sheep is 558 g of coarse quality which is used for preparation of coarse blankets and floor coverings. Half breds perform well under the routine management condition and crosses upto 75% exotic level have been found to perform well under organised farming system. It may be desirable to restrict exotic level preferably upto 50 per cent. Russian Merino, corriedale and Rambouillet rams have been used extensively in the state. Sheep and Goat devolvement board has 10 sheep and goat development farms with 2326 Hectares land. 5,863 indigenous and cross bred sheep and goat of valuable germ plasm available with the corporation. Nine sheep and goat management training centres, 12 shearing machines and 4 meat processing units equipped with modern machinery for exports are developed by state animal husbandry department. More than 2,000 co-operative societies of sheep and goats registered in the state. Animal Husbandry Division of the Nimbkar Agricultural Research Institute embarked upon a breeding program to increase the efficiency of lamb production of local Deccani sheep from the year 2000. This program intended to increase the prolificacy through the introduction of prolificacy governed by a single major gene from Garole sheep breed, combined with selection for a comprehensive breeding objective including all important traits of economic importance.

Kerala

Since the sheep population is very small there is no need of any specific breeding policy. More over the state is not suitable for increasing sheep production. The state may obtain its requirement of sheep for mutton from neighbouring states.

General Recommendations

- There is a need to strengthen the ram raring centres for production of superior breeding rams of indigenous meat type breeds involving local sheep farmers should be followed.
- Potential mutton type breed such as, Madras Red, Mandya, Nellore, Deccani, Mecheri, etc., may be restricted to selective breeding by using superior rams of the respective breeds for enhancing mutton production.
- For profitable meat production, we need to improve a breed, which must have fast and efficient growing lambs. The ewes should be prolific and must have good mothering ability. Fast growth and early maturity are the most important characteristics for meat production.

- In case of crossbreeding, implementation must be considered situation specific i.e. resources needed for crossbreds must be available. Therefore, breeding of sheep should be in a way that matches the available and accessible resources.
- It has been found that lambs should preferably be weaned at two month age for putting them under intensive feeding to harvest economic market weight around 5 - 5½ month of age.
- Fecundity may be improved by proper selection for high prolificacy and adopting appropriate management to support it. Cross breeding of prolific native Garole sheep with other indigenous sheep of southern India subject to availability of sufficient milk with the ewes to sustain twin/triplet should be attempted so that their reproductive efficiency could be increased.
- Apparel wool production may be intensified only in the temperate areas such as Nilgiri and Kodai hills of southern region. In these areas 3/4th crosses of Rambouillet or Merino including Bharat Merino may be propagated and annual clips may be obtained to meet the requirements for apparel manufacture and also to meet the demand of mutton.
- Farmers need to be impressed upon use of superior breeding rams.

State Wise Recommendations

The Department of Animal Husbandry and Dairying, Govt. of India, set up a "Task Force" to take stock of the present status of sheep, goat and rabbit production in the country and make recommendations. The state wise breeding policies as recommended by "Task Force (1996)" for sheep improvement in Southern India are given as here under:

Andhra Pradesh : Selective breeding of Nellore and Deccani in the respective native tract for improving body weight and prolificacy.

Karnataka : Selective Breeding of Mandya, Bellary and Deccani in its breeding tract. Crossbreeding of Deccani, Hassan and Bellary may be undertaken with Rambouillet or Bharat Merino rams and the exotic inheritance may be stabilized at 50 % or lower level.

Kerala : There is no need of any specific breeding policy. More over the state is not suitable for increasing sheep production. The state may obtain its requirement of sheep for mutton from neighbouring states.

Tamil Nadu : Selective breeding of locals namely, Madras Red, Mecheri, Ramnad white and Vembur in their respective tract. Mecheri sheep which produces skins of excellent quality and texture having good market value may be improved further through selection. Cross breeding of Coimbatore with Rambouillet, Bharat Merino and stabilizing exotic inheritance at or below 50 per cent level is recommended. Nilagiri Synthetic (Sandyno) should be propagated further to increase the population in the hills. The area would serve as nucleus for availability of improver ram. In Nilagiri hills and Kodai hills there is scope for production of apparel wool by multiplying the Sandyno and Bharat Merino in the hill region.

Pondichery	will follow the breeding policy of adjacent area.

REFERENCES

Acharya, R.M. 1982. Sheep and goat breeds of India. Rome. F.A.O. Animal Production and Health paper No. 30.

Arora, A. L. 2007. Sheep Development Programme for South India. In Compendium Interactive Meet on Sheep Production for Meat and Wool and Its Utilization in the Plains of Southern Region of India 29th May 2007 Organized by Organized by Southern Regional Research Centre, CSWRI, Mannavanur, Kodaikanal, TN.

Arora, A.L. and Prince L.L.L .2004. Contribution and significance of Indigenous sheep in India. In Compendium of Lectures, Training Programme on " Characterization and conservation of Animal Genetic Resources" conducted at NBAGR, Karnal during 2-11th August 2004 pp 95-103.

Arora, AL, Mishra AK and Prince, L L L .2007. Sheep genetic resources in India: Their conservation and improvement. Lead paper in Proceedings of National symposium on role of animal genetic resources in rural livelihood security at Ranchi during Feb, 8-9,2007.

Census Report. 2003. Seventeenth Livestock Census Report . Department of Animal Husbandry and Dairying, Krishi Bhavan, New Delhi.

DAHD Report .2006. Basic Animal Husbandry Statistics. Department of Animal Husbandry and Dairying, Krishi Bhavan, New Delhi, AHS Series-10.

DARE report. 2005. DARE & ICAR Annual Report, Indian Council of Agricultural Research, New Delhi.

Khan, B.U., Arora, A.L. and Sharma, R. C. 2001. Enhancing Productivity of Sheep: Technology Dimensions in 2002. Technology options for sustainable livestock production in India: proceedings of the Workshop on Documentation, Adoption, and Impact of Livestock Technologies in India. edited by Birthal, P. and Parthasarathy Rao, P. pubs by ICRISAT-Patancheru, India.

Narayanaswami, M., Balaine, D.S. and Chopra, S.C .1975. Effect of non genetic factors on body weights in Mandya sheep. *Indian Journal of Animal Research* 37 133-135.

Nimbkar, C., Ghalsasi, P.M., Swan, A. A., Walkden-Brown, S. W. and L. P. Kahn. 2003. Evaluation of growth rates and resistance to nematodes of Deccani and Bannur lambs and their crosses with Garole. *Animal Science* 76, 503-515.

Rajapandi, S., Ramesh Saravanakumar, Richard Jagatheesan, P.N. and Viswanathan, K. 2006.Distribution pattern and management practise of Coimbatore sheep. *Indian Veterinary Journal* 83, 891-892.

Raman, K. S., Sundararaman, M. N., Haribhaskar, S and Ganesakale, D.2003. Biometrics and breed characteristics of Madras Red sheep. *Indian Journal of Small Ruminants* 9, 6-9.

Richard Jagatheesan, P.N., Arunachalam, S., SIvakumar, T. and Selvaraj, M.2003a. Performance of Mecheri sheep in its breeding tract. *Indian Journal of Animal Sciences* 73, 909-912.

Richard Jagatheesan, P.N., Arunachalam, S., SIvakumar, T. and Selvaraj, M.2003b. Performance of Mecheri sheep in farm and farmers' flock. *Indian Journal of Animal Sciences* 73, 1055-1057.

Shakyawar, D.B., Parthasarathy, S. and Swain, N. 2007. Performance of Bharat Merino & its cross in Kolar and future prospects. In Compendium Interactive Meet on Sheep Production for Meat and Wool and Its Utilization in the Plains of Southern Region of India 29th May 2007 Organized by Organized by Southern Regional Research Centre, CSWRI, Mannavanur, Kodaikanal, TN.

Swain, N., Sureshkumar, S., Rajapandi, S. and Parthasarathy, S. 2007. Performance of Bharat Merino and Avikalin sheep at Southern Temperate Region of Kodai Hills. In Compendium Interactive Meet on Sheep Production for Meat and Wool and Its Utilization in the Plains of Southern Region of India 29th May 2007 Organized by Organized by Southern Regional Research Centre, CSWRI, Mannavanur, Kodaikanal, TN.

Terminal Report .1986. Terminal report of AICRP on Sheep Breeding, Project Coordinator Sheep Breeding, CSWRI, Avikanagar.

Terminal Report .2007. Terminal report (1994-2006) of Network Project on Sheep Improvement, Project Coordinator Sheep Breeding, CSWRI, Avikanagar.

Viroji Rao, S. T., Ravindra Reddy and Veerabramahiah, K. 2002. Comparative study on pre weaning and post weaning growth performance in two strains of Nellore sheep. *Indian Veterinary Journal* 79, 1315-1316.

□□□

20

Malpura: A Promising Breed for Enhancing Mutton Production

A.K. MISHRA
Central Sheep and Wool Research Institute, Avikanagar - 304 501, Rajasthan

Introduction

India is endowed with rich sheep genetic resources having 42 descript breeds of sheep and account for 5.50% of world sheep population. A static sheep population around 40 million has been there from post independence to late seventies and after that an erratic trend was observed. However, in last decade sheep population substantially increased and presently 58.2 million of sheep are available in India. Sheep constitute an important species of livestock in rural agrarian economy and numerically, Rajasthan is a leading state in sheep. It has one fourth of the total sheep of the country. Malpura is a famous mutton breed of Rajasthan and well adapted to hot climatic conditions of semi-arid regions. It produces about one kilogram of white coarse wool annually. The animals are fairly well built with long legs and face is of light brown with short and tubular ears.

Habitat and Distribution

This breed derives its name from the small town "Malpura" located at about 90 km from Jaipur in Rajasthan. India has been divided into four distinct sheep raising geographical regions namely North temperate, North western, Eastern and Southern Peninsular. Malpura sheep had been classified in Northwestern ecozone of India and predominates in Jaipur, and Tonk districts, however it is also marginally distributed in Swai Madhopur, Ajmer, Chittorgarh and Bhilwara districts of Rajasthan. The area is exposed to hot-dry winds from April to June with lot of dust in the atmosphere. Relative humidity starts rising in the late part of June or early July.

The animals of Malpura breed are well built with long legs and having square compact body. The face is of light brown color. Ears are short and tubular, with a small cartilaginous appendage on the upper side. Both sexes are polled. Tail is medium to long and thin. Fleece is white and extremely coarse and hairy. Belly and legs are devoid of wool. Udder is well developed and ewes have the sufficient milk to sustain and accelerate the growth of their neonates. Averages for body measurements such as body length, height at withers and chest girth in adult males were 72.12±0.70, 71.93±0.68 and 83.62±0.89 cm and in adult females the corresponding figures were 63.91±0.14, 64.04±0.22 and 70.11±0.21cm, respectively (Acharya, 1982). The body measurements viz. length, height, girths', length of the ear and tail of different age groups as reported by Arora *et al.* (1975) are given in table 1.

Table 1. Body measurements of Malpura sheep.

Parameters (cm)	Milk teeth (N=28)	2-teeth (N=134)	4-teeth (N=88)	6-teeth (N=46)
Length	62.18	63.60	64.99	64.51
Height	62.21	63.80	64.43	64.13
Heart girth	67.25	67.70	69.06	70.80
Paunch girth	62.69	69.60	70.76	72.01
Head length	24.93	25.40	25.76	26.29
Eye's distance	14.28	14.00	14.50	14.64
Ear length	6.38	6.20	6.21	6.28
Tail length	22.89	24.60	24.34	23.58

Production Characteristics

Growth Performance

Malpura is a medium to heavy sized breed and on an average its adult body weight is 40 kg in males and 30 kg in females. At organized farm, average birth weight, 3-month weight, 6-month weight and 12- month weight ranged from 2.90 to 3.07 kg, 12.86 to 14.60 kg, 19.24 to 21.00 kg and 24.04 to 27.63 kg (Sharma et al. 1999; Arora *et al.* 1999; Arora et al. 2001). The flock of Malpura sheep raised at CSWRI, Avikanagar since long back was analyzed in terms of growth profile traits of lambs for a period of about 30 years and results are given in table 2. In farmers' flock average birth weight ranged from 2.77 to 3.50 kg and 6-month body weight ranged from 16.4 to 18.94 kg (Kaushish *et al.*, 1990; Sharma *et al.* 2002). Based on 482 records, Sharma *et al.* (1999) reported the average daily gain (ADG) to be 128.0 g in pre-weaning stage (0-3 months) and 52.0 g in post-weaning stage (3-12 month) in Malpura lambs. In another similar study, Arora *et al.* (1999) reported these ADG to be 117 and 60 g respectively.

Table 2. Least-square means of body weight (kg) of Malpura at Avikanagar

Particulars	Birth weight	3-Month weight	6-Month weight	12-M weight
Overall	2.86 (3586)	12.22 (2852)	18.36 (1819)	24.47 (1535)
Period				
1972-79	2.51 (1329)	9.54 (1058)	14.70 (727)	21.32 (361)
1980-89	3.12 (763)	14.40 (609)	20.85 (478)	26.64 (354)
1990-99	2.95 (1494)	12.70 (11850	19.52 (1171)	25.46 (820)

Values in parentheses are number of observations

Least square means (irrespective of year of lambing and sex of lamb) for birth, 3, 6 and 12-month body weight were 3.09, 14.16, 21.44 and 28.00 kg respectively (Table 3). Differences due to year of lambing were significant for body weights at all the ages. The body weight at six month reached up to 24.50 kg in year 2003 which is an excellent achievement in routine management system indicating that selection was in right direction. No definite trend over the years was observed and it might be because of differential availability of inputs in terms of feed/fodder, physical environmental conditions and other managemental factors prevailing in different years..Male lambs were significantly heavier at all the ages compared to their counterpart females. Results of body weights revealed that Malpura, lambs have the genetic potential for increased body weights and consequently mutton production. The overall least squares means of average daily body weight gain between 0-3, 3-6 and 6-12 months were 123.5±0.8 , 79.3±0.8 and 32.8±0.5 grams, respectively (Table 4). Ewe weight significantly influenced the birth, 3-month and 6-month body weight of lamb.

Table 3. Growth performance of Malpura sheep (kg)

Particulars	B-weight	3-M weight	6-M weight	12-M weight
Overall	3.09 (1082)	14.16 (995)	21.44 (926)	28.00 (684)
Year				
2002-03	3.08 (156)	12.09 (145)	20.62 (142)	32.73 (97)
2003-04	2.98 (225)	15.85 (215)	24.50 (205)	28.65 (173)
2004-05	2.84 (246)	13.42 (234)	20.41 (220)	26.97 (167)
2005-06	3.15 (216)	14.30 (193)	20.72 (167)	26.47 (132)
2006-07	3.00 (239)	14.33 (208)	20.97 (192)	25.86 (115)
Sex				
Male	3.09 (552)	14.84 (508)	22.85 (461)	30.43 (294)
Female	2.93 (530)	13.48 (487)	20.04 (465)	25.57 (390)

Values in parentheses are number of observations

Table 4. Least square means for average daily gain (g) in Malpura.

Particulars	Average daily gain (g)		
	0-3 month	3-6 month	6-12 month
Overall	123.5 (995)	79.3 (918)	32.8 (668)
Year			
2002-03	108.7 (145)	84.0 (140)	56.7 (97)
2003-04	142.7 (215)	94.5 (205)	23.0 (166)
2004-05	117.2 (234)	76.4 (219)	31.5 (163)
2005-06	123.6 (193)	68.4 (166)	27.9 (128)
2006-07	125.4 (208)	73.2 (188)	25.1 (114)
Sex			
Male	130.3 (508)	86.6 (459)	36.2 (284)
Female	116.8 (487)	72.0 (459)	29.5 (384)

Values in parentheses are number of observations

Reproductive Performance

Reproductive efficiency of sheep is an important component of flock productivity and its augmentation is essential for improving the low reproductive performance of native sheep breeds. Reproductive performance of Malpura ewes during the year 2001-202 to 2005-2006 is presented in Table 5. Tupping percent in Malpura sheep was about 96%. The over-all lambing % and lambing rate on ewes available basis was 88.86 and 92.04 %, respectively. The litter size was 1.05 whereas weaning rate was 0.99, indicating the good survival of weaner. The twining percentage ranges between 2.30 to 7.27 % in Malpura flock. On an average Malpura ewe produces 5.19 % twins.

Table 5. Reproductive performance of Malpura sheep.

Year	Tupping %	Lambing % (available basis)	Litter size	Lambing rate (available basis) in %	Weaning rate (%)
2001-02	96.22	85.36	1.04	89.43	1.00
2002-03	98.37	93.34	1.02	94.47	0.97
2003-04	99.23	90.16	1.08	96.72	1.04
2004-05	98.93	91.18	1.06	90.44	1.03
2005-06	90.66	86.02	1.04	90.06	0.91
Pooled	96.52 (1324)	88.86 (1172/1319)	1.05 (1108/1053)	92.04 (1214/1319)	0.99 (998/1007)

Within parentheses are number of observations; Lambing rate =Number of lambs born in a year/Number of ewes available

Ewes Productivity Efficiency

The ewe productivity efficiency in terms of kg weaned per ewe lambed or weaning efficiency (W.E.) of the ewe is an extremely important characteristic of economic mutton production program because it covers two important factors: (i) Mortality of the lambs up to weaning stage (90 days) and (ii) Weaning weight of available lambs. Therefore, to improve the W.E. of the ewe due care has to be given to both the factors because if the contribution of the either factor or both is low then automatically weaning efficiency will be affected adversely. The ewes' productivity efficiency of Malpura sheep in terms of kg of lamb produced per ewes lambed was significantly affected by year and parity. The Malpura ewes weaned 14.31 kg of lambs which was reached up to 26.47 kg in 12 months of age (Table 6). The ewes' productivity efficiency in terms of kg of lamb produced per kg body weight of ewes was 0.101, 0.447, 0.671 and 0.815 kg at birth, weaning, 6 and 12 months of age respectively and in 6th parity the Malpura ewes produces lambs equal to her body weight. The results indicated that Malpura females are more efficient in later parities.

Table 6. Ewes Productivity Efficiency of Malpura ewes.

Particulars	N	Birth	3-Month	6- Month	12- Month
Overall	859	3.21	14.31	21.44	26.47
Year					
2002	100	3.14	13.43	21.31	30.89
2003	165	3.31	17.28	26.19	30.57
2004	196	3.07	14.08	21.62	26.39
2005	205	3.38	13.81	18.89	23.33
2006	193	3.13	12.94	19.23	21.18
Parity	859				
1	331	3.02	13.05	19.36	23.91
2	209	3.15	14.36	20.95	26.14
3	160	3.12	13.48	19.96	24.89
4	98	3.25	15.04	22.51	27.04
5	49	3.26	13.66	21.10	24.32
6	12	3.44	16.27	24.79	32.54

Feedlot Performance

Series of nutritional manipulation experiments (Arora and Karim, 1995; Karim and Santra, 2000; Karim *et al* 2002; Santra *et al* 2002;)have been conducted on pre and post weaner lambs at CSWRI, Avikanagar. Results indicated that Malpura lambs are quite suitable for fat lamb production under intensive and semi-intensive production systems. Unlike Dorset/Suffolk x Malpura and Awassi x Malpura crossbreds, Malpura lambs did not deposited the undesired carcass fat under intensive feeding system at varying level of concentrate to roughage ratios such as 70:30, 60:40 and 50:50. It has been possible in Malpura lambs at the Institute farm that by providing ad-lib compound ration (60 concentrate: 40 roughage) weaner lambs of about 2 ½ month age could achieve 25 kg body weight at about 5 month of age (Bulletin, 1998). Under intensive feeding, 25 kg finishing weight could be achieved in a total age of 151 days with a protocol of 60 days weaning and 91 days of intensive feeding with 135 g and 16% feed efficiency. Malpura lambs could achieve 150 g ADG with 1:5 to 7 feed efficiency ratio under intensive feeding on 50:50 roughage (pala leaves) and concentrate based ration (Anonymous, 1998).

Milk Yield in Malpura Sheep

Sufficient milk availability in sheep is one of the critical factors determining the lambs' growth and its survival during pre-weaning stage in particular and overall growth increment of the lambs in general as it provides essential nutrients to the lambs in their early stage of life. On an average Malpura ewe produces 531.63 g milk per day. The effect of year of lambing and week of lactation significantly affected milk yield. The ewe parity had significant affect on the ADMY and was higher in the 2nd parity in Malpura, which indicated that mammary gland, secreted more milk after 1st parity of the ewe. A decreasing trend in milk yield over the ten-lactation week was observed. Maximum milk yield was obtained in 3rd week of lactation among all the weeks included in the study, which indicated that ewes secreted relatively more milk in their mammary glands up to 3rd week of lactation compared to other latter weeks. The results reveal that Malpura ewes have sufficient milk to sustain growth of their neonates.

Utility of Malpura Sheep

Valuable Germplasm

An elite flock of pure Malpura sheep has been developed through selective breeding at CSWRI, Avikanagar and has been serving as resource center for providing limited number of breeding rams. Improved breeding rams of Malpura are being produced every year and supplied to the interested farmers, State Govt. and development agencies for genetic improvement in farmers' flocks. As per the feedback received, the breeding rams supplied were performing well under farmers' management conditions.

Fat Lamb Production Technology

Production of fat lamb is a promising commercial mutton production program that can prove as a boon for meat industry. Malpura weaner lambs (2 months age) are capable of achieving 25 kg body weight at about 5 months of age under intensive feeding on composite ration of the ratio of 60 concentrates and 40 roughage. The major advantage of this technology is that sheep owners need to rear the animals only for about 5 months and not for the whole year. Further, farmers will get the handsome profit after 5 months and this profit would be at par or even more than when the lambs are reared for 12 months under extensive grazing system. By considering all the inputs required for raising the lambs to attain finishing weight of 25 kg it is estimated that net profit of about Rs. 370/- per lamb on slaughter basis and Rs. 100/- on live animal sale basis can be obtained. These figures are based on prevailing prices in local market. This technology would be helpful to the clientele for reducing the time period from 12 to 5 months in getting almost same profit per animal in addition to avoiding the mortality risk and unnecessary rearing of lambs up to yearling stage.

Economics of fattening lambs for mutton production

	Cost (Rs)
A. Cost of items	
1. Cost of 2 months old lamb	
(@ Rs.40/- per kg live weight) for weaning weight 11 kg)	440
2. Fattening cost (i.e. 25 kg live weight)	
(Required body wt. gain = 14 kg	391
Av. F.E. = 18.0% or 5.55 kg feed required for 1 kg gain in body weight	
Total feed required 14 x 5.55 (77.7 kg)	
(Cost of feed @ Rs. 503 per quintal)	
3. Labour cost (Rs. 70/- per day for 90 days).	150
One labour for lambs	42
4. Health cover	5
5. Interest on capital investment per lamb	22
Total rearing expenditure per lamb	900
B. Returns from meat production (through slaughter)	
1. Meat @Rs. 80/- per kg (13.0 kg meat; dressing % = 52)	1040
2. Skin + Wool (500 g)	120
3. Head, Hooves and Offals	80
4. Manure	10
5. Stomach/Intestine	20
Total return per lamb	1270
Net profit per lamb	370

C. Income (through sale of live animals)	
1. Return per lamb (25 kg body weight @ Rs.40/- kg)	1000
2. Expenditure per lamb	900
Net profit per lamb	100

Research and Development

Malpura sheep has been extensively used as dam breed under cross breeding programs of CSWRI, Avikanagar in evolving Avikalin (for carpet wool), Avimaans (for mutton), Bharat Merino (for fine wool) and Indian Karakul (for pelt). The exotic rams breed used were Rambouillet for Avikalin, Dorset /Suffolk for Avimaans, Rambouillet/Russian Merino and Karakul. The inheritance of Malpura breed in these developed strains varied from 25 to 50%. The lambs of the Avimaans strain were capable of achieving higher growth rate and market weight compared to Malpura lambs. However, the lambs were having tendency to deposit fat under intensive feeding system. The reproductive efficiency of Avimaans ewes was also poor. Rams of Awassi breed were crossed with Malpura during 1995 to improve live weight gain and milk production. The lambs of the Awassi x Malpura halfbreds exhibited 18.70, 27.05, 19.56 and 14.90 % improvement to their counterpart Malpura lambs at birth, 3, 6 and 12 month of age. The crossbred ewes produced 29% more average daily milk yield than its counterpart Malpura ewes. Further, these crossbreds produced 37% more 1st six monthly GFY than Malpura and survivability was almost at par with native Malpura. In 1997, Malpura rams were purchased by Orissa state government (Eastern state of India) under Integrated Livestock Development Project for distribution among farmers' sheep flocks, on the basis of performance Das *et al.* (2001) reported that Malpura sheep could be used in Eastern India particularly in Orissa for production of mutton. Recently, Garole a prolific sheep was introduced to enhance the reproductive efficiency of Malpura sheep and still is continuing.

Conclusion and Recommendations

Malpura is a famous mutton breed of Rajasthan and well adapted to hot climatic conditions of semi-arid regions. It produces about one kilogram of white coarse wool annually and mainly found in Jaipur and Tonk districts. The animals are fairly well built with long legs and face is light brown with short and tubular ears. This breed has been found worthwhile as dam parent in crossbreeding programs undertaken by Central Sheep and Wool Research Institute, Avikanagar and other development agencies of State Government. In future the more emphasis will be laid on enhancing the lambing rate and economic returns of the Malpura sheep. Improved breeding rams of Malpura sheep produced through selective breeding are being provided to the sheep farmers of breeding tract and other development agencies for genetic improvement in farmers' flocks. There is need to establish the ram rearing centers of Malpura breed to meet out the demand of farmers and to avoid the intermixing of the rams of other breeds.

REFERENCES

Acharya, R.M. 1982. Sheep and goat breeds of India. Animal Production and Health Paper 30, FAO, Rome, Italy.

Acharya, R.M. and Mohan M. 1979. Genetic consideration in crossbreeding for evolving new breeds of sheep-Results of crossbreeding experiment. Indian *Journal of Animal Genetics and Breeding* 1, 37-45.

Anonymous 1998. 35 years of Research, CSWRI, Avikanagar, Rajasthan, India.

Arora, A. L. and Karim, S. A. 1995. Growth response of intensively fed lambs. *Indian Journal of Animal Sciences* 65, 1046-1048.

Arora, A.L., Sharma, R.C. and Kushwaha, B.P. 2001. Breeding technologies for increasing mutton production in sheep breeds of Rajasthan. Invited paper In: 4th National Seminar of ISSGPU on scope and strategies for increasing meat production from small ruminants and rabbit in new millenium, November 9-11, Bharathiar University, Coimbatore India, 13-19.

Arora, A.L., Sharma, R.C., Narula, H.K. and Kumar, R. 1999. Comparative performance of Avikalin and Malpura sheep. *Indian Journal of Small Ruminants* 5, 4 -8.

Arora, C.L, Acharya, R.M., Badashiya, B.S. and Dass, N.C. 1975. Characterization of Malpura breed in Rajasthan and future prospects for its improvement. *Indian Journal of Animal Sciences* 45, 843-848.

Bohra, S.D.J, Kumar Satish, Bapna, D.L. and Singh, R.N. 1984. Wool attributes in Malpura, Sonadi and their F_1 crosses with Dorset and Suffolk sheep. *Cheiron* 13, 287-291.

Das, S, Mohanty, A and Sahoo, G. 2001. Performance of Malpura sheep and its crosses under hot and humid climatic conditions. *Indian Journal of Animal Production and Management* 17, 20-21.

Karim, S.A. and Santra, A. 2000. Growth performance of Malpura and crossbred rams under intensive feeding. Small Ruminant Research, 37: 287-291.

Karim, S.A., Santra, A. and Verma, D.L. 2002. Growth, feed conversion efficiency and carcass characteristics of Malpura and Malpura x Awassi crossbred lambs in a hot semi-arid environment. *Asian-Australasian Journal of Animal Science* 15, 377-381.

Kaushish, S.K., Rawat, P.S. and Sharma, S.C. 1990. Performance of native sheep (Malpura) and its crosses with Avikalin under semi arid condition. *World Review of Animal Production* 25, 43-46.

Mishra, A. K., Arora, A. L., Kumar Sushil, Sharma, R. C. and Singh V.K. 2005. MALPURA: A Mutton type sheep breed. CSWRI, Avikanagar, India.

Santra, A., Karim, S. A. and Chaturvedi, O.H. 2002. Effect of concentrate supplementation on nutrient intake and performance of lambs of two genotype in a semi-arid rangeland. *Small Ruminant Research* 44, 37-45.

Sharma, R.C., Arora, A.L., Mishra, A.K. and Sushil Kumar. 2003. Production performance of Malpura sheep in an organized farm of Rajasthan. *Indian Journal of Small Ruminants* 9, 119-121.

Sharma, R.C., Arora, A.L., Kumar, R. and Narula, H.K. 1999. Impact of genetic and non-genetic factors on growth profile in Malpura and Avikalin lambs. *Indian Journal Animal Sciences* 69, 820-822.

❑❑❑

21

Advanced Reproductive Technologies for Improving Reproductive Efficiency of Sheep

S.M.K. NAQVI, ANIL JOSHI, S. SAHA

Central Sheep and Wool Research Institute, Avikanagar - 304 501

Introduction

Sheep husbandry plays an important role in rural economy of India by contributing wool, mutton, manure, skin and milk for mankind. It also generates employment for about six million marginal and landless labourers. A large segment of landless and marginal farmers in our country rear them for their livelihood under different agro-climatic conditions at a low cost on natural vegetation, grazing lands, wastelands and uncultivated lands. A major proportion of sheep population (>70%) thrives well and maintains its production in arid and semi arid areas of western region and southern plateau (southern peninsular region). As per livestock census (2003), only five states of the country i.e. Andhra Pradesh, Rajasthan, Karnataka, Tamil Nadu and Jammu and Kashmir cover about 60% of the sheep population. It is clearly indicated that sheep is densely populated in hot arid and semi arid regions. However, despite a large population of sheep in these ecological zones, their productivity is low due to inadequate feeding, management and breeding. The productivity and profitability of sheep farming are directly proportional to reproductive efficiency, which is one of the major factors affecting the total output of mutton/ wool from sheep. The reproductive efficiency of Indian sheep in general, and in breeds of arid and semi-arid regions particular, is relatively low. Most of the sheep breeds available in arid and semi-arid areas have late maturity (>650 days age at first service), long lambing interval (about 1 year) and low fecundity (< 1 lamb/year). Therefore augmenting rate of reproduction is generally desirable and economical in farm animal production especially in meat producing animals. There has been steady increase in the domestic consumption/demand of meat due to rapid urbanization, awareness of quality diet regulation, improvement in economic status and living standard of especially middle class population. The estimated requirement of mutton is 687 M Kg against the current availability of only 239 M kg. The mutton output generated from per Indian sheep is also much below the world's average despite the fact that percent of sheep slaughtered are similar (32- 36 %). This is mainly due to low carcass yield of 10-12 kg as against 17 kg of world's average.

In order to measure reproductive performance of sheep, parameters like fertility, prolificacy, fecundity, age at first parturition and parturition interval are considered. Reproductive efficiency of sheep is governed by three factors: a. fertility, b. prolificacy and c. the lamb survival. The fertility is measured by percent of ewes lambed outof ewes bred; prolificacy, measured by percent of lamb born out of ewes lambed; lamb liveability, measured by percent of lambs born alive of total lambs born and percent of lambs weaned of lambs born alive. Thus overall measure of reproductive performance is the percent lambs weaned (or market) of ewes bred. This emphasizes the importance of death or loss of lambs as a source of loss in reproductive efficiency.

Martin and his colleagues (2004) described some of the measures to improve reproductive efficiency.

Artificial Control of Reproductive Events

(1) Control of oestrous cyclicity and lambing so as to harvest maximum benefit from opportunities in seasonal markets and in seasonal peaks in pasture supply. Lambing may be avoided in unfavourable (extreme winter or summer) environment.

(2) Control of time of ovulation and parturition so we can provide precisely timed nutritional supplements.

(3) Maximise the number of progeny:

  a. Through maximising the ovulation rate

  b. Through minimising post-fertilisation wastage

  c. Accelerating lambing: Three lambs in two years, five lambs in three years

In general most Indian sheep lamb once in a year. Accelerating lambing refer to more than one lambing per year. Intensive lambing can be practised either by natural means or artificial control using exogenous hormone. Accelerated lambing provides uniform supply of lambs through out year and reduces overall maintenance cost of sheep flock since total non-productive days (no pregnancy and weaning) are reduced. The ewes in annual production system occupy 145-155 days for pregnancy, engaged in weaning for 60-90 days and remain idle (non-productive, except some wool is grown) for about 120-140 days.

Sheep in tropics are in general non-seasonal polyestrus with intermittent acyclicity. Studies have shown (Prasad et al, 1982) that Marwari sheep in arid tropical region exhibit estrus through out year if nutrition is not a limiting factor. There is possibility to breed such ewes every 8 months (5 months pregnancy + 3 months weaning) when nutrition is adequate and thermal stress is avoided. However, some potential disadvantages which may be encountered include lower fertility, lower prolificacy and smaller birth weights.

Nutritional Regulation of Reproduction

There are ample evidences available that nutrition regimen (high or low plane of nutrition), under feeding or over feeding influence the reproductive efficiency of sheep (Robinson, 1996). The positive effect of nutritional supplementation on reproduction is well documented. Even short duration supplementation (4-6 days) effect the ovulation rate (Oldham and Lindsay, 1984). Feed flushing prior to breeding season has been in practice since long to enhance lamb crop. However, intricate physiological mechanism through which nutrition influence the reproduction are not well understood. Both high and low plane of nutrition are detrimental to growth and survival of embryos. High level of nutrition increase the heat increment and also attributed to reduction in plasma level of progesterone (Wallace *et al.*, 1994). Under-nutrition limits the supply of amino acids and glucose to developing embryos/foetus and alters the composition of uterine milk/fluid which in turn results in impaired or retarded embryonic foetal development (Parr et al, 1982), production of small sized lambs (Rhind et al., 1980; McDonald et al., 1981). Maurya and his colleague (2004) experimentally induced poor body condition in Malpura ewes through restricted feeding for a period of 14 weeks followed by re-alimentation. They showed that restricted feeding reduced the duration of estrus (15 h versus 26 h) and increased the estrus interval (31.5 days versus 18.6 days) compared to control. Animal subjected to restricted feeding required more services per conception than those of control (1.3 mating/ewe versus 1.9 mating/ewe).

Literature available on the effect of feed intake level on the ovulation rate in temperate sheep is confusing and contradictory (Rhind, 1992). Gunn (1983) suggested that ovulation rate is dependant on short-term flushing only in ewes, which are with in the certain intermediate range of body condition. McEvoy, *et al.* (1995b) showed that flushing superovulated ewes before mating did not increase ovulation rate. Information on the effect of feeding level on the superovulatory response and embryo yield is not available on sheep reared in tropics. Naqvi and colleagues (2002a) examined the superovulatory response and embryo recovery in Bharat Merino ewes maintained on variable dietary level in tropical environment. Dietary flushing i.e. supplementation of concentrate feed @ 150 g/

ewe/d of Bharat Merino sheep prior to superovulation and mating did not influence the estrus response, superovulatory response, embryo recovery and yield of transferable embryos. The extra nutrition or flushing did not influence the interval to onset of estrus and duration of estrus in superovulated ewes (Naqvi et al., 2002a). McEvoy *et al.* (1995a) also reported that diet with variable energy level (0.6 and 2.3 x maintenance energy) did not influence the interval to onset of estrus in the ewes treated for superovulation.

It is known that ovulation rate in ewes with moderate body condition, increased on feeding above their energy requirement (flushing) during the few weeks before mating (Rhind, 1992; Molle *et al.*, 1995; 1997). In contrast to above reports where flushing during the periovulatory period was considered to be beneficial for enhanced ovulation rate at naturally occurring estrus in normal ewes, the same may not be case when ewes were superovulated using exogenous gonadotrophins. Flushing of ewes prior to superovulation with additional quantity of concentrate did not increase the ovulation rate (Naqvi *et al.*, 2003; McEvoy 1995b). Superovulatory responses were not affected in this study presumably because individual potential for ovulation was maximized as a result of exogenous gonadotrophin treatment. The key determinant of Prolificacy is ovulation rate, which may be improved by nutrition (Scaramuzzi, *et al*,2006). Since the mechanism by which nutrition promotes an increase in ovulation rate is still obscure, a better understanding of follicular dynamics during the oestrous cycle may help to elucidate the mechanism.

The effect of nutrition on metabolic and reproductive hormones, which influences the ovarian activities (folliculogenesis and ovulation) has been described elaborately by Scaramuzzi *et al.* (2006). The effect of energy balance on metabolic and reproductive hormones is widely reported. The negative energy balance primarily effect the hypothalamo-hyphophyseal gonadal axis (Wade and Jones, 2005) and characterized by hypoglycaemia, hypoinsulaneamia, suppressed plasma IGF-1 and elevated plasma GH. These alterations are associated with inhibition of GnRH pulsality, anovulation and anoestrus (Kiyama *et al.*, 2004). High plane of nutrition (positive energy balance increase the peripheral leptin and insulin level and glucose intake by the ovary which in turn is associated with enhanced folliculogenesis and ovulation rate. Overfeeding during early pregnancy alters hepatic metabolism of progesterone and thereby reduces the peripheral progesterone concentration and pregnancy rate in sheep (Parr *et al.*, 1993). Three major metabolic systems reported for influencing ovarian activities are a. Insulin –glucose b. IGF and c. Leptin. All these three systems are having complex interaction in mediating follicular response to nutrition (Scaramuzzi *et al.*, 2006).

Somachitac *et al.* (2007) studied the short term nutritional supplementation of ewes and suggested that local ovarian actions of nutrients i.e. increase in the concentration of glucose and decrease in the concentration of progesterone in follicles, have a role in mediation of nutritional influences on folliculogenesis.

Balanced nutrition of the ewe from post-conception to delivery of lamb is one of the factors influencing the embryonic and foetal development and extrapolated to postnatal development.

Physiological Regulation

It is becoming essential to improve the reproductive potential of the Indian sheep by augmenting ovulation rate with the application of exogenous hormones (Naqvi *et al.*, 2001, 2002a), immunization of inhibin peptides (Naqvi et al., 2004) or introgression of *FecB* gene from Garole sheep into non-prolific large size sheep (Naqvi et al., 2002b). Conservation of sheep genetic resources in the form of spermatozoa, ova and embryos assume significance if a gene of major interest or effect is detected. Garole rams produce good quality semen (Joshi *et al.*, 2003), which can be stored for short-term and long-term (Joshi *et al.*, 2001, 2005). Our recent study has shown evidence of production of a large number of in vivo embryos from a single flush of a Garole ewe (Naqvi *et al.*, 2006a) and birth of Garole lambs of higher birth weight following transfer of *in vivo* derived Garole embryos into large size non-prolific ewes (Naqvi et al., 2006b). There is an ample scope to exploit the *FecB* gene carrier rams (Kumar *et al.*, 2006) for production of fecundity gene carrier progeny born through non- prolific sheep. The establishment of a prolific homozygous sheep flocks with prolific Garole inheritance and large body size of non- prolific will act as a precursor to incorporate prolificacy traits in vast and diverse non- prolific sheep genetic resources available in our country. It will also serve as a new option for significantly increasing lambing rates of non- prolific sheep breeds in their natural habitat.

One of the basic questions that created lot of interest in research was how ovulation rate and litter size are controlled in sheep. It is now established that ovarian follicular development and ovulation rate are determined by highly co-ordinated level of endocrine and paracrine signalling. The endocrine signalling or regulation implies role of ovaries, hypothalamus and pituitary gland while paracrine regulation involves signalling with in follicular cells i.e. "oocyte, cumulus, granulosa and theca cells" (Matzuk *et al.*, 2002; Shimasaki *et al.*, 2004; Juengel *et al.*, 2004; Juengel and Mc Natty, 2005; Mc Natty *et al.*, 2003, 2005a). Recent advent of new molecular techniques and discoveries of genes regulating fertility in small ruminants, particularly in sheep has led to new insights in reproductive biology. Adopting a strategy that can maximize ovulation rate and minimize post- fertilization wastage may enhance the litter size, which is mainly determined by the ovulation rate.

It is now evident that ovarian follicular development and ovulation rate are determined not only by neuro-endocrine process involving hypothalamo-hyphophyseal gonadal axis but also by paracrine signals between the oocyte and its adjacent somatic cells (cumulus, granulosa and theca cells). The oocytes through specific secreted proteins (TGFâ super family) regulate follicular growth from primordial to preovulatory follicles and control number of ovulating follicles. These discoveries have lead to a paradigm shift in understanding the mechanism involved in regulation of ovarian function and also forming basis for development of new therapeutic reagents for regulating the fertility/ovulation rate. The physiological mechanism involved in folliculogenesis from establishment of primordial follicles at birth to ovulation at puberty is reasonably well understood in livestock species. Based on the knowledge of endocrine control of ovulation, ovulation rate in small ruminants can be enhanced through the administration of exogenous hormones. The intricate physiological mechanisms that control the number of ovulating follicles or apoptosis of maturing follicles are not still fully understood. During the last two decades physiologists limited their studies on the neuro-endocrine regulation of folliculogenesis and ovulation through the involvement of reproductive axis (hypothalamus, pituitary and ovaries) and compared mechanism of action between non-prolific and prolific sheep breeds. Research efforts are being directed to unveil intricate physiological regulation that determines ovulation rate in sheep (Fabre *et al.,* 2006) and that may lead to develop novel therapeutic reagents for fertility regulation.

Natural Methods to Control/Regulate Reproductive Events

Martin and his colleagues (2004) described three strategies to improve reproductive efficiency in small ruminants with out use of exogenous hormones/ pharmaceutical agents, which may lead to "green, clean and ethical small ruminant reproduction". These are

1. Male effect to induce estrus synchronization and ovulation
2. Focussed feeding to improve reproductive efficiency
3. Maximize offspring survival through a. management b. nutrition c. genetic selection for behaviour (temperament)

The sudden introduction of fertile male into female flock can induce the estrus / ovulation in females (Martin et al., 1986). Although, the male effect has not been reliable in all breeds of sheep and goat. This approach has not been explored systematically in Indian breeds as in exotic i.e.: Merino, Suffolk etc. The male effect response in female is altered with nutrition and photoperiod (Blache et al., 2003). The mechanism involved in Male effect on female reproduction has not yet been fully understood. Limited knowledge is available how all exoreceptors and genetic inputs interact each with each other in brain centre and effect the reproductive events (Boukhiq et al.,1999). Research efforts are warranted for unravelling the intricate mechanism of action through which Male effect is perceived in female. Strategic and economical use of feeding resources for animal has been the utmost concern for animal industry especially in drought prone areas. What quantity and when to use feeding supplement that may provide maximum benefit in overall productivity including reproductive efficiency is always considered as a management tool. Feed supplement for male for a period of 8 weeks before mating has been practiced to ensure optimum testicular size and sperm production (Martin and Walkden-Brown, 1995). Overfeeding and lack of exercise are detrimental to male performance (Combrink and Schoeman, 1993). Further research is required to ascertain that concept developed on other breeds may be extended to Indian breeds of sheep. It is also not known whether maintaining rams at constant body weight throughout the year or providing supplementation feed for 8 weeks prior to mating is relatively economic and effective.

Role of Assisted Reproductive Technologies

Assisted reproductive technologies (ARTs) have been developing since last 2-3 decades and tendering/expecting to revolutionize the animal production. Some reproductive technologies are already developed and being applied while others are at the stage of development. These technologies include artificial insemination, estrus synchronization, estrus induction, synchronized lambing, semen freezing, superovulation and embryo transfer, in vitro fertilization, in vivo and in vitro embryo production, embryo cryopreservation, embryo splitting, cloning, production of transgenic animals etc. ARTs are being applied for introduction, improvement and preservation of elite and endangered germplasm and to increase the reproductive efficiency of small ruminants (Baldassarre and Karatzas, 2004). Application of ARTs has shown impact on the genetic improvement of sheep for growth and the enhancement of reproductive efficiency. ARTs have also facilitated the development of methods to transfer desired single genes or, alternatively, the entire genome from desirable individuals or embryos.

Artificial insemination and estrus synchronization are the most widely used reproductive technologies to upgrade the germplasm and enhance the reproductive efficiency. Synchronized mating and lambing could be achieved with the utilization of these technologies to produce lambs at the desired time. Frequent or accelerated lambing can be induced for increasing the reproductive rate of sheep.

A. Oestrus induction and synchronization in sheep

Synchronization of estrus in sheep is generally achieved either with the use of prostaglandin F_2a (PGF) or intravaginal sponges (Naqvi et al. 2001). Progesterone impregnated vaginal sponges may be considered as a treatment of choice due to low cost and efficiency. Artificial control of oestrus and ovulation in sheep can offer three major advantages for (a) Synchronizing oestrus in donor and recipient in ET programme, (b) Reducing inter-lambing period and compact lambing in conducive environment and (c) Performing large scale AI programme. Estrus among native and crossbred sheep could be induced and synchronized during the summer and winter seasons when incidence of estrus activity among sheep especially in crossbred is poor. Progesterone (0.35g) in combination with PMSG is a suitable treatment for estrus synchronization in acycling ewes. Indigenously developed intra-vaginal sponges loaded with progesterone were kept *in situ* for 12 days in vagina and PMSG (200 IU) given on day of sponge withdrawal to induce estrus in anoestrus sheep (Naqvi et al., 1996a; 1996b). Estrus in Indian native and crossbred ewes could also be synchronized with substantial efficiency (80-100%) with the use of PGF (Mathur et al, 1987; Naqvi and Kalra, 1990). At our institute, we have been using two doses of PGF 10.0 mg each 10 days apart in donor and 7.5 mg in recipient ewes to synchronise estrus and achieving 80-100% estrus response (Naqvi et al.1998d). A single dose of PGF @7.5 mg also could be effective for synchronization of estrus in native sheep when administered between 8 to 11 of the cycle (Das et. al., 1999).

B. *Artificial insemination using fresh and frozen semen*

Artificial insemination (AI) in farm animals is the first generation of reproductive biotechnologies, which has been used extensively in almost every country of the world for improvement and up gradation of livestock species especially cattle (Thibier and Guerin, 2000). Globally more than 3.3 Million sheep and 0.5 million goats are artificially inseminated annually.

Garole and Kendrapada sheep possesses *FecB* gene responsible for prolificacy Introgression of Fec *B* gene for prolificacy in non-prolific breed of sheep is exemplary to increase the reproductive rate. A large number of rams of proven status for fecundity/prolificacy and homozygous for *FecB* gene are required to cover the existing population for improvement in lambing rate. Extensive exploitation of these rams through artificial insemination and their wider distribution to farmers may produce progenies with improved fecundity character for significantly improving the reproduction of low producing sheep breeds. At Avikanagar, it was possible to produce 235 Garole x Malpura (GM) lambs through AI of 438 Malpura ewes using fresh diluted semen obtained from Garole rams. The overall conception and lambing rate achieved in six years AI programme were 63.5 and 53.7%, respectively. It is highlighted that AI could be successfully utilized to produced F1 crossbred lambs from small sized Garole sires and ewes of 2-3 times in size compared to the sire.

The widespread use of AI and the realization of its full potential depend essentially on the success rate in term of the lambing. Extensive use of elite rams is possible through AI with frozen semen but the fertility after cervical AI with frozen semen is low. However, acceptable fertility could be achieved with fresh diluted ram semen. *Ex situ* conservation in the form of spermatozoa of breeds having fecundity genes will be useful for inducing prolificacy in non- prolific breeds that are best adapted to their natural habitat.

C. *Multiple ovulation and Embryo transfer (MOET)*

Embryo transfer technology has been integrated in small ruminant improvement programme by providing faster multiplication, transportation and conservation of germplasm. As per the report of International Embryo Transfer Society (2005), about 25,000 sheep and 7000 goat embryos were transferred worldwide (Givens *et al.*, 2007). Relative risk of transmission of infectious pathogen via embryo transfer is likely insignificant as with transfer of live animals. Like cattle, MOET is not practiced /accepted extensively despite the recent improvement in the embryo collection and transfer procedures in sheep

MOET in livestock is considered an important tool to:

(i) Increase the reproductive rate of elite female.

(ii) Generation of large number of offspring from superior female that can increase the female selection differential within the breeding programme.

(iii) Increase the rate of genetic gain (50-100%) by reducing the average ages of breeding females and so the generation interval.

(iv) Increase the selection among breeding females .It is being used in various countries to introduce new breeds and multiply expensive germplasm at a faster rate.

A large number of rams of proven status for fecundity/prolificacy are required to cover the existing population for improving the lambing rate. There is a possibility to produce such rams within a short period by increasing the lambing rates of ewes through MOET. Extensive exploitation of these rams through use of frozen semen and AI technology and their wider distribution to farmers may produce progenies with improved fecundity character.

Despite of these advantages, MOET still lacks acceptance as a breeding tool for sheep improvement. The major factors responsible are: (i) high cost-benefit ratio, (ii) complicated nature of technology, (iii) involvement of surgical/ laparoscopic procedures for embryo collection and transfer and (iv) low product price. With the recent development and/or further refinement in techniques of embryo production, collection and transfer, it may be possible to invite the attention of small ruminant industry towards the use of MOET for increasing meat, milk and wool production. MOET is an expensive technique as the cost estimated per lamb was $ 260 with success rate of 2.13 and 3.45 lambs borne per flush (Nicholas, 1996). In our programme the arbitrary calculated cost per lamb produced ranges from Rs 800-1000 depending upon the success rate. The use of MOET will only be worthwhile if high value offsprings are produced and considered for AI or natural mating. Great efforts have been made to simplify the technology and to refine the protocols for superovulation, estrus synchronization and methods of collection and transfer of embryos. Several research workers have reviewed the recent advancement made in sheep through intensive research in ETT (Naqvi *et al.*, 2001; Cognie *et al.*, 2003). Despite the recent improvement in the ETT in sheep, cost of MOET is high as compared to that in cattle. The value of individual animal and the cost is more than the benefits gained through faster multiplication. McClintock and Nicholas (1991) concluded that MOET is unlikely to be profitable for Australian meat sheep industries. Similarly, impact of ETT on genetic gain in dairy sheep reported to be low (Barril *et al*, 1997). The cost of a lamb born under MOET is ten times the cost of one born from AI. MOET has considerable potential for genetic improvement of sheep having low reproductive rate (Cloete *et al.,* 1998). The Indian sheep breeds have poor reproductive performance (Arora and Garg, 1998) and the mean number of lambs reared per ewe averages between 0.6-0.9. It is possible to augment the reproductive rate of these sheep for genetic gain through use of MOET and AI technologies.

Although the cost benefit ratio for production of lamb through embryo transfer is very high but still this technology is effective for widespread use in breed improvement programme. The use of MOET will only be worthwhile if valuable offspring of high genetic merit are produced and considered for AI or natural mating under sheep improvement program in the country. A large number of rams of proven status for fecundity/prolificacy are required to cover the existing population for improving the lambing rate. There is a possibility to produce such rams within a short period by increasing the lambing rates of ewes through MOET. Extensive exploitation of these rams through use of frozen semen and AI technology and their wider distribution to farmers may produce progenies with improved fecundity character.

Conclusion and Recommendations

The reproductive efficiency of Indian sheep is relatively low due to late maturity, long lambing interval, low fecundity and lamb survival. Feeding, management and breeding govern the reproductive efficiency of sheep. Artificial and natural methods have been developed to control the reproductive events to enhance the reproductive efficiency. The use of natural methods for increasing reproductive efficiency in sheep is cost-effective to increase the productivity, especially mutton production through accelerated lambing and uniform supply of lambs throughout the year. High and low plane of nutritional supplementation for either short or long-term duration have a moderate to significant effect on ovulation rate, fertility and growth of foetus and newborn. Artificial control of cyclicity may be applied in Indian sheep for fixed time AI, induction of estrus during anestrus or non-breeding season, synchronized lambing in sheep for enhancing reproductive efficiency. Further development and refinement of protocols for ram semen freezing, multiple ovulation and embryo transfer, induction of twining through nutritional and physiological manipulations are required before application of these technologies in field.

REFERENCES

Arora, C.L. and Garg, R.C. 1998. Sheep production and breeding. International Book and Distribution Company. Lucknow.

Baldassarre, H. and Karatzas, C.N. 2004. Advanced assisted reproduction technologies (ART) in goats. *Animal Reproduction Science* 82-83, 255-266.

Barrilet, F.1997. Genetics of milk production. In Piper L., Ruvinsky,A. editors. The Genetics of sheep. CAB International: wallingford,U.K..p.539-564.

Blache, D., Zhang, S. and Martin, G.B., 2003. Fertility in males: modulators of the acute effects of nutrition on the reproductive axis of male sheep. *Reproduction* Suppl.61, 387-402.

Boukhiq, R., Goodman, R.L. Berriman, S.J., Adrian, B. and Lehman, M.N. 1999. A subset of gonodotropin-releasing hormone neurons in the ovine medial basal hypothalamus is activated during increased pulsatile lutenizing hormone secretion. *Endocrinology* 140, 5929-5936.

Cloete, S.W.P., van Niekerk, F.E. and Rust, J.M. (1998). Application of embryo transfer for the improvement of multiple rearing ability in medium wool Merino ewes. *Sheep and Goat Research Journal* 14, 173-179.

Cognie,Y., Baril,G., Poulin,P. and Mermillod, P. 2003. Current status of embryo technologies in sheep and goat. *Therigenology* 59: 171-181.

Combrink, G.C. and Schoeman, S.J. 1993. The influence of exercising rams on the lambing performance of a Merino flock. *South African Journal of Animal Science* 23, 24-25.

Das, G.K., Naqvi, S.M.K., Gulyani, R., Joshi, Anil and Mittal, J.P. 1999. Effect of two protocols of PGF2á treatment for synchronization of estrus in a tropical sheep. *Theriogenology* 50, 283.

Fabre, S., Pierre, A., Mulsant, P.,Bodin, L., Pasquale, E.D., Persani, L., Monget, P.M. and Monniaux. D. 2006. Regulation of ovulation rate in Mammals: Contribution of sheep genetic models. Reprod. Biol. Endo, 4:20 doi: 10.1.1186/1477-7827-4-20. Article URL http:// www. rbej.com/ content/4/1/20.

Givens, M.D., Gard, J.A., and Stringfellow, D.A. 2007. Relative risk and approaches to biosecurity in the use of embryo technologies in livestock. Theriogenology 68, 298-307.

Gunn, R.G., 1983. The influence of nutrition on the reproductive performance of ewes. In: W.Haresign (Editor), Sheep Production, Butterworth,London pp 99-110.

Joshi, A., Bag, S., Naqvi, S.M.K., Sharma, R.C. and Mittal, J.P., 2005. Effect of post-thawing incubation on sperm motility and acrosomal integrity of cryopreserved Garole ram semen. *Small Ruminant Research* 56, 231-238.

Joshi, A., Bag, S., Naqvi, S.M.K., Sharma, R.C., Rawat, P.S. and Mittal, J.P. 2001. Effect of short-term and long-term preservation on motion characteristics of Garole ram spermatozoa: A prolific microsheep breed of India. Asian-Australasian *Journal of Animal Science* 14, 1527-1533.

Joshi, A., Naqvi, S.M. ., Bag, S., Dang, A.K., Sharma, R.C., Rawat, P. ., Mittal, J.P., 2003. Sperm motion characteristics of Garole rams raised for a prolonged period in a semi-arid tropical environment. *Tropical Animal Health and Production* 35, 249-257.

Juengel J.L, Hudson, N.L., Whiting, L and Mc Natty, K.P. 2004. Effects of immunization against bone morphogenetic protein 15 and growth differentiation factor 9 on ovulation rate, fertilization, and pregnancy in ewes. *Biology of Reproduction* 70, 557- 561.

Juengel, J.L and Mc Natty, K.P. 2005. The role of proteins of the transforming growth factor-â superfamily in the intraovarian regulation of follicular development. *Human Reproduction Update* 11, 144- 161.

Kiyama, Z., Alexander, B.M., Van Kirk, E.A., Murdoch, W.J., Hallford, D.M.and Moss, G.E. 2004,Effects of feed restriction on reproductive and metabolic hormones in ewes. Journal of Animal Science 82, 2548-2557.

Kumar, D., Joshi, A., Naqvi, S.M.K., Kumar, S., Mishra, A.K., Maurya, V.P., Arora, A.L., Mittal, J.P. and Singh, V.K. 2006. Sperm motion characteristics of Garole X Malpura sheep evolved in a semiarid tropical environment through introgression of FecB gene. *Animal Reproduction Science* doi: 10.1016/j.anireprosci. 2006.07.004.

Martin, G.B. and Walkden-Brown, S.W.1995. Nutritional influences on reproduction immature male sheep and goats. *Journal of Reproduction and Fertility* (Suppl.) 49, 437-449.

Martfn,G.B.,Milton,J.T.B.,Davidson,R.H.,Banchero,Hunzikar,G.E.,Lindsay,D.R. and Blache,D.2004.Natural methods for increasing reproductive efficiency in small ruminants. Animal Reproduction Science 82-83: 231-245.

Martin. G.B., Oldham, C.M., Cognie, Y. and Pearce, D.T. 1986. The physiological responses of anovulatory ewes to the introduction of rams - a review. *Livestock Production Science* 15, 219-247.

Mathur, A.K., Srivastava, R.S., Singh, G. and Kalra, D.B. 1987. Synchronization of oestrus and fertility in ewes treated with prostoglandin F_2 alpha. *Indian Journal Animal Science* 57, 709-710.

Matzuk., M.M., Burns, K.H., Viveiros, M.M and Eppig, J.J. 2002. Intercellular communication in the mammalian ovary: oocytes carry the conversation. *Science* 296, 597- 641.

Maurya, V.P., Naqvi, S.M.K. and Mittal, J.P. 2004.Effect of dietary energy level on physiological responses and reproductive performance on Malpura sheep in the hot semi-arid regions of India. *Small Ruminant Research* 55, 117-122.

Mc Natty K.P., Juengel, J.L., Wilson, T., Galloway, S.M., Davis, G.H., Hudson, N.L., Moeller, C.L., Cranfield, M., Reader, K.L., Laitinen, M.P., Groome, N.P., Sawyer, H.R and Ritvos, O. 2003. Oocyte- derived growth factors and ovulation rate in sheep. *Reproduction* (Supplement) 61, 339- 351.

Mc Natty, K.P., Juengel, J.L., Reader, K.L., Lun, S., Myllymaa, S., Lawrence, S.B., Western, A., Meerasahib, M.F., Mottershead, D.G., Groome, N.P., Ritvos, O. and Laitinen, M. P. 2005. Bone morphogenetic protein 15 and growth differentiation factor 9 cooperate to regulate granulosa cell function in ruminants. *Reproduction* 129, 481- 487.

McClintock,A.E. and Nicholas,F.W. 1991. The implication of advanced breeding techniques. Australian Meat and Livestock Research and Development Corporation Project Report US.016.

McDinald, I. Robinson, J.J. and Fraser,C. 1981. Studies on reproduction in prolific ewes.VII. Variability in the growth of individual foetuses in relation to intra-uterine factors. *Journal of Agricultural Sciences* (Camb.),96,187-194.

McEvoy, T.G., Robinson, J.J., Aitken, R.P., Findlay, P.A., Palmer, R.M. and Robertson, I.S. 1995a. Dietary induced suppression of pre-ovulatory progesterone concentrations in superovulated ewes impairs the subsequent in vivo and in vitro development of their ova. *Animal Reproduction Science* 39, 89–107.

McEvoy, T.G., Robinson, J.J., Aitken, R.P., Findlay, P.A. and Robertson, I.S., 1995b. Relationship between pre-ovulatory feed intake, progesterone priming and the subsequent in vitro development of ova collected from superovulated ewes. *Theriogenology* 43, 276.

Molle, G., Branca, A., Ligios, S., Sitzia, M., Casu, S., Landau, S. and Zoref, Z. 1995. Effect of grazing background and flushing supplementation on reproductive performance in Sarda ewes. *Small Ruminant Research*. 17, 245–254.

Molle, G., Landau, S., Branca, A., Sitzia, M., Fois, N., Ligios, S. and Casu, S. 1997. Flushing with soyabean meal can improve reproductive performances in lactating Sarda ewes on a mature pasture. *Small Ruminant Research* 24, 157–165.

Naqvi , S.M.K., Palta , P. , Joshi , A., Gulyani , R. , Paul , V. , Maurya ,V.P. and Manik , R.S. 2003. Increase in ovulation rate after immunization of Malpura ewes against a synthetic peptide sequence of the á-subunit of bovine inhibin. *Reproduction, Fertility and Development* 16, 226-227.

Naqvi, S.M.K., Gulyani, R., Joshi, A., Maurya, V.P., Kumar, D., Mittal, J.P. and Singh, V.K. 2006a. Production of 24 transferable embryos in one flush by a Garole microsheep. *Veterinary Record* 158, 414- 415.

Naqvi, S.M.K. and Kalra, D.B. 1990. Estrus synchronization and superovulation in crossbred ewes in semi-arid tract of Rajasthan. *Cherion* 19, 13-16.

Naqvi, S.M.K., Gulyani, R. and Mittal, J.P.1998d. Estrus synchronization response in Kheri ewes treated with prostaglandinF_2á. *Indian Journal of Animal Science* 68, 564-565.

Naqvi, S.M.K., Gulyani, R. and Pareek S.R.1996a. Estrus synchronization with Progesterone impregnated vaginal sponges during summer season. *Indian Journal of Animal Reproduction* 17, 15 -16.

Naqvi, S.M.K., Gulyani, R., Joshi, A., Das, G.K and Mittal, J.P. 2002a. Effect of dietary regimens on ovarian response and embryo production of sheep in tropics. *Small Ruminant Research* 46, 167- 171.

Naqvi, S.M.K., Joshi, A., Das, G.K and Mittal, J.P. 2001. Review. development and application of ovine reproductive technologies: an Indian experience. *Small Ruminant Research* 39, 199- 208.

Naqvi, S.M.K., Joshi, A., Gulyani, R., Kumar, D., Kolte, A.P., Kumar, S., Maurya V.P., Saha, S. Mittal, J.P. and Singh V.K. 2006b. Production of prolific microsheep by embryo transfer into large non- prolific sheep. Veterinary Record 159: 522- 526.

Naqvi, S.M.K., Mathur ,A.K. and Kalra, D.B. 1996b. Estrus synchronization using prostaglandin F_2 alpha with and without progestagen in ewes treated for superovulation. *International Journal of Animal Science* 11, 139-141.

Naqvi, S.M.K., Maurya, V.P., Joshi, A., Sharma, R.C. and Mittal J.P., 2002b. Production of crossbred lambs through artificial insemination of non- prolific medium size Malpura and Avikalin ewes using fresh diluted semen of prolific micro size Garole rams. *Asian-Australasian Journal of Animal Science* 15, 633- 636.

Naqvi, S.M.K., Palta, P., Joshi, A., Gulyani, R., Paul, V., Maurya, V.P. and Manik, R.S. 2004. Increase in ovulation rate after immunization of Malpura ewes against a synthetic peptide sequence of the á-subunit of bovine inhibin. *Reproduction Fertility and Development* 17, 226- 227.

Nicholas, F.W. 1996. Genetic improvement through reproductive technology. *Animal Reproduction Science* 42, 205- 214.

Oldham, C.M. and Lindsay, D.R. 1984. The minimum period of intake of lupin grain required by ewes in dry to increase their ovulation rate when grazing dry summer pasture. In: Lindsay, D.R. and Pearce, D.T. (Eds.). Reproduction in sheep. Australian Academy of Science, Canberra pp. 274-276.

Parr R.A., Davis I.F., Miles M.A. and Squires T.J. 1993.Feed intake affects metabolic clearance rate of progesterone in sheep. *Research in Veterinary Science 55, 306-310.*

Parr, R.A., cummings, I.A., Clark, I.J. 1982. Effect of maternal nutrition and plasma progesterone concentration on survival and growyh of sheep embryo in early gestation. *Journal of Agricultural Science* (Camb.) 98, 39-46.

Prasad, S.P. 1982. Optimum season of breeding for sheep production in arid climate. Final report of ad-hoc scheme submitted to ICAR, New Delhi.

Rhind, S.M., 1992. Nutrition: its effects on reproductive performance and its hormonal control in female sheep and goats. In: Speedy, A.W. (Ed.), Progress in Sheep and Goat Research. CAB International, Oxford, pp. 25-51.

Rhind, S.M., Robinson, J.J. and McDinald, I. 1980. Relationship among uterine and placental mfactors in prolific ewes and their relevance to variation in foetal weight. *Animal Production* 30, 115-124.

Robinson, J. J. 1996. Nutrition and reproduction. *Animal Reproduction Science 42,25-34.*

Scaramuzzi, R.J., Campbell, B.K., Downning ,J.A. Kendall, N.R., Khalid, M., Munoz-Gutierrez,M. and Somchit, A. 2006. A review of the effect of supplementary nutrition in the ewe on the concentration of reproductive and metabolic hormone and the mechanism that regulate folliculogenesis and ovulation rate. *Reproduction Nutrition Development 46, 1-16.*

Scaramuzzi,R.J., Campbell, B.K., Dowling,J.A., Kendall, N.R., Khalid, M. and Munoz-Gutierrez, M. 2006. A review of the effect of supplementary nutrition in the ewe,on the concentration of reproductive and metabolic hormones and the mechanism that regulate folliculogenesis and ovulation rate. *Reproduction Nutrition Development* 46, 339-354.

Shimasaki, S., Moore, R.K., Otsuka, F and Erickson, G. F. 2004. The bone morphogenetic protein system in mammalian reproduction. *Endocrine Review* 25, 72- 101.

Somachitac, A., Campbell, B.K., Khalid, M., Kendall, N.R. and Scaramuzzi, R.J. 2007. The effect of short-term nutritional supplementation of ewes with lupin grain (Lupinus luteus), during the luteal phase of estrous cycle on the number of ovarian follicles and the concentration of hormones and glucose in plasma and follicular fluid. *Theriogenology* 68, 1037-1046.

Thibier, M. and Guerin, B. 2000. Hygienic aspects of storage and use of semen for artificial insemination. *Animal Reproduction Science* 62,233-251.

Wade, G.N. and Jones, J.E. 2005. Neuroendocrinology of nutritional infertility. *American Journal of Physiology Regulation Integration and Comparative Physiology* 287, R1277-R1296.

❑❑❑

22

Nutrition Reproduction Interaction in Farmer's Sheep Flock

DAVENDRA KUMAR, V.P. MAURYA, SAJJAN SINGH
Central Sheep and Wool Research Institute, Avikanagar - 304 501, Rajasthan

Introduction

Nutrition and reproduction are interrelated phenomena in the evolutionary shaping of living creatures; if nutrition is compromised, reproductive capacity is diminished resulting in lowered commodity output. A complete reproductive cycle of ovulation, conception, pregnancy and lactation is one of the most energetically expensive activities that a female mammal can undertake. In nature breeding patterns have been evolved to ensure that abundant food supply coincides with late pregnancy and lactation. In semi-arid environments the pattern of food supply varies greatly from year to year due to unpredictable nature of the rainfall. Majority of small ruminants in India are range managed, exposed to wide seasonal variation in feed resources and environmental conditions. Many research reports have clearly demonstrated that intake of energy, protein, vitamins and minerals are the most important nutritional factors affecting reproduction. This is particularly evident for sheep in the tropics where feed is often inadequate due to which most of the sheep cannot build sufficient body reserves (particularly energy and protein). As a direct consequence of feed inadequacy, sheep of most smallholder resource-poor farmers are often in poor body condition (emaciated). Depleted body condition during period of energy deficiency reduces heat tolerance (Naqvi, 1987 and Hooda and Naqvi, 1990). The average reproductive efficiency of sheep in farmers flock range from 75-80% with 5-8% twinning (Singh et al., 2005). In farm-bred sheep maintained under grazing on protected rangeland and concentrate supplementation in critical physiological stages, the average reproductive efficiency ranged from 80-85 % in different genotypes (CSWRI, 1998). During summer the pasture vegetation has as low as 5% CP while after monsoon season the lush pasture vegetation has 10-12 % CP. It is also a fact that even during summer the animals maintain the CP content of the diet at constant 10-12 % by intensive selection on plant parts rich in protein. It is evident that the quality and quantity of protein is not limiting in range managed small ruminant, as the animal consuming 500 g DM/ day are able to meet their protein requirement. Hence, the energy availability is the major limiting factor for reproduction losses of animals under semi arid and arid environment.

In ewes, reproduction is controlled by the hypothalamus area of the brain, the pituitary gland and the ovaries. As the amount of daylight decreases over time, this is sensed by the hypothalamus, which in turn signals the pituitary gland via gonadotrophin releasing hormone. The pituitary then stimulates the ovaries by releasing follicle stimulating hormone and luteinizing hormone. Then, the ovary secretes estrogen and a heat cycle results. This photoperiodic effect is more pronounced in temperate climate. Once sheep cycle, they will continue to come into estrus every 15 to 17 days until successfully bred or until they reach the anestrous period. Gestation in ewes lasts for 144 to 152 days, depending on breed and season of lambing. The nutritional demand for ovulation, fertilization and early pregnancy are small and easily met whereas the nutritional demand increases sizably during late pregnancy more so in early lactation. However, the stages of reproductive cycle, particularly the process of ovulation and conception will have serious implication under low plane of nutrition. The consequences of low plane of nutrition in drought conditions could be total shut down of reproductive process in the form of anovulation and anoestrus in ewes and failure to manifest estrus in young animals (Maurya *et al.,* 2004). Nutrition influences reproduction directly by supplying needed nutrients for the processes of oocyte and spermatozoa development, ovulation, fertilization, embryo survival and the establishment of pregnancy and indirectly through its impact on the circulating concentrations of the hormones and other nutrient-sensitive metabolites that are required for the success of these processes. Although there are species differences in the particular physiological and behavioural mechanisms mediating nutrition-reproduction interactions, two findings are consistent: (1) Reproductive physiology and behaviours are sensitive to the availability of oxidizable metabolic fuels and (2) When reproductive attempts are made, ovarian hormones play a major role in the changes in ingestion, partitioning, and utilization of metabolic fuels (Wade and Schneider, 1992).

Effect of Nutrition on Puberty

Puberty, the age at first ovulation, occurs approximately at 7 to 9 month of age in sheep. Besides genetic and environmental factors, plane of nutrition is the main factor, which influence puberty. Adequate nutrition is necessary for the proper pubertal development. High plane of nutrition advances the puberty considerably. Undernutrition is known to delay the onset of puberty. Restricted feeding during early life alters the rate of growth of lambs which, delay the onset of puberty (Dyrmundsson, 1987 and Foster *et al.,* 1988). Inadequate nutrition may influence pituitary function i.e. pattern of synthesis and/or release of gonadotrophins and response of target organs to either gonadotrophins or gonadal hormones (Lamming, 1969). Tyrosine may function as a nutritional signal influencing the central nervous system control on GnRH release (Hall *et al.,* 1992).

Effect of Nutrition on Fertility

The practice of flushing, short period of nutritional improvement, is widely used by the farmers to increase fertility (Boland *et al.,* 2001). Sheep farmers of Western part of the country supplement freshly harvested ripe Babool pods to their sheep with the belief that it acts as a flushing agent. Sehgal and Vijay Gopal (1991) also observed that nutrient intake and body weight gain improved on supplementing babool pods to ewes maintained on basal Cenchrus hay diet. In another study evaluating the link between flushing in form of concentrate supplementation @ 1.5 % of BW one month prior to breeding season on reproductive efficiency of ewes it was observed that the conception rate increased from 66.7 % in control to 79.2 % in supplemented group (Chaturvedi et al., 2006). Whereas Redmer *et al.,* 2005 have reported that placental growth and pregnancy outcome are severely compromised in adolescent ewes over nourished to promote rapid maternal growth. In the adolescent, maternal over nutrition has a negative impact on midgestation placental angiogenic factor/ receptor expression. This may impact placental vascularity and explain why uteroplacental mass, blood flow, and nutrient uptake are compromised in late pregnancy, resulting in low-birth-weight offspring. Proper feeding, mineral supplementation, Vitadae injection induced estrus in 45.16% (14/31) post partum anestrus Malpura ewes under farmer's field (Kumar *et al.,* 2007).

The critical role of trace elements/ micronutrients in the fertility of ewes has shown that inadequacies impair conception by reduced ova implantation, early embryonic loss and foetal death. They play an important role in the metabolic processes of sheep, as they are involved in the action of several enzymes and co-enzymes. Thus, their deficiency, excess or imbalance often leads to sub-optimal growth and fertility (Smith and Akinbamijo, 2000). Among micronutrients folic acid, selenium (Se), zinc (Zn), iron (Fe), molybdenum (Mo), cobalt (Co), copper (Cu) and the antioxidant vitamins A and E are of particular importance during reproductive processes. Se is the cofactor of glutathione peroxidase enzyme system responsible for regulation of extra and intra cellular hydroperoxidase (Burk and Hill, 1993). Zn is essential in the production of many of the sex hormones including testosterone and gonadotrophin releasing hormone (Hambidge et al., 1986) and carbohydrate and protein metabolism, hence affects reproduction (Masters and Fels, 1980 and Masters and Moir, 1983). Oral supplementation of 5 to 25 mg Se as sodium selenite or as a selenium/ ion powdered metal bolus 1 month before lambing to ewes with a low blood Se concentration (0.026 µl/ ml) increased lambing percentage (Godwin et al., 1970) and lambing and weaning weights (D'yachenko and Levinski, 1991), while Cu/ Se supplementation increased proportion of twin born (Hill et al., 1970). Dietary Co supplementation one month before mating drops lamb loss due to abortion and probable reabsorption between mid- pregnancy and birth and causes less neonatal death (Fisher and Mac Pherson, 1991). Supplementation of 100 ppm Zn in addition to a normal 23 – 25 ppm Zn diet one month before mating to lambing, increased feed intake of ewes by 15 %, and ewes had a higher fertility rate, were more prolific and produced heavier lambs at birth and weaning (Ali et al., 1998). Zn deficient diets in ewes may cause prolonged labour, pregnancy toxaemia and retained placenta (Apgar *et al.*, 1993). Oral supplementation additional 45 mg Mn and 20 mg Zn per day to the ewes consuming herbage of marginal Mn and Zn (Mn 45 mg and Zn 20 mg/ kg DM) status increased lambing percentage. A multi- trace element bolus containing 5.3 g Cu, 50 mg Se, 90 mg Co, 100 mg I, 3.2 mg Mn and 4.7 g Zn with daily release rate of 15 mg Cu, 0.1 mg Se. 0.2 mg Co, 0.3 mg I, 9.4 mg Mn and 13 mg Zn increased proportion of twin born and higher pregnancy rates (Hemingway *et al.*, 2001).

Consumption of some pasture legumes also affects fertility in sheep. These legumes contain chemicals called phyto-oestrogens, which can affect the reproductive system of sheep. Many subterranean clover varieties and other clovers, such as red clover, contain high levels of the phytooestrogen formononetin. In the long-term, ewes grazing green pastures with high formononetin levels suffer reduced fertility, while short-term effects can include a prolapsed uterus, difficult lambing and a declining lambing percentage. While the introduction of new low-oestrogen pasture varieties and more effective flock management have reduced the incidence of sheep infertility, continued monitoring for signs of clover infertility is still required. Medics and lucerne can contain high levels of the phytooestrogen coumestrol, particularly in response to disease stress. In areas where subterranean clover is oestrogenic it would be wise to buy rams from a breeder (in a similar area) with high levels of flock fertility. These rams should produce progeny with resistance to the development of clover infertility.

Effect of Nutrition on Ovarian Function

Regulation of ovarian activity is an integrated process encompassing both extraovarian signals and intrafollicular factors (Webb et al., 2004). Initiation of primordial follicle growth and the early stages of folliculogenesis can occur without gonadotropins, but FSH may affect the rate of preantral follicle growth. Antral follicle development from 1 to 4 mm in sheep is completely gonadotropin dependent. These recruited follicles express a range of mRNA encoding steroidogenic enzymes, gonadotropin receptors, and local regulatory factors and their receptors. As follicles continue to mature, there is a transfer of dependency from FSH to LH, which may be part of the mechanism involved in selection of follicles for continued growth. Locally produced growth factors, such as the IGF and members of the transforming growth factor-ß superfamily, work in concert with gonadotropins throughout the follicular growth continuum and can have significant effects on follicle selection. Changes in nutrition have an effect on ovarian activity without significant variation in circulating gonadotropin concentrations and can be correlated with changes in circulating concentrations of metabolic hormones, including insulin, IGF-I, GH, and leptin. Ewes fed at a level of twice the maintainance requirement a week prior to scheduled ovulatory period showed increased concentrations of glucose, insulin and leptin in the blood reflecting increased number of follicles growing in ranging in size of 2-3 mm (Vinoles *et al.*, 2005). Nutritional stimulation increase blood concentrations of insulin, leptin and FSH, decreases growth hormone and oestadiol while IGF-I appears unaffected (Rex *et al.*, 2006). Nutrition can also affect the expression of mRNA encoding components of the ovarian IGF system to regulate the sensitivity/ response of follicles toward gonadotropins. Hence, the roles of growth factors in follicular development and survival depend on gonadotropin status and differentiation state of the follicle. It is the integration of these extra ovarian signals and intrafollicular factors that determine whether a follicle will continue to develop or be diverted into atretic pathways. The stimulation of these intra-follicular systems by improved nutrition leads to a suppression in follicular oestradiol production. The consequence of these direct actions on the follicle is a reduced negative feedback to the hypothalamic-pituitary system and increased FSH secretion that leads to a stimulation of folliculogenesis. Allison (1977) reported greater number of follicle of > 2 mm diameter and larger mean follicle diameters in ewes of high live weight compared with ewes of low live weight. Higher population of large, estrogenic and potentially ovulatory follicles have been reported in high condition animals compared to low condition animals (Mc Neilly *et al.*, 1987)

The ability of nutrition to alter ovulation rate of ewes is well known. A rapid improvement in body condition is usually associated with greater number of large sized follicles (Abecia et al., 2006), an increased ovulation rate (Naqvi et al., 2002) and litter size. 2.5-3.0 % increase in ovulation rate for each 1.0 Kg increase in live weight of ewe has been reported (Cumming, 1977). Alterations in ovulation rate may be related to cell-entry-rate of glucose in animals on a high plane of nutrition. Dietary supplements containing high energy and protein have been shown to increase ovulation rate in ewes (Downing *et al.*, 1995a). Similarly, increase in ovulation rate was reported when glucose was infused directly (Downing et al., 1995b). Thus it is likely that short term energy supply is directly involved in follicle recruitment and perhaps also in follicle growth. However, this effect may be of short duration when diet level is altered. Infusion of branched chain amino acids over a five-day period in the late stage of estrous cycle increases ovulation rate in sheep (Downing et al., 1990). Several report indicate that feeding fat altered the growth pattern of follicles and this effect is somewhat independent from energy (Mattos et al., 2000). Recent observation by Mitchell et al. (2004) suggest that Co/Vitamin B_{12} status of donor ewes for carrying out multiple ovulation and embryo transfer programmes, is important as it affected ovulation rate. In this study ewes with mean circulating B_{12} concentrations of 1483 and 182 pmol/l had mean ovulation rates of 14.4±1.28 and 9.9±1.55, respectively, following super ovulation. Dietary restriction has been shown to alter follicle growth characteristics in superovulated sheep (Yaakub et al., 1997). Under feeding (0.5M) than feeding at 1.5 of the maintenance requirement resulted in increased rate of ova wastage (Rhind et al., 1989). Under nutrition, which results in lower ovulation rate is associated with decreased LH pulse frequency, which is likely due to inadequate hypothalamic GnRH secretions. However, McEvoy et al. (1995) reported that low energy diet produced high proportion of viable ova from ewes as compared to high energy diet.

Effect of Nutrition on Embryo Quality, Development and Survival

Studies have been carried out to determine the relationship between level of feed intake during early pregnancy, concentrations of peripheral progesterone and embryo survival in sheep. Feed intake in sheep can influence the concentration of progesterone with a strong negative correlation between dietary intake and progesterone concentration (McEvoy *et al.*, 1995). This effect is likely due to an increase in the rate of catabolism of progesterone and in hepatic circulation at higher feeding levels (Parr *et al.*, 1987; 1993). Progesterone is believed to play an important role in oocyte maturation and in early embryo development. However, as steroids are selectively stored in fat, any dietary regime that results in fat mobilization will result in the release of stored progesterone. This may account for some of the increased progesterone evident in animals on low dietary intakes. Concentration of progesterone and embryonic interferone-*tau* have been reported to be positively correlated. Thus minor changes in maternal progesterone concentration during the initial period of embryo development may alter the secretion of this antiluteolytic agent and may be critical to embryo survival. In sheep overfeeding which reduced circulating progesterone concentration also reduced pregnancy rates and decreased both the rate of development and viability of embryos (Parr *et al.*, 1987). Peripheral concentrations of progesterone on day 0 and 1 relative to the LH peak are important for embryo survival in sheep. This presumambly modifies follicular maturation and oocyte quality. However, Abecia *et al.* (1997) have suggested that the effect of progesterone on embryo development in acting primarily through the effect of progesterone in uterus.

Although, the effect of extremes of dietary intake on embryo development is evident but the point at which this change occurs is still unknown. When ewes were infused with high levels of glucose and superovulated, yield of good quality embryos was reduced (Yaakub *et al.*, 1997). Glucose infusion has also been reported to reduce pregnancy rate. While the reason for such an effect of glucose on embryo development is not clear, it may be due to unusually high plasma glucose concentration interfering with cell signalling mechanisms during preovulatory follicle growth, oocyte development or early embryo development. It is known that high glucose concentration is deleterious to embryo development *in vitro* (Moley *et at*., 1996). Blastocyst formation is well recognised as a key developmental process in the growth and development of an embryo. The blastocoel cavity forms as a consequence of fluid transport across the trophectoderm. This process is partially facilitated by Na/K-ATPase; m-RNA for this enzyme has been identified in day 7 bovine embryos. Dietary intake and diet type can alter the expression of transcripts of genes involved in early embryo development, such as Na/K-ATPase and Cu/Zn SOD (Wrenzycki *et al*., 2000). Abecia *et al* (1997) reported a decrease in the *in vitro* secretion of interferon-*tau* in day 15 embryos from undernourished ewes. An increase in the *in vitro* secretion by endometerial tissue of PGF2μ was evident in the same amount. Thus, nutrient requirements for optimum follicle growth and embryo development may be quite different. This highlights the importance of diet around the time of mating and in particular the significance of extreme overfeeding or underfeeding post mating, in regulating pregnancy rate.

Undernutrition (Below 50% of maintenance) and over feeding (Above 150% of maintenance) provoke reduced rates of embryonic survival (Karim 2006). Micronutrients also play a key role during development and survival of embryos. The consequences of micronutrient imbalance on the developing conceptus may not be apparent at the time of the nutritional deficiency, but may be manifested later in development. However, supplementary micronutrients provided later in gestation or during postnatal life cannot completely reverse the detrimental effects caused by earlier micronutrient imbalance. Deficiency of a specific micronutrient, such as Zn, during pregnancy can result in a greater incidence of fetal malformation and resorption than general malnutrition. Micronutrients affect the number of developmental stages that are susceptible to micronutrient status and that leads even alterations in maternal and conceptus metabolism, as a consequence of their essential role in enzymes and transcription factors and through their involvement in signal transduction pathways that regulate development. Micronutrient-induced disturbances in the balance between the generation of free oxygen radicals and the production of antioxidants that scavenge free radicals may provide an additional mechanistic explanation. The detrimental effects of many micronutrient deficiencies, particularly Zn and Cu, can be alleviated by supplementary antioxidants, whereas deficiencies of antioxidant vitamins A and E are likely to reduce defence against free radical damage. In Se deficiency condition free radicals accumulate and disrupt the processes of postaglandins and steroid synthesis (Hemler and Lands, 1980; Staats *et al*., 1988) and the development of embryo (Alvarez and Storey, 1989; Goto et al., 1992). The response of giving injection of 10 mg Se and 136 IU Vitamin E before mating was interesting (Segerson and Ganapathy, 1981) as these ewes produced more fertile ova, more uterine contractions and more ewes with large number of sperm attached to the zona pellucida due to improved sperm motility.

Maternal and fetal circulating levels of both energy and protein have influence over fetal growth. Fetal glucose diminishes when maternal glucose is low and the fetus tries to make up the energy balance by it own glucogenic and gluconeogenic process (Lindsey *et al.*, 1993). The effect of protein levels on embryonic development indicate that either it has no effect (Abecia *et al.*, 1997) or that an excess of rumen degradable protein in ewe diet elevated plasma urea and ammonia levels in uterus with an associated increase in embryo mortality (Mc Evoy *et al.*, 1997). Fetal ovary is remarkably sensitive to maternal nutrition during its early development with lifetime effects on ovulation rate (Robinson *et al.*, 2006). Undernutrition affects the concentration of oogonia in the ovine ovary as early as day 47 of fetal life (Borwick *et al.*, 1994), postpone the arrest of ovarian metabolic activity on day 62 (Borwick *et al.*, 1995) and reduces subsequent litter size (Robinson, 1990).

It is important to differentiate between optimum conditions for follicle growth, (both in terms of number of follicles and paracrine environment), and optimum conditions for embryo survival. Nutritional conditions may not be similar for both. While the significance of increased glucose on increasing ovulation rate has been identified above, it is possible that increased glucose is also having a deleterious effect of embryo development. Preimplantation mouse embryo expressed to hyperglycemia have delayed development. While direct infusion of glucose to ewes can increase ovulation rate (Downing *et al.*, 1995b), embryo quality can be dramatically reduced (Yaakub *et al.*, 1997). It is not clear why high concentration of glucose have such deleterious effects, but it is possibly due to interactions between insulin, glucose and glucose transport proteins during early embryo development. Before blastocyst formation, metabolic reliance of embryos is on lactate and pyruvate rather than glucose. Indeed it is likely that pre-morula embryos can not metabolise glucose. Glucose transport was increased in mouse embryos in response to glucose deprivation, and both IGF-1 and insulin can stimulate glucose uptake (Pantaleon and Kaye, 1996). Moley et al. (1996) suggest that elevated concentrations of glucose may alter activity of the polyol pathway and kreb's cycle, resulting in retarted embryo development. These types of observations indicate the significance that changes in the glucose-insulin-IGF-1 axis may have an early embryo development.

Effect of Nutrition on Foetal Growth and Development

Despite the spectacular effect of nutrition on ovulation rate, normal nutritional fluctuations experienced in the field during the first 100 days of pregnancy have less effect on the survival of embryos than does nutritional change in late pregnancy. However, as discussed earlier, gross over nutrition during the first month of pregnancy is thought to cause losses due to a reduction in a hormone, progesterone, necessary for the maintenance of pregnancy. This counteracts any benefit gained from an increased ovulation rate due to high nutrition at mating. In cases of extreme under nutrition during the initial 100 days of pregnancy, embryo survival can also be affected. The effects of extreme under nutrition during this early stage of pregnancy are usually more severe in ewes experiencing their first gestation than in older ewes. During the final third of pregnancy from 100 days until term, the fetus grows rapidly. Nutrition has a dramatic effect on fetal growth, birth weight, ewe milk production and lamb survival following birth. This is more pronounced in ewes carrying multiple fetuses. During the early stages of pregnancy, the placenta grows much faster than the fetus; however, at about day 90, placental growth slows while fetal growth rate accelerates. Adequate ewe nutrition early in pregnancy ensures that the placenta can grow sufficiently to perform the role of supplying maternal nutrients and oxygen to the fetus. It is usual to allow ewes that are fat at mating (body condition score 3.5 to 4.0) to decline gradually in body condition to score 3.0 during the first two-thirds of pregnancy (14 weeks), as slow growth of the fetus during this period does not place excessive demands on the ewe. Caution must be observed, however, because rapid weight loss in ewes can trigger some fetal deaths. The grazing and feeding strategy should be planned in advance to ensure adequate nutrition for ewes during the final third of pregnancy. The cheapest way of providing this nutrition is by grazing green pastures. Mating, therefore, should be planned so the critical period of pregnancy coincides with a flush of feed in late winter and early spring. This is especially important in high-performance flocks where there is a large demand for protein and energy in multiple- bearing ewes. An added advantage of late summer/autumn mating and winter/spring lambing is an improvement in ewe fertility. If ewes, especially those bearing twins, do not receive adequate nutrition to maintain body condition at or above score 2.5 during the final stage of pregnancy, they are likely to succumb to pregnancy toxaemia or hypoglycaemia. Pregnancy toxaemia can be caused by poor nutrition, sudden stress such as yarding, or through depressed appetite by having ewes over fat. This can result in total loss of the ewe and fetuses. In addition, adequate nutrition and body condition of the ewes during the final third of pregnancy is essential to ensure; satisfactory birth weight of the lambs, better chance for survival of the lambs, better development of the mammary tissue and potential milk production of the ewe, early onset of lactation and supply of colostrum, satisfactory development of wool follicles in fetal skin and future fertility of ewe lambs.

Effect of Nutrition on Male Reproduction

In sexually mature rams, nutrition affects neuroendocrine system thereby controls testicular activity (Martin *et al.,* 1994), testicular growth and sperm production (Oldham *et al.,* 1978). High food intake stimulates GnRH-LH pulse frequency for only 10-20 days, but continues to promote testicular growth over several months in mature male sheep. Short- term dietary treatments lead to changes in concentrations of glucose, fatty acids, insulin and leptin in blood and cerebrospinal fluid. Amino acids affect GnRH-LH secretion directly in sheep. The addition of fatty acids to the diet also increases LH pulse frequency, but does not affect the concentrations of insulin or leptin in the CSF. It appears that acute responses to changes in nutrition involve a range of alternative pathways, possibly including interactions among insulin, leptin and energy substrates. Effects of long-term dietary treatments on testicular size are only partly dependant on the GnRH-LH system and so must also be depend on other unknown pathways (Blache et al., 2000). With the exception of severe undernutrition, the effects on spermatogenic function are not accompanied by similar changes in endocrine function of the testes, as measured by the production of testosterone or inhibin. The energetic components of the diet, rather than the protein content, seem to be responsible for affecting gonadotrophin secretion in rams. Using mature male sheep to study the reproductive endocrine responses to acute and chronic changes in diet, Blache et al., 2002 have shown that glucose does not appear to be involved directly, but fatty acids can stimulate GnRH-dependent pathways that initiate changes in testicular function. Their work also indicates that GnRH-independent (perhaps also neuroendocrine) mechanisms may be involved. In the brain, it seems likely that intracerebral insulin and orexin are important mediators of the GnRH response to nutritional changes, but it is unlikely that leptin plays a role, at least in mature animals. Where these substrates act and whether they are also involved in the gonadotrophin-independent pathways is not clear. In the gonads, the gametogenic tissue responds rapidly to changes in nutrition, but the endocrine compartments are less affected (Martin and Walkden-Brown, 1995).

The adequate nutrition and optimum supply of specific micronutrients during two months prior to breeding is generally advocated (Saunders, 2003). Micronutrient influence both quantity and quality of sperms. The Zn, Co, Se and Mo play important role in male reproduction. Zn deficiency causes reproductive failure in rams, because the testes are more sensitive to dietary Zn than the other body tissues and sub clinical Zn deficiency reduces testicular growth well before other more overt symptoms appear (Underwood and Somers, 1969; Martin and White, 1992). Zn deficiency also reduces testicular diameter during pubertal development in young lambs (Martin and White, 1992). The zinc dependent pathway participating in the testis formation may alter (Martin and White, 1989) as Zn is involved in cell differentiation and replication. Zn supplementations increase sperm production (Underwood and Somers, 1969). Beneficial effects of micronutrients (Zn, Co and Se) on sperm motility, percentage of live sperm and sperm membrane integrity in ram lambs have been observed by Kendall *et al.* (2001). An improved antioxidant status in the form of increased concentrations of glutathione peroxidase in seminal plasma was attributed to Se. Se deficiency affects the morphology and motility of spermatozoa (McKenzie *et al.*, 1998). In Se deficiency, there is an increase in free oxygen radicals, which in turn decrease fertility, as free oxygen radicals attack the sperm membranes and decrease their viability (Irvine, 1996). Increased Se status will increase the antioxidant glutathione peroxidase activity thus leading to an increase in male fertility. Supplementation of minerals using glass bolus that contained Zn (1.52%), Co (0.5 %) and Se (0.15%) increased both percent live sperm and percent motile sperms (Kendall *et al.*, 2000). A beneficial effect of Se supplementation (50 mg as barium selenate by s/c injection) to rams on the viability and motility of their semen has also been observed by Anderson *et al.* (1996). Interestingly the beneficial effect in terms of the proportion of normal sperm was more pronounced following freezing and thawing than in fresh semen. Kendall *et al.* (2000) observed an optimum seminal plasma Zn (41 μm per ml) concentration and glutathione peroxidase activity (2.32 U per ml) for better fertility in male. Zn supplementations reduce the proportion of abnormal spermatozoa (Underwood and Somers, 1969). Zn is also important for the attachment of head to tail in spermatozoa and also required for the production of an antibacterial compound released from the prostate gland into semen (Abdel-Rahman *et al.*, 2000).

Nutritional Management for Optimizing Reproduction

Three stages of the production cycle can be managed nutritionally by strategic feeding to optimize reproduction. These periods are; pre-breeding, late gestation and lactation. Although, strategic use of nutritional supplements has long been an important management tool in production systems but, in the light of recent research, we can extend the list of points of focus: a) boost sperm production before mating; b) maximize ovulation rate; c) avoid early embryo loss; d) program the future productivity of the developing fetus. If these periods of supplementation are focussed with respect to both composition and duration, they can be cost effective for extensive management system mostly practiced by farmers in semiarid environments.

Boost sperm production

It should now be common practice to feed supplements to males for 8 weeks before mating to ensure that their testicular size and sperm production are maximal. Unfortunately, it also seems to be common practice to overfeed rams in the interests of maximizing fertility and body size, yet there is little doubt that this, as well as a lack of physical exercise, is detrimental to their performance.

Maximize ovulation rate

An advantage of small ruminants is that we can increase their ovulation rate by manipulating their diet. For example, we can feed lupin grain to ewes for as little as 4 days in the final stages of the estrous cycle to induce a 20-30 per cent increase in the frequency of twin ovulation.

Avoid early embryo loss

In sheep embryo loss varies from 7 to 46 per cent. Severe undernutrition is one of the many factors suggested as a cause but overfeeding in the first few weeks after fertilization can cause problems. This is apparently due to an increase in the clearance of progesterone and thus a degradation of the uterine environment. Exogenous progesterone at this time can also improve fetal growth. Therefore, supplements that are used to increase ovulation rate may need to be stopped precisely within a few days after ovulation so that they do not intrude into the period of early embryo development.

Program the future productivity of the developing fetus

In extensive management systems, ewes are often forgotten between mating and the final few weeks of pregnancy because it is generally considered that there are no serious demands on their reserves until the fetus begins to reach a significant size. However, this ignores the nutritional requirements for the development of the placenta as well as any critical periods during the development of the fetal organs between days 60 and 120 of pregnancy. The consequences of effects of maternal undernutrition on the fetus may not become evident until much later, after birth or even after sexual maturity, by which time the link between cause and effect will have been clouded. Three aspects of sheep production are affected by nutrition during pregnancy.

(i) The initiation and development of secondary follicles in the skin of ovine fetus is reduced by undernutrition of the mother during the final third of pregnancy; this effect is permanent so will decrease wool production and increase fibre diameter in later life.

(ii) Muscle fibres are formed during the fetal period and the process is complete by birth, so the inhibitory effects of undernutrition during gestation have long term consequences for growth and carcass quality.

(iii) The development of reproductive axis is affected with undernutrition reducing the number of sertoli cells in new born ram lamb, delaying the development of ovarian follicles in fetal ewes and reducing lifetime reproductive performance in ewes.

Feeding supplements for a short, strategically chosen period near the end of pregnancy can have profound effects on both the ewe and her lamb. Lupin and maize fed to ewes in the last week of pregnancy can improve the amount of colostrum available to the lamb even in ewes grazing high quality pasture. The extra colostrum at birth can translate into better survival of lambs and perhaps also into faster growth rates. The short duration of feeding needed to elicit the colostrum response means that there is little or no effect on the birth weight of the lambs, which, in theory, obviates concern about increasing the incidence of dystocia. It has long been acknowledged that overfeeding ewes in the final weeks of pregnancy may cause excessive fetal growth leading to dystocia and an increase in lamb mortality. The situation may be more complex than a simple incompatibility of fetal size and size of birth canal. Progesterone stored in adipose tissue of over fat ewes may disrupt the process of parturition and lead to hypoxia-induced damage to the central nervous system of their lambs. Whatever the mechanism, we need accuracy in the timing of the supplements at this point in the reproductive process, as we do in the period that covers ovulation and embryo survival. Research is also needed into the types of diet (duration, amount, content) and the timing of any supplements that are to be fed.

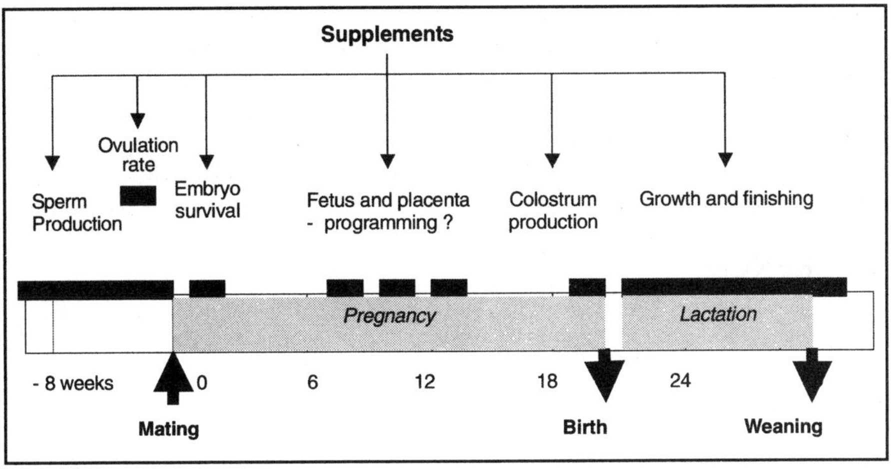

Fig. Periods during the reproductive process of small ruminants when supplementary feeding may affect the reproductive success of the flock (Martin *et al.*, 2004).

Conclusion and Recommendation

The interaction between nutrition and reproduction is a topic of increasing importance. Both over and under nutrition have profound and sometimes persistent effects on reproduction of male and female in sheep. Although the final manifestation of a detrimental effect of nutrition on fertility may be the death of the embryo, it is not totally clear if nutrition affects embryo quality through changing the developmental capacity of oocyte or through changes occurring during embryo development. Thus, the relationship between nutrition and reproduction is complex and responses are often quite variable and inconsistent. Three stages of the production cycle can be managed nutritionally by strategic feeding to optimize reproduction. These periods are; pre-breeding, late gestation and lactation. The pre-breeding period can positively influence production by flushing both the rams and ewes with increased energy. Increasing energy to ewes should increase body weight and number of eggs ovulated, thus, increasing lambing rate. Usually, flushing is accomplished by feeding grains high in energy. Nutrition during late gestation affects lamb size at birth, lamb vigor and survival, ewe mammary system development and ewe body fat reserves. All of these factors are crucial in increasing productivity. Proper nutrition during lactation affects the milking ability of ewes and subsequent production. Within the farming system, the aim should be to only use supplementation with nutrients where it is necessary; that is, where an actual or likely deficiency has been diagnosed. Further, the supplement used should be cost-effective and appropriate to the farming system in question. It is therefore perceived that improvement of sheep diets on smallholder farms will greatly improve reproductive performance and therefore flock productivity on smallholder farms. In the long run, this will also impact positively on the economic status of the household. There is need for intensified research efforts, farmer education, quality control of vitamin-mineral pre-mixes and development of community grazing land, in order to improve sheep nutrition, and, consequently, the reproductive performance and overall productivity of farm animals in developing tropical countries. Given a choice, the priorities for feeding should be for ewe flushing and during lambing (lactation). Early pregnancy, (beginning from as early as the second cycle to day 70) may be a period where feed could be restricted, to help build up pasture cover for lambing.

REFERENCES

Abecia, J. A., Lozano, J. M., Forcada, F. and Zarazaga, L. 1997. Effect of level of dietary energy and protein on embryo survival and progesterone production on day 8 of pregnancy in Rasa Argonesa ewes. *Animal Reproduction Science* 48, 209-218.

Abecia, J. A., Sosa, C., Forcada, F. and Meikle, A. 2006. The effect of under nutrition on establishment of pregnancy in ewes. *Reproduction Nutrition and Development* 46, 367-378.

Abdel- Rahman, H.A., Belely- El, M.S., Qarawi- Al, A.A. and Mougy- El, S.A. 2000. The relationship between semen quality and mineral composition of semen in various ram breeds. *Small Ruminant Research* 38, 45- 49.

Ali, H.A., Ezzo, O.H. and El-Ekhnawy, K.E. 1998. Effect of zinc supplementation on reproductive performance of Barki ewes under practical field conditions. *Veterinary Medicine Journal of Giza* 46, 77-87.

Allison, A.J. 1977. Effect of nutritionally induced live weight differences on the ovulation rates and the production of ovarian follicles in ewes. *Theriogenology* 8, 19-24.

Alvarez, J.G. and Storey, B.T. 1989. Role of glutathione peroxidase in protecting mammalian spermatozoa from loss of motility caused by spontaneous lipid peroxidation. *Gamete Research* 23, 77-90.

Anderson, J.M.L., Ap Dewi, I. and Axford, R.F.E. 1996. The effect of selenium supplementation on fresh and frozen ram semen. *Animal Science* 62, 672.

Apgar, J., Everett, G. A., Fitzgrald, J. A. 1993. Dietary zinc deprivation effect parturition and outcome of pregnancy in the ewes. *Nutrition Research* 13, 319- 330.

Blache, D., Chagas, L.M., Blackberry, M.A., Vercoe, P.E. and Martin, G.B. 2000. Metabolic factors affecting the reproductive axis in male sheep. *Journal of Reproduction and Fertility* 120, 1-11.

Blache, D., Adam, C. L. and Martin, G. B. 2002. The mature male sheep: a model to study the effects of nutrition on the reproductive axis. *Reproduction (Supplement)* 59, 219-33.

Boland, M.P., Lonegran, P. and Callaghan, D. 2001. Effect of nutrition on endocrine parameters, ovarian physiology and oocyte and embryo development. *Theriogenology* 55, 1325-1340.

Borwick, S.C. Rhind, S.M. and McMillen, S.R. 1995. Effect of under nutrition from the time of mating on the ovarian; development in foetal sheep at 62 d of gestation. *Journal of Reproduction of Fertility*, Abstract Series No. 15, p 52.

Borwick, S.C. Rhind, S.M. and McMillen, S.R. 1994. Ovarian steroidogenesis and development in fetal Scottish Blackface ewes, undernourished in utero from conception. *Journal of Reproduction of Fertility*, Abstract Series No. 14, p 14.

Burk, R.F., Hill, K.E., 1993. Regulation of selenoproteins. *Annual Review of Nutrition* 13, 65-81.

Chaturvedi, O.H., Bhatta, R., Verma, D.L. and Singh, N.P. 2006. Effect of flushing on nutrient utilization and reproductive performance of ewes grazing on community rangeland. *Asian-Australasian Journal of Animal Science* 19, 521-525.

CSWRI. 1998. 35 years of research, Central Sheep and Wool Research Institute, Avikanagar.

Cumming, I. A. 1977. Relationship in the sheep of ovulation rate with live weight, breed, season and plane of nutrition. *Australian Journal of Experimental Agriculture and Animal Husbandry* 17, 234-241.

Downing, J. A., Joss, J., Connell, P, and Scaramuzzi, R.J. 1995a. Ovulation rate and concentrations of gonadotrophin and metabolic hormones in ewes fed lupin grain. *Journal of Reproduction and Fertility* 103, 137-145.

Downing, J.A., Joss, J. and Scaramuzzi, R.J. 1995b. Ovulation rate and concentrations of gonadotrophin and metabolic hormones in ewes infused with glucose during the luteal phase of the estrous cycle. *Journal of Endocrinology* 146, 403-410.

Downing, J.A., Scaramuzzi, R.J. and Joss, J. 1990. Infusion of branched chain amino acids will increase ovulation rate in the ewes. *Proceedings Australian Society of Animal Production 18, 472.*

D'yachenko, L.S. and Levinski, V.A. 1991. Balancing trace elements in diet given to ewes. *Ovtsevodstvo* 5, 29- 30.

Dyrmundsson, O.R. 1987. Advancement of puberty in male and female. In: *New Techniques in sheep production.* Eds. I.F.M. Marai and J.B. Woven. Butterworths, London. Pp 65-76.

Fisher, G.E.J. and Mac Pherson, A. 1991. The effect of cobalt in pregnant ewes on reproductive performance and lamb viability. *Research of Veterinary Sciences* 50, 319- 327.

Foster, D. L., Ebling, E. B. J., Vannerson, L. A. and Veenvilet, B.A. 1988. Modulation of gonadotrophins secretion during development by nutrition and growth. Proceedings of the XXXV Easter School in Agricultural Sciences University of Nottinghom, London, Butterworths, pp 99-100.

Godwin, K.O., Kuchel, R.E., Buckley, R.A. 1970. The effect of selenium on infertility in ewes grazing improved pastures. *Australian Journal of Experimental Animal Husbandry* 10, 672- 678.

Goto,Y., Noda, Y., Narimoto, K., Umaoka, Y. and Mori, T. 1992. Oxidation stress on mouse embryonic development in vitro. *Free Radical Biology Research* 13, 47-53.

Hall, J. B., Schillo, K. K., Hileman, S. M. and Boling, J. A. 1992. Does tyrosine acts as a nutritional signal mediating the effect of increased feed intake on leutinizing hormone patterns in growth restricted lambs. *Biology of Reproduction* 46, 573-579.

Hambidge, K.M., Casey, C. E., Krebs, N. F. 1986. Zinc. In: Mertz, W. D. (Ed), Trace Elements in Human and Animal Nutrition. 2 Academic Press, Orlando, FL, pp. 1-138.

Hemingway, R.G., Parkins, J.J. and Ritchie, N.S. 2001. Enhanced reproductive performance of ewes given a sustained release multi trace- element/ vitamin ruminal bolus. *Small Ruminant Research* 39, 25- 30.

Hemler, M.E. and Lands, W.E.M. 1980. Evidence of peroxide-initiated free radical mechanism of prostaglandin biosynthesis. *Journal of Biology and Chemistry* 225, 6253-6261.

Hill, M.K., Walker, S. and Taylor, A.G. 1970. Effect of marginal deficiency of copper and selenium on growth and productivity in sheep. *Newzeland Journal of Animal Research* 12, 261- 270.

Hooda, O.K. and Naqvi, S.M.K. 1990. Effect of thermal load and feed restriction on relative adaptability of Malpura and Avikalin sheep in semi-arid region. *Indian Journal of Animal Sciences* 60, 608-611.

Irvin, D.S. 1996. Glutathione as a treatment for male fertility. *Review of Reproduction* 1, 6-12.

Karim, S.A. 2006. Nutrition and reproduction interactions in small ruminants. National Seminar on Innovations and Recent Advances in Reproduction for Augmenting Small ruminant Production, December 28-30, 2006, pp 161-167.

Kendall, N. R., Mackerzie, A. M. and Telfer, S. B. 2001. The effect of a copper, cobalt and selenium sulphate soluble glass bolus given to grazing sheep. *Livestock Production Science* 68, 31- 39.

Kendall, N.R., McMullen, S., Green, A. and Rodway, R.G. 2000. The effect of zinc, cobalt and selenium soluble glass bolus on trace element status and semen quality of ram lambs. *Animal Reproduction Science* 62, 277-283.

Kumar, D., Saha, S., Mann, J. . and Singh, V.K. 2007. Induction of estrus in ewes under field conditions of semi-arid region of Rajasthan. In: National Symposium on Animal Resource Development through Physiological, Nuclear, Genetic and Biotechnological Interventions, January 10-12. pp 132 (Abst No EB 2)

Lamming, G.E. 1969. Nutrition and reproduction. In: The Science of Nutrition of Farm Livestock. Ed. Cuthbertson Pt. I. Pergamon, London. Pp. 411-453.

Lindsey, D.R., Martin, G.B. and Williams, I.H. 1993. Nutrition and reproduction. In: World Animal Science Series. Reproduction in domasticated animals. Ed. G. J. King, Elsevier Science Publishers, Amsterdam.

Mattos, R., Staples, C.R. and Thatcher, W.W. 2000. Effects of dietary fatty acids on reproduction in ruminants. *Reviews of Reproduction* 5, 38-45.

Martin, G.B., Milton, J.T.B., Davidson, R.H., Hunzicker, G.E.B., Lindsay, D.R. and Blache, D. 2004. Natural methods for increasing reproductive efficiency in small ruminants. *Animal Reproduction Science* 82-82, 231-246.

Martin, G.B., Tjondronegoro, S. and Blackberry, M.A. 1994. Effects of nutrition on testicular size and the concentrations of gonadotrophins, testosterone and inhibin in plasma of mature male sheep. *Journal of Reproduction and Fertility* 101, 121-128.

Martin, G.B. and Walkden-Brown, S.W. 1995. Nutritional influences on reproduction in mature male sheep and goats. *Journal of Reproduction and Fertility (Supplement)* 49, 437-49.

Martin, G.B. and White, C.L. 1989. The effect of zinc deficiency on reproduction in the ram. *Journal of Trace Element Experimental Medicine* 2, 170 (Absr).

Martin, G.B. and White, C.L. 1992. Effect of dietary zinc on gonadotrophin secretion and testicular growth in young male sheep. *Journal of Reproduction and Fertility* 96, 497-507.

Masters, D.G. and Fels, H.E. 1980. Effect of zinc supplementation on the reproductive performance of grazing Merino ewes. *Biology of Trace Element Research* 2, 228-281.

Mc Evoy, T.G., Robinson, J.J., Aitken, R.P., Findley, P. A., Palmerm, J.J. and Robertson, I.S. 1995. Dietary induced suppression of pre-ovulatory progesterone concentration in superovulated ewes impair the subsequent in vivo and in vitro development of their ova. *Animal Reproduction science* 39, 89-107.

Mc Evoy, T.G., Robinson, J. J., Aitken, R.P., Findley, P.A. and Robertson, I.S. 1997. Dietary excess of urea influence the viability and metabolism of pre implantation sheep embryos and may affect fetal growth among survivors. *Animal Reproduction Science* 47, 71-90.

Masters, D. G. and Moir, R. J. 1983. Effect of zinc deficiency on pregnant ewes and developing fetus. *British Journal of Nutrition* 49, 365-372.

Maurya, V.P., Naqvi S.M.K. and Mittal, J.P. 2004. Effect of dietary energy level on physiological responses and reproductive performance of Malpura sheep in the hot semi-arid regions of India. *Small Ruminant Research* 55, 117-122.

McKenzie, R.L., Rafferty, T.S. and Beckett, G.J. 1998. Selenium: an essential element for immune function. *Immunology Today* 19, 342-345.

Mc Neilly, A.S., Jonassen, J.A. and Rhind, S.M. 1987. Reduced ovarian follicular development as a consequence of low body condition in ewes. *Acta Endocrinologica* 115, 57-83.

Mitchell, L.M., van der Ven, W., van Soest, W., Watt, R.G., Ashworth, C.J., Rooke, J.A. and McEvoy, T. G. 2004. Effect of cobalt/Vitamin B12 status on ovum recovery from superovulated ewes. *Reproduction* Abs. Ser. No. 31, p. 8.

Moley, K.H., Chi, M. M-Y., Manchester, J.K., McDougal Jr., D.B. and Lowry, O.H. 1996. Alterations of intraembryonic metabolites in preimplantation mouse embryos exposed to elevated concentrations of glucose: A metabolic explanation for the developmental retardation seen in preimplantation embryos from diabetic animals. *Biology of Reproduction* 54, 1209-1216.

Naqvi, S.M.K. 1987. Physiological adaptation of sheep during energy crisis and thermal stress. Ph.D. Thesis submitted to Kurukshetra University, Kurukshetra, India.

Naqvi, S.M.K., Gulyani, R., Joshi, A., Das, G.K. and Mittal, J.P. 2002. Effect of dietary regimens on ovarian response and embryo production of sheep in tropics. *Small Ruminant Research* 46, 167-171.

Oldham, C.M., Adams, N.R., Gherardi, P.B., Lindsay, D.R. and MacKintosh, J.B. 1978. The influence of level of feed intake on sperm-producing capacity of testicular tissue in the ram. *Australian Journal of Agricultural Research* 29, 173-179.

Pantaleon, M. and Kaye, P.L. 1996. IGF-1 and insulin regulate glucose transport in mouse blastocysts via IGF-1 receptor. *Molecular Reproduction and Development* 44, 71-76.

Parr, R.A., Davis, I.F., Fairclough, R.J. and Miles, M.A. 1987. Overfeeding during early pregnancy reduces peripheral progesterone concentration and pregnancy rate in sheep. *Journal of Reproduction and Fertility* 80, 317-320.

Parr, R.A., Davis, I.F., Miles, M.A. and Squires, T.J. 1993. Liver blood flow and metabolic clearance rate of progesterone in sheep. *Research in Veterinary Science* 55, 311-316.

Redmer, D.A., Aitken, R.P., Milne, J.S., Reynolds, L.P. and Wallace, J.M. 2005. Influence of Maternal Nutrition on Messenger RNA Expression of Placental Angiogenic Factors and Their Receptors at Midgestation in Adolescent Sheep. *Biology of Reproduction* 72, 1004–1009.

Rex J.S., Bruce K.C., Jeff A.D., Nigel R.K., Muhammad, K., Minerva, M.G. and Anongnart, S. 2006. A review of the effects of supplementary nutrition in the ewe on the concentrations of reproductive and metabolic hormones and the mechanisms that regulate folliculogenesis and ovulation rate. *Reproduction Nutrition Development* 46, 339-354.

Rhind, S.M., Mc Kelvey, W.A.C., Mc Millen, S., Gunn, R.G. and Elston, D.A. 1989. Effect of restricted food intake, before and after mating, on the reproductive performance of grey face ewes. *Animal Production* 48, 149-155.

Robinson, J.J. 1990. Nutrition in the reproduction of farm animals. *Nutrition Research review* 3, 253-276.

Robinson, J.J., Ashworth, C.J. Rooke, J.A. Mitchell, L.M. and McEvoy, T.G. 2006. Nutrition and fertility in ruminant livestock. *Animal Feed Science and Technology* 126, 259-276.

Saunders, P.T.K. 2003. Germ cell-somatic cell interactions during spermatogenesis. *Reproduction (Supplement)* 61, 91–101.

Segerson, E.C. and Ganapathy, S.N. 1981. Fertilization of ova in selenium/ vitamin-E treated ewes maintained on two plan of nutrition. *Journal of Animal Science 51,* 386- 394.

Sehgal, J.P. and Vijay Gopal. 1991. Utilization of babool pod in sheep ration. *Annual report*, CSWRI, Avikanagar, Rajasthan, pp. 38.

Singh, V.K., Bhagwan, P.S.K., Arora, A.L. and Karim, S.A. 2005. Terminal report, Jai Vigyan Project on Improvement in migratory sheep production program for tribal farmers of North West, CSWRI, Avikanagar.

Smith, O. B. and Akinbamijo, O. O. 2000. Micronutrients and reproduction in farm animals. *Animal Reproduction Science* 60-61, 549-560.

Staats, D.A., Lohr, D.P. and Colby, H.D. 1988. Effects of tocopherols depletion on the regional differences in adrenal microsomal lipid peroxidation and steroid metabolism. *Endocrinology* 123: 975-980.

Underwood, E.J. and Somers, M. 1969. Studies of zinc nutrition in sheep. 1. The relation of zinc to growth, testicular development and spermatogenesis in young rams. *Australian Journal of Agricultural Research* 20, 889-897.

Vinoles, C., Forsberg, M., Martin, G.B., Cajarville, C., Repetto, J. and Meikle, A. 2005. Short term nutritional supplementation of ewe in low body conditions affects follicle development due increase in glucose and metabolic hormones. *Reproduction* 129, 299-309.

Wade, G.N. and Schneider, J.E. 1992. Metabolic fuels and reproduction in female mammals. *Neuroscience Biobehaviour Review* 16, 235-272.

Webb, R., Garnsworthy, P.C., Gong, J.G. and Armstrong, D.G. 2004. Control of follicular growth: Local interactions and nutritional influences. *Journal of Animal Science* 82, E63-E74.

Wrenzycki, C., de Sousa, P., Overstrom, E.W., Duby, R. T., Herrmann, D., Watson, A J., Niemann, H., O'Callaghan, D. and Boland, M P. 2000. Variations of relative mRNA abundances and pyruvate metabolism in bovine embryos are associated with superovulated heifer diet type and quantity. *Journal of Reproduction and Fertility* 118, 69-78.

Yaakub, H., O;Callaghan, D., O'Doherty, J.V. and Boland, M.P. 1997. Effect of dietary intake on follicle numbers and oocyte morphology in unsuperovulated and superovulated ewes. *Theriogenology* 47, 182.

□□□

23

Preservation of Ram Semen Recent Trends

ANIL JOSHI, DAVENDRA KUMAR, S.M.K. NAQVI
Central Sheep & Wool Research Institute, Avikanagar- 304501, Rajasthan

Introduction

Semen preservation enables widespread use of superior rams for artificial insemination and *ex-situ* conservation of elite or endangered breeds for preserving genetic diversity. The two main methods of storage are either in the form of liquid state at low temperature for short-term preservation or in a frozen state at sub-zero temperature for long-term preservation of semen. The life span of spermatozoa is prolonged by preservation of semen either through reduction (short-term preservation) or arrest (long-term preservation) of their metabolic activity. The principal advantages of using liquid semen are that fertility is possible with low numbers of spermatozoa and the storage is economical. However, during short-term preservation of ram semen there is a gradual decline in motility, change in morphological integrity and a rapid decline in fertility beyond 24 h of storage (Maxwell and Salamon, 1993). The limitation of short-shelf life period of liquid semen after dilution has necessitated the development of a viable technology for cryopreservation of ram spermatozoa. Frozen storage has the advantage of being able to preserve important genes for future use, provide insurance against loss of a particular sire, permit the transport of semen over long distances and allow large numbers of females to be inseminated over extended periods of time or at different times of the year.

Historical Aspects of Research

Landmark breakthroughs in semen preservation first came into prominence with the discoveries of the protective agents in egg yolk for cooling (Phillips, 1939) and glycerol for freezing (Polge *et al.*, 1949). Freezing of ram semen was first performed by Emmens and Blackshaw (1950) with satisfactory sperm survival after thawing but the success was not encouraging in terms of fertility. The research on ram semen freezing got impetus later on by the work of Salamon and Maxwell done in Australia and Dauzier, Colas and Corteel in France (Salamon and Maxwell, 1995 a,b; Maxwell *et al.*, 1999). The consequences of sperm cryoinjury caused by the cryopreservation of ram spermatozoa are impaired transport and poor survival in the female reproductive tract. Concerted research efforts are going on worldwide to improve the methods of cryopreservation and insemination for obtaining good post-thaw recovery and fertility with frozen semen. The progress made on various aspects of ram semen freezing and AI such as diluent composition, processing, freezing and thawing methods, cryoinjury to sperm plasma membrane and fertility has been extensively reviewed elsewhere from time to time (Sadler, 1966; Visser, 1974; Graham *et al.*, 1978; Watson, 1979; Maxwell 1984; Evans 1988; Salamon and Maxwell 1995 a,b; Watson 1995; Maxwell and Watson 1996; Bailey *et al.*, 1999; Watson 2000; Salamon and Maxwell 2000; Naqvi *et al.*, 2001; Evans *et al.*, 2004; Gillan *et al.*, 2004).

Status of Research

Although ram semen has also been frozen in ampoules but the most common packages for its storage are pellets and polyvinyl chloride (PVC) straws. Ram semen can be successfully stored in pellets (Salamon *et al.* 1996) and straws (Salamon and Maxwell, 2000) for prolonged period of time with no detrimental effect on the subsequent survival of spermatozoa after thawing. Noteworthy results on cryopreservation of ram semen have obtained by workers from different countries (Fiser and Fairfull, 1989; Abdelhakeam *et al.* 1991; Mathur *et al.*, 1991; Sanchez-Partida *et al.*, 1992; Payne *et al.*, 1994; Gonzalez *et al.*, 1996; Valcarel et al., 1996; Soderquist *et al.*, 1997; Joshi *et al.*, 1998a; Aisen *et al.*, 2000; Gil *et al.*, 2000; Joshi *et al.*, 2001; Paulenz *et al.*, 2004). The mean percent post-thaw motility achieved among them ranged from 41 to 70%. The Australian in pellet freezing (Sanchez-Partida *et al.*, 1992) while Argentinean (Aisen *et al.*, 2000), Swedish (Gil *et al.*, 2000) and our group in straw freezing have reported more than 70% post-thaw motility of ram spermatozoa. The wide variation in this may be contributed by the adoption of different freezing protocol by each laboratory. Pellet freezing is advantageous because of easy handling and simple equipment required for it. However, there is no control on freezing rate and variability exists even between pellets of the same sample. Moreover, there is also a difficulty in maintaining identity of stored pellets, as they cannot be coded distinctly like straws. Programmable cell freezers have been designed for the controlled cooling and subsequent warming of biological specimens specially sperms and embryos. They provide variable cooling and heating rates within the temperature range of

+ 30 to –180 °C. The operation of the freezer relies upon the controlled influx of atomised liquid nitrogen into the work chamber by a microprocessor-based programmer with an associated temperature controller allowing pre- set linear cooling rates to be obtained. A resistance temperature probe situated within the work chamber detects temperature and the signal is fed to the controller and also to a digital display. A second temperature probe, also within the chamber, provides signal to the chart recorder for maintaining permanent record of the freezing cycle.

In order to overcome the disadvantage of proper identification of stored pellets and lack of repetition in freezing conditions, further studies were done in our laboratory by freezing ram semen in French plastic mini and medium straws. The protocol for cryopreservation of ram semen in straws is based on programmable freezing and computer-assisted sperm analysis (CASA) techniques (Joshi and Mathur, 1996; Mathur and Joshi, 1996; Joshi et al., 1998a; Bag et al., 1999; Joshi *et al.*, 2001; Bag *et al.*, 2002a, 2002b, 2004; Joshi *et al.*, 2005). The main advantage of using this protocol is complete automation of the freezing and evaluation processes. The results were consistent and reproducible giving an average post-thaw recovery of 70% motile spermatozoa using rams of exotic (Joshi *et al.*, 1998b), crossbred (Bag *et al.*, 1998; Joshi *et al.*, 1998a; Bag *et al.*, 2002a; 2004) or native breeds (Mathur and Joshi, 1996; Bag *et al.*, 1999; Joshi *et al.*, 2001; Bag *et al.*, 2002a, 2004; Joshi *et al.*, 2005). The high post-thaw recovery may be attributed to the criteria of processing only those ejaculates for cryopreservation which have thick consistency, rapid wave motion, 90% initial motility and >3000 x 10^6 spermatozoa per ml.

New concepts of Research

In a recent study it has been shown that ram sperm can be held at 5 degrees Celsius for up to 48 hours prior to freezing with no injurious effects on motility, membrane integrity, or fertilizing potential (Purdy, 2007). The purpose of this study was to determine the fertilizing capacity of ram semen collected at distant locations while being shipped to semen laboratory for cryopreservation. In our laboratory we have started programmable freezing from the ambient temperature itself at a slow rate simulating the slow cooling upto 5°C in a cold handling cabinet prior to the use of cell freezer. This step will allow the elimination of cold handling cabinet and large number of semen doses can be preserved under controlled rate cooling and freezing using a large capacity cell freezer. Moreover, the complete run of the freezing protocol in the cell freezer can make this reproducible at any time of the year and any research laboratory. For this step we have added one step in the existing protocol that allows the controlled rate of cooling upto 5°C. Our results indicated that cooling of ram spermatozoa under controlled conditions, compared to conventional cooling prior to programmable freezing, did not influenced the post-thaw survival but influenced the acrosomal integrity of frozen-thawed ram spermatozoa.

In Israel a novel freezing technology based on "multi- thermal gradient (MTG) directional solidification for freezing bulk quantity of semen has been developed and patented (Arav, 1999). The semen in the test tube is moved in a constant velocity through a linear temperature gradient so that the cooling rate and ice front propagation are precisely monitored. The MTG apparatus is designed to control ice crystal propagation by changing the thermal gradient or the liquid- ice interface velocity and so it can optimise the ice- crystal morphology during freezing of semen (Arav *et al.*, 2002a,b; Gacitua and Arav, 2005). Although acceptable post-thaw motility of ram spermatozoa has been reported with MTG protocol but it will be most applicable in future for banking frozen semen in large volumes. The use of large volume has advantages over that in small volumes in straws because it reduces wastage of semen during packaging and also allows insemination of large group of animals. However, freezing or thawing large volumes had problems associated with heat transfer that tend to cause varied cooling rates during cooling and recrystallization during warming. To overcome the problem of recrystallization during warming, the Israeli workers have also developed a new container (Hollow- Tube ™, IMT, Israel) that works by causing high temperature water to flow inside the tube causing uniform warming in spite of large volume (Gacitua and Arav, 2005).

Advances in Evaluation of Cryopreserved Ram Semen

Sperm motility is one of the most important characteristics for evaluating the fertilizing potential of ejaculated and frozen-thawed spermatozoa. Although traditional semen quality evaluation procedures based on visual assessment are simple but the subjective nature of the method leads to variability of results between observers. Sperm motility assessed microscopically is not well correlated with either *in vivo* or *in vitro* fertilization due to human and technical limitations of the method for assessing detailed sperm movement characteristics. The objective methods of evaluating sperm motility based on CASA technique give better prediction of fertility (Saacke, 1982) and the results are ideal for comparison within and between research laboratories (Dunphy *et al.*, 1989).

The three main factors those can provide erroneous CASA results are settings of the semen analyzer on the basis of sperm dimensions, sperm concentration and image digitization. CASA technique is accurate for objective measurements of sperm kinematics, if care is taken when preparing the semen samples and proper setting of the instrument. The results of CASA are inaccurate if the sperm count for analysis if below 20 X 10^6 sperm/ ml or above 50 X 10^6 sperm/ ml. Hence, it is necessary that all the samples should be diluted to approximately to the recommended range and the digitization rate along with the settings of the instrument are kept constant prior to CASA analysis for all the observations so as to ensure reliability and repeatability of the results. It is very important to identify the dilution medium that should not have any adverse effect on the sperm motion characteristics. Therefore, before interpretation of any CASA results it must be ensured that dilution medium does not influence any sperm motion attributes. Moreover, the time lapse between semen dilution and the CASA analysis should be short thereby enabling the spermatozoa to survive until completion of analysis. Apart from these three, there are several other factors, which can also decide on the accuracy of the CASA results. The role of all these factors is very important for meeting the operational standards and comparability of automated semen analyzers (Davis and Katz, 1992, 1993).

Validated CASA procedures are now being applied in most laboratories during freezing of ram spermatozoa. Apart from identifying motile and static spermatozoa CASA can also categorize spermatozoa on the basis of velocity of each motile sperm. It also measures the mean sperm velocity and related sperm track dimensions (Edward *et al.*, 1995; Holt and Palomo, 1996; Joshi *et al.*, 2003; Kumar *et al.*, 2007). Moses *et al.* (1995) by CASA determined that sperm kinematics parameters changed significantly after freezing and thawing of ram spermatozoa. Our studies based on CASA observations have confirmed these observations. It has been observed that prior to freezing the sperm motility is high and almost all spermatozoa are rapidly motile. After thawing even if the motility may be ≥ 70 % and majority of the frozen-thawed sperms are rapidly motile but there are significant number of medium and slow moving sperms. CASA can also provide valuable information of the effect of post- thaw incubation on sperm kinematics of ram spermatozoa (Bag et al., 2002a,b; Bag *et al.*, 2004; Joshi *et al.*, 2005). Computer automated sperm- head morphology (ASMA) can also be performed with the analyzer that is equipped with additional morphometry software. The morphology option provides morphometry measurements for total, normal and abnormal sperms, number and percent of total, normal, bent and coiled tail sperms and number and percent of total, distal and proximal cytoplasmic droplets (Gravance *et al.*, 1997; Rijsselaere *et al.*, 2004).

Factors Affecting Functional Integrity of Semen on Cryopreservation

The combinations of storage temperature, chemical composition of extender and the hygienic control are the key factors that affect the life span of spermatozoa (Yoshida 2000). Unlike some mammalian species, ram spermatozoa are susceptible to various stresses during freezing and thawing. Cryopreservation of ram semen also depends upon several interrelated factors (Fiser and Fairfull 1986, 1989; Schmehl *et al.*, 1986; Pontbriand *et al.* 1989; Joshi and Mathur, 1994; Joshi *et al.*, 1998a, 1998b, 2000; Bag *et al.*, 1998, 2002a, 2002b, 2004; Joshi *et al.*, 2005) but their adverse effects are manifested on thawing (Holt *et al.* 1992; Holt and North, 1994). The problem, still, is that even with the best preservation a technique to date post-thaw survival is restricted to about 50 % of sperm population (Watson 1995). The complex anatomy of ewe cervix (Halbert *et al.* 1990) is also a major factor affecting the fertility with frozen-thawed semen (Eppleston and Maxwell 1993, Naqvi et al., 1998, 2001). Despite thorough research, progress in the preservation of ram semen has reached a plateau. As evidence of this, a recent study by Salamon et al. (2004) showed that ram semen cryopreserved

35 years ago yielded fertility results after laparoscope-aided intrauterine AI that are within today's acceptable standards. Nevertheless, over the past 30 years there has been much knowledge gained on the functional behaviour of the spermatozoa after preservation.

Recent studies have revealed that preservation techniques do not merely suspend spermatozoa in time, but induce sublethal damage that affects their functional capacity making them behave very differently to fresh spermatozoa. Many studies aiming to improve the fertility of preserved spermatozoa have addressed the issue by increasing the number of motile spermatozoa inseminated (Gillian *et al.*, 2004). Spermatozoa reach to a state of suspended quiescence for an indefinite period on cryopreservation but can be resuscitated on thawing. Frozen-thawed ram spermatozoa are immediately capable of penetrating oocytes whereas fresh spermatozoa require pre-incubation (Garde *et al.*, 1993). Watson (1995) to suggest that the spermatozoa emerge from frozen storage in a state resembling that of capacitation having by-passed some of the normal development process. The observations of Gillan et al. (1997) that a high proportion of ram spermatozoa were capacitated after freezing and thawing compared to fresh spermatozoa confirmed the assumptions of Watson (1995). The premature capacitation as a consequence of freezing and thawing curtails the life span of frozen-thawed spermatozoa and they have a short time to achieve fertilization compared to fresh spermatozoa. There is also increased embryonic loss after insemination with preserved semen, which decreases the fertility. Furthermore, the advanced stage of development of embryos fertilized by spermatozoa that have bypassed normal capacitation process may result in abnormal embryo development (Gillan *et al.*, 1997). An attempt to alter the membrane state was made by Maxwell *et al.*, (1999) by post-thawing addition of seminal plasma to ram spermatozoa that resulted in less capacitation.

Irrespective of the type of diluent or method of storage used, damage occurs to a significant portion of spermatozoa resulting in a reduction in motility, a decline in viability and decreased fertility as a consequence of frozen storage. Moses et al. (1995) by computer-aided semen analysis (CASA) determined that sperm kinematics parameters changed significantly after freezing and thawing of ram spermatozoa. Our studies based on CASA confirmed these observations (Joshi *et al.*, 1998 a,b; Bag *et al.*, 2002 a,b; 2004; Joshi *et al.* 2005). It has been reported that there is a disruption to the axonemal elements of the tail (Watson, 1995) and this may not only reduce the number of motile spermatozoa but also affects the type of motility. The architecture of the mitochondria in the midpiece is altered during freezing and thawing resulting in a loss of structure and this damage has been shown to decrease the respiratory rate of spermatozoa (Windsor and White, 1995; Windsor, 1997). Using the fluorochrome rhodamine 123 (R123), which accumulates in the mitochondria of actively respiring spermatozoa, Gillan and Maxwell (1999) reported that the mitochondrial respiration of frozen-thawed spermatozoa decreases more rapidly than that of fresh spermatozoa when subject to thermal resistance test at 37°C. Also using R123, Windsor (1997) observed the effect of cryoinjury on mitochondrial respiration and demonstrated that mitochondrial respiration is important for ram spermatozoa to be able to successfully penetrate the cervix. Although at present it is not known how damage to mitochondria can be prevented, but it would be expected that methods that can avoid direct damage to spermatozoa or minimize the sublethal effects due to cryoinjury might be beneficial. There are reports that the proportion of double-stranded decreases after freezing and thawing (Gillian and Maxwell, 1999) and the surface area of the sperm head also decreases significantly possibly owing to the over condensation of the chromatin in the nucleus (Gravance *et al.*, 1998). Freezing and thawing disrupts the stability of ram sperm chromatin, suggesting that the reduced fertilizing efficiency of cryopreserved semen *in vivo* might partly be due to abnormal DNA structure (Peris *et al.*, 2004). Variation between individual animals in their ability to withstand the preservation process may be owing to the susceptibility of their DNA to the freeze-thawing induced damage to spermatozoa. This can be used to visualize differences in DNA sequences between individual rams classified as having poor, average and good post-thaw recovery of motile spermatozoa as assessed by CASA, plasma membrane integrity and acrosome integrity (Gillan *et al.*, 2004).

Advances in Separation and Preservation of Sorted Sperms

The development of sperm sexing technique has added new dimensions on semen processing for the associated uses of the semen (Evans *et al.*, 2004). It is well documented that the only marker in sperm that can be effectively used for the separation of X and Y- bearing chromosome is DNA and the method used worldwide is a patented technique known as the 'Beltsville Sperm Sexing Technology'. The technique involves staining of DNA in the sperm nucleus with a DNA- permeant fluorescent dye bizbenzimide (Hoechst 33342) followed by excitation by a UV laser. The sperms are passed under pressure through an orienting nozzle in single file and the degree of fluorescence is measured optically as the sperms are sorted in which X chromosome bearing sperm fluoresce more brightly than Y chromosome bearing sperms (Johnson et al., 1989; Johnson *et al.*, 1999). The complete sperm sorting process is not only time consuming but also factors like dilution, high pressure and mechanical stresses alter the functional integrity of sperms, while staining and UV exposure affect the integrity of sperm DNA (Garner, 2001; Evans *et al.*, 2004). The sorting technique requires two stages of extensive semen dilution depending on the species. The first involves a 20-1000 fold extension of the neat semen for staining and the second involves further final dilution of 200- 30000 folds after passage through the flow cytometer, when the sorted spermatozoa combine with the sheath fluid in the collection tube. The overall influence of these factors can have an adverse effect on the viability and function of sorted sperms (Siedel and Garner, 2002; Hollinshead *et al.*, 2003).

Freezing of sorted sperms require an additional step of reconcentration by centrifugation to enable packaging thereby imposing additional stress to the sperm. In spite of these steps samples of sexed sperm with high proportion of motile and acrosome intact sperm can be obtained from frozen ram sperm (O'Brien *et al.*, 2003). Standard freezing protocols have been modified for post-sorting cryopreservation of sperm (Evans *et al.*, 2004). For sorted ram sperm no difference was observed in post-thaw motility or pregnancy rate of inseminated ewes after freezing sorted sperm in either Zwitterion or Test-yolk-buffered freezing media (Hollinshead *et al.*, 2002). The ability of these samples to bring about conception *in vivo* has not been reported but lambs have been born after frozen semen has been thawed, sorted, refrozen, thawed and used following *in vitro* fertilization and transfer of resultant embryos to surrogate dams (Hollinshead *et al.*, 2002; Hollinshead *et al.*, 2004a, b). The high cost of the sperm flow cytometry sorter and skilled personnel presently restrict the widespread use of this technology in small ruminants. The central location of any such facility will require transport of semen in special media for which short-term liquid storage of semen before and after sorting holds promise. The future use of this technology lies in conjunction with MOET, AI and associated reproductive technologies those are expected to provide a large scope for more efficient use of sexed sperm for specialized research program (Maxwell *et al.*, 2004). High-purity sorting of frozen-thawed spermatozoa for immediate use or re-cryopreservation for later use would enhance the commercial and conservation applications of sperm sexing technology.

Conclusions and Recommendations

Cryopreservation adversely affects many sperm attributes, such as motility, respiratory activity, membrane state and DNA quality. Consequently, the viability is reduced leading to poor fertility. Therefore, it is pre-requisite to enhance the post-thaw survival and minimize the membrane damage to ram spermatozoa for achieving acceptable lambing rates by using frozen-thawed semen. CASA offers a tremendous scope for rapid and precise evaluation of sperm motion characteristics of ram spermatozoa frozen under controlled conditions in a programmable cell freezer. However, continuous research and development efforts are still required to provide reliable relationship and correlations between sperm motion and morphology parameters that can be used as an index to predict the fertilizing potential of frozen-thawed ram spermatozoa. By thorough evaluation of preserved spermatozoa it may be possible in future to identify molecular markers for resistance to the detrimental effects of preservation and reduced fertility after preservation. Further research efforts are therefore required to develop better freezing protocols that can minimize freeze-thaw induced adverse effects to ram spermatozoa and improve the fertility. Preservation and AI of sexed semen in sheep holds promise for future but that will require strategies to prevent destabilization of sperm membranes, continued improvement in the high- speed sorting equipments along with sorting efficiency and development of efficient procedures for handling and processing of flow cytometrically sorted spermatozoa.

REFERENCES

Abdelhakeam, A. A., Graham, E. F., Vazquez, I. A. and Chaloner, K. M. 1991. Studies on the absence of glycerol in unfrozen and frozen ram semen. *Cryobiology* 28, 43-49.

Aisen, E. G., Alvarez, H. L., Venturino, A. and Garde, J. J. 2000. Effect of trehalose and EDTA on cryoprotective action of ram semen diluents. *Theriogenology* 53, 1053-1061.

Arav, A. 1999. Device and methods for multigradient directional cooling and warming of biological samples. *US Patent 5 873 254.*

Arav, A., Yavin, S., Zeron, Y., Natan, D., Dekel, I and Gacitua, H. 2002 a. New trends in gametes cryopreservation. *Molecular and Cellular Endocrinology* 187, 77- 81.

Arav, A., Zeron, Y., Shturman, H., Zeron, Y. and Gacitua, H. 2002 b. Successful pregnancies in cows following double freezing on large volume of semen. *Reproduction Nutrition Development* 42, 1- 4.

Bag, S., Joshi, A., Mathur, A.K. and Mittal, J.P. 1998. Controlled freezing and automated evaluation of frozen-thawed ram spermatozoa packaged in 0.5 ml straws. *Indian Journal of Animal Reproduction* 19, 136-139.

Bag, S., Joshi, A., Mathur, A.K., Rawat, P.S. and Mittal, J.P. 1999. Effect of thawing temperature on motion characteristics of frozen-thawed ram spermatozoa. *Indian Journal of Animal Science* 69, 16-18.

Bag, S., Joshi, A., Rawat, P.S. and Mittal, J.P. 2002 a. Effect of initial freezing temperature on the semen characteristics of frozen-thawed ram spermatozoa in a semi-arid tropical environment. *Small Ruminant Research* 43, 23-29.

Bag, S., Joshi, A., Naqvi, S.M.K., Rawat, P.S. and Mittal, J.P. 2002 b. Effect of freezing temperature, at which straws were plunged into liquid nitrogen, on the post-thaw motility and acrosomal status of ram spermatozoa. *Animal Reproduction Science* 72, 175-183.

Bag, S., Joshi, A., Naqvi, S.M. and Mittal, J.P. 2004. Effect of post-thaw incubation on sperm kinematics and acrosomal integrity of ram spermatozoa cryopreserved in medium-sized French straws. *Theriogenology* 62, 415-424.

Bailey, J. L., Bilodeau, J. F. and Cormier, N. 1999. Semen cryopreservation in domestic animals: a damaging and capacitating phenomenon. *Journal of Andrology* 21, 1-7.

Davis, R.O and Katz, D. ., 1992. Standardization and comparability of CASA instruments. *Journal of Andrology* 13, 81- 85.

Davis, R.O and Katz, D.F., 1993. Operational standards for CASA instruments. *Journal of Andrology* 14, 385- 394.

Dunphy, B.C., Kay, R., Barratt, C.L.R and Cooke, I.D. 1989. Quality control during the conventional analysis of semen, an essential exercise. *Journal of Andrology* 10, 378- 385.

Edward, A.Y., Windsor, D.P., Purvis, I.W., Sanchez-Partida, L.G. and Maxwell, W.M.C. 1995. Distribution of variance associated with measurement of post- thaw function in ram sperm. *Reproduction Fertility and Development* 7,129- 134.

Eppleston, J. and Maxwell, W.M. C.1993. Recent attempts to improve the fertility of frozen ram semen inseminated into the cervix. *Wool Technology and Sheep Breeding* 41, 291-302.

Emmens, C.W. and Blackshaw, A.W. 1950. The low temperature storage of ram, bull and rabbit spermatozoa. *Australian Veterinary Journal* 26, 226-228.

Evans G. 1988. Current topics in artificial insemination of sheep. *Australian Journal of Biological Sciences* 41, 103-116.

Evans, G., Hollinshead, F.K. and Maxwell, W.M.C. 2004. Preservation and artificial insemination of sexed semen in sheep. *Reproduction Fertility and Development* 16, 455-464.

Fiser, P. S. and Fairfull, R. W. 1986. Combined effects of glycerol concentration, cooling velocity and osmolality of skim milk diluents on cryopreservation of ram spermatozoa. *Theriogenology* 25, 473-484.

Fiser, P.S. and Fairfull, R.W. 1989. The effect of glycerol-related osmotic changes on post-thaw motility and acrosomal integrity of ram spermatozoa. *Cryobiology* 26, 64-69.

Gacitua, H. and Arav, A. 2005. Successful pregnancies with directional freezing of large volume buck semen. *Theriogenology* 63, 931- 938.

Garde, J., Gutierrez, A., Artiga, C.G. and Vazquez, I. 1993. Influence of freezing process on *in vitro* capacitation of ram semen. *Theriogenology* 39, 225.

Garner, D.L. 2001. Sex-sorting mammalian sperm: concept to application in animals. *Journal of Andrology* 22, 519- 526.

Gil, J., Soderquist, L and Rodriguez-Martinez. 2000. Influence of centrifugation and different extenders on post-thaw sperm quality of ram semen. *Theriogenology* 54, 93-108.

Gillan, L., Evans, G. and Maxwell, W.M.C. 1997. Capacitation status and fertility of fresh and frozen-thawed ram spermatozoa. *Reproduction Fertility and Development* 9, 481-487.

Gillan, L. and Maxwell, W. M. C. 1999. The functional integrity and fate of cryopreserved semen in the female tract. *Journal of Reproduction and Fertility* (Suppl) 54, 271-283.

Gillan, L., Maxwell, W.M.C. and Evans, G. 2004. Preservation and evaluation of semen for artificial insemination. *Reproduction Fertility and Development* 16, 447-454.

Gonzalez, C.I.M., Oba, E., Bicudo, S.D. and Souza, M.I. . 1996. Cryopreservation of semen of ideal rams with egg yolk-milk extender. *13th International Congress on Animal Reproduction*, Sydney, Vol 3, p 2-9.

Graham, E.F., Crabo, B.G. and Pace, M.M. 1978. Current status of semen preservation in the ram, boar and stallion. *Journal of Animal Sciences* 47 (*Suppl*), 8-103.

Gravance, C.G., Vishwanath, R., Pitt, C., Garner, D.L and Casey, P.J. 1998. Effects of cryopreservation on bull sperm head morphometry. *Journal of Andrology* 19, 704-709.

Gravance, C.G., White, C., Robertson, K.R., Champion, Z.J. and Casey, P.J. 1997. The effects of cryopreservation on the morphometric dimensions of caprine sperm heads. *Animal Reproduction Science* 49, 37- 43.

Halbert, G.W., Dobson, H., Walton, J.S. and Buckrell, B.C. 1990. The structure of the cervical canal of ewe. *Theriogenology* 33, 977-992.

Hollinshead, F.K., Gillan, L., O'Brien, J.K., Evans, G. and Maxwell, W.M.C. 2003. *In vitro* and *in vivo* assessment of the functional capacity of flow cytometrically sorted ram spermatozoa after freezing and thawing. *Reproduction Fertility and Development* 15, 351- 359.

Hollinshead, F.K., O'Brien, J.K., Maxwell, W.M.C and Evans, G. 2002. Production of lambs of pre-determined sex after the insemination of ewes with low numbers of X- or Y-chromosome bearing spermatozoa. *Reproduction Fertility and Development* 14, 503- 508.

Hollinshead, F.K., Evans, G., Evans, K.M., Catt, S.L., Maxwell, W.M.C. and O'Brien, J.K. 2004 a. Birth of lambs of a pre-determined sex after in vitro production of embryos using frozen-thawed sex-sorted and re-frozen-thawed ram spermatozoa. *Reproduction* 127, 557- 568.

Hollinshead, F.K., O'Brien, J.K., Gillan, L., Meyers, M., Maxwell, W.M.C and Evans, G. 2004 b. Liquid storage of flow cytometrically sorted ram spermatozoa, *Theriogenology* 62, 587- 605.

Holt, W.V, Head, M.F. and North, R.D. 1992. Freeze-induced membrane damage in ram spermatozoa is manifested after thawing: Observation with experimental cryomicroscopy. *Biology of Reproduction* 46, 1086-1094.

Holt, W.V and North, R.D. 1994. Effects of temperature and restoration of osmotic equilibrium during thawing on the induction of plasma membrane damage in cryopreserved ram spermatozoa. *Biology of Reproduction* 51, 414-424.

Holt, W.V and Palomo, M.J. 1996. Optimization of a continuous real- time computerized semen analysis system for ram sperm motility assessment and evaluation of four methods of semen preparation. *Reproduction Fertility and Development* 8, 219- 230.

Johnson, L.A., Flook, J.P. and Hawk, H.W. 1989. Sex preselection in rabbits: live births from X and Y sperm separated by DNA and cell sorting. *Biology of Reproduction* 41, 199- 203.

Johnson, L.A., Welch, G.R. and Wim, R. 1999. The Beltsville sperm sexing technology: High- speed sperm sorting gives improved sperm output for *in vitro* fertilization and AI. *Journal of Animal Science* 82, 213- 220.

Joshi, A. and Mathur, A.K. 1991. Influence of freezing rate on post thaw motility and velocity of ram spermatozoa. In: Proceedings of the 7th International Symposium on Spermatology, Cairns, Australia, pp 9.3-9.4.

Joshi, A. and Mathur, A.K. 1996. Assessment of sperm kinematics of frozen-thawed ram spermatozoa. *Indian Journal of Animal Reproduction* 17, 42-44.

Joshi, A., Bag, S. and Mittal, J.P. 1998 a. Freezability of ram spermatozoa packaged in mini and medium size straws. In: Proceedings of 8th World Conference of Animal Production Seoul, Korea, OR 1-13, pp 258-259.

Joshi, A., Bag, S., Rawat, P.S. and Mittal, J.P. 1998 b. Influence of storage period on motion characteristics of frozen-thawed Awassi ram spermatozoa. *Indian Journal of Animal Science* 68, 1032-1033.

Joshi, A., Bag, S., Mathur, A.K., Rawat, P.S. and Mittal, J.P. 2000. Effect of osmolality of Test-yolk-glycerol extender on motion characteristics of frozen-thawed ram spermatozoa. In: Proceedings of the 14th International Congress of Animal Reproduction, Stockholm, Sweden, 2, 132.

Joshi, A., Bag, S., Naqvi, S.M.K., Sharma, R.C., Rawat, P.S. and Mittal, J.P. 2001. Effect of short-term and long-term preservation on motion characteristics of Garole ram spermatozoa: A prolific microsheep breed of India. *Asian-Australasian Journal of Animal Science* 14, 1527-1533.

Joshi, A., Bag, S., Naqvi, S.M.K., Sharma, R.C. and Mittal, J.P. 2005. Effect of post-thawing incubation on sperm motility and acrosomal integrity of cryopreserved Garole ram semen. *Small Ruminant Research* 56, 231- 238.

Joshi, A., Naqvi, S.M.K., Bag, S., Dang, A.K., Sharma, R.C., Rawat, P.S. and Mittal, J.P. 2003. Sperm motion characteristics of Garole rams raised for a prolonged period in a semi-arid tropical environment. *Tropical Animal Health and Production* 35, 249- 257.

Kumar, D., Joshi, A., Naqvi, S.M.K., Kumar, S., Mishra, A.K., Maurya, V.P., Arora, A.L., Mittal J.P. and Singh V. K. 2007. Sperm motion characteristics of Garole X Malpura sheep evolved in a semiarid tropical environment through introgression of *FecB* gene. *Animal Reproduction Science* 100, 51-60.

Mathur, A.K., Srivastava, R.S., Joshi, A. and Kalra, D.B. 1991. Cryopreservation of ram semen in Egg yolk lactose raffinose citrate glycerol extender. *International Journal of Animal Science* 6, 79-81.

Mathur, A.K. and Joshi, A. 1996. Freezability of ram spermatozoa in organic amine-based semen diluents. In: Proceedings of13[th] International Congress of Animal Reproduction, Sydney, Australia, June 30th to July 4th, P 2-10.

Maxwell, W.M.C. 1984. Current problems and future potential of artificial insemination programmes. In: *Reproduction in Sheep*. (Eds Lindsay D R and Pearce D T), Cambridge Press, Cambridge, pp 291-298.

Maxwell, W. M. C., Evans, G., Hollinshead, F.K., Bathgate, R., De Graaf, S.P., Eriksson, B. M., Gillan, L., Morton, K.M. and O,Brien, J.K. 2004. Integration of sperm sexing technology in to the ART toolbox. *Animal Reproduction Science* 82-83, 79-95.

Maxwell, W.M. ., Evans., Mortimer, S.T., Gillan, L., Gellatly, E.S. and McPhie, C.A. 1999. Normal fertility in ewes after cervical insemination with frozen-thawed spermatozoa supplemented with seminal plasma. *Reproduction Fertility and Development* 11, 123-126.

Maxwell, W.M.C. and Salamon, S. 1993. Liquid storage of ram semen: A review. *Reproduction Fertility and Devlopment* 5, 613-638.

Maxwell, W.M.C. and Watson, P.F. 1996. Recent progress in the preservation of ram semen. *Animal Reproduction Science* 42, 55-65.

Moses, D.F., de las Heras, M.A., Valcarel, A., Perez, L and Baldassarre, H. 1995. Use of computerized motility analyser for the evaluation of frozen-thawed ram spermatozoa. *Andrologia* 27, 25-29.

Naqvi, S. M. K., Joshi, A., Bag, S., Pareek, S.R. and Mittal, J.P. 1998. Cervical penetration and transcervical AI of tropical sheep (Malpura) at natural estrus using frozen-thawed semen. *Small Ruminant Research* 29, 329-333.

Naqvi, S.M.K., Joshi, A., Das, G.K. and Mittal, J.P. 2001. Development and application of ovine reproductive technologies: an Indian experience. *Small Ruminant Research* 39, 199-208.

O'Brien, J.K., Hollinshead, F.K., Evans, K.M., Evans, G. and Maxwell, W.M.C. 2003. Flow cytometric sorting of frozen- thawed spermatozoa in sheep and non-human primates. *Reproduction Fertility and Development* 15, 367- 375.

Paulenz, H., Soderquist, L., Adnoy, T., Nordstoga. A., Gulbrandsen, B and Berg, K. A. 2004. Fertility results after different thawing procedures for ram semen frozen in minitubes and mini straws. *Theriogenology* 61,1719-1727.

Peris, S. I., Morrier, A., Dufour, M. and Bailey, J. L. 2004. Cryopreservation of ram semen facilitates sperm DNA damage: Relationship between sperm andrological parameters and sperm chromatin structure assay. *Journal of Andrology* 25, 224-233.

Phillips, P.H. 1939. Preservation of bull semen. *Journal of Biological Chemistry* 130, 415.

Payne, S.R., Oliver, J.E. and Upreti, G.C. 1994. Effect of antifreeze proteins on the motility of ram spermatozoa. *Cryobiology* 31,180-184.

Pontbriand, D., Howard, J.G., Schiewe, M. C., Stuart, L.D. and Wildt, D.E. 1989. Effect of cryoprotective diluent and method of freeze-thawing on survival and acrosomal integrity of ram spermatozoa. *Cryobiology* 26, 341-354.

Polge, C., Smith, A.U. and Parkes, A.S. 1949. Revival of spermatozoa after vitrification and dehydration at low temperatures. *Nature* 164, 666.

Purdy, P.H. 2007. Preserving ram semen: New science shows it can be stored for up to two days before freezing. *Sheep Magazine* March 1, 40-41.

Rijsselaere, T., Soom, A.V., Hoflack, G., Maes, D and de Kruif, A. 2004. Automated sperm morphometry and morphology analysis of canine semen by the Hamilton-Thorne analyser. *Theriogenology* 62, 1292- 1306.

Saacke, R. G. 1982. Components of semen quality. *Journal of Animal Science* 55 (Suppl 2), 1- 13.

Sadler, R.M.F.S. 1966. The preservation of mammalian spermatozoa by freezing. *Laboratory Practicals* 15 , 413.

Salamon, S. and Maxwell, W.M. . 1995 a. Frozen storage of ram semen .I. Processing, freezing, thawing and fertility after cervical insemination. *Animal Reproduction Science* 37, 185-249.

Salamon, S. and Maxwell, W. M. C.1995 b. Frozen storage of ram semen .II. Causes of low fertility after cervical insemination and methods of improvement. *Animal Reproduction Science* 38, 1-36.

Salamon, S and Maxwell, W.M.C. 2000. Storage of ram semen. *Animal Reproduction Science* 62, 77-111.

Salamon S, Gillan L, Evans G and Maxwell W M C. 1996. Fertility of ram semen after long-term frozen storage. *Proceeding of the 13th International Congress on Animal Reproduction*, Sydney, Australia, Vol 3 , P 24-25

Salamon S, Gillan L, Evans G and Maxwell W M C. 2004. Fertility of ram semen after 35 years of frozen storage. In:*Proceedings of the 15th International Congress on Animal Reproduction*, Porto Segura, Brazil.

Sanchez-Partida, L.G., Maxwell, W.M.C. and Palez, L.G. 1992. Proline and glycine betaine in cryoprotective diluents for ram spermatozoa. *Reproduction Fertility and Development* 4, 113-118.

Schmehl, M.K., Anderson, S.P., Vazques, I.A. and Graham, E.F. 1986. The effect of dialysis of extended ram semen prior to freezing on post-thaw survival and fertility. *Cryobiology* 23, 406-414.

Seidel, G.E.Jr and Garner, D.L. 2002. Current status of sexing mammalian spermatozoa. *Reproduction and Fertility* 124,733- 743.

Soderquist, L., Madrid-Bury, N. and Rodriguez-Martinez. 1997. Assessment of ram sperm membrane integrity following different thawing procedures. *Theriogenology* 48,1115-1125.

Valcarel, A., de las Heras, M.A., Moses, D.F., Perez, L.J. and Baldassarre, H. 1996. Comparison between Sephadex G-10 and percoll for preparation of normospermic, asthenospermic and frozen/thawed ram semen. *Animal Reproduction Science* 41, 215-224.

Visser D. 1974. Recent advances in the deep-freeze preservation of ram semen. *South African Journal of Animal Sciences* 4, 275-288.

Watson P.F. 1979. The preservation of semen in mammals. In: *Oxford Reviews of Reproductive Biology* (Ed Finn C A) Clarendon Press, Oxford, London, U.K., Vol I, and pp 283-350

Watson P. F. 1995. Recent developments and concepts in the cryopreservation of spermatozoa and the assessment of their post-thawing function. *Reproduction Fertility and Development* 7, 871-891.

Watson, P.F. 2000. The causes of reduced fertility with cryopreserved semen. *Animal Reproduction Science* 60-61, 481-492.

Windsor, D.P. 1997. Mitochondrial function and ram sperm fertility. *Reproduction Fertility and Development* 9, 279-284.

Windsor, D.P and White, I.G. 1995. Mitochondrial injury to ram sperm during procedures associated with artificial insemination or frozen storage. *Animal Reproduction Science* 40, 43-58.

Yoshida, M., 2000. Conservation of sperms: current status and new trends. *Animal Reproduction Science* 60-61, 349-355.

❑❑❑

24

Thermal Stress and Sheep Production in Semi-arid Tropics

V.P. MAURYA, S.M.K. NAQVI, R. GULYANI

Central Sheep and Wool Research Institute, Avikanagar - 304 501, India

Introduction

Sheep raised in the semi-arid tropics, usually exhibit low productivity because of shortages of feed and harsh climatic conditions (Abi Saab and Sleiman 1995). Environmental factors, animal factors and thermoregulatory mechanism have a significant impact on the energy exchange between the animal and the environment (Nienaber *et al.* 1999). Heat stress is typically evidenced by an increase of mean and amplitude of body temperature (Hahn, 1995). The decrease in feed intake and changes in body weight and condition (Robertshaw, 1987) are indicators to monitor prolonged heat stress in grazing livestock. Heat also affects the comfort level of animals, which in turn adversely influence the productivity of animals (Armstrong, 1994). In arid and semi-arid tropical environment sunshine hours are longest (10-12 h) and ambient temperature is elevated up to 48°C during summer day when most of the sheep are grazed in field and therefore exposed to high intensity of thermals and nutritional stress (Maurya *et al.* 2004). Animals exposed to hot environment manifest a significant increase in physiological responses to counter balance heat stress (Naqvi, 1987; Hooda and Naqvi, 1990). The reproductive functions of the sheep are drastically affected by thermal stress because it has deleterious effect on estrus incidences, estrus intensity and embryo production (Naqvi *et al.* 2004). However the reproductive efficiency of different breeds of sheep inhabiting in the arid and semi-arid tropical regions is relatively low due to high ambient temperature conditions and feed scarcity (Arora and Garg 1998). Thermal stress is known to reduce the circulating estradiol–17β concentration with consistent increase in adrenocorticotropin secretion, which in turn blocks the estradiol induced estrus behavior (Doney *et al.* 1973). The reproductive behavior of the animals in the environment in which they are kept should be fine-tuned to produce maximum number of offspring most efficiently (Lindsay 1996). Heat stress is revealed by the inability of animals to cope with its environment, a phenomenon that is often reflected to achieve full genetic potential (Dobson and Smith 2000).

Thermal Stress and Physiological Responses

Sheep maintained in a semi-arid region have to adapt by special physiological means to maintain their thermal equilibrium. Controlled vasoconstrictor tone of skin blood vessels, upper respiratory tract evaporative cooling and alterations in coat insulation (Copenland, 1983). Kuthe (1976) observed that high ambient temperature results in loss of water via skin and lung of sheep leading to an increased water intake. The onset of sweating is the next rapid reaction of animals to heat exposure and increases linearly with the increase in ambient temperature (Kamal, 1975). Animals exposed to hot environment manifest a significant increase in physiological responses and reduction in productive potential (Hassanin *et al,* 1996). The numerous physiological mechanisms for coping with thermal stress have been reviewed (Blackshaw and Blackshaw, 1994). These include sweating, more rapid respiratory rate and greater vasodilatation with increased blood flow to the skin surface.

Table 1. Effect of high temperature stress on physiological responses of Malpura sheep.

Attributes	Shed	Hot Chamber
Respiration rate (breath/min.)	54.4 ± 4.08^{b}	88.6 ± 3.08^{a}
Pulse rate (beat/min.)	89.9 ± 1.33^{b}	98.8 ± 1.11^{a}
Rectal temperature (°C)	38.9 ± 0.14^{ab}	39.3 ± 0.04^{a}
Sweating rate (gm/m²/h)	120.8 ± 6.06^{b}	288.7 ± 20.03^{a}

^{ab}means with different superscript in same row differs significantly ($p<0.05$).
Source: Maurya et al., 2007

During the thermal stress a significant increase in the respiration rate, pulse rate, rectal temperature and sweating rate have been reported (Maurya *et al* 2007). Animal, which can maintain their physiological responses within normal limits under stressful environmental condition, may be considered adapted to that environment and hence, may be worth rearing commercially (Johnson, 1991).

Thermal Stress and Reproductive Efficiency of Ewes

Thermal stress markedly impairs reproductive success by disrupting those events occurring at the time of ovulation. The pre ovulatory release of LH and the expression of estrus behaviour seems to be especially sensitive to stress (Wolfenson et al. 1995; Dobson *et al.*, 2001). The thyroid gland also plays a role in decreasing the reproductive activity during thermal stress (Farghaly, 1984) with a decrease in the level of triiodothyronine (T3) and thyroxin (T4) hormone. Season of the year influence the superovulatory response of ewes (Naqvi *et al.*, 1998). It is noted that a proportion of ewes treated with gonadotrophin had good superovulatory response (<4-5 CL) but did not exhibit estrus symptoms and therefore could not be bred/ mated naturally. This leads to wastage of drug and increase in cost of embryo production.

Heat stress is reported to increase incidence of weak estrus, silent heat, ovulation failure and number of service per conception in buffaloes (Mohamed, 1974). Estrus behaviour in sheep is a part of a continuum of interaction between the sexes, which includes mutual sniffing and licking as well as motor pattern that are conducive to mating act (Fabre-Nys *et al.* 1993). The behavioural manifestation of estrus (circling, tail fanning, head turning, standing and approaching to ram) ranges from a phase of low intensity to a phase of high intensity occurring 8 to 12 hour after commencement and ends with another phase of low intensity with gradual disappearance as ewes enter in the diestrus stage (Maurya *et al.*1998 and 2005). The incidence of estrus and receptivity of ewes are important factors, which can affect over all embryo production (Tompkins and Bryant, 1974). The proper assessment of sexual behaviour and receptivity before breeding of ewes treated for super ovulation may increases the possibility of fertile mating which can influence the overall success rate of embryo transfer performance.

The embryo's survival depends on the maternal environment i.e. physiological regulation of oviduct and uterine function. Exposure of pregnant females to heat stress during the embryonic period leads to embryonic loss (Ealy et al.,1993). One cause of embryonic death is the disruption of embryonic development caused when the embryo is exposed to elevated temperature (Hansen, 2002). The thermal stress influenced the quality of embryos produced from the sheep. Naqvi *et al* (2004) reported higher proportion of embryonic abnormality in ewes exposed to thermal stress as compared to the ewes kept under shed. Thermal stress is reported to decrease follicular growth (Badinga *et al.*, 1993) and 17β estradiol concentration in follicular fluid (Wolfenson *et al.*, 1997) in dairy cattle. These consequences may lead to disruption in follicular development and influence the oocyte competence for either fertilization or embryo development. The effect of heat stress on oocyte competence can affect the yield of quality embryos.

Table 2. Effect of thermal stress on superovulatory response and embryo production and quality in Bharat Merino ewes.

Attributes	Shed	Hot chamber
Duration of estrous (h)	37.7±1.59[a]	31.7±3.57[b]
Interval to estrous (h)	25.4±1.12[b]	30.6±1.16[a]
Mean ovarian response (CL+LF)	8.86±1.27(7)	8.0±1.12(7)
Mean ovulation rate (No. of CL/ewe)	8.14±1.17(7)	6.71±1.10(7)
Transferable embryo No.(%)	21[a] (84.0)	14[b] (45.2)
Poor quality embryos	4[b] (16.0)	17[a] (54.8)

[a,b] with different superscript in same row differ significantly ($p<0.05$).
Source Naqvi et al 2004, Small ruminant research

The thermal stress during the pregnancy badly affects the growth of fetus (Mc Crabb, 1996). The thermoregulation capability of ewes also depends on their productive potential. In different world's leading laboratories, it has been found that placental growth is restricted in sheep, which do not thermo regulate well when exposed to a hot summer gestation. These sheep which are susceptible to placental restriction can be identified from the measurement of body temperature made in the previous year (Henson *et al*, 2001). In this context there is an urgent need for developing a embryonic drugs or antioxidants to overcome deleterious consequences on embryonic survival are that use thermoregulatory molecules to reduce the magnitude of hyperthermia caused by heat stress.

Body Condition Score and Thermal Stress

Scores are based on the amount of muscle and fat over and around vertebrae in the loin region. Sheep usually are scored from 1 to 5, 1 being extremely thin, 5 extremely fat, and 3 averages. Condition scoring should be "hands-on" rather than by eye. Hands-on scores give a better measure of condition mainly for sheep because the body of sheep generally hidden by wool. Body condition scoring in sheep can be useful means for assessing the flocks at mating time (Gordon, 1997). During the thermal stress the feed intake in the sheep reduces and which results in the decrease in the body condition score of sheep. The ewes of lower body condition used to have less duration of estrus and more estrus interval as compared to ewes having high body condition scoring (Maurya *et al*, 2006). The ewes with low body fat reserves have fewer large ovarian follicles and therefore less number of follicles become capable of undergoing the final stage of development and ovulation and has greater possibility of reproductive failure and reduction in production level (Britt, 1992, Michels *et al.*, 2000). The incidence of estrus and receptivity of ewes are important factors, which can affect over all embryo production (Tompkins and Bryant, 1974) reported an early negative energy balance during folliculogenesis could impair follicular development.

Consequently the production of defective follicle during the breeding period may result in low plasma concentration of progesterone and be associated with low fertility. The lamb born from the ewes of lower body condition had significantly lower body weight than of lambs born from the ewes of good body conditions (Maurya *et al,* 2007).

Thermal Stress and Reproductive Efficiency of Ram

The most important factor for over all productivity of flocks are Rams. A highly fertile and adapted ram will settle a greater number of ewes early in the breeding season, and will also be more likely to fertilize a higher proportion of eggs than less adapted and rams with poor fertility. Sexual performance of rams is highly variable (Price, 1987, Maurya et al 1999) and strongly influences flock fertility (Matos and Thomas, 1992). Rams that exhibit relatively rapid ejaculation rates are capable of inseminating a greater number of ewes per unit of time than males of having poorer libido or mating technique (Perkins et al., 1992). Tests for ranking rams on sexual performance are repeatable (Snowder *et al.,* 2002) and reliable predictors of sexual performance under field conditions (Ibarra *et al.,* 2000), potentially allowing breeders to evaluate the mating competence of individual males before they are employed in a breeding programme. In the fertile male harmonious interaction of gametogenic and endocrine testicular activity ensures the production of spermatozoa and seminal plasma in an integrated fashion. Under the thermal stress one of these functions may be disrupted more seriously than the other, thus effecting spermatogenesis and steroidogenesis to different degrees. The stage of spermatogenesis that is most susceptible to elevated temperature is the primary spermatocytes although damage to B spermatogonia can occur in bulls and prolonged exposure to heat can damage dividing spermatocytes and spermatids. Heat stress occurs when the scrotum is not able to reduce the temperature of the testicles below normal body temperature. During thermal stress, Merino rams with heavy skin wrinkle have a relatively poor ability to control testicular temperature resulted in loss of motility, increased proportion of abnormal sperm, decreased concentration of sperm and ultimately cessation of spermatogenesis (Tiwari and Sahni 1975; Marai *et al* 2006). In addition to this there were a higher percentage of abnormal spermatozoa, acrosomal damage. Stress also caused the increase in temperatures of rectum, body skin and scrotal skin. Blanc et al 1981 reported that heat stress on scrotum drastically affects the function of sertoli cells and release of androgen.

Mating behavior is an important aspect of male reproductive function as it has a direct bearing on the number of females mated. Alterations in sperm production caused by heat stress commonly include decreased sperm numbers, decreased sperm motility and increased numbers of abnormal sperm (Meyerhoeffer *et al.,* 1985). If weather is extremely hot during the breeding season, rams may be removed from the ewes during the day and kept in a cool place until returning to the ewe flock at night for breeding. Greater emphasis should be placed on the adaptation of animals to their environment and on their behavioural response to different stresses. Genetic selection for certain traits could result in the development of breeds and strains of sheep better able to survive and thrive in such systems.

Options for Improving Productivity of Sheep

Even highly specialized, well-adapted type of sheep can produce with maximum efficiency unless some means are taken to modify or reduce environmental stress. Shearing of fleece is usually during comfortable months of the year. Shearing before lambing is advantageous. By doing this the breakage and weakness in wool may be avoided and ewes will take less space when they will closely confined under shed for lambing. Shearing of rams or ewes to facilitates cooling for higher fertility, needs to be carefully timed to obtain maximum benefit.

Breeds of have been developed for specific purposes often in areas with common environmental conditions. Probably the largest single factors in adaptation of breeds is the available supply of feed. Besides the genetic potential, feed play an important role in exploiting the growth and production potential of animals. Low productivity of sheep is mainly due to poor nutrition. Dietary nutrition promotes the programming and expression of the metabolic pathways that enable animals to achieve their genetic potential for reproduction. Nutrition has an important impact on the reproductive performance in sheep. So there is an urgent need for increasing the feed availability both in terms of quantity and quality.

Sunshades may be desirable in hot humid weather when natural shade is not available. The provision of shed for growing lambs is desirable under hot summer condition. Feed intake may also be reduced with out shade and losses from sunstroke are possible. Housing is more necessary in the wetter, more humid climates. Shelter and care at lambing time can materially affect neonatal mortality.

To maintain health, reproductive function and productive capacity ewes must have adequate amount of body fat, which is an indicative of energy store (Atti *et al.*, 2001). The ewes with appropriate body condition score (BCS) and that maintain it through out gestation and after lambing, have higher productivity than ewes of poor BCS (Russel 1969). The routine assessment of body condition score of sheep at every stage of productive cycle can facilitate appropriate dietary changes to correct any deficiencies. Optimum nutrition and appropriate body condition of ewe will ensure full exploitation of genetic potential of the flock.

Conclusion and Recommendations

During thermal stress, a provision of shed is necessary to minimize the fluctuations in the physiological responses and changes in blood constituents. A simulated micro-environment against thermal stress may be created which will able to alter the onset of estrus, sexual behaviour, conception and lambing rates of ewes reared in the semi-arid tropical environment. There is a need to develop suitable strategies to reduce the effect of thermal stress, which may prevent against the high occurrence of thermal stress induced infertility in sheep and economic losses to the farmers.

REFERENCES

Abi Saab, S. and Sleiman. F.T. 1995. Physiological responses to stress of filial crosses compared to local Awassi sheep. *Small Ruminant Research* 16, 55-59.

Armstrong, D. V. 1994. Heat stress interaction with shade and cooling. *Journal of Dairy Science* 77, 2044-2050.

Arora, C.L., Garg, R.C., 1998. Sheep production and breeding. International Book and Distribution Company, Lucknow.

Atti N, Theriz M and Abdennebi L. 2001. Relationship between body condition at mating and reproductive performance in the fat tailed Barbarine breed. *Animal Research* 50, 135-144.

Badinga, L., Thatcher, W.W., Wilcox, C.J., 1993. Effect of environmental stress on follicular development and steroidogenesis in lactating Holstein cows. *Theriogenology* 39, 797-810.

Blackshaw, J.K. and A.W. Blackshaw. 1994. Heat stress in cattle and the effect of shade on production and behaviour: a review. *Australian Journal of Experimental Agriculture* 34,285-295.

Blanc MR, Hochereau-de Reviers MT, Cahoreau C, Courot M, DacheuxJL. Inhibin: effects on gonadotropin secretion and testis function in the ram and rat. In: Franchimont P, Channing CP (eds.), Intragonadal Regulation of Reproduction. Toronto, ON: Academic Press; 1981, 299-326.

Britt, J.H.1992. Impacts of early postpartum metabolism on follicular development and fertility. Procc. Anim.Conven.American Assoc. Bovine Practitionars. 39-43.

Copeland, R. S. 1983. *Responses to heat stress.* In: T.N. Edey (Editor) A Course Manual in Tropical Sheep and Goat Production. AUIDP, pp. 122-125.

Doney J M, Gunn R G and Griffiths J S. 1973. The effect of pre-mating stress on onset of estrus and ovulation rate in Scottish black face ewes. *Journal of Reproduction and Fertility* 35, 381-384.

Dobson H and Smith, R.F. 2000. What is stress and how does it affect fertility. *Animal Reproduction Science* 60, 743-752.

Dobson, H., Tebble, J.E., Smith, R.F. and Ward, W.R., 2001. Is stress really all that importance. *Theriogenology, 55,* 65-73.

Ealy, A.D., Drost, M. and P.J. Hansen. 1993. Developmental changes in embryonic resistance to adverse effects of maternal heat stress in cows. *Journal of Dairy Science* 76, 2899–2905.

Fabre-Nys, C., Poindron,P. and Signoret, J.P.,1993. Reproductive behaviour.*In reproduction in domesticated animals.* 89,147-149. (ed.G.J.King), *Elsevier,Amsterdam.*

Farghaly, H.A., 1984. *Thyroid and some ovarian hormones and their relation to reproduction in Egyptian water buffalo.* MSc. Thesis, Faculty of Agriculture Cairo, University, Cairo.

Gordon, I., 1997. Embryo Transfer and Associated techniques in sheep. In: Controlled Reproduction in Sheep and Goat, CAB International, Oxon, UK.

Hahn, G. L. 1982. *Housing for cattle, sheep and poultry in the tropics*. In Animal Production in the Tropics. (Ed) Yousef, M.K.Praeyer Publisher, New York.

Hansen, P.J. 2002.Embryonic mortality in cattle from the embryo's perspective. *Journal of Animal Science* 80(E. Suppl. 2, E33–E44.

Henson P.J. Drost M. Rivera R.M., Paulalopes F,F., Al-Katanani Y.M., Kríninger C.E. and Chas Jr. C.C. 2001. Adverse impact of heat stress on embryo production. Causes and stratagies for mitigation. *Threriogenology* 55, 91-103.

Hooda, O K. and Naqvi. S.M.K. 1990. Effect of thermal load on feed restriction on relative adaptability of Malpura and Avivastra sheep in semi-arid region. *Indian Journal of Animal Science* 60, 608-611.

Hassanin S.H., Abdulla E.B., Kotby E.A., Abd-Elaziz, A.M.S. and El-Fouly, M.A. 1996. Efficiency of asbestos shading for growth Bakri rams during hot summer. *Small Ruminant Research* 20, 199-203.

Antunovic, Z., Sencic D., Seprnada, M. and Liker, B. 2002. Influence of the season and the reproductive status of ewes on blood parameters. *Small Ruminant Research* 45, 39-44.

Ibarra, D., Laborde, D., van Lier, E., 2000. Repeatability and relationship with field mating performance of a serving capacity pen test in rams. *Small Ruminant Research* 37, 165–169.

Johanson, K.G. 1991. Body temperatures and respiratory rates of free ranging Merino sheep in and out of shed during summer. *Australian Journal of Agriculture Research* 42, 1347-1357.

Lindsay D R. 1996. Environmental and reproductive behaviour. *Animal Reproduction Science* 42, 1-12.

Mari, Ibrahim F. M., El-Darawany, Abdel-Halim, A., Ismail, El-Sayed A.-F. and Abdel-Hafez, Mohamed A. M. 2006. Tunica dartos index as a parameter for measurement of adaptability of rams to subtropical conditions of Egyptian *Animal Science Journal* 77, 487-494.

Mc Crabb, G. J. and Bortolussi, G. 1996. Placental growths and the ability of sheep to thermoregulate in hot environment. *Small Ruminant Research* 20, 121-127.

Matos, C.A.P., Thomas, D.L. 1992. Physiology and genetics of testicular size in sheep: a review. *Livestock Production Science* 32, 1-30.

Maurya V P, Naqvi S M K and Mittal J P. 1998. Reproductive behaviour of Malpura and Bharat Merino ewe lambs under semi-arid condition. *Indian Journal of Animal Sciences* 68: 533-535.

Maurya V P, Naqvi S M K and Mittal J P. 2004. Effect of dietary energy level on physiological responses and reproductive performance in Malpura sheep in hot semi-arid region of India. *Small ruminant Research* 55,117-122.

Maurya V P, Naqvi S M K, Gulyani R, Joshi A and Mittal J P. 2005. Effect of thermal stress on sexual behaviour of superovulated Bharat merino ewes. *Asian Australasian Journal of Animal Sciences* 18,1402-1405.

Maurya, V.P., Naqvi, S.M.K., and Joshi, A.., Mittal, J.P. 1999. Dominance , libido and serving capacity of Garole rams under semi-arid region of Rajasthan *Indian Journal of Animal Sciences* 69, 773-775.

Maurya V.P., Naqvi S.M. ., and Joshi, A. 2007. Effect of thermal stress on physiological responses and blood profile of native Malpura sheep reared in hot semi-arid region of India. *Indian Journal of Animal Sciences* (in press)

Maurya V. P., Naqvi, ,S.M.K., Kumar, D., Gulyani, R., Joshi, A. 2006. Annual Report CSWRI, Avikanagar. PP 45.

Maurya, V.P., Kumar, S., Kumar, D., Gulyani, R., Joshi, A., Naqvi, S.M.K., Arora, A.L. and Singh, V. K. 2007· Effect of body condition score on reproductive performance of chokla ewes. *Indian Journal of Animal Sciences* (Accepted)

Meyerhoeffer, D.C., Wetteman, R.P., Colleman and Wells, M.E.1985. reproductive criteria of beef bulls during and after exposure to increased ambient temperature. *Indian Journal of Animal Sciences* 60, 352-357.

Michels H, Decuypere E and Onagbesan O. 2000. Litter size, ovulation rate and prenatal survival in relation to ewes body weight: genetic review. *Small Ruminant Research.* 38, 199-209.

Mohamed, A.A. 1974. *Physiological changes in the reproductive organ of buffalo from particular to conception.* PhD Thesis faculty of agriculture Al Ajhar university, Egypt.

Naqvi, S.M.K., Maurya, VP Gulyani, R., Joshi, A. and Mittal, J.P. 2004. Effect of thermal stress on superovulatory response and embryo production in Bharat Merino ewes. *Small Ruminant Research* 55, 57-63.

Naqvi, S.M.K., Joshi, A., Das, G.K. and Mittal, J.P. 2001. Development and application of ovine reproductive technologies: An Indian experience. *Small Ruminant Research* 39, 199-208.

Naqvi, S.M.K., Gulyani, R., Das, G.K. and Mittal J.P., 1998. Superoulatory response of ewes during different seasons. *Indian Journal of Animal Sciences,* 68, 249-250.

Naqvi, S.M.K. 1987. Physiological adaptation of sheep during energy crisis and thermal stress. Ph.D. thesis submitted to Kurukshetra University, Kurukshetra, India.

Nienaber, J.A., G.L. Hahn and R.A. Eigenberg. 1999. Quantifying livestock responses for stress management: a review. *International Journal of Biometeorology* 42, 183-188.

Perkins, A., Fitzgerald, J.A., Price, E.O., 1992. Sexual performance of rams in serving capacity tests predicts success in pen breeding. *Indian Journal of Animal Sciences* 70, 2722–2725.

Price, E. O. 1987. Male sexual behavior. *Vet. Clin. North Am. Food Anim. Pract.* 3:405–422.

Robertshaw, D. 1987. Heat stress. Pages 31-35 *in* Proc. Grazing Livestock Nutrition Conference, Jackson Hole, WY, July 23-24, 1987. Univ. Wyoming, Laramie, WY. Stricklin, W. R., and C. C. Kautz-Scanavy.

Russel A J F, Doney J M and Gunn R G. 1969. Subjective assessment of body fat in live sheep. *Journal of Agricultural Science (Cambridge)* 72, 451-454.

Snowder, G.D., Stellflug, J.N., Van Vleck, L.D. 2002. Heritability and repeatability of sexual performance scores of rams. *Journal of Animal Science* 80, 1508-1511.

Tiwari, S.B. and Shani, K.L. 1975. Semen quality of Rambouillet tams and their crosses (with native) during hotter part of the year. *Indian Veterinary Journal* 52, 614-619.

Thompkins, T. and Brayant, M.J., 1974. Oestrus behaviour of ewes and the influence of treatment with progesterone. *Journal of Reproduction and Fertility* 41, 121-132.

Wolfenson, D., Thatcher, W.W., Badinga, L., Savio, J.D., Meidan, R., Lew, B.J., Brawtal, R. and Berman, A., 1995. The effect of heat stress on follicular development during the estrus cycle in lactating dairy cattle. *Biology of Reproduction* 52, 1106-1113.

Wolfenson, D., Lew, B.J., Thatcher, W.W., Graber,Y., Meidan, R., 1997. Seasonal and acute heat stress effects on steroid production by dominant follicles in cows . *Animal Reproduction Science* 47, 9-19.

□□□

25

Measures to Amleorate Heat Stress Under Arid Conditions

S.K. KAUSHISH

Central Arid Zone Research Institute, Jodhpur 342003, Rajasthan

Introduction

India is a rich repository of sheep and goat germ plasma, possessing 20 recognised goat (Acharya, 1982) and 57 sheep breeds (World watch 95). It is a multi crop species and produces meat and fiber besides manure and skins. They also produce milk. The sheep and goat population of India is 61.8 and 120.1 million, respectively (Livestock census, 2003). Goat is considered to be the principal meat animal of India as it contributes 40.7 percent of the total meat produced in the country. It is a highly adaptable and prolific species. In spite of heavy depletion owing to the sacrifice of 39.10 million goats (42,7% of the total population) annually, there has been an increase of 26.4 percent in the goat population over the last survey (91.0 million).

Generally, the term "Stress" is used to describe the detrimental effects of variety of factors on the health and performance of the animal. Under the stress conditions, there is redistribution of body resources including energy and protein at the cost of decreased growth, reproduction and health. The different types of stress are: environmental, nutritional, physiological, physical, psychological, anaesthesia and surgery, exercise, noise, starvation, altitude and transportation.

Stress is one of the major factors, which affects the physiological functions of the body and production of livestock. The domesticated animals are protected against the environmental hazards so that their production is influenced to the minimum possible extent. The domestic livestock are homeothermic animals, which can maintain their body temperature and are therefore, able to keep a fine balance between the heat of metabolism and the heat of environment. But for keeping that balance and to maintain a homoeothermic condition, the animals have to do a lot of efforts and also have to spend energy if the animals are kept above the upper critical temperature or lower than the lower critical temperature.

Homoeothermic animals are able to maintain a balance between heat of metabolism and heat of environment to avoid hyperthermia or hypothermia. The thick hair coat or wool in sheep and goat acts as an insulator for environmental temperature. As ambient temperature exceeds body temperature the hair or wool coat protects the skin and cutaneous blood flow from warming effects of solar radiation and thus help prevent excessive increase in core body temperature. Goats are less susceptible to environmental stress because of their unique water conservation capacity, higher sweating rate, lower basal heat metabolism, higher respiration rate, higher skin temperature and constant cardiac output and water output (Rai et al., 1979; Lu, 1989). However heat stress do occur in these animals when they are exposed to ambient temperature above critical temperature zone.

Heat stress is caused primarily due to high air temperature but can be intensified by high humidity, thermal radiation, low air movement and physiological status of the animals. Like other animals, small ruminants undergo several physiological and behavioral adjustments to cope with the heat stress. Increased physiological reactions, decreased voluntary feed intake, feed digestibility, thyroid function, blood insulin and increase in cortisol etc have been observed during heat stress in small ruminants (Bhattacharyya and Hussain, 1974; Joshi *et al.*, 1977; Kaushish *et. al.*, 1995,1997. The consequences of these changes are the decreased production and reproductive functions of animals under stress (Kaushish and Sahni, 1977). Apart from decreased growth rate in young, heat stress lengthens estrous cycle, affects ovulation and decrease conception rate in the female goat and sheep, and in male libido and semen quality are reduced. Further exposure of ewes to heat during pregnancy results in damage to embryo, decreased birth weight and viability of lamb. Amelioration of heat stress in animals involves breeding, management and environment modification. Breeding and management irlc1udes selection of heat tolerance breed, use of crossbreeding, nutritional management and control of disease and parasites. Environmental modification may include provision of shades, enhancement of evaporative cooling and increase air movement in the shed etc. However to alleviate heat stress in small ruminant, not much studies have been done.

Provision of Shelter

The shelter should be conducive to the expression of normal behavior by the animal (Nagpaul and Kataktaware, 2003). The microclimate of shade in broad sense is the climate in side shade/surrounding the animal. The housing environment facilitates normal behavior, avoid stress anr let animal arrive at a high age with high and stable performance (Brade 2001). Compared to the sheep maintained in shade, the animals while grazing or enroute migration are exposed to direct solar radiation, which amounts to several times higher impinged heat on animal body. Further on exposure to heat, the crossbred animals experience greater heat stress than the natives (Singh, *et. al.*, 1980), which is due to their different mode of body heat dissipation. The respiration rate, rectal temperature, skin temperature and fleece temperature 1 and 2 cm above the skin were higher in solar exposed animals as compared to the animals maintained in shed (Kaushish *et. al.*, 1989).

Providing shed during intense sunshine to sheep and goat reduces the heat load of these animals and reduces the adverse effect of heat stress. Sheep having access to shed show better growth rate, feed conversion efficiency and improved fertility than control one (Moose and Rose, 1964).

The physiological responses viz respiration rate and heart rate were lower in shaded and nutritionally restricted or ad lib feeding sheep than exposure in sun (Hooda and Naqvi, 1990). Kaushish et. al., (1987,1993) studied the effect of heat and water restriction on physiological norms, serum electrolytes and body water compartments in goats Heat stress during pregnancy also significantly affects the growth rate of the fetus and therefore provision of shed has been recommended during this period. For hot humid tropical India, CIRG, Makhdoom have designed an effective and low cost housing for stall fed goats. The high (3-5m height) 'A' shaped roof shed covered with fire proof material and the main axis running east-west provides a cooler environment than housing with hay thatches and asbestos sheet. The painting the sidewalls white outside reduces the surface temperature inside the shed by 12-22°C. Further water can be sprayed periodically on the floor and roof of shelter to reduce the heat load on the animals during peak hour. For loose housing system they have also designed the suitable housing system. Effect on physiological parameters: Sharma (1983) observed that extremes of temperature did not show appreciable effect on the physiological reaction whereas humidity affected pulse as well as respiration rate in crossbred animals in individual stalls. However, Kishonti and Adam (1985) found increased respiration rate and body temperature in heat stressed cattle.

Evaporative Cooling

Evaporative cooling is probably the most practical means of cooling livestock. Sheep do not have water on their skin and the effect of air humidity might have less effective on surface evaporation. The tropical Merino sheep lose 90% of total evaporative heat loss by non respiratory means. The enhancement of non respiratory evaporative cooling through provision of wallowing, sprinkling and fine misting of water, dry and wet cooler have been used for reducing heat load in different domestic animal. However little study has been made in this aspect in small ruminant. Partly because sheep and goat are grazing animals and providing water showering, misting of water or air cooler during heat stress may not be feasible. However providing these means during peak hour of sunshine may reduce the heat load of these animals. In lamb, mist cooling in conventional shed resulted in a significant improvement in live weight gain.

Nutritional Alleviation of Stress

Nutritional management in combination with other methods such as genetic selection of heat resistant breeds, housing and management modification can be most effective in combating heat stress in small ruminants. Declining feed intake and feed energy units to production unit has been identified as a major cause of reduced production in animals during heat stress. Therefore improvement or maintaining the production should be possible through altered diets that either promote higher intake or improve the biological efficiency of converting feed units into production unit. However in small ruminant balancing nutrient requirement under biological limitations of heat stress is the conservative approach. In this approach productive limits of animals under heat stress and limiting nutrients for productive functions are both recognized.

Reducing dietary forage to grain ratios decreased heat load in sheep and goat during heat stress. This is due to greater heat of rumen fermentation of forages compared with concentrate. Total heat production in goat fed a high grain diet is less than those fed a high forage diet, which gives the animal the capacity to balance some heat gradients between animal and environment. Compared to goat consuming only forage diet, goat fed a 50:50 forage to grain ration are capable of maintaining homeostasis under heat stress which was evidenced by smaller changes in rectal temperature, feed intake, blood metabolites and blood enzymatic activity. However care must be taken about the negative effect of the high concentrate diets on production efficiency and rumen function.

The physiological responses elicited by heat-exposed animals during summer to counter the imposed stress (respiratory ventilation and evaporative cooling) increase their maintenance energy requirement. Further even with efficient thermolysis in sheep and goat the core temperature remain in a higher plane also adding to maintenance requirement due to Van't Hoff effect. Therefore provision of additional better nutrition is essential to cope with the heat stress. Nutritionally stressed sheep experience more heat stress than those on nutritionally adequate diets (Hooda and Naqvi, 1990). Alternatively sheep with higher plane of energy level show better heat tolerance than sheep with low plane of nutrition (Hooda and Naqvi, 1997). The wool growth rate was shown to be unaffected by high environmental temperature in sheep on a high quality diet, but was lower in low quality diet.

Role of Minerals and Electrolytes

The calcium requirement of the animals also increases by feeding high fat in diet as calcium forms soaps with fats and it is not available to the animals. Thus the supplementation of calcium in diet should be increased above the requirement of the animals.

Sodium bi carbonate supplementation in diet is beneficial for the production of animals under stress. Milk production and feed in take improves with addition of 0.85% sodium bi carbonate in heat stressed dairy cows. Heat stressed lactating goats and cattle prefer to drink water to over hydrate them to protect against the stress.

Role of microbial feed additives

The effect of microbial feed additives in the diet of diseased or stressed animals is much higher as compared to healthy animals. In ruminants the beneficial effect of microbial feed additives can be summarized as

1. Protection from diarrhea
2. Improved growth rate in farm animals
3. Improved nutrient utilization
4. Improved milk production

Dietary Supplementation

Feeding some fat also reduce the heat stress in sheep and goat. Because the heat of increment of fat is less than that of carbohydrate and protein in ruminants. As a result of feeding supplementary fat dietary energy density is increased and heat load of the animal is reduced during heat stress. Supplementary feeding of fat have been shown to improved daily gain in heat stressed lamb and have no effect on growth hormone, cortisol, rectal temperature and respiration rate. But feeding hydrolyzed animal fat changes the milk fat composition and increase the calcium requirement (NRC, 1981).

Supplementation of sodium bicarbonate may be beneficial for the production of ruminants under heat stress. This is based on the based on the hypothesis that buffering capacity via saliva secretion is lowered due to a rapid loss of carbondioxide during accelerated respiration of heat stressed animals. Dietary supplementation with NaHC03 also can correct low ruminal pH (Tripathi et al., 2004) resulting from high concentrate feeding and less rumination in heat stressed animals. Feed intake and milk production has been shown to increase in heat stressed animal fed with $HC0_3$.

Providing ascorbic acid to heat stressed lamb lower the respiration rate and rectal temperature than control lamb and prevents the changes in PCV, Hb and serum total protein. It also increases serum thyroxin and decreased cortisol-indicating improvement of heat tolerance in lamb. Use of natural shed and iodine supplementation improved the fertility of tropical Merino ewes and this effect is most pronounced when ambient temperature is most pronounced. Similarly supplementation of iodine for the first 7-10 days after lambing in shaded ewes have higher pregnancy rate than un shaded and un supplemented ewes.

Deficiencies in Na, K, Cl and to a lesser extent for Ca and Mg may occur in heat stressed animals. This is attributed to increase in rate of loss and decreases in intake during hyperthermea. Since K, Na and Cl are essential in maintaining osmotic balance of heat stressed animals, supplementation above NRC requirement may be increased.

Thermoregulatory Behavior

The behavioural adjustments include posturalchanges, qualitative and quantitative changes in food intake in food intake and water consumption, and diurnal and nocturnal patterns of activity and characteristics patterns in parental behaviour.

Behavioral pattern and thermoregulation

	High temperature		**Lower temperature**
Activities	Seeking shade	Activities	Shivering
	Licking body surface	Patterns	Huddling
	Night grazing		Increased locomotor activities
	Eating succulent feeds		Seeking microclimate with higher temperature
Patterns	Anorexia		
	Body extension		
	Group dispersion		
	Extra drinking		

Behavior of Cattle Under Heat Stress

Since management is a major factor in the well being of cattle in a hot climate, it is worth describing the symptoms that indicate heat stress in cattle. A number of these are valuable guide and indicate when it is necessary to take steps to alleviate the strain. In approximate order of ascending severity they are as follows:

1. *Refusal to lie down:* This is a common phenomenon and is presumably caused by the need to expose as much of the body as possible to the atmosphere. It was noted that animals that became recumbent involuntarily had the greatest difficulty in avoiding hyperpyrexia and needed intermittent housing or sprinkling.

2. *Huddling:* Contrary to expectation, cattle under heat stress huddle together and will disperse only when the temperature drops. This is presumably an endeavor to create own microclimate.
3. *Body splashing:* Strenuous effects may be made by cattle to wet their coats either by pawing water back over themselves or by beating their heads into the water in their through. They do not, however, use their saliva for this, purpose, as do some animals.
4. *High respiratory rates:* These become serious only when second phase breathing occurs, which the deep inspiratory movement of can recognize by the movement of flanks. This condition must always be alleviated.
5. *High rectal temperature:* Any persistent temperature higher than 41°C must be promptly treated.
6. *Open mouth breathing*: This is also indicative of advanced stress and is characterized by the head being extended, the tongue protracted and profuse salivation, taking place.

Selection of Heat Tolerant Breed

To avoid the effect of high environmental heat stress, heat tolerant breeds of sheep and goat should be selected for tropical environment. Small ruminant of tropical origin are more tolerant than breeds of temperate origin (Singh, et al., 1980; Singh *et al.,* 1982; Gupta and Acharya 1987). Sheep breeds like Chokla, Malpura and Nali are more heat tolerant than purebred Merino or Rambouillet (Gupta and Acharya, 1987). Crossbred are also fairly comparable to native or in between native and exotic sheep in their heat tolerance (Gupta and Acharya, 1987). Similarly goats like Sirohi, Beetal and their crosses with Alpine, sannen or toggenburger performs well in the tropic.

Parasitic Diseases

The prevalence of snail born diseases like fasciolosis, amphistomiasis and schistosomiasis have been reported more during and after rainy season. They suck the blood and cause weakness in the animal. To check these various drugs can be used.

Other Managemental Practices

Follow of light managemental practices reduces the heat load of sheep and goat. Grazing increases heat production and contribute to the heat load of the animal. Therefore grazing of sheep and goat may be restricted during the daily peak hour of heat stress. Animal should be provided ad lib and preferably cooled drinking water as water intake increases in sheep and goat during heat stress (Joshi *et al.,* 1977; Gupta and Acharya, 1987). Care should be taken for pregnant ewes and doe. Because of increasing metabolic activity, pregnancy imposed extra heat load to animal during heat stress. Further heat stress affects fetal development during late gestation in the ewe and fetal growth is influenced due to level of nutrition at this stage (Brown *et at.*, 1977).

Therefore provision of shed and better nutrition is essential in pregnant sheep and goat during heat stress. Shearing of lamb may be done because. it improved daily gain and serum growth hormone level significantly and decreased respiration rate and rectal temperature in heat stressed lamb. Weak animal is more susceptible to heat stress than healthy animal. Therefore control of diseases and parasites in small ruminant needs to be taken care during summer season.

Conclusion and Recommendations

Although small ruminant are resistant to high environmental temperature, but they do suffer from heat stress when exposed to upper critical temperature. Although several methods are followed to alleviate heat stress in cattle, buffalo and pig, but little such measures have been studied in small ruminants. However selection of breed of heat resistant, provision of shed, nutritional manipulation and modification of management may greatly reduced the heat stress of small ruminants.

REFERENCES

Acharya, R.M. 1982. Sheep and Goat Breeds of India, FAO, Rome

Bhattacaryya, A.N. and Hussain, F. 1974.Intake and utilization of nutrients in sheep fed different levels of roughage under heat stress. *Journal of Animal Science* 38, 877-886.

Brade, W. 2001. Animal welfare aspects of dairy farming: definition, conditions and criteria. *Practische Tierarzt 82*, 588-94. (*Dairy Science Abstract 2002* 3,6507).

Brown, D.E., Harrison, P.C. and Hinds, F.C. 1977. Heat stress effects on fetal development during late gestation in the ewe. *Journal of Animal Science* 44, 442-446.

Gupta, U.D. and Acharya, R.M. 1987. Heat tolerance in different genetic groups of sheep in semi arid conditions. *Indian Journal of Animal Science* 57, 1314-1318.

Hooda, O.K and Naqvi, S.M.K. 1990. Effect of thermal load and and water restriction on the reletive adaptability of Malpura and Avikaline sheep in semi arid region. *Indian Journal of Animal Science* 60, 608-611.

Hooda, O.K. and Naqvi, S.M.K. 1997. *International Journal of Animal Science* 17, 201-204.

Joshi, B.C., Aravindan, M., Singh, K.S. and Bhattacharyya, N.K. 1977. Effect of high environmental stress on the physiological responses of bucks. *Indian Journal of Animal Science* 47, 200-203.

Kaushish, s. K. 1993. Serum electrolyte and water compartment responses to acute heat and water stress in Beetal and Black Bengal goats. *Indian Journal of Animal Sciences* 63,1 277-1279.

Kaushish, S.K, Karim, S.A. and Rawat, P.S.1989 Responses of native and cross bred sheep to solar exposure during summer in semi arid region. *Indian Veterinary Journal* 66,852-56

Kaushish, S.K. Sengupta, B.P. and Georgie, G.C. 1997 Effect of thermal stress and water restriction on cortisol level of Beetal and Black Bengal goats. *Indian Journal of Animal Sciences* 67, 1104-5.

Kaushish, S.K., Karim, S.A. and Rawat, P.S.1995 Physiological responses of unshorn and shorn native and crossbred sheep under heat exposure *Indian Journal of Animal Sciences* 65, 714-17.

Kaushish,S.K, and Sahni,K.L. 1977 Reproductive performance of Merino sheep during autumn and spring seasons under semi arid conditions. *Indian Journal of Animal Sciences*. 47, 18-22

Kaushish,S.K. ,Georgie, G. C.and Sengupta, B.P. 1987. Effect of heat and water resttiction on physiology responses of Beetal and Black Bengal goats. *Indian Journal of Animal Sciences*. 57, 461-465

Kishonti and Adam. 1985. Effect of heat on dairy cows and use of water sprinkling to compensate for heat stress. *Allattenyesztes-es-Takarmanyozas* 34, 521-29

Moose, M.G. and Ross, C.V. 1964. Effects of shades and concentrate levels on lambs *Journal of Animal Science* 23, 1201(Abst).

Nagpaul, P.K. and Kataktaware, M.A. 2003. shelter management for dairy animals. All India Dairy Husbandry Officer's Workshop on *Transferable Technologies for enhancing Milk Production* from Sept. 5-6, NDRI, Karnal

NRC 1981. National Academy of Sciences press, WD pp152

NRC, 1981. National Academy of Sciences press, WD, pp 152.

Rai, A.K., Mehta, B.S., Gour, D and Singh, M.1979. Sweating in in sheep and goats. *Indian Journal of Animal Science* 49, 546-548.

Sharma, R.N. 1983. Feeding and milking behaviour of crossbred cattle. M.Sc. Thesis submitted to Zkurukshetra University, Haryana.

Singh, M., More, T. and Rai, A.K. 1980. Heat tolerance of different genetic groups of sheep exposed to elevated temperature conditions. *Journal of Agricultural Sciences* 94, 63-67.

Singh, M., More, T., Rai, A.K. and Karim, S.A. 1982. A note on the adaptability of native and crossbred sheep to hot summer conditions of semiarid and arid areas. *Journal of Agricultural Sciences* 99, 525-528.

Tripathi, M.K., Santra, A., Chaturvedi, O.H. and Karim, S.A. 2004. Effect of graded levels of $NaHCO_3$ supplementation on ruminal fluid pH, feed intake, nutrient utilization, growth performance and blood biochemical constituents of lamb on high concentrate diets. *Animal Feed Science and Technology* 111, 27-39.

❏❏❏

26

Shelter Management for Improving Small Ruminant Production

RAJEEV GULYANI
Central Sheep and Wool Research Institute, Avikanagar, Rajasthan - 304 501

Introduction

Sheep (*Ovis aries*) have been domesticated for over 10,000 years and figure prominently in the story of civilization and human survival, as evidenced by religious practices and symbols and cultural rituals that involve this species, not to mention their importance to agriculture over several millennia. Sheep are widespread across the world and vary widely in their capacity to heat and cold. Like all mammals, sheep are forced to increase heat production to maintain core body temperature as environmental temperature drops. The critical temperature at which this occurs varies from 0°C for adult sheep in full fleece to 20-25°C for newly shorn animals to 30-36°C for newborn lambs. On the other hand, sheep are farmed for meat and wool at very low ambient temperatures in temperate regions and peak daily temperatures of 49°C in arid and semi-arid regions. Coping mechanisms are thus stretched to extremes and ultimately adversely affect the overall productivity of the animal, even though the animals are adapted to the region over several centuries.

Sheep have a most important role in the utilization of vast areas of grazing lands throughout our country. These lands, whether arid, semi-arid or hilly have characteristics, which preclude cultivation and crop production. However, these regions permit pastoral agriculture using the grazing animal like sheep and goat to convert the natural vegetation into products useful to man. The natural characteristics of sheep and the diversity of their functions, adaptability and performance have combined to establish their wide distribution throughout the country. Besides their well-recognized contribution to human needs for meat, fibre and milk, they provide a living and a way of life for millions of landless and marginal farmers of our country.

India is endowed with 40 breeds of sheep, which are fine wool type, carpet wool type or hairy meat type. Though these breeds are well adapted to the environmental conditions of their specific regions, their productivity is influenced by the extremes of the environment. Prolonged exposure to high ambient temperature, wind velocity and solar radiation are some of the important climatic factors, which directly or indirectly influence the productivity of these animals. The natural climate is not continuously optimum for livestock production throughout the year. It is, therefore, necessary to protect the animals from adverse climatic conditions by providing comfortable housing system for exploiting their genetic potential to the fullest. Sheep do not require any elaborate housing, but require full protection against wind, rain, extremes of temperature and humidity. Provision of suitable shelter to the sheep during adverse environmental conditions will buffer the extremes of these climatic factors and reduce the stress on the animals. The basic principle of shelter is that it should alter or modify the environment for the benefit of the animal.

On the basis of geographical conditions, India can be divided into four zones: the humid (North Eastern region, coastal states like Goa, Kerala and coastal regions of the country), sub-humid (U.P Plains, Bihar, Maharashtra and parts of Gujarat and M.P), temperate (Jammu & Kashmir, Himachal Pradesh and Uttranchal) and arid and semi-arid (Rajasthan, North Gujarat, Haryana, Punjab, Andhra Pradesh, Tamil Nadu, Karnataka and parts of M.P). Physiologists prefer to classify the climate according to its effect on physiological functions. It can be hot (above40°C), warm (20-30°C), temperate (10-20°C) and cool (below 10°C) on the basis of ambient temperature and can be humid (over 15 mm Hg) and dry (below 15 mm Hg) on the basis of vapor pressure.

In spite of a number of sheep development activities in operation in different states of the country, sheep rearing still continues to be nomadic and thus mostly concerned to poor and landless people. As suitable grazing lands are scanty in most of the states, the shepherds keep on migrating their flocks over extensive areas within or in the neighbouring states. In Rajasthan 59% of the flocks are penned in open fields away from human habitat and the rest penned in kutcha courtyards near the house. In the peninsular India 34-41% of the flocks are penned in enclosures near the house and the rest in open fields. During extreme summer months of the year, the flocks are grazed during the cooler hours of the day. The grazing starts in the late hours of the day and the animals are brought to the water points sometime in the noon hours of the following day. The animals are usually rested during the hotter part of the day under the shade of trees or under covered shelters usually made of locally available thorny bushes and tree branches.

Our surveys of shelter systems in Rajasthan under NATP have revealed that larger flocks are mostly housed in open enclosures with a boundary or fencing of thorny bushes reinforced with tree branches, 5-6 feet in height to protect against predators. This enclosure is usually built around shady trees to provide shade during noon hours. In extreme arid regions where tree cover is scarce, a covered shelter made up of thorny bushes is provided in one corner of the enclosure to provide protection during noon hours. Smaller flocks are usually at the back of the house in a separate enclosure with a hut provided for protecting animals from cold and rain.

Looking into the information available on the basis of various surveys of sheep shelter management systems in the country, it is clear that sheep are mainly reared by the poor and landless in our country that mostly lead a nomadic way of life. Such flocks are provided only minimal shelter. Sheep shelter management systems on modern lines can be observed in Central and State Sheep farms, with some NGO's supported by government grants and in a few private commercial farms.

Like most other ruminants, sheep is a social animal that needs the presence of other conspecifics in order to experience a state of well-being. Sheep should therefore be housed in a social setting whenever possible. Behavioural changes such as augmentation of aggression, incomplete and abnormal reproductive behaviour, decreased foraging, decreased stretching, teeth grinding, increased respiration rate, unusual posturing, reduced alertness, anorexia are indicators of a sheep kept in isolation. Sheep are conspicuously vigilant and their natural response to fear inducing situations is to instantaneously flock and flee. Sheep have a strong need to flock and this species-typical behaviour requires at least three animals. Hence the saying that " one sheep is not a sheep." Free ranging animals always maintain uninterrupted visual contact with at least one other member of the flock. Indoor-housed animals should therefore always be positioned in a way that they can see at least one other conspecific. Visual isolation is extremely stressful for sheep and under no circumstances should they be housed by themselves. Sheep establish social hierarchies and so it is preferable to establish groups of animals that are already familiar with one another and are of the same age and sex while housing them. Sheep are relatively fearful by nature. Any abrupt changes in their surroundings makes them apprehensive but they adjust if they are in the presence of their familiar and experienced conspecifics.

Normally sheep do not require elaborate housing facilities, but providing minimum housing requirements will definitely increase productivity. High investments in physical facilities for sheep are seldom needed or justified. It is only sedentary flocks and more intensive system of production that involves housing consideration. It is essential to provide shelter to young lambs, pregnant ewes and breeding rams against inclement weather conditions and predation.

Optimum performance requires some protection from inclement weather conditions and predation. Decisions must be made relative to the needs for and benefits from housing and equipment, and how to provide it economically. If the sheep are to perform their function of profitably producing meat and wool, the following are the basic requirements of building and equipment for sheep:

- Protection from cold, rainy weather.
- Shade in the summer.
- Good drainage to ensure dry conditions.
- Security from predators.
- Storage for feed and equipment.
- Convenient, efficient working area for handling sheep.

The overall objective of providing physical facilities is to enable the operator to perform, efficiently and effectively, the necessary production practices with the flock. Because physical equipment may be very expensive, it is necessary to spend some time thinking about providing an additional facility beyond the bare essentials for each unit.

Since the animal husbandry practices are dependent on geographic, climatic and economic factors, it is extremely difficult to recommend an ideal model of housing system to suit the requirements of sheep throughout the country. One general model can be recommended for organized farms and other for rural areas. These models can suitably be modified according to the prevailing conditions of the area in which these animals are to be reared. The number of units of each type to be provided in the yard would depend upon the number of animals to be looked after. The area in each unit shall depend upon the minimum floor space specified for each animal and number of animals to be kept in each unit.

Shelter Management for Sheep in Organized Farms

The sheep shed is the most expensive of all the physical facilities for sheep. Therefore it is necessary to know what is needed, what is available, and whether or not it is most advantageous to remodel an existing building or to construct a new one. Not only should the present needs be considered, but also the building should be designed to permit expansion or alteration if the situation warrants it. In addition, it may be advantageous to visit existing sheep housings to get feedback about the advantages and disadvantages of their system. The following are considerations and suggestions for planning and constructing a shed.

1. *Topography of the land:* The building should be situated at a higher elevation than the surrounding to offer a good slope for drainage of rain water, urine, manure and waste products.
2. *Soil type:* Soil composed of sand and gravel is best suited for foundation. Clay soil which swells during rainy season may not be suitable.

3. *Exposure to sun and protection from wind:* Shed should be constructed to obtain maximum sunlight. The long axis should always be in north-south direction. For protection from strong winds fodder trees should be planted at least 10 feet away from the sheds.
4. *Accessibility:* The sheds should be connected by good roads for movement of farm machinery and equipment.
5. *Water supply:* Fresh and clean water supply should be ensured.
6. *Durability and attractiveness:* Sheds should be constructed of durable material and designed in a manner to give it a scenic view.
7. *Surroundings*: Areas infested with wild animals, poachers and industries should be avoided.
8. *Labour*: Farm should be at an approachable distance from the locality to ensure availability of labour in time.
9. *Marketing:* The market should be at an easy and approachable distance to get a good price for the farm produce.
10. *Fodder production:* There should be reasonable area surrounding the shed for fodder cultivation and its preservation as silage and hay.
11. *Scope for future extension:* Sufficient area should be available around the shed for future expansion.

Types of Sheds

Generally three types of shed systems are advocated for housing sheep. These are:

Open shed: In this system there is no special construction except for roof and fencing. This type of shed is good for housing adult flock throughout the year. However during severe winter and summer, artificial screens made of grass or bamboo may be provided on the sides.

Semi-open shed: In this one side of the shed is open and may be provided with a paddock, which is fenced.

Closed shed: This type of shed is fully enclosed with controlled environment, light and ventilation.

Sheep are also housed as per the type of animal.

Flock shed: The flock shed is used for housing ewes kept for breeding purpose. A shed of 15x4x3 m dimension can comfortably accommodate 60 sheep.

Ram shed: The ram shed is used for housing rams kept for breeding purpose. A shed of 4x2.5x3 m dimension can accommodate 8-10 rams. The shed can be partitioned to form equal compartments of not more than one-meter height.

Lamb shed: This shed is used for housing lambs until they attain maturity and are weaned and also those animals, which are weaned and have not attained maturity. A shed of 7.5x4x3 m can accommodate around 75 animals easily. The shed can also be partitioned if needed into two compartments breadth wise of 5x4 m dimension to keep un-weaned lambs and another compartment of 2.5x4 m to keep weaned and immature animals.

Lambing shed: This shed is used as maternity shed for ewes which are about to lamb. This shed is usually of 1.5x1.2x3 m dimension. A manger for holding feed and hay and a bucket for water is provided in the shed.

Sick shed: This shed is used for segregating sick and disabled animals. The dimensions of this shed are 3x2x3 m. A door of one metre width and two metre high is provided whose upper half has a wire net covering.

Table 1. Floor space requirements of an animal unit

Type of animal	Floor space/animal (sq. m)
Dry ewes	1.0
Rams	1.9-2.8
Ewe with lambs	1.5
Lambs	0.4

Other details of a sheep shed

Floors: Dirt/ moorum (kutcha) floors are very satisfactory for holding pens and outside lots. They are cheapest to build and are quite satisfactory if properly maintained each year when the shed is cleaned. In dirt floors it is easy to drive steel posts and set up various pen arrangements. It is preferable for the floors in the inside of the shed to be built higher than the outside ground level to prevent surface water from entering. It should be sloped toward the front. The other types of floors being used in sheep sheds are concrete and wooden battens. A concrete floor may be advantageous as it is more permanent, requires less maintenance, does not develop mudholes and may be dryer. On the other hand it must be bedded and it is initially more expensive than dirt floor. Where the rainfall is quite heavy, wooden batten floors are preferred. The width of each plank is 75-100 mm and the thickness 25-40 mm. The clearance between two planks is 10-15 mm. for disposal of dung and urine. The wooden batten flooring is constructed at a height of at least one metre above the ground level.

Walls: Walls are built for protection and proper control of animals upto two thirds of the height from the burnt clay bricks. The upper one third of the wall may be of chain link net of galvanized steel wire supported by angle irons and brick masonry columns. In some cases only chain link fence supported by angle iron at regular intervals is raised from the ground level all around the perimeter upto 5-6 feet high to form the wall. In the case of lamb shed the upper one third part above the brick wall may be of wire netting supported by angle iron.

Roof: The roof is usually made up of corrugated steel sheet or of asbestos cement sheet. A thatch of 2-3 inches can be put over the roof for better insulation. Where the rainfall is not heavy the roof can be totally of thatch.

Gate: Each shed should be provided with one or more gates for easy accessibility and control of flock. The gate may be 0.8 m wide and one metre high and made up of angle iron frame and chain link net.

Ventilation: Sheep should not be housed in closed sheds or rooms unless it is necessary to save lambs at parturition or after shearing. Warm buildings with high humidity are usually conducive to respiratory problems. If sheep are sheltered from cold rain, wet snow and drafts, low temperatures are not harmful. Ventilation of sheep shed is required to remove moisture from inside the shed, provide fresh air for sheep, remove excess heat in hot weather and remove odors and gasses from the sheep and their offal. A shed with an open front to the south of the prevailing wind with some adjustable openings on the windward side should be satisfactory for air movement. Ventilation systems should be designed in such a way that drafts at animal level are avoided while adequate air movement above animal level is guaranteed.

Lighting: Sheep are relatively fearful when experiencing sharp contrasts in lighting and they are fearful of moving from light to dark areas. A uniform lighting pattern is therefore recommended. Sheep are seasonal breeders, with decreasing daylight initiating the onset of the reproductive cycle. Exposure to the natural light rhythm is therefore important for indoor housed animals who are assigned to breeding programmes.

Feeders: These may include a number of types like grain or concentrate troughs, hayracks, combination grain and hay bunks and fence bunks. The concentrate troughs should be 4-6 in. deep, 10-12 in. wide and 12-15 in. above the ground. They should be made of sturdy material like galvanized steel sheets to resist weather conditions and can be of portable type. The number of feed troughs may vary according to the number of animals. The hayracks may be constructed to feed from one or both sides with height of 12-15 in. from ground, width 14-16 in. for a one side feeder or 20-24 in. for both side feeders. In fence-line bunks there is no need for the operator to enter the shed. They are also well adapted to feeding mechanically. These may be constructed of concrete or wood. Such a bunk should be 6 in. deep, 12 in. wide at the bottom and 24 in. wide at the top.

Waterers: The simplest method of watering sheep is to use troughs or tanks. Metal or concrete troughs are satisfactory provided they are cleaned frequently. If tanks are used in summer, they should be shaded if possible so that the sheep will drink more. Also make the tanks shallow so that the lambs that fall into it can climb out easily.

Dipping tank: Depending upon the size of the flock, it may be prudent to plan for an area where sheep can be dipped. Dipping vats may be made of galvanized metal and portable or built permanently of concrete. If a galvanized steel tank is used, it should be well bedded down and the soil tight around it to prevent the sides from bulging when it is filled. The dipping tank should be built at one side of the yard.

Foot-bath: A foot bath made of galvanized steel sheet or brick in cement mortar should be provided at the entrance to the yard to protect the animals from foot rot disease. These baths may be embedded in the soil suitably.

Shearing and storeroom: The shearing room is meant for shearing sheep, storing wool and shearing equipment. The room may be of 6x2.5x3 m dimension or more depending on the flock size. The floor of this room is preferably made of wooden planks. The store room is meant to store feed and medicines and is usually made next to the shearing room, often separated from each other by a dividing wall. These rooms are provided with a door and two windows each.

Shepherd's house: The shepherd's quarter is meant for the caretaker and is located at a convenient place in the yard. This room can also be used to keep records, to revive chilled lambs and to do many other jobs in comfort. If a cot is provided, it may pay dividends for the shepherd to spend part of the night in the quarters to more closely monitor the ewes and to render assistance when needed. This room can also serve as an office and have telephones, etc.

Materials for shed construction: The common materials used for construction of sheds are asbestos and galvanized iron sheets, angle iron, tubular frame, cement, lime, sand and bricks. Bamboo and straw are also useful construction materials.

Cleaning and disinfection of shed: Cleaning of sheds is a problem and it should receive attention in the planning stage. If clear-span construction is used and partitions are only temporary, it should be possible to take them down and clean large spaces mechanically. Pests such as rodents, flies and birds should be considered in planning a shed. Mice and rats eat and spoil much feed and may also be carriers of disease. In slotted flooring the dung and urine are accumulated below, which may be cleaned once in a week. The rest of the flooring should be cleaned daily. Lime or bleaching powder may be sprinkled in the shed once in a week. The walls and roof may be disinfected with a flame gun once a fortnight. The walls may also be white washed with lime water to which 5% phenol is added. Sodium hydroxide at 4% may be used as disinfectant spray.

Sheep Equipment

The type of operation affects the kind and amount of equipment required. Larger operations may justify automation to save labour, while smaller ones may use more labour and be handled by members of the family. Labour may be in surplus and cheap in some areas, while it may be scarce and expensive in others.

Similarly cost of power may be prohibitive in some areas, while it may be relatively low and justify the use of more power and less hand labour.

The common equipments used in sheep management are spring balance, platform balances, ear-tags, tag imprinting die-set, hoof cutter, shearing scissors, shearing machine, measuring tape, feed and water troughs, hand trolley, axe, sickle, buckets, torch, wooden scrapper, tattooing forceps, hay racks, fencing rods and panels, etc.

Sheep Shelters for Rural Areas

Existing housing system: In the existing system of housing sheep, smaller number of sheep are generally housed together with goat, cattle and buffaloes in most parts of the country. In tropical climate they are kept in the open in the vicinity of the farmer's house. In temperate climate all these animals are either kept together in a separate house constructed adjacent to the house of the owner or are kept at ground floor and the farmer and his family live on the first floor of the house.

In warm climate small flocks of sheep are generally housed by the side of the house of the farmer taking advantage of the wall/ walls of the existing house. The other sides are enclosed by using either locally available fencing material like wood or dry thorny material or some katcha wall is constructed as enclosure. One gate either of wood or wrought iron is fixed at the entrance. In such traditional type of shelter no provision is made for construction of roof. As a result during rainy season animals get wet and their body resistance may go down and make them sick. Moreover, in such structures no provision is generally kept for extension and with increasing numbers there is overcrowding, which is a pre-disposing factor for outbreak of pneumonia. Chances of predation and theft are always there in this system. Large flocks of sheep are generally migratory in nature and are kept in the open. In north-temperate region the migratory flocks are generally taken into caves to protect them from heavy rains and snowfall. Some of the well to do and educated farmers are providing slightly better shelter to their animals. Using local material available region wise, they are providing thatch alongwith fencing of thorny material or thatch with kutcha walls and better thatch on the backyard of human houses. These housing systems after some alterations and modifications can be converted into improved houses where animals can be accommodated with comfort and under hygienic conditions.

Observations on Sheep Housing in the Field

Housing of Sheep in Arid Region: Observations in the field reveal that there is no dearth of space with the sheep owners in this region. They have one or more corrals for keeping their flocks. Corrals are usually surrounded by fence of thorny bushes of *Zizyphus nummularia* or the fence is made up of pile of phog wood (*Calligonum polygonides*). Hieght of fence ranges from 5-6 feet. In the open corral, on an average 78.9 sq feet/ sheep (range 28.3 to 250 sq feet) space is available for housing of sheep during the evening hours. For pregnant and newly lambed ewes 8.6 sq feet/ sheep (range 6.6 to 9.8 sq feet space is available in the corrals. Breeders usually house adult animals in the open space; thatched roof or covered area is used only for lactating ewes and lambs to protect them from inclement weather. Covered space is of round shape or rectangular. The dome shaped hut is 19-22 feet in diameter and 15-18 feet in height with slanting roof prepared by putting piles of bajra stems and kheemp above them and tying them with rope made of twisted thin stems of kheemp. Rectangular covered space provided for housing of lambs was 3.2 sq feet (range 1.4 to 3.9 sq feet), whereas open space provided to lambs was 3.5 sq feet/lamb. Walls of the covered huts are made of pucca bricks or mud with mud masonary and plastered with a mixture of cow dung and mud. The covered/ open boundary walls have ventilators. When the flocks are on migration the huts are used for storing dry fodder (bajra kuttar/ groundnut fodder/ lahsu-dry leafy part of *calligonum polygonides*. The mud floor is sandy in all the corrals except that for lambs in the covered areas where floor is plastered with a mixture of cow dung and mud. For drinking water, water channels at common watering point are available. In addition to this, some breeders have iron troughs for watering of lambs.

On the basis of survey of various shelter management systems pertaining to sheep and goat surveyed in western, eastern and southern Rajasthan, information was collected on land holding size, housing systems, feed management, migration, health management, reproduction management and production parameters. The main focus was on the type of shelter provided to the different categories and type of animals. The salient findings based on survey of sheep/ goat farmers are: -

Caste : Rebaris, Gujar, Jat, Rajput, Yadav and Muslims.

Land Holding Size : Mostly landless farmers but some have land holdings ranging from 2.5 to 6.0 acres.

Flock Size : The average flock size was 88 animals and it ranged from 39 to 259 animals. The number of males ranged from 1 to 7 and number of females from 38 to 255.

Breeds of Sheep : Marwari, Jaisalmeri, Kheri.

Breeds of Goat : Marwari, Sirohi.

Housing System : Sheep are mostly housed in open paddocks. For protection of animals, a protective boundary is made from thorny bushes 5 to 6′ in height. In sheep and goat mixed flocks of smaller size, usually a hut or thatched structure is provided in the centre/ corner of the paddock. In some cases lamb nurseries were also observed.

Orientation : The orientation of the paddocks was predominantly East-West but North-South orientation was also observed.

Duration of Housing of Animals : 10-12 hours only. In Jaisalmer region, the sheep are left to graze in the open pasture for upto 2-3 days before they are brought back for watering.

Watering : Sheep are watered once a day, usually in the afternoon when they rest. They drink water from the village pond or from common water points/ wells provided by the Govt.

Feeding : Animals are mainly maintained on grazing in forest area or community grazing lands. Lambs and lactating ewes are supplemented with pala leaves.

Sanitation : In arid regions, particularly Jaisalmer and Barmer, the paddocks are not swept daily but cleaned only after 6 month or simply the paddock is shifted to a new site. The droppings pile up and act as insulation during cold desert nights. Otherwise the paddocks are swept daily after the animals leave for grazing. The floor is kept kutcha and the urine is absorbed, but the mud is not changed nor lime, etc sprinkled.

Construction Material Used : Since the sheep paddocks are mostly of open type, they are provided with a protective boundary of either stone bricks/stone slabs or simply thorny bushes. Thorny bushes are most commonly found in the arid region. Huts and thatched structure are also made from locally available construction material like tree branches, grasses, etc.

Migration of Flock : Usually large flocks, encountered in Barmer/Jaisalmer regions, migrate to neighbouring states of Punjab, Haryana, U.P. and M.P. They are on migration for 4-5 months in a year.

Health Management : For health management i.e. for vaccination and deworming, the sheep farmers are in contact with the State Animal Husbandry Department. However, the farmers also know ethno-veterinary practices. The common ailments affecting sheep are Pneumonia and diarrhea, though disease outbreaks are also encountered from time to time.

Reproductive Management : The age of ram at first breeding is 1 to 1.5 yrs, while the age of ewe at 1st lambing is 1.5 yrs. to 2.0 yrs. Mostly single lambs are born. Rams are kept free with the flock during the breeding seasons. At other times their prepuce is tied to prevent them from breeding. The ewes, which do not lamb for 2 consecutive seasons, are culled. Among reproductive disorders, abortion is the most common.

Shearing of Animals : Done 2 to 3 times in a year on community or shared basis.

Revenue : Major earning of a sheep farmer is from the sale of live animals followed by sale of wool, milk and manure.

Conclusion and Recomendations

Under rural conditions for smaller units of sheep, elaborate housing system is not required. The cheapest form of shelter may be constructed against the side of an existing building. The length and width of the shed may vary according to the number of animals to be reared. The height of the shed may be at least 3 metre. In construction of this type of rural shelter, preference may be given to utilization of locally available materials like bamboo or cheap wood as supporting material, bamboo sticks, bamboo mats or thorny bushes as fencing material and kans, moonj, pani pula or palm leaves as roofing material. The shape of thatched roof should be slanting. The floor must be at a raised platform and should be made of mooram or simple mud and the mud should be replaced at suitable intervals. The important point is that space requirement / animal must be strictly followed to avoid overcrowding and care should be taken for proper ventilation. There should be a separate provision for keeping females before and after parturition for a few days. If males are housed with the females, their prepuce must be tied. At one corner of the enclosure a small area may be earmarked for segregation of sick animals.

For large flocks in rural areas, the sheep shelter may be on the lines described for organized farms. In such cases all the construction material used for organized farm may be replaced by cheaply available local material. Bamboo or hard wood pillars may be used in place of angle irons as supporting structures. Kutcha walls, bamboo sticks / mats, sarkanda or dried thorny bushes may be used as fencing material in place of chain-link net fencing. The asbestos sheets may be replaced by simple thatch made of locally available materials like kans, moonj, palm leaves, etc. Separate sheds may be constructed for accommodating males, females, sick animals, parturiting and lactating animals and neonates. For providing supplementation, mangers of suitable dimensions may be constructed using local material inside the sheds or a common manger can be constructed outside in the vicinity of these sheds. Similarly watering arrangement may be made for availability of clean and fresh water.

27

Significance of Body Condition Score on Carcass Traits of Sheep

V.P. MAURYA, V. SEJIAN, DAVENDRA KUMAR
Central Sheep and Wool Research Institute, Avikanagar, Rajasthan 304 501

Introduction

Body condition affects both sheep and lamb performance. Poor body condition is associated with increased post-partum interval, weak lambs at birth (Maurya *et al.*, 2007), lower weaning weights and reduced sale receipt per sheep (Rossi and Wilson, 2006). Increasing post-partum interval will result in a younger, smaller lamb at weaning and will also result in lower incomes if sold at weaning. Weak lambs at birth may not get adequate colostrum and are more susceptible to disease, reduced weaning weights, reduced feedlot performance and less desirable carcass traits. Research clearly shows that sheep in moderate body condition will have a shorter interval from lambing to first estrus than sheep in poor condition. Body condition score (BCS) is one of the most important factors in determining the desirable carcass traits (Apple, 1999). Condition scoring of sheep at slaughter has shown that, among animals of the same age and sex, live weights, carcass weights, and edible tissue yield are highly correlated with condition score. Thus those involved in marketing of live sheep would find condition scoring useful in estimating expected saleable sheep meat (Hassen *et al.*, 1999).

Meat from small ruminants accounts for almost 30 % of the meat consumed in the world (Okello *et al.*, 1984). In terms of palatability and delicacy, mutton is preferred to beef. It is possible that sheep could play a leading role in the provision of animal proteins to supplement other sources such as beef, chicken, pork and fish. The complex task of determining carcass value and, ultimately, live value is no longer based on the whole carcass, but rather on pieces (primal and sub primal cuts) from the carcass. However, very little information is available concerning sub primal cut yields from sheep carcasses. This information gap poses a major stumbling block in the valuation process of spent animals and culled breeding stock. Therefore, the objective of this paper was to assess the effect of BCS on carcass quality and cutability characteristics and sub primal cut yields from carcasses of culled sheep.

BCS and Live Weight

Association between live weight and body condition scores exist in sheep (Maurya *et al.*, 2006) and beef cows (Apple, 1999). As the animals increase in live weight, their body condition scores tend to increase and as sheep lose live weight their body condition scores tend to decrease (Hassen *et al.*, 1999)

What body condition scores feel like on a live cow

Body condition score	What you feel at each site			What the score means for meat production
	Backbone	Spinal processes	Eye muscle	
3	Prominent and sharp.	Sharp ends. Fingers easily pass under ends.	Very thin and feels hollow.	Very lean. Poor meat yield. Should be fed more. Further weight loss may result in death.
5	Prominent but smooth.	Smooth and rounded. Fingers pass under ends.	Some tissue present. Feels flat.	Lean. Moderate meat yield for adults. Too low for prime calves.
7	Smooth and round over the top but still elevated.	Smooth. Need pressure to feel ends.	Full coverage to end of spinal processes. Feels rounded.	Medium. Ideal for prime calves. May be too fat for adult sheep where a slightly lower score is often preferred.
9	Only detected with pressure.	Cannot be felt.	Well rounded.	Fat. Too much feed has been used. Fat has to be cut off meat when processed.

Source: Apple, 1999

Relationship Between BCS & Carcass Characteristics

Generally the carcass weight and live weight increased as the BCS increased (Gregory *et al.*, 1998). In goats carcass weight increases by 0.43 to 0.54 kg for every 1 kg increase in live weight (McGregor, 1990). However, carcass length did not showed this trend, suggesting that the increase in weight occurred in the soft tissues rather than in skeletal size. Animals having low condition score had limited body fat reserves. The extent to which the rate of increase in fatness increased as condition score increased. The dressing percentage of slaughter sheep may be influenced by age, weight, sex, body condition, amount of gut fill at slaughter, whether the carcass is weighed hot or cold and, of course, by the number of body components included in the yield calculation (Apple et al., 1999). Dividing the un shrunk live weight into the weight of the hot carcass with liver, heart and kidney but without skin, head, feet and viscera will yield dressing percent values in the 45- 52 % range (Gregory *et al.*, 1998). The interactions among the individual factors affecting dressing percentage are great, making it rather difficult to accurately predict carcass yield or quality by visual examination of the live animals. However the single factor which affect maximum the carcass characteristics is the body condition (Shelton *et al.*, 1984). Body condition score and carcass value were subjective assessment. Condition score reflects subcutaneous fat deposits, which is higher in sheep than in goat. It also reflects muscular development and coverage of the ribs. Similarly carcass value can be assessed according to visible subcutaneous fat coverage and leanness of the meat. Body condition score and carcass value were better correlated in sheep than in goats. Carcass quality can be assessed by measuring the fat deposits in the omentum, kidney and brisket as well as in the thickness of the *Longissimus dorsi muscle.*

BCS and Its Relation to Fat Content

The BCS has been related to the fat cover and carcass fat by many researchers (Tinker *et al.*, 1990). A BCS range of 3 to 4 will include most mutton sheep. In slaughter trials, researchers have found that sheep with body condition scores of 3, 3.5, and 4 had a carcass fat content of 7 to 9 percent, 15 to 18 percent, and 25 to 27 percent, respectively. Sheep with condition scores of 3.5 or higher show a smoother appearance across the ribs and back. The breed type of sheep can influence where fat is deposited. Some sheep show very little fat over the ribs, but will deposit fat over the hooks and pins. Other sheep show uniform deposits of fat across the ribs and back, with no patchy deposits around the tail head. Fat scoring is used in live stock market reports as a method of estimating Fat Classes (Delfa et al., 1991). Fat scores and fat classes range from 1 to 5

The relationship between cow fat classes and tissue depth at the GR site

Fat Class	Description	Tissue depth at GR site
1	Very lean	up to 3 mm
2	Lean	4 to 6 mm
3	Moderately lean	7 to 9 mm
4	Moderately fat	10 to 12 mm
5	Fat	over 12 mm

Source: Apple *et al*, 1999

Fat Classes are decided in the meat works by measuring the tissue depth of the carcass at the GR site (Schro derl and Staufenbiel, 2006). The GR site is on the second last long rib (12th rib) as a site 110 mm from the midline (ridge of the spine). The tissue depth at the GR site includes muscle and fat. The GR site is regarded as a good reference point as it:

- Provides a reliable indication of the meat and fat content of the carcass,
- Is easy to measure.

The GR measurement is recorded in mm. The idea with fat scoring is to use your hands to estimate the fat class into which the carcass will be assigned after the sheep is slaughtered. Fat value increased linearly and bone value decreased linearly as BCS increased from 2 to 5. Carcasses from BCS-5 sheep had the least valuable sub primal cuts from the chuck, loin, and round; therefore, within the Utility grade, carcasses from the fattest group (BCS of 5) of sheep had less total value than all other groups of sheep (Apple *et al,* 1999a).

Effect of BCS on Different Parts of Carcass

It is a general finding that live weight, hot carcass weight, dressing percentage, fat thickness, *Longissimus* muscle area, and numerical yield grade increased as BCS increased from 3 to 5 (Apple *et al.,* 1999 a; Funston et al., 2003). Carcasses from BCS 1- 2 sheep are graded as Cutter or lower, and carcasses from BCS- 3 and BCS- 4 sheep are graded as utility. Hides from BCS-1 sheep were of greater value than hides from cows assigned a BCS of 3 or higher. The value of the tongue, tripe, and feet decreased linearly as BCS increased from 2 to 5. Blood meal, liver, honeycomb- tripe, kidneys, heart, trachea, lungs, and bone meal from BCS- 2 sheep were more valuable than that of sheep assigned a BCS of 3 or higher. Body condition scores had no effect on values of the head meat, gullet, weasand meat, oxtail, or spleen (Shemeis *et al.,* 1994).

Effect of BCS on Different Carcass Cuts

BCS affects the values of sub primal cuts, minor cuts, and lean trimmings from utility carcasses (Apple *et al,* 1999b). Within the Utility grade, BCS had no effect on the value of the chuck roll, rib eye roll, strip loin, bottom sirloin butt, ank steak, or lean trimmings. The shoulder clod from carcasses of BCS- 1 and 2 sheep had more value than the shoulder clod from carcasses of BCS- 3 or 4 sheep. Moreover, carcasses from BCS- 3 sheep had more valuable chuck tenders than carcasses of BCS- 4 or more sheep. Top sirloin butts from carcasses of BCS- 1 and 2 sheep and tenderloins from carcasses of BCS- 2 sheep had greater value than those from carcasses of sheep assigned a BCS of 3, 4, or 5. Value of the knuckle was higher for carcasses from BCS- 3 sheep than carcasses of sheep assigned a BCS of 4 or 5. Inside rounds from BCS- 4 and 5 sheep were of less value than inside rounds from BCS-1 and 3 sheep. Carcasses of BCS- 5 sheep had less valuable outside rounds than carcasses of sheep assigned a BCS of 1, 2, 3, and 4, and less valuable eye of rounds than carcasses of cows assigned a BCS of 1, 2, 3, and 4.

Effect of BCS on Gross and Net Value of the Carcass

Absolute gross and net values of sheep were affected by the different BCS. Within the grade of Cutter, carcasses from BCS- 1 sheep had the lowest and carcasses from BCS- 4 sheep had the highest gross and net values. Among utility carcasses, the gross and net values of BCS-5cows were greater than the gross values of BCS- 1, 2, 3, and 4; however, the live value of cows producing utility carcasses was unaffected by BCS (Apple *et al.*, 1999b). Across both quality grades the gross and net value increased linearly as BCS increased from 1 to 4. Sheep designated with a BCS of 4 and 5 had greater gross and net values than Sheep assigned a BCS of 3 or lower. Moreover, BCS-3 sheep had higher gross and net values than BCS-1, and 2 sheep; BCS-1 sheep had the lowest gross and net values.

Effect of BCS on By-products Yield and Value

There is generally a correlation between BCS & by-products yield (Apple et al, 1999a). Even though BCS- 1 and 4 sheep had similar total by- product values, carcasses from BCS- 4 sheep were more suitable for fabrication into boneless, sub primal cuts for the food- service trade. Thus, these fatter sheep would be of greater value to both If producers and packers.

Impact of BCS and Carcass Quality on Economy

Although BCS-4 and BCS-5 sheep produced carcasses grading utility or higher and had the highest gross and net values, carcasses from these two scores had the lowest yields of trimmed or denuded sub primal cuts (Devendra, 1981). Additionally, carcasses from BCS- 4 and 5 sheep would incur the greatest labor costs associated with trimming sub primal cuts. Sheep assigned a BCS of 1 had the greatest by-product values. Moreover, BCS- 1 sheep produced the least valuable carcasses and also had the lowest gross, net, and live values among all BCS. Sheep designated with a BCS of 1 would require additional feed inputs to improve carcass quality traits, carcass value, and, ultimately, live value; however, feeding sheep can introduce a substantial amount of risk due to increased feed costs (Hanset et al., 1987). On the other hand marketing sheep at a BCS of 3 could optimize economic returns to both producers and packers. Feeding to, or marketing at, a BCS of 3 would result in a high percentage (73.3 %) of carcasses grading utility and maximize total lean product yields. Carcasses from BCS- 3 sheep were lean and would produce sub primal cuts requiring minimal trimming to achieve a marketable fat- trim level (Reed *et al.*, 1988).

Computer Tomography Index to Improve Carcass Traits

CT and ultrasonic measurements provide an effective means of selecting for improved carcass composition in sheep (Karamichou *et al.*, 2007). The ultrasonic and CT measured traits had greater heritablility than the subjectively assessed traits. The maternal environment, as expected, also affected many of traits. The ultrasonic and CT traits also showed high levels of genetic variation, which is the second requirement for making genetic progress. Judging by genetic correlations, selection on CT index achieved a moderate improvement in conformation, price and slaughter weight, and a decrease in fat class. These changes are in fact seen for fat class, conformation and live weight. Conversely, selection on live weight at scanning is expected to improve carcass and slaughter weight, and total price received, but have slightly deleterious impacts on carcass conformation score.

Effect of Weaning Age and Slaughter Weight on Carcass and Meat Quality of Lambs

Un weaned lambs raised on a grass- based diet with an *ad libitum* concentrate supplement exhibit greater production efficiency than their weaned counterparts; as they display greater growth, lower concentrate consumption and higher carcass dressing at slaughter (Caneque *et al.,* 2001). Following these same parameters, weaning at 65 days of age is more advantageous in terms of productivity than weaning at 45 days of age. Weaning represents a sudden change for lambs, reducing their growth rate and altering body condition with a drop in fatness and a proportional increase in the quantity of lean muscle. This effect is most pronounced in lambs weaned earliest. The greater fatness of the un weaned lambs, together with their lower total collagen values and the greater lightness of their meat indicates that the consumer may better accept their meat. From the human nutrition standpoint, the higher proportion of SFA in the meat of un weaned lambs is compensated for by its lower n-6/n-3 ratio confirming the idea that its consumption is healthier than that of weaned lambs in so far as lowering the risk of suffering cardiovascular disease. With regard to the higher-weight lambs, although their meat is more succulent due to the fact that it displays a higher water holding capacity (WHC), its texture may be affected by the lower solubility of its collagen. The meat of these lambs presents a lower PUFA/SFA ratio, which indicates that it is also inferior from a nutritional standpoint.

Effect of Crossbreeding on Growth and Carcass Characteristics in Sheep

Crossbreeding in sheep had lead to producing sheep of moderate body condition, which had better carcass characteristics (Dawson *et al.,* 2003). Crossbred ewes produced by crossing Texel sires with Rouge de l'Ouest dams and Rouge sires with Texel dams, relative to lambs from purebred Texel and Rouge ewes. The ewes were crossed with Texel or Rouge sires and the relative performance of the sires assessed in terms of lamb carcass quality. Ewe and ram genotype had a significant effect on conformation classification (Heelsum *et al.,* 2003). Lambs from Texel ewes had a greater conformation classification (4·0) than lambs from Rouge ewes (3.4) and Texel-sired lambs had a greater conformation (4.0) than Rouge sired lambs (3.6). Fat depth was significantly influenced by ewe genotype with rams from Texel X Rouge ewes having greater fat depths compared with lambs from Texel ewes (longissimus dorsi 2.3 and1.9

mm respectively). Chemical composition of the carcass and meat quality measurements were unaffected by ewe or ram genotype. Significant heterosis effects on lamb weight gain from birth to slaughter, age at slaughter, weight of perinephric and retroperitoneal fat and fat depth over *Longissimus dors*i and *Gluteus medius* were also been reported. The results from this study demonstrate that significant heterosis effects, particularly in terms of lamb growth rate, can be achieved by crossing Rouge and Texel breeds. In addition lambs from crossbred ewes have similar conformation characteristics to the Texel breed, which has been intensively selected for carcass characteristics. Duration of feeding also influence live weight and carcass characteristics in goats (Fehr *et al.*, 1976).

In another experiment conducted on cattle, the genotype of cows and bull have shown to influence the body confirmation and carcass quality (Marshall *et al.*, 1994). By using high index rams the leanness of the carcass of their crossbred offspring can be improved, without affecting live or carcass conformation and with only a small increase in size. The later is of great importance to crossbred dam line sheep, because the goal to keep the size of crossbred ewes unchanged is considered important to overall efficiency. High conformation rams gave slightly quicker finishing lambs with larger fat depths, but live and carcass conformation did not improve. This implies that selection on lean index will improve carcass quality, but the benefit of selection on live conformation remains unclear.

Management Strategies to Improve BCS and Carcass Traits

Several management strategies that may help maximize profit by improving the BCS and thereby increasing the carcass yield (Rossi and Wilson, 2006).

- Adjust stocking rate to ensure adequate forage during the stocking rate limiting months,
- Target a lambing season that fits the forage, supplements, marketing plan, and management,
- Cull open and poorly producing sheep,
- Control parasites and diseases,
- Provide a good mineral-free choice all year,
- Provide on daily basis supplemental crude protein for sheep grazing forages with less than 7 percent crude protein or that have a TDN to crude protein ratio above 7,
- Provide adequate protein for young ewes to improve rebreeding. This may require managing first and second lamb ewes in separate herds during the winter.
- Group sheep by age and nutritional needs. Separate groups for weaned ewes, yearling ewes, first lamb ewes and young ewes and adult ewes may be needed for part or all of the year,

- Provide higher quality forages or higher levels of energy supplements balanced with protein to reduce weight loss or improve BCS when sheep BCS is below 3.

Conclusion and Recommendations

Body condition at harvest affects products, carcass, and live values. Fatter sheep, with a body condition score (BCS) of 5 or 6, should have the greatest gross and net values because their carcasses should grade Utility or higher. Sub primal cuts from these carcasses would incur the greatest labor costs associated with trimming to a marketable fat-trim level. Thin-conditioned (BCS- 1 and 2) sheep will have less value and may require additional feed to improve their value. The mutton industry could optimize returns to producers and processors by marketing cull sheep with a BCS of 4. Carcasses from BCS- 4 sheep should be the most valuable, and the live value should be similar to BCS- 5 and 6 sheep. Information about carcasses from cull sheep is necessary as the mutton industry moves closer to a value-based marketing system.

REFERENCES

Apple, J.K. 1999. Influence of Body Condition Score on live and carcass value of cull beef cows. *Journal of Animal Science* 77, 2610-2620.

Apple, J.K., Davis, J.C. and Stephenson, J. 1999a. Infuence of Body Condition Score on by-product yield and value from cull beef cows. *Journal of Animal Science* 77, 2670-2679.

Apple, J.K., Davis, J.C., Stephenson, J., Hankins, J. E., Davis, J. R. and Beaty, S. L. 1999b. Influence of Body Condition Score on Carcass Characteristics and Subprimal yield from cull beef cows. *Journal of Animal Science* 77, 2660-2669.

Caneque, V., Velasco, S., Diaz, M., Perez, C., Huidobro, F., Lauzurica, S., Manzanares, C and Gonzalez, J. 2001. Effect of weaning age and slaughter weight on carcass and meat quality of Talaverana breed lambs raised at pasture. *Animal Science* 73, 85-95.

Dawson, L.E.R., Carson, A.F., McClinton, L.O.W., Kilpatrick, D.J. and Moss, B.W. 2003. Comparison of the carcass characteristics and meat quality of lambs produced from Texel and Rouge de l'Ouest ewes and their crosses. *Animal Science* 77, 57-65.

Delfa, R., Teixeira, A,, Blasco, I. and Colomer-Rocher, F. 1991. Ultrasonic estimates of fat thickness, C measurement, Longissimus dorsi depth, in rasa aragonesa ewes with same body condition score. Option *Mediterraneennes* 13, 25-30.

Devendra, C. and Burns M. 1983. Feeding and nutrition. In: *Goat Production in the Tropics*. CAB (Commonwealth Agricultural Bureaux), Farnham Royal, Slough, UK. pp. 90-115.

Fehr, P. M., Sauvant D., Delage J., Dumont B. L. and Roy G. 1976. Effect of feeding methods and age at slaughter on growth performances and carcass characteristics of entire young male goats. *Livestock Production Science* 3, 183.

Funston, R.N., Paterson, J.A., Williams, K.E. and Roberts, A.J. 2003. Effect of body condition, initial weight, and implant on feedlot carcass characteristics of cull cows. *The Professional Animal Scientist* 19, 233-238.

Gregory, N.B., Robins, J.K., Thomas, D.G. and Purchas, R.W. 1998. Relationship between body condition score and body composition in dairy cows. *New Zealand Journal of Agricultural Research* 41, 527-532.

Hanset, R., Michaux, R. and Stasse, A. 1987. Relationships between growth rate, carcass composition, feed intake, feed conversion ratio and income in four biological types of cattle. *Genetic Selection and Evolution 19,* 225-248.

Hassen, A., Wilson, D.E. and Rouse, G.H. 1999. Evaluation of carcass, live, and real-time ultrasound measures in feedlot cattle: I. Assessment of sex and breed effects. *Journal of Animal Science* 77, 273-282.

Heelsum, A.M., Lewis, R.M., Davies, M.H. and Haresign, W. 2003. Growth and carcass characteristics in wether lambs of a crossbred dam line. *Animal Science* 76, 43-53.

Karamichou, E., Merrell, B.G., Murray, W.A., Simm, G., and Bishop, S.C. 2007. Selection for carcass quality in hill sheep measured by X-ray computer tomography. *Animal* 1, 3-11.

Marshall, D.M. 1994. Breed differences and genetic parameters for body composition traits in beef cattle. *Journal of Animal Science* 72, 2745-2755.

Maurya, V.P., Naqvi, S.M.K., and Joshi, A. 2007. Effect of thermal stress on physiological responses and blood profile of native Malpura sheep reared in hot semi-arid region of India. *Indian Journal of Animal Science* (In press)

Maurya, V.P., Naqvi, S.M.K., Kumar, D., Gulyani, R. and Joshi, A. 2006. Annual report CSWRI, Avikanagar. PP 45.

McGregor, B.A. 1990. Boneless meat yields and prediction equations from carcass parameters of Australian cashmere goats. *Small Ruminant Research* 3, 465-73.

Okello, K.L., Ebong, C. and Opuda-Asibo, J. 1984. Effect of feed supplements on weight gain and carcass characteristics of intact male Mubende goats fed elephant grass (Pennisetum purpureum) ad libitum in Uganda.

Reed, J., Ebong, C., Tanner J., Gebru, G. and Akale Work, N. 1988. The nutritive value and use of feeds for fattening small ruminants in African highlands. A perspective of ILCA research. *Small Ruminant Research Network Newsletter* 12, 10-27. ILCA (International Livestock Centre for Africa), Addis Ababa, Ethiopia.

Rossi, J. and Wilson, T.W. 2006. Body condition scoring beef cows. In the bulletin published by university of Goergia.

Schro derl, U. J and Staufenbiel, R. 2006. Invited Review: Methods to Determine Body Fat Reserves in the Dairy Cow with Special Regard to Ultrasonographic Measurement of Backfat Thickness. *Journal of Dairy Science* 89, 1-14.

Shelton, Maurice., Gary Smolder and Figueiredo, E.A.P. 1984. Meat Production and Carcass Characteristics of the Goat. USAID Title XII. Small Ruminant Collaborative Research Support Program, Texas A & M University, San Angelo.

Shemeis, A. R., Liboriussen, T., Bech-Andersen, B and Abdallah, O.Y. 1994. Changes in carcass and meat quality traits of Danish Friesian cull cows with the increase in their age and body condition. *Meat Science* 37, 161–167.

Tinker, E.D., Buchanan, D.S., McCarter, M.N., Dolezal, H.G., Frahm, R.R. and Knori, L.W. 1990. Evaluation of body condition scoring for estimating carcass fat percentage of various cow types. Okla. Agric.Exp. Stn. Res.Rep. MP-129: 14–17. Oklahoma State Univ., Stillwater.

❑❑❑

28

Phytoestrogens in Mutton Production

BANDLA SRINIVAS, OM HARI CHATURVEDI, SHAIKH ABDUL KARIM

Central Sheep & Wool Research Institute, (via-Jaipur) Avikanagar - 304 501, Rajasthan

Introduction

The science of sheep nutrition, specifically for mutton production, gets increasingly involved with questions on how to supply nutrients that guarantee adequate physiological development, good health and reproduction because these animals are largely dependent on open system of feeding which means grazing in the pastures, grasslands etc (Holtenius, 2006). Failure of one breeding season in mutton production means a monitory loss of feeding expenses, loss of productive life time of ewe and net return on sale of 6-8 months old lamb. For profitable mutton production, functional conception rate of flock should be 70 per cent or above (Plant, 1981). Unlike like cultivated fodder, grazing/pasture lands contain different types of herbage with varied biochemical characteristics. Plants have co-evolved with predator populations of bacteria, insects, fungi and grazing animals, and have developed defence mechanisms which assist their survival. Leguminous trees and shrubs often have thorns, fibrous foliage and growth habits which protect the crown of the tree from defoliation. Many plants also produce chemicals which are not directly involved in the process of plant growth (secondary compounds), but act as deterrents to insect and fungal attack.

Secondary phytochemical compounds also affect animals (including humans) and the nutritive value of forages. The nature and action of toxins in plants has been the subject of several reviews (Duke, 1977, Rosenthal and Janzen 1979, Hegarty 1982, Seawright *et al.* 1985, Barry and Blaney 1987), in which attention was focused on pasture plants. Phytoestrogens are one among them drawn attention of the researcher recently for their full range of biological effects in farm animals.

Phytoestrogens

There is a lot of confusion about phytoestrogens whether these are beneficial or harmful both for animals and human beings in health and productivity spheres? Certain water soluble glucosides in plants bear close resemblance with oestradiol (a steroid hormone) and are knows as phytoestrogens. More than 800 types of phytoestrogens are present in nature. Plant origin oestrogens are important because a successful mutton production with high productivity, fertility and low incidence of production diseases is essential in order to compete in the market and fulfil food safety and ethical aspects for safe consumption. Many studies showed that regular consumption of phytoestrogens rich foods reduces risk or developing cancer but, notionally many livestock owners and personnel assume they are bad for health, specifically for reproduction (Owens, 2003). Consumption of phytoestrogens are higher in Asian countries and on vegetarian diets where traditional feeds/foods are still dominant on the dinning table and perhaps it is one of the reason for lower rates of mammary gland, prostate and colon cancer in these regions. Among forage crops, many legumes contain compounds that can affect the productivity of sheep, but not all of these are oestrogens. Any compound to qualify as phytoestrogens must have a chemical structure that permits the compound to bind with the oestrogen receptor and there by causing detectable effects.

Different types and their metabolism

Phytoestrogens are naturally occurring phytochemicals which are structurally and fundamentally similar to 17-â-estradiol. There are 3 principal classes of phytoestrogens, viz. isoflavones (e.g., lucerne, chickpeas, mung beans, whole grains, red clover, soya etc.), lignans (e.g., linseed, rye, buckwheat, millet, whole grains, sesame seeds, sunflower seeds, legumes, beans etc.) and coumestans (Alison *et al.*, 2003).Chemical structures of most common phytoestrogens present in plants is shown in figure 1 (Benassayag *et al.*, 2002). Isoflavones are the most potent form among three principle classes. They are thought to have a broad range of biological functions in plants, although these are poorly understood. However, two particular functions are recognised – (a) as phyto-alexin or stressor chemicals which are secreted by the plant in response to attack by parasites such as insects, fungi, viruses, etc., and which display activity against these parasites, and (b) chemicals which encourage colonisation of nitrogen-fixing bacteria on the roots of legumes. The biological functions in plants of the lignans and

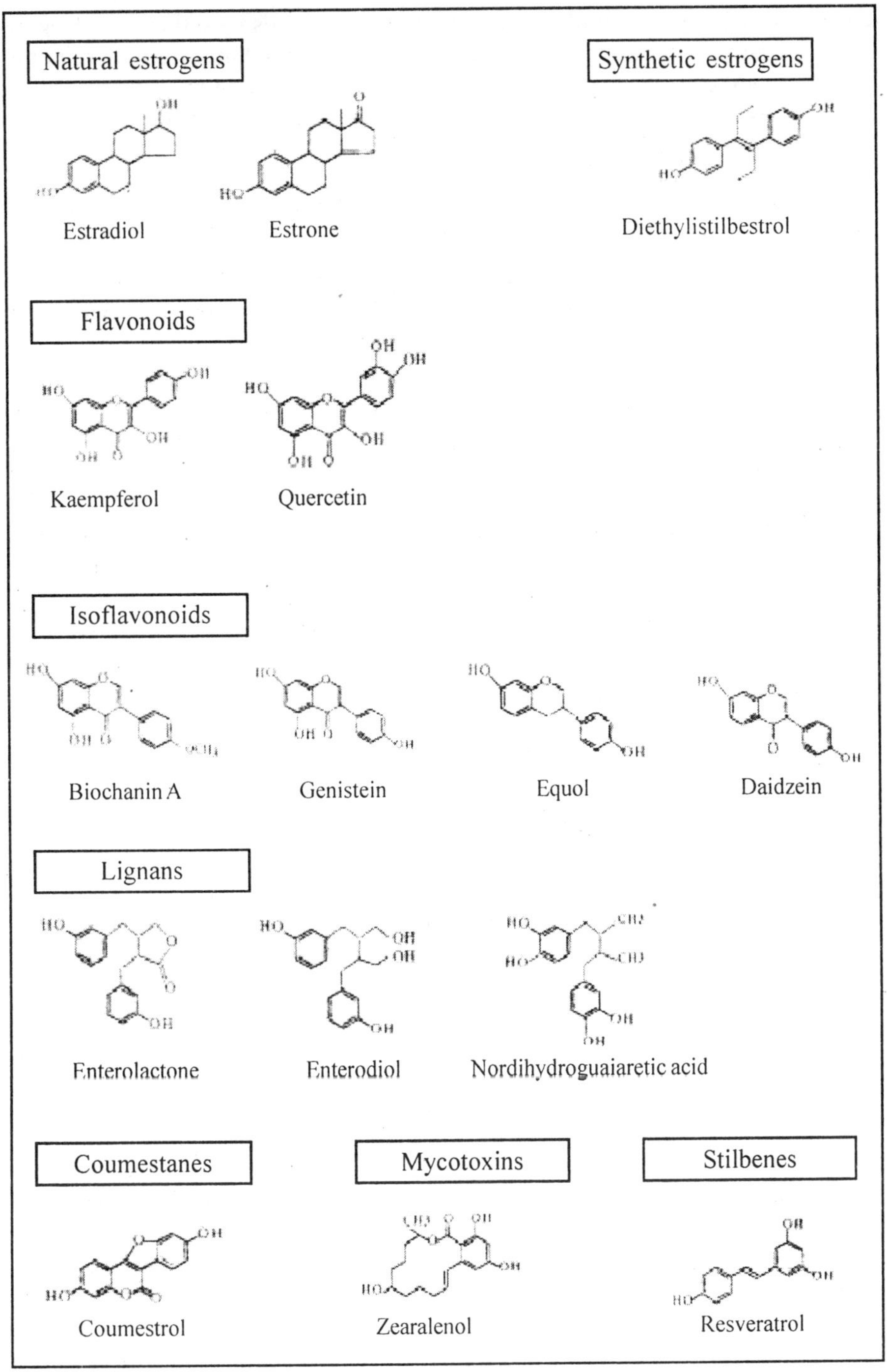

Fig. 1. Comparison of molecular structures of phytoestrogens and nature or synthetic estrogens.

coumestans is not generally understood. There are also synthetic oestrogens such as diethylstilboestrol used as drug in many animal and human ailments. Besides plant source of oestrogens, there is another oestrogen of fungal source such as zeralenone that has been elaborately investigated and reviewed (Morgavi and Riley, 2007), but falls outside the scope of this lecture.

The following are the 4 types of phytoestrogens (Nancy, 1995a,b; Fang et al., 2003):

- Type 1 – Isoflavones: Isoflavones appear to be widely distributed in the plant kingdom and over 700 different isoflavones are described. However, the isoflavones which display oestrogenic activity belong to a small sub-group and are restricted almost exclusively to the leguminosae family. The known oestrogenic isoflavones are daidzein, formononetin, genistein and biochanin A. In common animal/human foodstuffs such as soya, chickpeas, lentils and beans, the total levels of the oestrogenic isoflavones range between about 40 and 300 mg per 100 g dry weight. In the raw plant material, isoflavones occur principally as glycosides. Following ingestion by man and animals, the glycoside moiety is hydrolysed free by a combination of gastric acid hydrolysis and fermentation by intestinal bacteria. Some of the isoflavones in the aglucone form are absorbed directly and circulate in the blood, while the remainder are metabolised by intestinal fermentation to a variety of compounds which are also absorbed. The absorbed isoflavones and their metabolites appear to undergo little or no further metabolism in the body, being readily transported in the bloodstream, and ultimately being excreted in the urine.
- Type 2 – Lignans: Lignans are widely distributed in the plant kingdom. Over one hundred lignans are described and they are reported in common human foodstuffs such as cereals, fruits and vegetables. Oilseeds such as linseed have the highest known levels at 20-60 mg/100 g dry weight, while cereals and legumes have much lower levels at 0.3-0.6 mg/100 g, and vegetables have even lower levels at 0.1-0.2 mg/100 g. The most common lignan described is metairesinol. Dietary lignans also appear to be metabolised fairly efficiently within the gut by bacterial fermentation, yielding metabolites such as enterodiol and enterolactone which are absorbed into the bloodstream and excreted in the urine.
- Type 3 – Coumestans: Compared to isoflavones and lignans, oestrogenic coumestans appear to have a relatively restricted distribution in plants and generally occur at much lower levels. Lucerne, ladino clover and some other fodder crops such as barrel medic may have significant levels and have been reported to cause reproductive dysfunction in grazing animals. In the human diet, the important sources of coumestans are sprouts of soya and lucerne where levels up to 7 mg/100 g dry weight are reported. Whole soyabeans and other common foodstuff legumes contain levels of approx. 0.12 mg/100 g dry weight and most of that is concentrated in the

seed hull which commonly is removed in the preparation of human foodstuffs.

- Type 4 - Oestrogens: These are compounds closely related to animal oestrogens such as oestrone, oestradiol and oestriol. These have been described in plants such as liquorice, apple, French bean, pomegranate and date palm. Little is known of the metabolism and biological significance of these chemicals in humans and animals.

Ingestion and metabolism in sheep

Isoflavones are the principle source from legumes besides lignans and, to a much lesser extent, coumestans in sheep. Presence of cereals in concentrate supplements or feeding of kitchen vegetable wastes as practiced by livestock house holds would also boost the lignan intake (Manidool, 1985). However, commonly intake of isoflavones is far more than other types in sheep. A primary effect appears to be associated with their close structural relationship to naturally-occurring oestrogens which allows the phytoestrogens to mimic the effects of the endogenous oestrogens. The principle isoflavones biochanin A, daidzein, formonetin and geinstein are synthesized from phenylalanine in plats. Although formononetin has about half, one third and one quarter the oestrogenic activity of biochanin A, diadzein and genistein negative oestrogenic effects on sheep appear to be directly correlated to the formononetic content in common legumes, e.g., Lucerne (*Medicago sativa*), Berseem Clover (*Trifolium alexandrium*), Annual Medics (*Medicago* sp.), White clover (*Trifolium* repens), Subterranean clover (*Trifolium* subterraneum), red clover (*Trifolium* pratense), Soyabean (*Glycine* sp.) and pasture plants e.g., birdsfoot trefoil (*Lotus* corniculatus) etc. This is because rumen microorganism degrades genstein and biochanin A to *p*-ethylphenol and phenolic acids, thus destroying their oestrogenic effects. However, in sheep, rumen bacteria, anaerobic fungi are crucial in degrading plant compounds thereby making them important to performance of the host (Theodorou et al., 1996). Rumen microorganisms demethylate formononetin to daidzein, and then together with any daidzein directly consumed from the plants, these minor phytoestrogens are metabolized to equol which is a much more potent form of estrogen. Daidzein may not affect total volatile fatty acid (VFA) production but, quadaratically changes propionate proportion in the total VFA. This will result in change in acetate to propionate ratio in rumen. Increase *in vitro* gas production with rumen liquor from goat containing daidzein ranging from 5 to 50 mg/L is an indication of alteration in end products of fermentation (Mao *et al.*, 2007). Equol can absorb rapidly through the rumen wall. The amount of equol or the non-demethylated compound 4′ methoxy-equol excreted in urine can account to 70 per cent of the formononetic consumed by sheep (Cox and Braden, 1975). However, the rate of excretion of absorbed equol is rapid in the urine of cattle but not in sheep. Sheep is more susceptible to the estrogenic effects of isoflavones (Adams, 1995), although plasma concentration of isoflavones in cattle are comparable to those in sheep on same feed (Lundh *et al.*, 1990). Ruminal microorganism may take 6 to 10 days to adapt fully to these substrates, so genistein and biochanin A may produce oestrogenic effects in sheep during the first few days after introduction onto new pastures (Adams, 1995). However, only formononetin is considered important for long-term oestrogenicity. Coumestans are not greatly metabolized in the rumen. Isoflavones alter the responsiveness of the animal tissue to endogenous oestrogen. High serum levels of phytoestrogens may also inhibit the release of gonadotropic hormones from the pituitary gland and fill receptor sites in the cervix and uterus meant for endogenous oestrogens.

Biological Effects of Phytoestrogens

The action they have in the body is that of weak oestrogens because they bind to hormone receptors (Morley *et al.*, 1966). But they are also adaptogens, which means that they help to reduce high levels of oestrogen or boost low levels. Phytoestrogens have some 1/100 to 1/1000 the binding affinity of oestradiol for cellular receptors. Oestrogen receptor is a protein within certain cells which can bind to endogenous oestrogens in common but, phytoestrogens once coupled with its oestrogenic molecule, this protein communicates with other molecules and starts a process that ultimately leads to certain genes being switched or off. The genes regulate many hormone-influenced systems (Song *et al.*, 1999). Since phytoestrogens have an important role in helping to regulate hormones, they can be of use to relieve the unpleasant effects in reproduction, lactation and growth of sheep.

At high levels, dietary phytoestrogens can have profound physiological effects. An example of this is sheep and cattle grazing pastures containing a high proportion of subterranean clover or red clover which can contain levels of phytoestrogens as high as 5 per cent of the dry weight of the plant (Adams, 1986). Pasture leguminous and few non-leguminous herbage with an oestrogenic activity equivalent to daily injection with 4 ìg of diethylstilbestrol may decrease twinning rate in sheep (Bennett *et al.*, 1967). Lucerne containing 25 ppm coumestrol decreases ovulation rate in ewes (Smith et al., 1979). This occurs as a result of the competitively-inhibitory effect of the dietary phytoestrogens on endogenous oestrogen function in the hypothalamus. Male and female sheep and cows can develop androgenic symptoms. Such high dietary levels of phytoestrogens are rare as large population of sheep is habituated globally in arid and semi-arid regions where ingestion of leguminous forages by these animals is quite lesser than cattle. Permanent infertility in sheep is normally developed when exposed to high levels of oestrogenic activity for 4 to 5 months of the year (Adams, 1995). There is little doubt that low levels of oestrogenicity equalling 4 ìg of diethylstilbestrol to occur in many pastures without being recognized. It is far more common that most sheep diets contain low to moderate levels of phytoestrogens, and there is growing epidemiological evidence that such levels have a beneficial effect on sheep health which is contrary to the past belief. Purified isoflavones show a hypocholesterolaemic effect in sheep when animals artificially induced hypercholesterolaemia. Feeding of ewes with 700 g/d of a 60:30:10 mixture of lucerne hay, barley grains and soya bean meal for 14 days prior to ovulation increased the ovulation rate 2.22±0.16 against control with a ovulation rate of 1.50±0.16 (Abecia *et al.*, 1997). The ability of phytoestrogens to have both agonist and antagonist actions mean that it can be termed a selective oestrogenic receptor modulator (Owens and Ashby, 2002). Phytoestrogens influence the grcwth of the female sex organs and the development of the secondary sex organs. They may act directly on the ovary, uterus, vagina, mammary glands and prostate glands, or indirectly through the central nervous system. In the uterus they influence the proliferation of endometrium, in the male, sperm quality is affected. High levels of phytoestrogens may lead to infertility in ewes and difficulty in superovulation. They also influence non-reproductive organs, tissues and cells wherever oestrogen receptors are found, may have physiological effects (Table 1). Targets include bone, tumor cells, the cardiovascular system, the kidneys, liver, the immune system and brain (Owens, 2003) and there may be possibility of increase in organ weight of liver, kidney and spleen as is the case in most other plant toxins at higher doses (Srinivas *et al.*, 1994).

Table 1. Oestrogen receptor sites in sheep those can be afflicted by phytoestrogens.

Organ	Estrogen Receptor	
	α - Receptor Site	â- Receptor Site
Bone	Weak molar affinity present	Strong molar affinity present
Cardiovascular System	Weak molar affinity present	Strong molar affinity present
Central Nervous System	Weak molar affinity present	Strong molar affinity present
Gastro-intestinal Tract	Nil	Strong molar affinity present
Liver	Weak molar affinity present	Nil
Mammary Gland	Weak molar affinity present	Strong molar affinity present
Urogenital Tract	Weak molar affinity present	Strong molar affinity present

Cellular level interactions

Different types of phytoestrogens appear to act at the cellular level in a similar manner that is through interaction with cell surface oestrogen receptors. In the body, naturally-occurring oestrogens circulating in the blood largely exert their activity by interaction with oestrogen receptors on cell surfaces; such interactions then triggering a particular biological function of that particular cell (McEvoy et al., 2001). Phytoestrogens are able to bind to those oestrogen receptors because the structure of these compounds so closely resembles the endogenous oestrogens, but unlike the animal oestrogens, phytoestrogens only weakly activate the oestrogen receptor. As a result of phytoestrogens and endogenous oestrogens competing for the oestrogen-binding sites on cells, the more weakly oestrogenic phytoestrogens can be considered to have an anti-oestrogenic effect (Adams,

1995). This phenomenon is known as competitive-inhibition, by which is meant that the biological effect of an active substance is impaired by the competitive binding to a target receptor of a similar but less active compound. Thus a primary biological effect of phytoestrogens is held to be competitive inhibition of endogenous oestrogens.

Another more direct effect is the stimulation of synthesis of sex hormone-binding globulin (SHBG) in the liver, as occurs with orally administered synthetic steroidal oestrogens (Adams, 1986). High levels of dietary phytoestrogens are thought to be responsible for the higher SHBG levels seen in the sheep fed on more leguminous forages. SHBG plays an important role in binding and transporting the reproductive hormones (oestrogens, androgens) in blood so that the availability of reproductive hormones is regulated to a large degree by SHBG levels. Higher SHBG levels are considered beneficial in leading to reduction in both blood levels of unbound (and unregulated) reproductive hormones and metabolic clearance rates of the hormones. Although isoflavones are potent stimulators of SHBG synthesis, they only weakly bind to SHBG, so that the increased SHBG levels resulting from the dietary isoflavones are largely available for binding to endogenous oestrogens.

Phytoestrogens in Growth Activation and Mutton Quality

Unlike steroidal estrogens, the estrogenic isoflavones and coumestans found in berseem clover or lucerne seem to have little ability to enhance live weight gain in sheep (Trenkle and Burroughs, 1978; Morrison and Young, 1991). However, oestrogenic growth promotants may be additive with phytoestrogens in producing oestrogenic effect (Lookhart, 1980). Since oestrogen receptors can activate growth factor receptor signalling, phytoestrogens too can believe to have similar impact. However, phytoestrogens impact on activating growth factor receptor signalling may not be as strong as oestradiol (Levin, 2003). The required minimum level of phytoestrogens in the diet to activate growth, and a breakpoint of response in the growth of sheep to ingested phytoestrogens are not yet known. High levels of plasma phytoestrogens plays an important role in the metabolism regulation through reduced intake since central nervous system too have oestrogen receptors. These changes may interfere in the orchestrated or coordinated change in the metabolism necessary to support a physiological change (Bauman and Currie, 1980).

The U.S. Food and Drug Administration (FDA) has been thoroughly researching the effects of growth stimulating hormones for decades. Phytoestrogens are considered as non-nutrient feed/food components and levels are monitored in global market owing their after effects on consumption by human beings. Phytoestrogens draw important attention due to their principle effect on woman (Wilcox et al., 1990). Quantitatively phytoestrogens in feed are inextricably linked to animal health. Curative action for infertility or health, in general, is costlier. Hence, animal health management for toxic principles of plant origin needs to be integrated with feeding management. Phytoestrogens affect the profitability of mutton production through liver damage, mammary gland damage and reduction in fertility at mating. Chronic exposure to phytoestrogens may even affect the life time performance of ewes. The trade losses from phytoestrogens may be worked out with the following trade impact model (Wu, 2007):

- Total economic losses from phytoestrogens = Losses in oestrous cycle (variable and fixed costs for the extended period) + Loss in conception rate of stock + Loss in reproductive cycle in terms of lambs (single or twin) + Animal health loss and expenses incurred upon it + Losses due to shrinking of growth + Losses in carcass yield + Market reject loss/depleted cost for the mutton.

Any other miscellaneous losses may also be included as practically perceived by the farmer or entrepreneur. Animals grazed on oestrogenic pastures may accumulate isoflavones in their fat to a concentration approximately 20 folds greater than in plasma (Lindner, 1967). Nonetheless, these concentrations are very low compared with those in some vegetable products normally consumed by humans. Concentration of 1 to 5 ìg isoflavones/g of fat are approximately 1000 fold less than concentrations of phytoestrogens in foods derived from soyabeans. However, plant origin oestrogen levels in beef, and mutton are commonly well with in the range of natural levels of these hormones. To put this issue into perspective, it is important to remember that many foods contain significant levels of hormones. The fact is consumers are exposed every day to foods with far higher hormone levels than those found in any beef and mutton. For example:

1. Hormone levels (oestradiol equivalent) in beef or mutton are far less than those found in eggs. A hen's egg (about 50 grams) contains about 45 times as many oestradiol equivalents as 250 grams of steer or lamb meat,

2. Wheat germ and soybean oil contain phytoestrogens at several thousand times higher hormone equivalent concentrations than a serving steer or lamb meat.

The less harmful phytoestrogens indeed, recommended at a dose of 80 to 100 mg a day of isoflavones to women to combat menopause symptoms and cut the risk of oestrogen-dependent tumours such as breast cancer. Based on the more recent addition to the epidemiological knowledge of phytoestrogens which have found a relationship between dietary intake of isoflavones and lignans and lower rates of certain cancers, cardiovascular problems and menopausal symptoms, whimsical belief about their presence in the mutton is now changing and these substances are no more in the list of undesirable substances found in the mutton (Lagio *et al.*, 2003).

Conclusions and Recommendations

Low concentrations of phytoestrogens are widespread. There is controversy prevail on the effect of low to moderate levels of oestrogenicity of phytoestrogens on the health of sheep. The recently growing epidemiological evidence shows that they have a beneficial effect on sheep health as against past perceptions at low to medium levels. However, undoubtedly diets containing 25 ppm or more of phytoestrogens may affect the reproductive rate of sheep and non-reproductive organs, tissues and cells wherever oestrogen receptors are found, may have physiological effects. High levels of phytoestrogens effect mutton production by reducing the functional conception rate of ewes to less than optimum levels of 70 per cent. On the other hand low levels activate weight gains in sheep. A concentration of 1 to 5 ìg isoflavones/g of fat is thousand times lesser than the concentrations found in soy products and should not be an impediment to export mutton to the countries where population density is lesser and per capita mutton consumption is higher.

Detailed measurement of estrogenicity of the herbage/forage at low, moderate and high levels, accompanied by epidemiological studies, may be required to determine whether subclinical effects of phytoestrogens are important for health based animal nutrition in sheep to produce mutton that qualify international health standards. Threshold and critical levels of phytoestrogens in diets those are necessary to induce growth are also need to be determined.

REFERENCES

Abecia, J.A., Lozano, J.M., Forcada, F. and Zarazaga, L. 1997. Effect of levels of dietary energy on embryo survival and progesterone production on day eight of pregnancy in Rasa Aragonesa ewes. *Animal Reproduction Science* 48, 209-218.

Adams N. R. 1986. Measurement of histological changes in the cervix of ewes after prolonged exposure to oestrogenic clover or oestradiol-17β. *Australian Veterinary Journal* 63, 279-285.

Adams, N. R. 1995. Detection of the effects of phytoestrogens on sheep and cattle. Journal of Animal Science 73, 1509-1515.

Alison, M.D., William, R.P. and Mindy, S.K. 2003. Phyto-oestrogens. *Best Practice and Research Clinical Endocrinology and Metabolism* 17, 253-271.

Bauman, D.E. and Currie, W.B. 1980. Partitioning of nutrients during pregnancy and lactation: a review of mechanisms involving homeostasis and homeorhesis. *Journal of Dairy Science* 63, 1514-1529.

Bennett, D., Morley, F.H.W. and Axelsen, A. 1967. Bioassay responses of ewes to legume seards. II uterine weight responses frok swards. *Australian Journal of Agricultural Research* 18, 495-498.

Benassayag, C., Perrot-Applanat, M. and Ferre, F. 2002. Phytoestrogens as modulators of steroid action in target cells. *Journal of Chromatography B. Biomedical Sciences and Application* 777, 233-248.

Barry, T.N. and Blaney, B.J. 1987. Secondary compounds of forages. In: Hacker, J.B. and Ternouth, J.H. (eds), *The Nutrition of Herbivores,* Academic Press, Sydney, pp. 91-120.

Cox, R.I. and Braden, A.W.H. 1974. The metabolism and physiological effects of phyto-oestrogens in livestock. *Proceedings of Australian Society of Animal Production* 10, 122-128.

Duke, J.A. 1977. Phytotoxin tables. *CRC Critical Reviews in Toxicology* 5, 189-237.

Fang, H., Tong, W., Branham, W., Hong, H., Xie, Q. and Moland, C.L. 2003. Structure-activity relationships of 202 natural, synthetic and environmental chemicals for binding to the androgen receptor. *Chemcial Research and Toxicology* 16, 1338-1358.

Hegarty, M.P. 1982. Deleterious factors in forages affecting animal production. In: Hacker, J.B. (ed.), *Nutritional Limits to Animal Production from Pastures.* CAB, Farnham Royal, UK, pp. 133-150.

Holtenius, K. 2006. Editorial on Feed and animal health: Preface to the special issue. *Animal Feed Science and Technology* 126, 173-174.

Lagiou, P., Hsieh, C -C., Trichopoulos, D., Xu, B., Wuu, J., Mucci, L., Tamimi, R., Adami, H-O. and Cnattingius, S. 2003. Birth weight differences between USA and China and their relevance to breast cancer aetiology. *International Journal of Epidemiology* 32, 193-198.

Levin, E.R. 2003. Effect of oestrogen in activating growth factor receptor signalling. *Molecular Endocrinology* 17, 309-317.

Lindner, H. R. 1967. Study of the fate of phyto-oestrogens in the sheep by determination of isoflavones and coumestrol in the plasma and adipose tissue. *Australian Journal of Agricultural Research* 18, 305-312.

Lookhart, G.L. 1980. Analysis of coumestrol, a plant estrogen, in animal feeds by high-performance liquid chromatography. *Journal of Agriculture and Food Chemistry* 28, 666-671.

Lundh, T.J. -O., Pettrsson, H.I. and Martinsson, K.A. 1990. Comparative levels of free and conjugated plant estrogens in blood plasma of sheep and cattle fed estrogenic silage. *Journal of Agriculture and Food Chemistry* 38, 1530-1536.

Manidool, C. 1985. Utilization of tree legumes with crop residues as animal feeds in Thailand. In: *Relevance of Crop Residues as Animal Feeds in Developing Countries.* Proceedings of an International Workshop, Khon Kaen, Thailand, pp. 249-269.

Mao, S. Y., Zhu, W. Y., Wang, Q. J. and Yao, W. 2007. Effect of daidzein on *in vitro* fermentation by microorganisms fro the goat rumen. *Animal Feed Science and Technology* 136, 154-163.

McEvoy, T.G., Robinson, J.J., Ashworth, C.J., Rooke, J.A. and Sinclair, K.D. 2001. Feed and forage toxicants affecting embryo survival and fetal development. *Theriogenology* 55, 113-129.

Morgavi, D. P. and Riley, R. T. 2007. *Fusarium* and their toxins: Mycology, occurrence, toxicity, control and economic impact. *Animal Feed Science and Technology* 137, 199-200.

Morley F.H., Axelsen, W. and Bennet, D. 1966. Recovery of normal fertility after grazing on oestrogenic red clover. *Australian Veterinary Journal* 42, 204-211.

Morrison, D. and Young, J. 1991. Profitability of increasing lambing percentage in the Western Australian wheat belt . *Australian Journal of Agricultural Research* 42, 227-236.

Nancy, B. 1995a. Phyto-oestrogens and compounds that affect oestrogen metabolism, Part I. *Australian Journal of Medical Herbalism* 7, 11-16.

Nancy, B. 1995a. Phyto-oestrogens and compounds that affect oestrogen metabolism,Part II. *Australian Journal of Medical Herbalism* 7, 27-33.

Owens, J.W. and Ashby, J. 2002. Critical review and evaluation of the uterotrophic bioassay for the identification of possible estrogen agonists and antagonists: in support of the validation of the OECD uterotrophic protocols for the laboratory rodent. *CRC Critical Reviews of Toxicology* 32, 445-520.

Owens, W., Ashby, J., Odum, J. and Onyon, L. 2003. The OECD program to validate the rat uterotrophic bioassay: Phase 2: dietary phytooestogen analyses. *Environmental Health Perspectives* 111, 1559-1567.

Plant, J.W. 1981. *Field investigations of reproductive wastage in sheep*; In *Sheep*, Proc 58, Post-graduate Committee in Veterinary Science, University of Sydney, pp. 411

Rosenthal, G.A. and Janzen, D.H. 1979. *Herbivores - their Interaction with Secondary Plant Metabolites.* Academic Press, New York.

Seawright, A.A., Hegarty, M.P., James, L.F. and Keeler, R.F. 1985. Plant toxicology. *Proceedings of the Australia-USA Poisonous Plants Symposium.* Queensland Department of Primary Industry, Yeerongpilly, Brisbane.

Song, T.T., Hendrich, S. and Murphy, P.A. 1999. Estrogenic activity of glycitein, a soy isoflavone. *Journal of Agriculture and Food Chemistry* 47, 1607 - 1610.

Smith, J.F., Jagusch, K.T., Brunswick, L.F.C. and Kelly, R.W. 1979. Coumestans in lucerne and ovulation in ewes. *New Zealand Journal of Agricultural Research* 22, 441-446.

Srinivas, B., Somvanshi, R. and Biswas, J.C. 1994. Effect of feeding potato (*Solanum tuberosum*) plan on weight gain and haematology of guinea-pigs. *Indian Journal of Animal Nutrition* 11, 25-29.

Theodorou, M.K., Zhu, W.Y., Rickers, A., Nielson, B.B., Gull, K. and Trinci, A.P.J. 1996. Biochemistry and ecology of anaerobic fungi. In: The Mycota VI: Human and Animal Relationships. (Howard, Miller ed.). Springer Verlag, Berlin, Heildelberg, Germany, pp. 265-295.

Trenkle, A. and Burroughs, W. 1978. Physiological effects of estrogens in animal feeds with emphasis on growth of ruminants. In: Nutrition and Drug Interactions. (J. N. Hathcock and J. Coon, eds.), Academic Press, New York. p 577.

Wilcox, G., Wahlqvist, M.L., Burger, H.G. and Medley, G. 1990. Oestrogenic effects of plant foods in post-menopausal women. *British Medical Journal* 301, 905-906.

Wu, F. 2007. Measuring the economic impacts of *Fusarium* toxins in animal feeds. *Animal Feed Science and Technology* 137, 363-374.

❑❑❑

29

Scope and Prospects of Sheep Farming for Organic Meat and Wool Production in the Country

A.K. SHINDE
Central Sheep and Wool Research Institute, Avikanagar - 30450, Rajasthan

Introduction

In the recent year, large number of chemicals, antibiotics, hormones, steroids and growth promoters are used in animal rearing practices under intensive system for maximizing growth, meat and milk production. The residues of these hazardous chemicals appear in the meat and milk and its consumption in human being causes many health problems. In the recent years, the incidences of chemicals residues are of a great concern for ensuring food safety and protecting human health. The residues of pesticides in the milk were reported as early as in sixties in the country. Pesticides are commonly used during the production, storage, transport and distribution and processing of agricultural commodities for control of pest or unwanted plant species or spraying on the body of animal for control of external parasites. Pesticides enter animal body through consumption of contaminated food and generally deposited in the adipose tissue. As the concentration in the fatty tissue exceeds a threshold level, the excess residues are transferred into the lactating glands of animals and secreted in the milk. Other route for absorption of pesticides is through skin or digestive tract and deposition in the adipose tissue. Whenever, the animal mobilizes their body reserve to support growth or lactation, these pesticides enter the blood circulation and secrets in the milk (Dhami *et al.*, 2005). Pesticides don't generate immune response in body but when they bond with certain proteins, they become immunogenic and initiate the immune response against the body proteins, leading to generation of immunity against itself. They destructs the self-tissue and damage vital organs (Singhal and Chauhan, 2000). Organophosphate pesticides caused immuno-suppression, neuropathic and hepetopathic effects in human being when consumed through contaminated food above permissible limits. A wide range of antibiotics are used in treatment and prophylaxis of animals. The residues of some of the antibiotics are known to cause allergic reactions in human beings through consumption of contaminated food. On the other hand intake of low doses of antibiotics by the consumers through animal produce gradually replaced the sensitive gut micro flora by resistant form (Thakur *et al.*, 2000). Sex hormones in the recent years are also used in fattening of animals. The consumption of meat containing unacceptably higher levels of sex hormones causes many side effects in people, including breast enlargement and ovarian cysts. The feeding of GM feed byproducts like Bt cotton seed cake etc are now become part of animal ration in many cotton growing areas. The safe use of the GM byproducts need long term studies before wide scale use in animal feeding. The rearing of animals under organic farming appear to be the realistic approach for ensuring safe meat and milk production for human consumption to avoid the adverse effect of harmful chemical on human health through consumption of contaminated food.

Organic Animal Farming

Organic farming in animals reduces or eliminates the need for routine chemical treatments such as drenching or dipping for internal and external parasites, fly dressings, antibiotics, vaccines against many diseases, growth promoters for enhancing growth and mutton production, feed additives, hormones and rumen manipulators, use of synthetic fertilizers, pesticides, insecticides and other chemicals in pasture (Shinde et al., 2005). In organic sheep farming approach should be "*Prevention Is Better Than Cure*". A great demand for organic products is picking up for a much wider range of dairy, meat and wool products in the recent years because the consumers believe that these products are safe, high in quality and better for the environment as no synthetics or harmful chemicals are used in production and processing of product. The animal produces are contaminated through treatment of animals with chemical drugs, feeding of contaminated feed and fodder and also by use of chemicals during processing of produce. Some of the measures for production of organic produce in organic farming are :

1. Organic Treatment and Prophylaxis Measures

In the recent past more emphasis was given on food safety particularly on contamination of food by animal diseases. In the last decade the incidence of diseases in human being has been noticed due to consumption of meat and meat products infected by Bovine Spongiform Encephalopathy (BSE), Foot and Mouth Disease (FMD), Avian Influenza and Swine Fever. The incidence of health hazards has been also noticed by consumption of contaminated food items with toxins produced by microorganism. There are eight major diseases afflicting livestock according to OIE (an International Organization for Animal Health) classification that are serious and fast spreading in the country. These diseases include Foot and Mouth Disease (FMD), Contagious Bovine Pleuro-Pneumonia (CBPP) and Peste des Petits Ruminants (PPR), Blue Tongue (BT) etc. For controlling the incidence of diseases in the country strengthening of institutional and legal frame work, the regulatory mechanism for the import and export of livestock and livestock products with appropriate legal support is required but at present the absence of a uniform legislation for the control of movement of livestock across the country is proving to be a handicap for effective control of diseases. Animal diseases result in huge national economic losses. Loss of production on account of diseases alone amounts to more than Rs. 10000crore per year. Economic losses due to diseases occur in the form of direct losses on account of reduction of milk yield, quality and quantity of meat, work capacity and growth.

The organic farming regulations demand that animals should not be given any antibiotics in the few weeks before slaughter to reduce the antibiotic residues in the meat and meat products. The antibiotic residues in milk caused problem in people who are susceptible to an allergic reactions and more specifically with penicillin. Antibiotics, growth hormones, synthetic de-wormers and medicated feed additives including ionophores should be banned from regular use in organic livestock farming. The use of vaccines in some regulations is encouraged, while in others they are used under certain circumstances. The regulations in organic farming stressed that suffering animals must be treated with the most effective therapy available but animals that have been treated with synthetic medications or antibiotics should not be sold as organic animals for meat production. Sheep can be treated with certain drugs up to the third trimester of pregnancy and the lambs can be sold as organic. In organic farming the focus must be on the selection of naturally healthy animals and careful and management regarding stocking density, sanitation, and nutrition for ensuring animal health. These preventative measures improve resistance in animals against diseases and also help in maintaining parasite loads within normal range. A problem with enterotoxaemia can be resolved through efficient feeding management rather than with vaccination. The homeopathic drugs can be used as an alternative treatment measures for treatment of pneumonia and more enthusiastic approach may be initiated for trying other homeopathic remedies.

Parasites are potentially a major problem in sheep husbandry. The controlling of internal parasites without synthetic products by reducing stocking densities, rotational grazing and selecting grasses based on sward characteristics are identified management tools in the organic farming. The knowledge of parasite behavior is also a valuable tool. It has been recently learned that larvae are crawl to the tops of the blades when grasses are wet with dew. By moving sheep onto new pasture later in the day decrease ingestion of larvae and infection. Herbal de-wormers are now available in the markets and may be promoted in the treatments of animals against parasites. Their residue effect if any in meat and other produces required testing before wide spread use in the sheep flocks. Generally farmers in the region follows 3-4 times drenching with dewormers in year as a routine practice in their flocks. The dipping of sheep in chemicals is a routine practice in the organized farms as a prophylaxis measures for external parasites. Now it has been realized that conventional dips contain potentially dangerous organophosphorus compounds, which adversely affected the health of animals and human.

The residues of antibiotics persist for considerable time in the produce after treatment of animals with antibiotics (Thakur and Chauhan 2000). The withdrawal period of individual antibiotic varies greatly; therefore, it is essential to know the withdrawal period of antibiotics to avoid any adverse effect on the health of consumers through consumption of animal produce (Table 1).

Table 1. Withdrawal period of various antibiotics in animals

Antibiotics	Withdrawal period (days)	
	Meat	Milk
Procain penicilline	10	3
Benzathine penicilline G	30	-
Ampicilline	6	-
Amoxicillin	30	2
Benzathine Cephaprime	42	3
Sulphamethazine	10-28	4
Trimethoprim + Sulphadiazine	3	7
Oxytetracycline	15-22	15-22
Chlortetracycline (Long acting)	28	28
Erythromycin	14	2-3
Tylosin	21	4
Tilmicosin	28	-

There is limited information on adverse effect of antibiotics residues on health of human being after consumption of produce of treated animals. The residues of Penicillin, Sulphonamides and Neomycin antibiotics are known to cause allergic reactions in human. The intake of low doses of antibiotics residue by the consumers from animal produce gradually replaced the sensitive gut micro flora by resistant form (Thakur *et al.,* 2000). Most of the antibiotics are excreted in milk either unchanged or as a metabolite. Penicillin caused allergic reaction, chlorphenical caused aplasticanemia, granulocytopenia and liver damage, tetracycline inhibits growth and development of skeleton and chlortetracycline caused photosensitive reactions. Antibiotics are used in animals for treatment and prophylaxis of various diseases; animal produce of treated animals may be used after withdrawal period. These approaches can minimize the adverse effect of drugs on human health. FAO/WHO Codex Food Standards recommended maximum residues limits of different veterinary drugs in milk and tissues of sheep (Table 2). The milk or body tissues samples if contained more than permissible limits, then it must be removed from human consumption.

Table 2. Maximum residual limits of certain veterinary drugs in milk (mg/l) and tissues of sheep (mg/kg)

Drugs	Milk	Muscle	Kidney	Liver	Fat
Closantel	-	1500	5000	1500	2000
Diclazuril	-	500	2000	3000	1000
Dihydro-streptomycin /Streptomycin	-	500	1000	500	500
Fenbendazole/ Oxfendazole	100	100	100	500	100
Ivermectin	-	-	-	15	20
Levamisole	-	10	10	100	10
Moxidectin	-	50	50	100	500
Neomycin	-	500	10000	500	500
Oxytetracycline	-	100	600	300	
Spectinomycin	-	500	5000	2000	2000
Thiabendazole	-	100	100	100	100
Tilmicosin	50	100	300	1000	100
Triclabendazole	-	100	100	100	100

Source : FAO-WHO Alimentarius Codex Commission

2. *Organic Feed and Feeding Practices*

The presence of pesticide residues in various animal feeds and fodders has been reported from different parts of the country. The feed ingredients while storage are usually treated with pesticide for protection against damage caused by pests and insects. The residues of BHC (Benzene Hexachloride) and DDT (Dichlorodiphenyl trichloroethane) in straw samples up to 0.24ppm and 1.35ppm was reported by Singh *et al* (1988). The contamination of DDT and HCH (Hexachloran) in cattle feed (Kaphalia and Seth 1982) and in mustard, groundnut and sesame oil cakes (Dixit et al., 1989) was also reported from different parts of the country. Unnikrishnan *et al* (1998) also reported residues of BHC, DDT and endosulphan in cattle-feed and feed ingredients collected from farms located in Bangalore city. Agnihotri (1999) also found a higher levels of HCH and DDT in cereal straw. The HCH and DDT residues in dry fodders and concentrate mixtures from Jaipur city was also reported (Gupta et al., 1997, 2000). BHC, endosulphan, heptachlor, DDT and aldrin residues in concentrate feeds and roughage in villages around Karnal city was also reported (Prasad and Chhabra, 2001).

Monocrotophos, melathion, phosphamidon among the chlopyriphos; carbfuran among carbamates; cypermethrin among synthetic pyrethroids; cyhalothrin and fenvalerate among the other fenpropathrin pesticide residues were also detected in samples of concentrate mixture and dry fodder (Nag and Raikwar, 2002). The presence of quinalphos, parathion-methyl and ethion were detected from river Ganga. Melathion and monocrotophos were detected from ground water samples of Farukabad districts (Mahopatra *et al.*, 1994). Melathion and parathion were detected from irrigation and domestic wells and canals of Aligarh district (Bansal and Gupta, 2000). These indicate the spiralling incidence of contaminated feeds and fodders across the country and probable adverse effect on human health.

Organochlorine (OC) pesticides are DDT, lindane, dieldrin and aldrin and caused dizziness, nausea, twitching of arms and legs, tremors and convulsions and finally cessation of breathing in acute cases. They also caused fetal abnormalities and carcinogenic diseases (Robbins, 1991). Their half-lives have been found to be at least 20 years in both soil and water (Bhantnagar and Kumar, 1989). Because of their solubility in fats, this group of pesticides can accumulate and transfer from one food chain to another, e.g. from insects to birds, to fish, and thence to larger mammals, including humans (MAFF, 1989). DDT affected calcium metabolism, causes thinning of eggshell and death-in-chick embryo (Narayan, 2002). Organochlorine pesticides are not degraded in the body rather accumulate in adipose tissues. Organophosphate pesticides include parathion, melathion, sumithion, monocrotophos, quinaphos caused immuno-suppression, neuropathic and hepetopathic effects in human being (Hierons and Johnson, 1978). Organophosphorus pesticides are readily decomposed by physical and chemical and enzymatic process in the animal system. Some of the organophosphorus pesticides are resistance to degradation and appear in the milk (Chhabra 2005). 2,4-D (2,4-Delapon), the commonly used herbicide in agriculture. The main consequence of 2,4-D is higher incidence of lymphatic cancers in farmers who often sprayed the herbicide on their fields over many years (Saftlas, 1987). Rodenticids includes fluoroacetate derivatives thioureas, anti-vitamin K compounds. Such as warfarin contain anti-coagulants, caused bleeding to death in animals and humans. Rodenticids such as metal phosphides containig zinc and aluminum are dangerous for human health, if ingested (Jensen, 1983). Fungicides include inorganic and organo-mettallic compound, antibiotics, quinines and dithiocarbamates. They are usually based on compounds containing metals such as copper or mercury or hydrocarbons containing sulphur. Fungicides such as zineb and mancozeb are known to be carcinogenic in nature (Jensen, 1983). The pesticides and chemicals are used for pest control in agriculture and their adverse effect in human being could be minimized by use of correct dose, proper application tools and strict follow up of instructions and safe withdrawal period. The regular screening of feeds and fodders and animal produces for pesticide residues and withdrawal of contaminated feeds and fodders from animal feeding are some of the approaches for minimizing the risks arising from contaminated feeds in animal and human life.

Transgenic (Genetically Modified, GM) crops like BT-Cotton and Mustard are now grown in different parts of the country and their by-product in form of cakes, seed, bran etc are incorporated at various levels in the animal rations. At present stage, not much has been known about safe use of the GM feeds in animal feeding and its residues in milk and meat and effect upon human health. The agricultural by-products of GM crops may be used with caution in animal feeding. Singhal et al (2006) fed 2kg of non-GM cotton seed or 2kg of Bollgard cotton seed expressing the Cry 1 AC protein to each of 10 lactating cows. No difference in milk yield, body weight gain, milk composition, DM intake and nutrient digestibility was recorded. Similar observations in buffaloes was also recorded by Singh et al. (2002). Recently Hartnell and Hammond (2007) reported that Bt protein protect crops against a certain class of insects resulting in a reduction in the application of chemical pesticides benefiting the environment as well as the farmers. Bt crops or their by-products have been evaluated in feeding studies with lactating dairy cattle, lactating buffaloes, beef cattle, poultry, swine, sheep and fish at different location in the world. All the studies have concluded that Bt crops are as safe, nutritious and wholesome as their non-Bt counterparts.

3. *Organic Animal Produce and Processing*

The incidence of food borne diseases and food poisoning across the country are mainly due to improper handling of animal produce and products at different stages of processing from production to end users. *Pasteurellosis, Pseudotuberculosis, Leptospirosis, Listeriosis* and *Brucelosis* are some of the diseases spread through consumption of contaminated meat and meat products. *Trichinella* and *Teania* are the chiefly meat borne helminthes infection reported in the country. *Salmonella, Staphylococci, Streptococci, E. coli* and *C. botulium* bacteria infected meat and meat products caused food poisoning in human beings.

Pesticides residues have a greater affinity for fat molecules and generally not degraded in the body system rather it accumulates in adipose tissues. The higher contamination of organochlorine pesticides in meat samples collected from Hyderabad, Izatnagar and Ludhiana has been reported (Sastry, 1983). In the contrary, meat samples of sheep collected from Jaipur city indicated no presence of BHC, DDT, Aldrin, Dieldrin, Heptochlor,Endosulfan and Endrin organochlorine pesticides residues (Singh *et al.*, 2005). The problem is area specific and related to use of pesticides in crop cultivation. The potential risks of health hazards from contaminated feed and fodders has been examined by conducting experiments on animals. It has been observed that residues of organochlorine pesticides appears in animal produce after oral administration (Kapoor and Kalra, 1988, 1997; Singhal and Mudgal, 1990; Surendranath *et al.*, 1998). The study conducted at Environmental Research Institute, Lucknow under biological monitoring of pesticides in meat and meat products indicated that the countrywide variation in the overall toxic load of organochlorine pesticides in meat samples of different animal origin occurs within reasonable limit. The higher organochlorine pesticide residues in pork and buffalo as compared to sheep, goat and chicken was noticed. Sheep in the country are raised on natural pastures, unlike in the advance countries where they are grazed in pastures grown by application of pesticides and fertilizers. A study conducted at GB Pant Univ. of Agriculture and Technology indicated the presence of endosulphon, chlorpyriphos in liver, kidney and muscles of sheep and goats. All these studies confirmed that meat and meat products in the country are contaminated by the pesticide residues and required immediate attention for protection of animal produce from hazardous pesticides.

Wool are processed into finish product by application of chemicals at the different stages of processing from scouring to coloring and carding and many of them are harmful to human beings. Harmful chemicals and dyes cause allergic problem in human beings. Greasy wool is scoured in cleansing agents before being carded and spun. The chemical dyes and bleach materials are used in wool processing for manufacturing into fashionable style knitwear. Most of these chemicals are non-biodegradable and caused pollution of the environment. The knitwear developed by chemical process caused allergies and skin disorders in sensitive people.

Meat is another important produce of sheep, process into various products by chemicals, additives and enzymes for inhibiting the spoilage or preventing spoilage, improving the palatability and providing variety of products for trade. The present practices suggest that a large number of chemicals, drugs and hormones are used at various stages in animal husbandry. The excess use of these chemicals accumulate in animal's body and subsequently in the human through food chain and cause health problems.

In organic farming, animals shall not be slaughtered in a slaughterhouse in sight of other animals, animal shall not be administered any chemical, drug or hormone before slaughter except drugs for its treatment for any specific disease or ailment and meat of these animals should not be sold as organic. The regulatory standards on Good Animal Feed, Good Hygienic Practices (GHPs), Good Manufacturing Practices (GMPs) and Cold Chain System accomplishing quality; good processing and infrastructure facilities should be developed and strengthened in the country. The regular screening of feeds and fodders and animal produce for presence of pesticide residues and imposing ban on their use are some of the approaches for protecting the health of human beings.

Food Quality Regulations for Safe Production of Animal Produce

There are several regulations and agencies working in the country for administering food quality measures in the domestic and import produce market: Bureau of Indian Standards, Prevention of Food Adulteration Act and Rules, Export Inspection Council of India, Milk and Milk Products Order, Livestock Importation Order and Meat Food Products Order. In present days, major concern in sheep husbandry is to assure meat and wool and their products quality in an order to protect human or animal life and health from food additives, contaminants, toxins or disease causing organisms, zoonotic diseases, pests, trans-boundary diseases and residues of pesticides, chemicals, hormones, and veterinary drugs. The potential of export of meat and milk products cannot be exploited without meeting the International quality standards and sanitary and phyto-sanitary requirements. These require strengthening of institutional and legal framework for producing quality animal produce in the country. There are two binding agreements relevant for food regulations under WTO regime, which are of major concern to India are the agreements on Sanitary and Phytosanitary measures (SPS) and Technical Barriers on Trade (TBT) which stipulate quality assurance such as Hazard Analysis Critical Control Point (HACCP) based system, code of practices on Good Animal Feed Practices (GFPs), Good Hygienic Practices (GHPs), Good Manufacturing Practices (GMPs) and Cold Chain System. To accomplish quality of International Standards, good processing and infrastructure facilities need to be strengthened. The concept of producing meat and milk and their produce without harmful chemicals seem to be promising but, so far not much has been done in promoting organic farming in the livestock sector in the country. Few NGOs and other agencies like APEDA has taken lead in organic agricultural production, by and large, the livestock component of the system is virtually untapped. For ensuring organic livestock production, organic agriculture is a must. To capitalize on this niche market, a National Programme for organic production has been launched and the road map includes preparation of documents like National Standards, Accreditation, Criteria and Procedures for Accrediting, Inspection and Certification Agencies has been prepared. Most Standards in India are voluntary although health and safety regulations are mandatory for food products. The Bureau of India Standards (BIS) is responsible for formulating and setting National Standards has been harmonizing Indian Standards with International Standards since last decade. So far nearly 3500 Indian Standards have been harmonized with ISOTEC and EC Standards/ Regulations. Although BIS is the National Standards body of India, standards and certification schemes are also operated and enforced in certain specific sectors by other bodies. Director of Marketing and Inspection, Ministry of Agriculture regulate Sanitary and Phyto-sanitary measures in food items. The Ministry of Health and Food Processing deal with the quality and hygiene of process foods.

The *Codex Alimentarius Commission* (CAC) is an intergovernmental body to implement the Joint FAO/ WHO Food Standards Programme, which was established by an FAO conference resolution in 1961 and World Health Assembly resolution in 1963. Its principles objectives are to protect the health of consumers and facilitate the trade of food by setting International Standards on food (i.e. Codex Standards) and specific requirements covering pesticide residues, food additives, veterinary drugs, hygiene, food contaminants and labelling etc.

Table 3. Maximum residue limits of pesticides (mg/kg) in sheep and goat meat

Pesticides	MRL in meat	Pesticides	MRL in meat
2,4-D	0.05	Fenitrothion	0.05
Aldicarb	0.01	Fenthion	2.00
Aldrin and dieldrin	0.20	Fenvalerate	1.00
Azocyclotin	0.20	Heptachlor	0.20
Betazone	0.05	Methiocarb	0.05
Chinomethiona	0.05	Methomyl	0.02
Chlordane	0.05	Methoprene	0.20
Cyhexatin	0.20	Permethrin	1.00
Cypermethrin	0.20	Phorate	0.05
DDT	0.50	Pirimicarb	0.05
Deltamethrin	0.50	Pirimiphos-Methyl	0.05
Dichlorvos	0.05	Profenofos	0.05
Diflubenzuro	0.05	Propargite	0.10
Dimethipin	0.02	Propiconazole	0.05
Diquat	0.05	Propoxur	0.05
Dithiocarbamates	0.05	Triadimefon	0.05
Endosulfan	0.10	Triadimenol	0.05
Fenbutatin oxide	0.05		

Source : FAO/WHO Alimentarius Codex Commission

Animal Welfare for Safe and Quality Animal Produce

Humane treatment of animals is essential in organic farming it is believed that stresses caused by overcrowding; lack of feed or under-nutrition, improper handling during transport and other factors influenced the quality and quantity of milk, meat and their products. The sheep must have access to pasture having sufficient number of edible forage in it for grazing. They must also have access to clean water all the time or at regular interval. The supplementation of minerals and vitamins are also recommended. Animal cages for small animals and stables for large animals should be comfortable and avoid overcrowding. Besides better feeding, breeding and management of animals, other issues of preventing unnecessary pain or suffering of animals in generally, and more particularly when they are being transported from one place to another or when they are used as performing animals or when they are kept in confinement need attention in organic farming. There is an urgent need to make improvements in the design of vehicles so as to lessen the burden on draught animals. Most of the slaughter houses in the country are below standard prescribed by Indian Bureau of Standards and more efforts are required in the designing of slaughter-houses or in connection with slaughter of animals so that unnecessary pain of suffering, whether physical or mental, is eliminated in the pre-slaughter stages so far as possible, and animals are killed wherever necessary, in as humane a manner as possible under organic meat production.

Package of Practices for Organic Sheep Rearing

The three biggest concerns that consumers have are :

Genetic engineering of food products,

Use of chemical fertilizer and pesticides

Use of antibiotics in livestock, the feeding of animal by-products, and humane issues.

A holistic approach to organic farming aimed at preserving the health of the land, the people, and the animals should be taken.

- Native animals must be given due importance in breeding programs avoiding crossing breeding of native animals with exotic animals.
- Animals must be raised on organically grown feed. Transgenic (GMO) feedstuffs and animal by-products may be used after ensuring safe and wholesome and the majority of the diet for ruminants must be of green forage grown without use of pesticides and fertilizers.
- Antibiotics, growth hormones, synthetic de-wormers and medicated feed additives, including ionophores, are banned from regular use in organic livestock production. The use of vaccines in some regulations is encouraged, while in others they can only be used under certain circumstances.
- Humane treatment of animals is essential at various stage of production including housing, transport, slaughter and feeding management.
- Wool yarn should not be treated chemically throughout the entire production from the farm to the end garment. The natural, un-dyed and unbleached wool should be used for manufacturing fashionable style and quality garments without sacrificing and polluting the environment.
- The market for organic meat and wool is growing in the country. A number of organizations have made a major effort over the years in developing a market for organic lamb in European countries. Organic lambs in the markets has great demand, organic lambs received 20-30% higher premium than conventional lambs in European countries.

Conclusion and Recommendations

The demand of organic meat and wool is growing in the country as the consumers know about its beneficial effect in safe guarding the health of human being as well as for protection of the environment. Although not much has not been done in the organic animal production but attempts are being initiated for promotion of organic meat and wool production in the country. The sheep in the country are raised on natural vegetation, which is free from hazardous chemicals and may be promoted for organic meat and wool production. In general, organic farming in animals decreased the production and growth but improved the quality of animal produce and ensured supply of food free from agrochemical and veterinary drug residues for human use. The input costs for medication go down, this is offset by higher feed costs. In addition to raising cost, the certificate and license expenses add to higher cost of meat production. The consumer has to pay higher cost for organic produce and willing to accept it before wide scale propagation of organic sheep production in the country.

REFERENCES

Adiroubane, D. and Letchoumanae, S. 1996. HCH and DDT residues in bovine milk sampling from Karaikal region of Pondichery. *Pesticide Research Journal* 8, 115-118.

Agnihotri, N.P.1999. Pesticide safety evaluation and monitoring. All India Coordinated Research Project on Pesticides Residues, Indian Agricultural Research Institute, New Delhi, pp. 104-111.

Bansal, O.P. and Gupta, Rachana. 2000. Ground water quality of Aligarh district of Uttar Pradesh. *Pesticide Research Journal* 12, 188-194.

Bhantnagar, P. and Kumar, S. 1989. Update and bioconcentration of dieldrin, dimethoate and permethrin by *Tetrahymena pyriformis, Water, Air and Soil Pollution.* 40,345-349.

Chhabra, Aruna. 2005. Contaminants in animal feeds and their carry over to milk and milk products. National Seminar on Application of Nuclear and Non Nuclear Techniques in Detection of Residues in Food and Feed. Madras Veterinary College, 18-19 March, 2005.

Dhami, A.J., Trivedi, M.M., Nayak, J.B. and Brahmbhatt, M.N. 2005. Pesticide residues in animal feeds, milk and milk products -an overview. National Seminar on Application of Nuclear and Non Nuclear Techniques in Detection of Residues in Food and Feed. Madras Veterinary College, 18-19 March,2005.

Dixit, T.S.S., Kumar, S.N., Raizada, R.B. and Srivastava, M.K. 1989. Organochlorine insecticide residues in cattle feed. *Bulletin of Environment Contamination and Toxicology* 43, 691-696.

FAO/WHO. Codex Secretariat, Secretariat of the Joint FAO/WHO Food Standards Program, Food and Nutrition Division, Food and Agriculture Organization of the United Nations, Viale delle Terme di Caracalla, Rome

Gupta, A., Parihar, N.S. and Singh, V. 1997. HCH and DDT residues in bovine milk and milk powder. *Pesticide Research Journal* 9, 235-237.

Gupta, A., Singh, B., Parihar, N.S. and Bhatnagar, A. 2000. Monitoring of HCH and DDT residues in certain animal feed and feed concentrates. *Pestology* 24 (12): 47-49.

Hartnell, G.F. and Hammond, B.G. 2007. Safety and wholesomeness of genetically modified crops for livestock, poultry and aquaculture: focus on insect-protected crops in India. Proceeding of International Tropical Animal Nutrition Conference, NDRI, Karnal, 4-7 October, 2007.

Hierons, R. and Johnson M.K. 1978. Clinical and toxicological investigations of a case of delayed neuropathy in man after acute poisoning of organophosphorus pesticide. *Archives of Toxicology* 40, 279.

Jensen, A. A.1983. Chemical contamination in human milk. *Residue Review* 89,1-128.

Kaphalia, B.S. and Seth, T.D. 1982. Organochlorine pesticides contamination in some species of fodder grasses. *Environment Pollution Series* B.3. : 231-237.

Kapoor, S.K. and Kalra, R.L. 1988. Residues of HCH in milk after its oral administration or dermal application to Indian buffaloes, *Bubalus bubalis* (L). *Pesticide Science* 24, 193-204.

Kapoor, S.K. and Kalra, R.L. 1997. Transfer of HCH isomers from feed into milk of Indian Buffalo (*Bubalus bubalis* L.). *Pesticide Research Journal* 9, 72-78.

MAFF. 1989. Report of Working Party on Pesticide Residues: Food Surveillance Paper, 1982-86 (fruit and vegetables}, 1985-88 (animal products), 1987 (infant foods) HMSO.

Mohpatra, S.P., Gajbhiye, V.T. and Agnihotri, N.P. 1994. Organophosphorus insecticide residues in the aquatic environment of a rural area. *Pesticide Research Journal* 6, 157-160.

Nag, S.K. and Raikwar, M. K. 2002. First Annual Report of Lal Bahadur Shastri Young Scientist Scheme "Monitoring of pesticide residues in animal feed, fodder and milk with special reference to Bundelkhand region" submitted to ICAR, New Delhi.

Narayan, K.G. 2002. Veterinary public health. In: Hand Book of Animal Husbandry. Indian Council of Agricultural Research, New Delhi. pp 780.

Prasad, K.S.N. and Chabara, A. 2001. Organochlorine pesticide residues in animal feeds and fodders. *Indian Journal of Animal Science* 71, 1178-1180.

Robbins C.1991. Poisoned Harvest: A Consumers Guide to Pesticide use and Abuse. Victor Gollancz Ltd, pp 300-313.

Saftlas AF. 1987. Cancer and ocher causes of death among Wisconsin farmer. *American Journal of Industrial Medicine* 11, 119-129.

Shastry, M.S. 1983. Monitoring of pesticide residues in animal feeds and animal products. *Pesticide,* December issue. pp. 36-38.

Shinde A.K. 2004. Organic farming for producing quality mutton and wool from sheep under semi intensive system. Lecture delivered in Training programme on Advances in Semi-intensive Sheep Management held at CSWRI, Avikanagar December 21-30.

Shinde, A.K., Ravindra Kumar and Singh, V.K. 2005. Organic sheep farming: to protect human health from risks arising from agrochemicals and veterinary drugs in milk and mutton. National Seminar on Application of Nuclear and Non Nuclear

Techniques in Detection of Residues in Food and Feed. Madras Veterinary College, 18-19 March, 2005.

Singh V.K. and Shinde, A.K. 2003. Scope and Challenges of Organic Sheep Production in the Country. National Seminar on Organic Products and its Future Prospects. Sher-e-Kashmir University of Agricultural Sciences and Technology of Kashmir (SKUAST), Srinagar 7-8 October 2003.

Singh V.K., Shinde, A.K., Karim, S.A., Kumar Ravindra, Mann, J.S., Misra A.K., Swarankar C.P. and Gupta, N.P. 2005. Organic sheep farming for producing quality meat, milk and wool. Institute Project, CSWRI, Avikanagar.

Singh, J.D., Tiwari, D.P., Kumar, A and Kumar, M.R. 2002. Effect of feeding transgenic cotton seed vis-à-vis transgenic cotton seed on heamato-biological constituents in lactating Murrah buffaloes. X International congress of AAAP, New Delhi pp 257-258.

Singh, P.P., Battu, R.S. and Kalra, R.L. 1988. Insecticide residues in wheat grain and straw arising from their storage in premises treated with BHC and DDT under malaria control programme. *Bulletin of Environment Contamination and Toxicology* 40, 696-702.

Singhal, K.K. and Mudgal, V.D. 1990. Transfer of organochlorine pesticide (DDT) from feed to milk. *Indian Journal of Dairy Science* 43, 348-350.

Singhal, K.K., Kumar, S., Tyagi, A.K. and Rajput, Y.S. 2006. Evaluation of BT cotton seed as protein supplement in the ration of lactating dairy cows. *Indian Journal of Animal Science* 76, 532-537.

Singhal, Lokesh and Chauhan, R.S. 2000. Environmental pollution by agrochemicals and its impact on epidemics of animal diseases. *Pashudhan* 15, 1.

Surendra Nath, B., Unnikrishanan, V., Gayathri, V., Chitra, P.S. Preeja, C.N. and Rama Murthy, M.K. 1998, Organochlorine pesticide residues in animal tissue and their excretion through milk. *Journal of Food Science and Technology* 35, 547-548.

Thakur, D., Kumar, R. and Chauhan, R.S. 2000. Consideration of antibiotics usage in animal health and production. *Pashudhan* 17, 1.

Unnikrishanan, V., Bhavadasan, M.K., Surendra Nath, B. and Ram Chand. 2005. Chemical residues and contaminants in milk: A Review. *Indian Journal of Animal Science* 75, 592-598.

Unnikrishanan, V., Surendara Nath, B., Gayathri, V., Sampath. K.T. and Rama Murty, M.K. 1998. Organochlorine pesticide residues content in feed and feedstuff. *Indian Journal of Dairy and Biosciences* 9, 59-64.

□□□

30

Market Trends in Goat and Sheep Meat: National and International Scenario

M. K. AGNIHOTRI[1], V. RAJKUMAR[2]

[1]Ministry of Agriculture, Department of Animal Husbandry, Dairying & Fisheries, Government of India, New Delhi-1. [2]Scientist, Department of Meat Science and Technology, Madras Veterinary College, Chennai.

Introduction

There are about 61 million sheep and 124 million goats in the country. Their population has increased by 7.0 and 1.3 %, respectively during the inter-censal period from 1997 to 2003. Worldwide goat and sheep population, unlike most other livestock species, has increased steadily during the last twenty years. Their numbers are increasing much more rapidly in the developing countries than in the developed regions. This is mainly due to the ability of small ruminants to survive and produce on low cost feed, their adaptability to difficult (arid) environment and suitability to small, low-capital family farms that needs additional income and low cost nutrition.

The country, with about 11% of the world livestock population occupies a significant place numerically in respect of livestock wealth. It possesses 56.60 percent of buffaloes, 13.66 percent of cattle, 15.63 percent of goat, 6.03 percent of sheep and 1.48 percent of world's pig population. With such a high population of livestock, on meat production front, the contribution is meager 2.23 percent (6.10 million tones) to the world total meat production of 273 million tones (FAO,2006).The current meat production from goat and sheep is 0.475 and 0.239

million tones, respectively as against the world total goat and sheep meat production of 4.95 and 8.63 million tones(FAO,2006).The average carcass weight in case of goat and sheep is 10 and 12kg, respectively which is lower than even from Pakistan being 17kg and China being 13 and 15kg .

Meat eaters in India prefer sheep and goat meat and chicken. The small frame size of goats and sheep enables them to be slaughtered hygienically. Carcass size is sufficient to meet small family needs and does not pose marketing and distribution problems. Unlike cattle and pigs, meat from these species is acceptable across religious beliefs and faith. Meat from goat and sheep, though it is costliest, yet most preferred. Due to these factors their slaughter rate is high (48% for sheep,38% for goats) compared to bovine (about9%). The factors like outbreak of bovine spongiform encephalopathy (BSE) in the United Kingdom and its association with Creutzfeldt-Jakob disease (CJD) in humans had adversely affected the demand and beef consumption in Europe. Outbreaks of swine fever, followed by massive stamping out operations in Germany, Belgium and the Netherlands in recent years have influenced public perception about the pork eating. Outbreaks of Bird flu opened new market to the sheep and goat meat trading in India.

A recent FAO study found that livestock production is one of the major causes for the world's most pressing environmental problems, including global warming, land degradation, air and water pollution, and loss of biodiversity.Therefore, to increase meat production from small ruminants, mere increase in numbers is not going to be sustainable in a longer run. Their productivity per unit time has to be improved by adopting modern tools and techniques of breeding, feeding, management and disease control.

Consumption Patterns of Goat and Sheep Meat

In many parts of the developing world, rise in income, education and the growing consciousness for the balanced diet, has led to increase in non-vegetarianism and meat consumption including India. Tough the per capita consumption of meat in the developing world has more than doubled; still a person in the developing world consumes only one third quantity of meat in comparison to a person in the developed countries.

Indian scenario

Holiday calendar indicating demand for goat and sheep meat has been presented in Table 1. During these holidays demand for meat rises and the commercial goat and sheep farmers can schedule their production accordingly. Indian consumers prefer "skin off" goat / sheep carcasses. Whereas, in some North Eastern States, "skin on" carcasses are also in demand. There is regional variation among consumers perception over goat and sheep meat. In Northern India goat meat fetches more market prices whereas in Southern part, it is mutton. Meat from all breeds and age of goat and sheep has a value, but young and well-

Table 1. Holiday calendar (2007 -2010) indicating demand for goat and sheep meat

Holiday	2007	2008	2009	2010
Eid ul-Adha Festival of Sacrifice	December 20	December 8	November 28	November 17
Muharramn/Islamic New Year	January 20	January 10	December 8	
Mawlid al-Nabi Prophet's Birthday	March 31	March 20	March 9	February 26
Start of Ramadan Month of Fasting	September 13	September 2	August 22	August 11
Eid ul-Fitr Festival of Fast Breaking	October 13	October 2	September 21	September 10
Diwali	November 8	October 28	October 17	October 14
Pongal	January 14-17			
Christmas	December 25			

managed goats receive premium prices. Meat from young male is preferred over female. Goat meat is consumed in more than 40 countries. Goat and sheep meat in India is more in demand during winters. During summer demand is low and is probably the best period for exports (Sharma *et al.* 2003).

International scenario

In most of the countries where goat meat is in demand, live weight and carcass weights are the major criterion, instead of age (Table 2). Malaysian consumers require lean meat and prefer intact male carcasses. The Caribbean market requires skin-off, frozen carcasses between 10 – 18 kg.

Table 2. Consumer uses and preferences for goat meat in different countries

Market	Consumer	Weight preferences	Other considerations
Italy	Ethnic	Live weights: 14-18 kg Dressed weights: 7-9 kg	Weight preferences depend on time of year. Lighter kids preferred for Easter and Christmas seasons.
Greece			
Middle East			
Pakistan			
S.E. Asia			
Australia	Hotel and restaurant Industry	Dressed weights: 12 –15 kg	Premium quality carcass is required, particularly high value cuts such as racks and loins.
Taiwan	Middle to high income wage earners. Predominately male	Live weights: 20-30 kg Dressed weights: 10-16 kg	Prefer skin on carcasses. Meat must be extremely lean with strong flavour.

Contd...

Table 2. Contd...

Caribbean	Middle income earners	Dressed weights: 10-18 kg	Require skin off carcasses. Will use packages of cubed, bone-in-pieces from older, less conditioned goats.
Malaysia	Middle to high income earners	Live weights: 25-30 kg Dressed weights: 13-20 kg	Use both skin-on and skin-off carcasses. Prefer intact male carcasses and also prefer to slaughter their own animals.
	Indian and Malay		Skin –off wet dishes, curries and stews
	Chinese		Skin-on for traditional Chinese cooking
Singapore	Housewives from the Indian community	Live weights: up to 50 kg Dressed weights: up to 25 kg	Prefer skin-off carcasses. Prefer lean product
United States	Ethnic groups located primarily in metropolitan areas and health conscious individuals.	Varies depending on ethnic group.	Demand from Hispanic, Indian and West Indian communities round the year and increased demand from Chinese communities during winter.

Demand and Supply

International scenario

Increase in meat production and its demand is expected to take place mainly in developing countries in the near future. Between 1995 and 2020 they will account for almost 85% of the increase in global demand as against 25% in developed countries. A demand driven livestock revolution, is therefore, anticipated in developing world. The forecast ruminant and non-ruminant meat production and demand by the year 2010 indicates a deficit of ruminant meat production in India, Indonesia, Malaysia, Pakistan and Philippines except in China (Table 3). Worldwide, demand for meat is forecasted to rise by more than 55 % (Figure 1) between 1997 and 2020; the demand for pork and beef will decrease and would be replaced by the poultry (Figure2). No change in demand for goat and sheep meat is predicted. China alone would account for more than 40% of this increase, compared to India's 4 percent. Even though demand for meat will double in South Asia, South-East Asia, and Sub- Saharan Africa, per capita consumption of meat will not match with the developed world. This gap suggests that people in these regions will have the potential to consume considerably more meat. Poultry will account for 40 % of the global increase in demand for meat to 2020, far higher than 28 % it accounted for in 1997 reflecting a dramatic shift in taste from red meat to chicken (Figure 2).

Indian scenario

Considering 70 % population as non-vegetarian, the annual per capita availability of meat in India is roughly 5.0 kg as against the recommended intake

Table 3. Forecast ruminant and non- ruminant meat production and demand per person by 2010 in India and some selected Asian countries.

Country	Ruminant meat demand (kg/capita)	Ruminant meat production (kg/capita)	Non-ruminant meat demand (kg/capita)	Non ruminant meat production (kg/capita)
China	5.8-7.4	8.5	54.5-71.3	49.8
India	5.1-6.2	3.8	1.5-1.9	26.2
Indonesia	4.6-6.0	2.5	12.3-16.7	26.2
Malaysia	6.3-7.9	0.6	73.7-93.9	139.6
Pakistan	16.9-20.3	14.6	19.8-24.5	9.8
Philippines	4.4-4.9	2.7	27.8-31.7	29.8

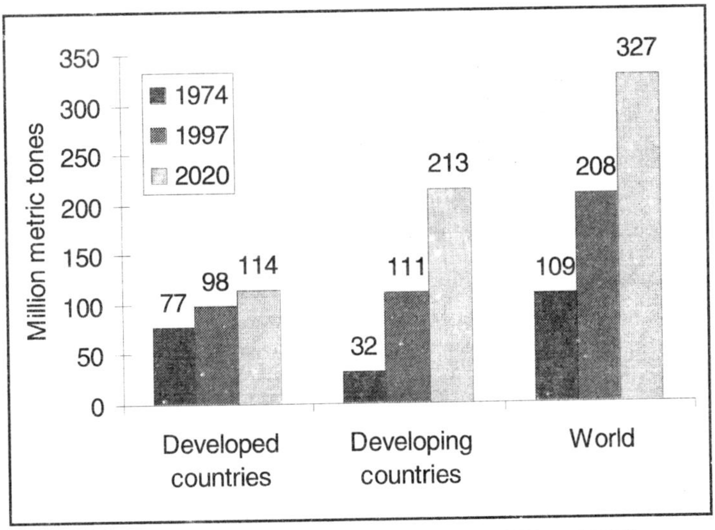

Fig 1. World demand for meat in 1974, 1997 and 2020

of 11.00 kg. The demand for meat and eggs in India is estimated to vary from 4.7 to 7.2 million tones in 2010 and 6.3 to 12.1 million tones in 2020 (Dastagiri 2004). More than 70 per cent meat in India is derived from cattle, buffalo and pigs for that preference is limited. For meeting domestic demand, dependency is on goat meat (chevon) and mutton. The annual per capita availability in developing countries is 26.0 kg and 76.0 kg in developed countries. The animal protein availability is only 9 g as against 16 g in Asia and 25 g the World average. The protein supply from meat in India is only 1.7 g / day as against 8.8 g in developing and 25.6 g in developed countries. Therefore, country needs to produce more not only for exports but also to feed the burgeoning human population. In spite of the progress made, a significant gap exists between demand and supply and goats and sheep could play a significant role in meeting this short supply.

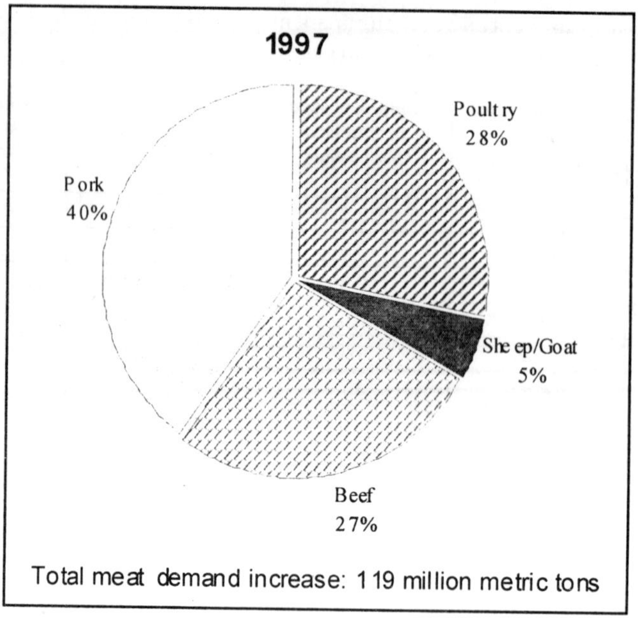

Increase in meat demand by the year 2020

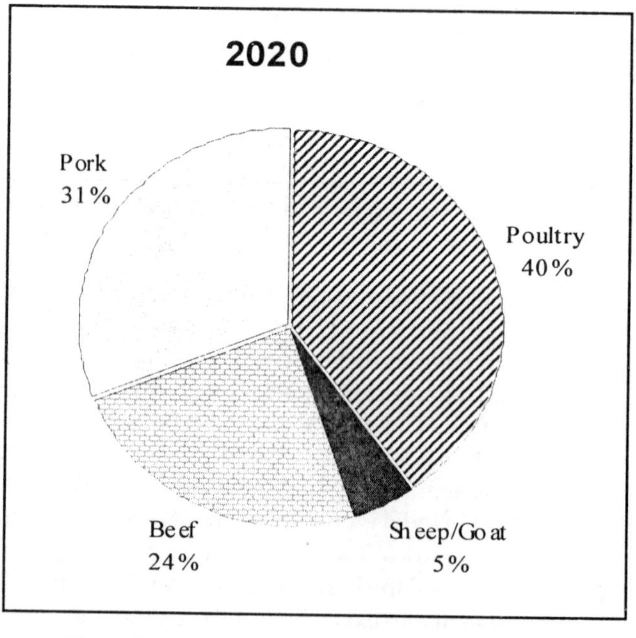

Fig. 2. Types of meat demanded globally in 1997

Trends in Consumption and Production of Goat and Sheep Meat Towards 2020

Although the differences are still large, the developing world is increasing its share in consuming animal foods (Table 4) as indicated by a remarkable increase in intake of calories and protein of animal origin. The developed world is not increasing its animal food intake (Delgado *et al.*, 1998) due to incasing risks of heart diseases and obesity. By 2020, the developing countries will eat 70 % more meat than the developed countries, although their per capita consumption will be only 38 % of developed country levels. In Table 5a the projected growth rates and total meat consumption are presented. The growth in demand in China will be a major force for meat and cereal production. In Table 6a predictions are given for meat production. Most of the growth in sheep and goat meat consumption and production is projected to occur in the developing countries (Table 5 b, 6b).

Table 4. Trends in consumption of calories and protein from animal products in the period 1973–2003

Region	% Calories					% Protein			
	1973	1983	1993	2003		1973	1983	1993	2003
Developing world	8	9	11	15		19	21	26	31
Developed world	28	28	27	34		55	57	56	59
India	4	5	8	12		8	9	12	18
China	6	8	18	24		12	14	28	32

Source: Zijpp (1999); and FAOSTAT

Table 5a. Consumption trends of meat from 1982 to the year 2020

	Annual growth of meat consumption (%)			Total meat consumption (million tones)		
	1982-1993	1993-2000		1983	1993	2020
Developing world	5.3	2.9		50	89	194
Developed world	1.2	0.5		88	99	113
India	1.8	2.4		1.2	3.5	8
China	8.3	3.2		17	39	89

Table 5b. Consumption trends of goat and sheep meat (million tones)

	1990	2000	2004	2005
Developing Countries	6.46	8.76	10.35	10.95
Developed Countries	2.90	2.30	2.18	2.16
India	0.59	0.68	0.70	0.71
China	1.10	2.82	4.16	4.61

Source: FAOSTAT (2007); and Delgado *et al.*, (1998).

Table 6a. Production trends of meat from 1982 to the year 2020

	Annual growth of meat production (%)		Total meat production (million tones)		
	1982-1993	1993-2000	1983	1993	2020
Developing world	5.2	2.9	50.0	89.0	194.0
Developed world	1.2	0.5	88.0	99.0	113.0
India	1.8	2.4	2.9	4.3	8.0
China	8.3	3.1	17.0	40.0	89.0

Source: FAOSTAT (2007); and Delgado *et al.*, (1998).

Table 6b. Production trends in goat and sheep meat (million tones)

	1990	2000	2004	2005
Developing Countries	6.32	8.51	9.92	10.47
Developed Countries	3.44	3.05	2.76	2.83
India	0.61	0.69	0.72	0.73
China	1.07	2.74	4.00	4.38

In Australia and USA, a significant increase in production cost of, both dairy and meat, is visualized due to steep increase in feed grain prices in the market. Owing to the increase in oil prices internationally, there is subsidy for production of ethanol and bio-diesel. The barren lands and the pastures are increasingly being converted in to the field for the cultivation of bio-diesel. Feed grains like maize and other coarse grains, otherwise used as an ingredient in the animal feed, are now used for the ethanol production. This new trend of land and feed grain utilization has resulted in the increased cost of production and might adversely affect the price competitiveness of animal products (meat and milk) in the global market. The meat industry in Europe is facing many challenges like protection of environment, animal welfare, food safety, animal diseases, nutrition (obesity, hyper-tension), increasing feed and labor cost, and it is visualized that by 2014 India and China would emerge as the major new stake holders in this sector.

Preference for the Indian Meat

There is a growing preference for meat exported from India due to certain inherent merits like lean character, relative freedom from toxic feed additive residues and near organic meat production. India is the 5th largest exporter of bovine (buffalo) meat in the World. Currently India has been exporting meat to more than 60 countries. Meat is exported in frozen bone less and de-glanded form and is free from FMD virus due to its ageing for minimum of 24 hrs at 2^{0}C to bring down the meat pH below 6.0.Meat is available at very competitive prices. The Indian buffalo and lamb meat has established itself in the markets of South East Asia, Middle East and African countries.

Trade and Changes in Meat Prices

There has been a significant increase in meat and meat products export from India. It has increased from Rs.408.25 crore in 1994-95 to Rs.1863.19 crores in 2004-05 (APEDA, 2005).Among the animal products exported from India, meat and meat products account for more than 90 % of the total exports volume. Remaining 10 % are dairy products. The major contribution to this was from export of buffalo meat, sheep/ goat meat and animal casings. Processed meat contribution to the total exports earning was meager Rs.4.79 crores.

The Saudi Arabia, UAE and Qatar are the major destinations for export of sheep and goat meat from India. The export of sheep and goat meat during 2003-04 was around 16821 tones worth Rs.110.39 lakh. During 2003-04, it has shown a quantum jump, when compared with 2001-02 and 2002-03 figures when it was only 3915 and 4974 tones, respectively. In MFPO licensed meat processing plants in the same period, 4045.78 tones worth Rs.5891 lakh meat food products were manufactured. The trends in processed meat food products for the last five years do not reveal significant increase in production (Agnihotri, 2006) and sheep and goat meats are not exceptions to it.

The estimated annual net exports from the developed world to developing countries in 1990 was 0.6 million tons (Delgado *et al*,. 1998) which is equivalent to 0.3 % of world meat production. In 2020, the developed countries will export 11.6 million tons of meat to the developing countries.

The market price of a kg of sheep and goat meat currently ranges from Rs. 120 in smaller towns and cities to Rs. 280 in metropolises of India. During the last 6-7 years, market has recorded almost 50 % increase in the cost of goat / sheep meat. Due to Bird flu scare in poultry sector, the cost of goat meat shot up to Rs.300 kg at the time of outbreak. Whereas, the international market price for lamb carcass (New Zealand and London market, FAO 2007) decreased from US$ 4430/ton in the year 2003 to US$ 4003/ton in the year 2007. The current month global market rate is US$ 4136/ton of frozen lamb carcass. Five years ago it was traded for 3366 US$/ton and 3217 US$/ton was ten year old price for frozen lamb (FAO, 2007).

Rules and Regulations for Export

(a) Standards have been laid down for export of meat and meat products under Export (Quality Control and Inspection) Act, 1963. The export of Raw Meat (chilled and frozen) shall be allowed subject to the provision specified to the gazette notification on Raw Meat (chilled and frozen) under Export (Quality Control and Inspection) Act, 1963 in January,1993. The notification lays down the Standards for abattoirs, meat processing plants and the products. Offal's too are subject to the same conditions of quality control and inspection.

(b) All consignments of Raw Meat (Chilled and Frozen) and export of canned meat products need to accompany with a Pre-shipment Inspection Certificate for which the Govt. has designated three agencies (i)All State Directorate of Animal Husbandry (ii) Export Inspection Agency (iii) Directorate of Marketing and Inspection, Government of India to carry out the inspection in accordance with either the standards prevalent in the exporting country or standards prescribed under the Meat Food Products Order, 1973 under Export (Quality Control and Inspection) Act, 1963 or orders made there under.

(c) Export of meat and meat products will be allowed subject to the exporter furnishing a certificate to the customs at the time of export that the above items have been obtained/sourced from an abattoir/ meat processing plant registered with Agricultural and Processed Foods Products Export Development Authority (APEDA).

Export Policy

The export of beef of cows, oxen and calf and offal's(includes heart, liver, tongue, kidneys and other organs) is prohibited and not permitted to be exported. Meat of buffalo (both male and female) and offal of buffalo except gonads and reproductive organs is free to export subject to the condition that exporter has to produce a certificate from the designated veterinary authority of the State, from which the meat or officials emanate, to the effect that the meat or officials are from buffaloes not used for breeding and milch purposes. The Quality control and inspection under (a) and (b) and (c) stipulated above under Rules & Regulations are required to be fulfilled. The meat and offal's of Indian sheep and goat is free for export subject to fulfilling the conditions (a) and (b) and (c) stipulated above.

Tallow, fat and/or oils of any animal origin excluding fish oil is prohibited for export.

Issues to be Addressed

- Because of the isolation due to odd geographic locations and the fact that small ruminants are generally maintained by resource poor illiterate farmers, there is limited access to the market and market information.They are not able to sell their products at the competitive prices. This needs to be looked into.

- The major production of sheep and goat meat in the country is from small farms. Since the livestock holding size in India is variable, small and the population is scattered and also migratory type in certain states, establishment and maintenance of traceability which is required not only for the purpose of export of various livestock products but also for the food safety, is a necessity and also a challenge.

- Animal food production will become crop based. Concentrations are likely to occur near centers of urbanization. The change to mixed crop-livestock and industrialized system has implications for the environment and health. Land degradation, including soil fertility, soil erosion and water pollution require measures worldwide including India.

- Free trade is the basis for the projected changes in patterns of consumption and production. Control of animal diseases is an important requisite for free trade. Industry should also think of playing a more active role in creation of more number of disease free zones, animal welfare issues, and the removal of deficiencies in the quality feed supply chain.

- On private sector we have modern integrated abattoirs where meat is produced adopting OIE guidelines and international quality standards for exports. State Government's need to take some bold decisions for modernization of municipal slaughterhouses as well others which caters to the need of domestic consumers. Environment and animal rights issues, by doing so, can be addressed more effectively.

- Strengthening of quality testing and monitoring laboratories to carry out the testing and monitoring of meat and meat product samples for pesticide residues, toxic metals, veterinary drug residues and mycotoxins etc has to be taken-up on priority.

- To remain competitive in the world market, information about trade linked animal diseases, veterinary drug residues and other harmful residues in feed and products of animal origin, their maximum permissible limits set by Codex and consequences of their presence beyond permissible limits to human health, trade and industry needs to be disseminated to farmers, feed manufacturers, field veterinarians, animal product processors and the consumers by the States.

- The consumers needs to be educated through print and electronic media to remove their apprehensions about hygiene, sanitation and safety of meat and meat products and also importance of modernization of abattoirs in producing safe and wholesome meat and healthy environment.

- Processing of meat in to value added products, establishment of cold chains, modern storage facilities and packaging of products in durable and attractive packaging materials are the other important areas which will require greater attention by the industry in near future to meet the growing demand of quality ready-to-eat meat and meat products. Not more than 2% of the total meat produced is processed into value added meat products. Sheep & Goat meat exports mainly comprise of chilled/frozen meats. The industry should seriously think of establishing backward linkages at primary production level with the farmers and process meat in to value added products to achieve the real "pink" revolution.

- There are hardly a few reputed manufacturers of meat processing equipments indigenously. Imported plants and machinery is subjected to high import duty. Waving off can reduce the cost of production.
- No taxes are imposed on fresh meat. For branded and packed meat products sales tax and excise duty is levied which needs to be revisited as it discourages the processing and encourages sale of unprocessed meat.
- There is need for harmonization of standards to meet the CODEX standards.

REFERENCES

Agnihotri, M.K. 2006. Regulatory requirements for hygienic meat production and export. Souvenir-cum-proceedings. National workshop-cum-seminar on commercial goat and Sheep farming and marketing: Farmer-Industry-Researcher interface, March 4-5, 2006. CIRG,Makgdoom, Farah, Mathura.pp.120-125.

APEDA. 2005. Agricultural and Processed Food Products Export Development Authority.

Export Statistics. 2005. Ministry of Commerce and Industry, Govt. of India, New Delhi.

Dastagiri, M.B. 2004.Demand and supply projections for livestock products in India. Policy Paper (21), National Centre for Agricultural Economics and policy Research(ICAR), New Delhi. pp.43.

Delgado, C.L., Courbois, C.B., Rosegrant, M.W. 1998. Global food demand and the contribution of livestock as we enter the new millennium. In: Paper presented at the BSAS-KARI Conference: Food, Lands and Livelihoods: Setting Research Agendas for Animal Science, Nairobi, Kenya.

FAOSTAT. 2006,2007. www.fao.org.

Sharma, N., Rajkumar, V. and Agnihotri, M.K. 2003. Goat Meat Production in India: Emerging Trends and Export Potential. Souvenir, Invited papers. Interaction Seminar on Meat Production and Export Potential: Impact of Globalization and Quality Concepts, 17th September, 2003, CIRG, Makhdoom. pp 42-50.

Zijpp, A.J. 1999. Animal food production: the perspective of human consumption, production, trade and disease control. *Livestock Production Science*. 59, 199-206.

❑❑❑

31

Scope of Value Addition and Commercialization in Sheep and Goat Meat

A.R. SEN[1], S.A. KARIM[2]

[1]National Research Centre on Meat, Chengicherla, P.B.No.19Uppal P.O., Hyderabad -500 039, A.P., [2]Central Sheep and Wool Research Institute, Avikanagar, Rajasthan 304 501

Introduction

The contribution of small ruminants in rural economy of the country is evident from the fact that they are closely associated with poorest of the poor. Sheep and goat rearing is traditional occupation of small and marginal farmers, landless laborers and the nomads. Both sheep and goat are sold for slaughter by the farmers having surplus animals in excess of the breeding requirement (mostly young lambs/kids) or adult animals after their effective productive life. Sheep and goat meat is consumed by all segments of society without any religious taboos. A small proportion of mutton and chevon produced in the country are converted to value added products which need be addressed to improve its earning potentiality. Most of meat production in India is under unorganized sector. Successful commercialized rearing is not possible unless the organized meat sector development takes place with modern hygienic slaughter practices. Because of the poor post harvest infrastructure and management, the highly valuable by products are either not utilized or under utilized and value addition is urgently required for these products to make Indian sheep and goat meat sector economically viable.

Sheep and Goat Meat Production and Export Potential

India ranks third in chevon production, next to China and Pakistan. Annual production of chevon is more than 475 thousand MT which contribute about 13 % to total chevon production in the world (Table 1). The reported annual slaughter rate for goat is about 38 %. The annual mutton production has crossed 239 thousand MT contributing around 3 % to total mutton production in the world. The reported annual slaughter rate for sheep in India is 33 %.

Table 1. Meat production of India from sheep and goat as compared to world

Year	Production (mMT)			
	sheep		Goat	
	India	World	India	World
2002	0.233	7.658	0.470	4.022
2003	0.236	7.879	0.473	4.252
2004	0.239	8.203	0.475	4.366
2005	0.239	8.470	0.475	4.562

FAOSTAT, 2006

The export of sheep and goat is less which is in Table 2. The export of small ruminant meat (Sheep and goat) has started picking up during the last three years. The fall in export of small ruminant meat is due to the reduction in the import by Saudi Arabia. Now that the market has opened up for Indian meat, the export of meat from sheep and goat is likely increased substantially. Indian chevon because of leanness is preferred in the world market. But continuing demand for goat meat in the domestic market limits the availability of chevon for export. However, the amount of meat being exported has increased recently to around 12 % of total chevon production.

Value Addition in Production and Processing

Value addition is an endeavor for increasing consumer acceptability and market demand which in turn provides better economic returns. Most of the sheep and goats are slaughtered in unregistered slaughter houses in small numbers ranging from 2- 10 by the individual butchers who generally sell of fresh meat on the same day. In our country, 10-12 months old sheep/goats are preferred for production of fresh meat. About 30 % of spent animals are also slaughtered at a very old age producing tough meat which is not preferred as fresh meat by the consumers. Meat processing by comminution and producing emulsified products has great advantage for this category of tougher meat. Toughness is reduced by mincing the meat and processing it to tasty and convenient meat products. The goat meat regardless of breed can be used to manufacture processed products of acceptable sensory quality, particularly when a spicy formulation is used. The low acceptability of goat meat by some consumers may be due to lack of tenderness, particularly from meat of older animals. Tenderness should not be a problem for comminuted meat products or products those undergo slow cooking/retorting (curries and stews). Another aspect in value addition is proper packaging of the finished products and favorable conditions for increasing freshness and shelf life of meat and meat products. Vacuum packaging and modified atmospheric storage can increase shelf life of such products. Packaging of chevon and mutton products with appropriate plastic films and laminates retain quality and increases shelf life during frozen and refrigerated storage conditions.

In India, only 2- 3 % of fresh meat produced is converted into processed meat products while in developed countries it is above 60 %. Without value addition as processed products, this sector can not be fully tapped to its full potential. The weakened image of beef resulting from the recent BSE scare, combined with the taboos associated with pork and beef in certain segments of the society could increase the appeal of processed meat products prepared from sheep and goat.

Development of Value Added Meat Products

The production of processed products involve study of meat and its functional properties as a raw material, binders, seasonings and additives, practical formulation, processing equipment and importance of public health regulations of meat and meat products. Technologically, the major consideration is selecting various meat ingredients for processed products those have ability to bind or emulsify fat and retain moisture. Moreover, the ingredients are also selected for color, flavor, tenderness and other quality attributes. Processed meat products are defined as those in which fresh meat is modified by various methods to yield certain desirable and palatable products. Meat processing originated in prehistoric times and no doubt developed soon after people became hunters. The original basis for meat processing was preservation by inhibiting or deterring microbial decomposition. In addition to this, preservation also resulted in flavor full and nutritious products. Before developing processed meat product it must be kept in mind about the desirable taste among the consumers. The western products may not be successful in Indian conditions.

There are numerous different processed meat products which have their own specific characteristics and method of preparation. They all can be classified as comminuted and non-comminuted products. Typical non-comminuted products include hams of all types, bacons, Canadian bacons etc. In the meat industry, many of these products are commonly referred to as smoked meats. Their distinguishing characteristics are that they are prepared from whole, intact cuts of meat. These products usually are cured, seasoned, heat processed and smoked and often they are molded or formed. Comminuted meat products involve subdividing the raw materials so that the product consists of small meat pieces, chunks, chips and slices. Most comminuted products may be classed as sausages.

Sausages are comminuted, seasoned meat products that may also be cured, smoked, molded and heat processed. Some sausages like frankfurters and bologna are referred to as emulsified products. In these products the meat can be so finely subdivided that the mix is viscous with many characteristics of an emulsion. Besides this, the emulsion can be molded in different shape and cooked viz. meat balls, nuggets, patties, kabab etc.

Meat pickles, like fruit and vegetable pickles have the potential of becoming a ready to eat, highly acceptable and shelf stable, convenient meat product of indigenous origin. Sheep and goat meat pickle can also be easily prepared by rural people and can be developed as a small scale cottage industry. Canned meat products are mainly enjoyed by the majority of people in the developed country. In India two types of canned products from sheep and goat meat are generally produced: meat canned in brine and meat curried and canned.

Enrobing is another promising technology for value addition of sheep and goat meat. It has become a means of offering consumers a new product. Some of the enrobed meat products developed is croquettes, cutlets and samosa using sheep and goat meat. Commercial enrobing provides an opportunity of value addition at a low cost to the processors. The composition of enrobing medium especially of carbohydrate and protein sources is important for final color of the product.

Recently fermented meat products are likely to play an important role in the development of processed meat products. Dry (Cervelats) and semi-dry (Thuringer) sausages are coarse ground fermented products. These are prepared using starter cultures (*Pediococcus or Lcatobacillus*) or can be held under specific conditions that preferentially promote the growth of organisms those impart, flavor, texture and preservative qualities. During fermentation the decline in pH not only imparts the tangy flavor and chew texture, but also helps in drying process.

Another shelf stable product through retort pouched technology has a great potential in Indian market. Nowadays number of working women are increasing in major Indian cities and the demand for such type of products are also increasing. The technology development for value addition and commercialization for this type of shelf stable product is a great challenge.

Value Addition in By-Products

The economics of the world meat industry demand that animal by products are to be utilized so that the livestock industry can stay economically competitive. Slaughtered and dressed animals produce 60- 70 % by products. Proper end use of these materials would contribute to considerable profit if they are properly collected and processed. In the last two decades there is considerable progress in India in this direction by way of development of useful technologies for better utilization of raw and fresh animal by products.

Non- utilization or underutilization of by products from slaughter animals has been a major concern in developing countries, since this contributes the source of economic returns by sale of animal based by products. If animal by products are not effectively utilized a valuable source of potential revenue will be lost and increasing cost of disposal of these products will be incurred by the industry. By products have multi facet uses- human food, medicinal and pharmaceuticals, leather goods, garments, feeds etc. Indigenous technologies are available for production of value added products which generate employment, value addition and removal of pollution. Technologies are also available for production of bio gas, which is a cheaper source of energy and also to removes pollution. Large export oriented mechanized state of the art slaughter houses produce large quantity of blood and paunch from sheep and goat. Presently these are utilized for manure purpose but could be utilized for value addition. The inedible portion are utilized for production of meat-cum-bone meal, tallow, bone chips etc.

Commercialization in Production and Processing

Indian sheep/goat breeds have tremendous genetic potential for higher meat production, which is possible through improving feeding practices and commercialization in farming. It has been reported that nutritional manipulation for meat animal production of the country can be increased by 300 % through improved and economic feeding of these animals. But our farmers are not very receptive to the developed technologies because of their poor socio economic condition and educational status. It has suggested that encouraging progressive farmers and entrepreneurs to enter in to venture of fattening of weaner lambs/ kid procured from existing production system for commercial meat production will go a long way as a economic venture. Moreover, farmers in one or two states have already taken up commercialization of goat farming. The problem arises here due to involvement of middlemen and the farmers' share is only 60 % in the total realization from the meat production. This can be solved through integration and concerted efforts from rearing to production of small ruminants.

We have 3600 registered slaughter houses throughout the country. Yet most of them are unhygienic, overcrowded and management is very poor with primitive infrastructure. Without modernization, value addition in quality and commercialization is incomplete. The recent trend in India is to establish latest state-of-the-art abattoirs cum meat processing plants. The industry has already established 10 most modern mechanized meat processing plants in various states based on slaughtering buffaloes and sheep. These plants are environment friendly, where the entire slaughter house by products is utilized. They are also adopting appropriate technologies to obtain value added products. The plants follow all the sanitary and phytosanitary measures required by International Animal Health code of OIE. Entrepreneurs have to be encouraged for production of value added meat products. In poultry products, a couple of renowned firms are coming up.

In the same line, mutton and chevon products have to be encouraged by such company with bigger advertisement and publicity.

Table 2. Export of sheep/goat meat from India (000MT)

Year	Export quantity (000MT)	
	Sheep	Goat
2001	3.468	0.026
2002	4.843	0.130
2003	16.54	0.28
2004	8.829	0.056

FAOSTAT 2006

Conclusion and Recommendation

Conversion of harvested meat to value added product would go a long way in value addition and further enhance profitability generating employment avenues for the unemployed. Sizeable migrated population of Indian origin in Middle East countries are prospective consumers for value added hygienically packed heat and eat meat product. The required infrastructure for processing and packaging has to be created in private sector for commercializing the venture. Value added meat and by products need to be developed both for domestic and export market in order to efficiently utilize the product from spent animals and low value cuts. Being a developing country if India needs to remain in competition in global meat trade, it has to come long way from production to consumption. Therefore, pragmatic and concerted efforts are required to build this value chain starting from farm to fork.

❑❑❑

32

Food Safety Regulations with Special Emphasis On Organic Residues On Meat

NAVNEET SAXENA, D. LAL, B.S. PUNIA

Central Institute for Research on Buffaloes, Hisar 125 004, Haryana, India

Introduction

People have the right to expect the food they eat to be safe and suitable for consumption. Food borne illness and food borne injury are generally unpleasant and sometimes they can be fatal. But there are other consequences also. Outbreaks of food borne illness can damage trade and tourism, and lead to loss of earnings, unemployment and litigation. Food spoilage is wasteful, costly and can adversely affect trade and consumer confidence. Food safety is an issue of growing importance due to several worldwide trends in food systems. The growing movement of people, live animals, and food products across borders, rapid urbanization in developing countries, changes in food handling, and the emergence of new pathogens or antibiotic resistance in pathogens all contribute to increasing food safety risk. These issues are recognized in international trade negotiations under the WTO and in the FAO's Committee on World Food Security (Unnevehr and Hirschhorn, 2000).

International food trade, and foreign travel, is increasing, bringing important social and economic benefits. But this also makes the spread of illness around the world easier. Eating habits too, have undergone major change in many countries over the last two decades and new food production, preparation and distribution techniques have developed to reflect this. Effective hygiene control, therefore, is vital to avoid the adverse human health and economic consequences of food borne illness, food borne injury, and food spoilage. Everyone, including farmers and growers, manufacturers and processors, food handlers and consumers, has a responsibility to assure that food is safe and suitable for consumption. In India, the quality and safety issues have been very well understood and there is regulatory framework for its effective implementation:

- *Prevention of Food Adulteration (PFA) Acts and Rules:* This is mandatory regulation for production, storage and sale of food products.
- A new Food Safety and Standards Bill 2006 have been passed by the Parliament recently and are awaiting endorsement by the President.

Institutional and Legal Framework

The multilateral trade agreements to be effective and functional enshrined in itself a robust institutional and legal framework. The sanitary and phytosanitary (SPS) agreement concerns the application of food safety, animal and plant health regulations and is integrated with the agreement on agriculture. Having recognized that many member countries would face exogenous roadblocks and hindrances on account of food science and food safety considerations among others, each Article of SPS Agreement suggests a foundation for formulation and application of regulatory mechanism within the national territorial boundaries. Therefore, issues of harmonization, equivalence, transparency, technical assistance and processes for dispute settlement have been specifically addressed albeit on the existing institutional and legal endowments in a member country.

Indian Institutions

The Ministry of Food and Consumer Affairs is the main Government agency dealing with product standards for consumption in the domestic market, although each Ministry/Department also has its own system of framing and notifying product standards (Mehta and George, 2003). The Government has been laying emphasis on export promotion since late 1960s. One result of this emphasis has been proliferation of Export Promotion Councils and Authorities. Some of these agencies, notably, Marine Product Export Development Authority (MPEDA) and Agriculture Produce Export Development Authority (APEDA) have been actively involved in implementation of policies relating to product standards.

The Bureau of Indian Standards (BIS) is the main standard setting body in India for all domestic market requirements. It sets voluntary standards that can be acquired to indicate the quality of the product by the use of "ISI" mark. However, BIS is also the guiding organisation behind most of the mandatory standards set by Government agencies. Notably, BIS is also the enquiry point of India under the WTO Agreement on Technical Barriers to Trade (TBT). International standards Organisation (ISO) has published ISO-22000:2005 – Food Safety Management

Systems - requirements for any organization in the food chain, with a view to provide framework for internationally harmonized requirements for systematically managing safety in food supply chains. Consequent to publication of ISO 22000, BIS has adopted this international standard as IS/ISO 22000:2005. BIS has launched Food Safety Management System (FSMS) certification IS/ISO 22000:2005 scheme which envisages grant of FSMS Certification license to organisations according to IS/ISO 22000.

The Export Inspection Council (EIC) is the chief enforcement body for exports, and the one relevant for our research study. The EIC was set up by the Government of India in order to ensure sound development of export trade of India through quality control and inspection. Keeping in view the presence of a large number of agencies, an Integrated Food Law is being considered by the Government of India. The domestic institutional and legal framework in India is evolving itself into a web of interlinkages. Be it the commercial interests or the health safety issues at the national level, a set of legal framework is already in place. In association with industry specific associations and commodity specific export promotion boards, financial and environmental issues too are being addressed.

International Institutions

World trade organisation (WTO)

The General Agreement on Tariffs and Trade (GATT) has been the beacon of the multilateral trading system since 1948. Through the results of the Uruguay Round of multilateral trade negotiations, the World Trade Organisation (WTO) came into being in 1994. The World WTO agreements on Sanitary and Phytosanitary Measures (SPS) and Technical Barriers to Trade Agreement (TBT) establish basic rules for international trade. The Agreement on the application of SPS is an elaboration of GATT rules as they relate to measures necessary to protect human, animal or plant or life of health. Under this Agreement, member countries are required to base their SPS measures on scientific principles and refrain from maintaining measures without sufficient scientific evidence. The Agreement encourages harmonisation of SPS measures and considers the standards set by three international standard setting bodies as acceptable standards. These are the Codex Alimentarius Commission, the International Office of Epizootic and the International Plant Protection Convention. Equivalence is encouraged and conformity assessment guidelines are laid down. Special and more favourable treatment provisions exist for developing countries, but in name only. The Agreement on Technical Barriers to Trade (TBT) allows members to apply standards (both mandatory and voluntary) for protection of human health or safety, animal or plant or life of health, or the environment.

International standards organisation (ISO)

ISO is the most important of international standard setting organisations. It is a world federation of 123 national standards bodies, an international non-governmental organisation with, however, a majority of its members coming from the public sector. ISO develops standards through 200 technical committees split into about 650 sub-committees and 2000 working groups. It also develops guide for standard setting. In preparing these, ISO interfaces with specific users of standards including those in the private sector. All its standards and guides are voluntary in nature. However, given its credibility as the most internationally accepted organisation, ISO standards have considerable trade affects due to their wide use in international trade. While ISO 9000 series is the general quality certification standard of ISO, there is an environmental management standard also, viz. ISO 14000 series.

Codex alimentarius commission

The Codex Alimentarius Commission, a UN body, compiles agreed-upon standards, guidelines and other recommendations into the Codex Alimentarius (the Codex Food Code, CFC). The CFC attempts to create harmonised standards. Prior to the SPS Agreement, the CFC could be adopted, applied and /or ignored at the discretion of a government. However, the CFC has now been adopted within the SPS Agreement as the benchmark. Codex Alimentarius standards, guidelines and recommendations, being internationally agreed legislative standards for measures applied to protect human health from risks arising from contaminants and toxins, etc. in foods, are mentioned in the SPS Agreement as the basis for regulations on contaminants in food in international trade. The Codex standards, guidelines and recommendations are used as the basis for settling disputes in international trade in food. The Codex Committee on Food Additives and Contaminants (CCFAC) is the forum within the Codex system for discussions concerning contaminants. Within CCFAC, the Codex General Standard for Contaminants and Toxins in Food is intended to provide a framework for the control of contaminants in traded foodstuffs.

The Indian Meat Industry

The meat production in India has been estimated as 6.4 Million Tonnes (FAO 2005 estimates). The value of meat produced accounts for Rs.21, 900 crores and meat products for Rs.828 crores. Meat production has increased at the rate of 4.1% annually during the last five years. India exported 306970 and 8885 metric tonnes of buffalo and sheep/goat meat respectively in 2004-05 valued at Rs. 1616 and Rs. 79 crores respectively. The total export earning from animal products during 2005-06 has touched Rs. 2647 crores against Rs.2252 crores during 2004-05 (Table 1). Livestock population, slaughter rate and meat production data indicate that buffalo population of 96 million produce 1.2 million tonnes of meat. This is due to effective culling practiced in buffaloes for both domestic and international markets.

International Trade in meat depends on a combination of factors which should be taken into account to ensure unimpeded trade, without incurring unacceptable risks to human and animal health. On account of likely risks the internal trade and marketing under the WTO environment requires the exporting country to comply with certain responsibilities. An exporting country is required to supply the certain information to importing countries on request such as information on the animal health situation and national animal health information systems to determine whether the country is free or has free zones of listed diseases, including the regulations and procedures in force to maintain its free status, regular and prompt information on the occurrence of transmissible diseases, and details of the country's ability to apply measures to control and prevent the relevant listed diseases.

Table 1. Export of Livestock Products from India 2001 -2005

(Quantity in Tonnes and value in Rs. Crores)

	2001-02		2002-03		2003-04		2004-05	
Animal Products	Qty	Value	Qty	Value	Qty	Value	Qty	Value
Buffalo Meat	243355.6	1144.4	297897.3	1305.5	343817.1	1536.8	306970.8	1615.6
Sheep / Goat Meat	3915.1	33.1	4973.6	40.0	16820.5	110.4	8885.3	79.4
Poultry Products	19876.0	130.1	26450.0	156.5	415228.2	202.4	264607.5	154.1
Dairy Products	24774.1	182. 5	21439.86	153.6	15882.7	155.2	48426.8	389.1
Animal Casings	464.3	9.6	8296.2	140.3	732.8	12.4	552.3	12.6
Processed Meat	267.1	1.3	669.5	4.8	986.1	7.6	107. 5	1.6
Total for Animal Products	292652.2	1500.9	359726.3	1800.5	793467.4	2024.8	629550.20	2252.3

Source: APEDA

Safety Regulations for Organic Residues in Meat Industry

In the early 1980's, there was a heightened interest in the threat of human health hazards from agricultural chemicals such as pesticides and from chemicals used as drugs (Anon. 1997). The findings of surveys suggested that the general public thought that chemical residues found in the edible tissue of food animals posed a greater risk than other hazards associated with foods (FSIS, 2005). The primary regulatory responsibility for controlling residues in edible tissues of food animals and poultry is jointly shared by the Food Safety and Inspection Service (FSIS) of the U.S. Department of Agriculture, the Food and Drug Administration, and the Environmental Protection Agency. EPA is responsible for the approval of pesticides and the establishment of their tissue tolerances in food animals. FDA is responsible for the approval of animal drugs, including feed additives, and the establishment of tolerances for animal drugs and environmental contaminants. FSIS is accountable for ensuring that the meat, poultry, and processed egg supply is safe, wholesome, and truthfully labeled. To improve the current meat and poultry inspection program, FSIS commissioned the National Research Council (NRC) in 1983 to evaluate the scientific basis of its inspection program, including its residue control programs. In United States, the Food and Drug Administration (FDA) is responsible for regulating medicated animal feeds and most animal health products. It approves the product, sets tolerance levels for residues in edible tissues, and determines how drugs are to be used. The FDA has published a Compliance Policy Guide (CPG 7125.37) outlining the responsible use of animal health care products.

An essential aspect of food safety is the control of residues that may result from the use of animal drugs and pesticides, or from incidents involving environmental contaminants. The codex document prescribes certain maximum residue levels for almost all the food commodities including animal products. An example of some pesticides in meat of sheep, poultry and cattle is presented (Table 2). Maximum permissible levels hav e also been established for residues that are the result of environmental contamination, such as cancelled pesticides that are no longer approved for use but persist in the environment (e.g., DDT), industrial chemicals (e.g., PCBs), and heavy metals. Tolerances for industrial chemicals and heavy metals are established by FDA and published in 21 CFR.

Table 2. Maximum residue levels for certain commonly used pesticides in meat

Name of Chemical contaminant	Maximum Residue Limits (in mg/Kg)		
	Sheep Meat	Poultry meat	Cattle Meat
Bifenthrin	-	-	0.5
Carbendazim	-	0.05	0.05
Chlorpropham	-	-	0.1
Chlorpyrifos	1	0.01	1
Chlorpyrifos-Methyl	-	-	0.05
Cyromazine	0.05	0.05	-
Spinosad	-	0.2	3.0
2,4-D	-	0.05	-
Aldrin and Dieldrin	-	0.2	-
Imidacloprid	-	0.02	-
Deltamethrin	-	0.1	-
Triazophos	-	-	0.01

Under the Federal Meat Inspection Act (FMIA), the Poultry Products Inspection Act (PPIA), and the Egg Products Inspection Act (EPIA), FSIS acts to ensure that USDA-inspected meat, poultry and egg products do not contain illegal levels of chemical residues. The FSIS National Residue Programme (NRP) provides a variety of sampling plans to prevent violative residues from entering the food supply, and develops national data on the occurrence of chemical residues to support risk assessment, enforcement and educational activities. The range of chemical compounds evaluated for inclusion in the various NRP testing programs is comprehensive in scope. It includes approved and unapproved pharmaceutical drugs and pesticides known or suspected to be present in food animals. It also includes any other xenobiotic or naturally occurring compounds that may appear in meat, poultry, and egg products and that may pose a potential human health hazard.

Food Animal Residue Avoidance Databank

Useful information can be obtained through the Food Animal Residue Avoidance Databank (FARAD). This is a program developed by pharmacologists and toxicologists at the University of California, University of Florida, North Carolina State University, and the University of Illinois, and supported by the USDA Cooperative State Research Education and Extension Service. FARAD is a repository of comprehensive residue avoidance information. FARAD maintains an up-to-date computerized compilation of current label information, withdrawal times, official regulatory tolerances, screening tests for detecting residues, and data on the fate of chemicals in food animals.

Food Safety Management Systems

Food safety is related to the presence of physical, chemical or microbial hazards in food at the point of consumption. Food reaches to consumers via supply chains that may link many different kinds of organizations. One weak link can lead to unsafe food and may be detrimental to health. As food safety hazards can occur at any stage of the food chain, adequate control throughout the chain is essential. All food businesses should have a full understanding of the products they produce, manufacture, sell and/or distribute; and shall have systems in place to identify and control hazards which are significant to the safety of food. This can be achieved by implementation of prerequisite requirements, by applying Hazard Analysis Critical Control Point (HACCP) principles and by using guides to good hygiene practice and assurance schemes (CAC/RAP, 2003). These are applicable to all types of food business and at all stages of the food chain, but allow for flexibility depending on the nature of the products involved and the size and complexity of the business. This standard integrates the principles of Hazard analysis and Critical Control Points (HACCP) system developed by Codex Alimentarius Commission and combines the HACCP plan with Prerequisites Programmes (PRPs) and is fully compatible with Quality Management Systems as per ISO 9001:2000 (TUV, 2006). Food safety management systems should be documented and updated as necessary. Written records should be kept for inspection by the local authorities. All companies operating in the food industry, irrespective of size, should have access to appropriate expertise. This expertise may be available in-house or through a consultancy arrangement.

Prerequisite Requirements

Prerequisite requirements are the basic environmental and operating conditions in a food operation that are necessary for the production of safe food. They control generic hazards and form part of good manufacturing and hygiene practices (GMP, GHP). The following list gives examples of the types of prerequisite requirements covering three key areas: product, premises and personnel. Not all of these are relevant to all types of food business, but they provide a checklist of the types of controls that operators should consider depending on the size and complexity of their business. The prerequisite requirements need to be in place before a HACCP system is developed. This will enable the HACCP system to focus on the significant product and process food safety hazards that require specific control to assure food safety. Prerequisite programs should be documented and records maintained.

Hazard Analysis Critical Control Point

The Hazard Analysis Critical Control Point (HACCP) system is being increasingly used as a food safety system all over the developed world. HACCP is not the magic bullet that solves all food safety problems. It is, when properly applied, a set of preliminary steps and principles that gives a systematic method for identifying significant hazards and properly applying preventive measures so that food borne hazards are prevented, eliminated or reduced to an acceptable level. With emerging international and national agreement on HACCP principles, their application would create commonality of understanding of the development, implementation and maintenance of a food safety system. Having these commonly understood principles, many food processors, for example, require their suppliers to have a HACCP system for production of ingredients that they supply. Knowing that a source of food borne hazards can be from a particular point there will be more attention given to that for implementation of effective, documented systems that eliminate or reduce the likely occurrence of food borne hazards. Application of HACCP offers widely understood principles for identifying significant risks and their control.

It is essential that there be scientifically documented steps and preventive measures. If this criterion cannot be met, then a HACCP system cannot be developed. A HACCP system can only be developed through proper application of the preliminary steps and principles of HACCP. An essential prerequisite to HACCP is the adoption of Good Manufacturing Practices (GMPs). The biggest problem in HACCP Plans is the lack of true CCPs. The HACCP philosophy is embedded in the residue avoidance programs (RAPs). As slaughter and processing establishments develop and implement HACCP programs mandated under new food safety regulations, they will need reliable information about the background of incoming animals. Under HACCP systems, slaughter plants must address significant chemical, physical, and microbial hazards associated with such animals. Section 417.2 of the rule requires that plants conduct a hazard analysis. The preamble of the rule describes the potential hazards that plants will need to consider during a hazard analysis. These hazards may arise from physical hazards, natural toxins, zoonotic bacterial and parasitic organisms, other microbial contaminants, chemical contaminants, including pesticides and drug residues, food additives, and hazards from food decomposition. These requirements will have an impact throughout the food animal production chain.

Good Agricultural Practices (GAP)

Good Agricultural Practices is a multidisciplinary area of work in FAO which is attracting a significant and growing demand for assistance from members. GAP work in FAO is focusing on provision of information, technical assistance and capacity building to help developing countries cope with changing and globalizing food systems and the proliferation of GAP standards developed mainly by the food industry, while not compromising their sustainable development objectives. Risks due to the accumulation in foods and feeds of natural contaminants such as mycotoxins (toxic substances produced by moulds), are widely recognised. Some agricultural practices, including reductions in pesticide and use of organic farming, could have an influence on the levels of natural contamination reached during the production and storage of food produces (in particular, the effects of changes in agricultural practices on the occurrence of moulds and the production of mycotoxins). Other important classes of potential contaminants include organic pollutants, pesticide residues, pathogenic microorganisms (e.g. Listeria as a water contaminant) genetically modified organizations (GMO) and GMO products, and beneficial microorganisms, such as those used for biological control. By definition, the good agricultural practices require to maintain the levels of contaminants below the toxicological limits, but often this is hampered by the lack of data on the quantitative relationships between agricultural practices, crop physiology and contaminant accumulation.

Grazing livestock should not be put at risk on lands exposed to agricultural spray drift or other industrial or naturally toxic events or activities which could introduce toxins or diseases. Field and laboratory testing of forages and water analysis plus a thorough knowledge of the land to be grazed should be a routine part of animal production in any geographical location.

Risk Analysis, Assessment and Management

Animal Production Food Safety activities concentrate on the link between animal (livestock and poultry) production and slaughter and processing operations. The live animal segment of the farm-to-table continuum holds enormous promise for developing food safety preventive programs based on the

HACCP approach. Through sound scientific studies and collaborative efforts among public and private stakeholders, the efforts for the identification of risk reduction strategies in animal production that have the potential to reduce chemical, physical, and microbial public health risks is being promoted. The scientific principles of risk assessment provide excellent tools for preparing a strategy to promote animal production food safety (ACS, 1998).

Risk assessment is the science of identifying and evaluating hazards. The two basic factors associated with risk assessment are the likelihood of an event and the consequences if the event occurs. It provides a good analytical tool to support decisions related to food safety. Risk assessment involves the development and use of models. A model is basically a framework of components that are followed to assess risks for any given commodity or process. Risk assessment is successfully used to evaluate chemical hazards such as residues in foods. In fact, the science of risk assessment is well suited to the analysis of chemical food hazards. Risk is based on various factors: level of exposure, route of exposure, length of exposure, etc. Through drug elimination study, the characteristics of each compound are known; the potential chemical hazard usually decreases with time. The FDA has established withdrawal times and violative tissue limits for all drugs approved for use in animal production. It is important to schedule slaughter in accordance with these withdrawal times. At the time of slaughter, random screening tests are performed to ensure that FDA tolerance limits are met.

Traceability and Authenticity

The main objective of traceability is to minimise any adverse health effects by a quick and complete recall. The recall should cover lots of product that do not meet the safety criteria, lots that are known to have caused adverse effects and others that are suspected of being capable of such effects (precautionary principle). For an adequate recall it is necessary that all animal products/meat and all of the ingredients used in producing the meat are traceable at all stages of production, processing and distribution. This needs to be carried out within a short period of time. One of the first considerations is to define a lot. Ideally, a lot is a quantity of feed produced and handled under uniform conditions. In practice, this usually means material produced within a limited period of time. The less uniform the conditions of production, the shorter that period of time should be. To apply such a principle implies knowledge about the uniformity of conditions.

The Sheep Residue Avoidance/Quality Assurance Program

The American Sheep Industry Association, in conjunction with the USDA, conducted a quality assurance audit to examine all phases of sheep production. Some of the recommendations address the prevention of drug residues (Anon. 1997). The sheep industry's quality assurance program goal is to produce high-quality, safe products. They focus on best management preharvest practices that are required for high product quality and safety. For the sheep producer, a pre-harvest quality assurance/product safety program includes a preventive flock health management program and knowledge of all aspects of proper sheep husbandry and care. Proper feeding management, sanitation, facilities, handling and transportation, and shearing, all contribute to a high quality product.

Conclusions and Recommendations

Consumers are increasingly concerned about the safety of their food, and this concern has led the animal industry to look carefully at the food it produces, which are the live animals (livestock or poultry) and other animal products. The organic contaminants have been perceived as a major threat in addition to the pathogenic microorganisms, particularly due to increased and indiscriminate use of agricultural chemicals in agricultural fields, large scale storage and during transportation. In view of the WTO agreements on SPS measures and TBT where the basic rules have been established for international trade and Codex Alimentarius standards, guidelines and recommendations, which are internationally agreed legislative standards as measures applied to protect human health from risks arising from chemical hazards, in food, it is a challenge for the meat industry to meet these stringent standards, especially in the current agricultural and animal management practices being followed in India at present. The meat industry will have to adopt Good (or Best) Management Practices, Total Quality Management (TQM) or Quality Assurance Programs (QAPs). Further, there is a need to develop proper HACCP plan and follow the general principles of hygiene practices laid by FAO/WHO. Integrating the good agricultural, animal keeping, grazing and manufacturing practices with TQM, QAP, HACCP plan will certainly help in achieving the international food safety standards so far as organic contaminants in meat are concerned.

REFERENCES

ACS. 1998. Understanding risk analysis. Available at www.rff.org/misc.docs/risk-book.htm

Anon.1997. Animal Production food safety - an overview for FSIS employees. www.kvaliteta.inet.hr/haccp/docs/ haccp7.htm.

CAC/RAP. 2003 Recommended International code of practice - General Principles of food hygiene, Rev. 4, 1-1969. www.codexalimentarius.net/download/standards /23/cxp_001e.pdf

FSIS. 2005. National residue Program data, the Red Book. www.fsis.usda.gov/Science/ 2005

Mehta, R. and George, J. 2003. Institutional and legal framework of food safety regulations in India. Workshop of Australian Centre for International Agricultural Research (ACIAR), March 19, Canberra.

TUV. 2006. Comprehensive study of food regulations and standards, food testing infrastructure, codex resource system, HACCP/ ISO/ organic products Certification/ Accreditation system. mofpi. nic. in/ projectprofiles/ TUVSA/ food_reg.pdf

Unnevehr, L. and Hirschhorn, N. 2000. Food safety issues in the developing world. World Bank Technical Paper no. 469.

❑❑❑

33

Meat Processing Technologies for Value Addition and Higher Income Generation

S. SURESHKUMAR, D.SANTHI

Southern Regional Research Centre, Central Sheep and Wool Research Institute, Mannavanur, Kodaikanal - 634 103 Tamil Nadu

Introduction

Sheep and Goat farming is practiced worldwide, with their products having a favorable image. The number of sheep and goats has increased globally, even in countries with high and intermediate incomes, despite major changes in agriculture due to industrial mergers, globalization, and technological advances in developed countries (Cheeke, 2004). Competitive advantages sustain production and marketing within the constraints of financial and strategic risks. Consumer food trends include convenience requirements, more meat, innovative dairy products, and growth in ethnic foods and one-dish meals, which include sandwiches, bowls, or cups as the entree. Convenience may mean less time for shopping, speed or ease of preparation, speed or ease of consumption, ready-to-eat or no preparation, or portability. Consumers often are willing to pay higher prices for convenient versions of their favorite products. Value added products are a means to provide convenience and economic profitability. Value-added products have been changed in form, function, or grouping to increase their economic value and/or consumption appeal (USDA, 2004). Value may be added to final products by decreasing costs or improving relative value (price received) of the final product. Costs may be decreased by using fewer market channel steps, lowering production or distribution costs, and by using uniform or consistent groups for inventory and display control. Relative value or price received is increased through products that are more palatable, usable, or available in a different form or time after sorting or processing. The relative value, and thus the amount of value that has been added or is available to the producer of the commodity or product, is ultimately determined through amounts and prices of purchases by consumers. The differences in attitudes and behaviors of various ethnic groups are important to businesses in terms of the goods and services that are offered. Ethnic groups may be defined as having a common and distinctive racial, national, religious, linguistic, or cultural heritage. Increased urban demographic growth has increased the demand for meat even though low innovation has contributed to the difficulty in preparing and cooking meat by urban peoples. Opportunities exist for sheep and goat meat because of its ecological image, dietetic and health qualities, and association with religious holidays, along with the tendency of consumers toward natural foods. Trends toward healthier diets could increase the demand for value-added products from nontraditional meat sources, such as sheep and goat meat, to supply products with decreased fat, lower cholesterol, and less sodium (Dawkins *et al.*, 1999). The potential for ethnic, value added, and convenient items is very high (Sloan, 2005), presenting justification for adding value to sheep and goat meat and their products to stimulate growth and profitability of meat animal production.

The market for processed foods in 2014-2015 in comparison to 2003-2004 is expected to go up 15 times for poultry and 4.3 times for buffalo meat. The constraints of the traditional meat marketing system can be eliminated by producing shelf-stable meat products by adopting specific processing techniques that can be stored under ambient temperatures, handled safely and also reach the consumers in rural and other areas far from food processing plants and abattoirs. Processed meat products provide tasty, convenience and designer foods to the meat consuming population with exceptional level of satiety. Also the quality and price value of low value meat and other edible byproducts can be upgraded through appropriate processing. It is obvious, therefore, that good-quality products can be produced with cheaper raw materials, while the more expensive carcass parts can be profitably sold to customers who can afford higher-priced meat or meat products (FAO, 1995).

Prices for meat products can be further decreased by using meat extenders, which may include plant or crop products such as flour and starches. Dairy products and fish, which can partially replace expensive meat while keeping the products at an adequate level of nutritional quality (Karmas, 1976).

Meat Consumption – Global scenario

As globalization has made the developing world richer, it has also whetted the appetite for more protein dense foods, namely meat. Hence there is a critical need to meet this protein demand. In the global scenario, the contribution of livestock to food supplies in developing countries is increasing at a higher rate than that of cereals. FAO and other institutions suggest that global meat production and consumption will rise to 300 million tonnes by 2020. Developing countries now account for almost half of the total world meat production (FAO, 1994a; FAO, 1994b). In developed countries a certain saturation and decrease in the overall consumption of livestock products can be seen but the demand for livestock products will continue to rise in developing countries as a result of their rapid expanding populations and higher incomes.

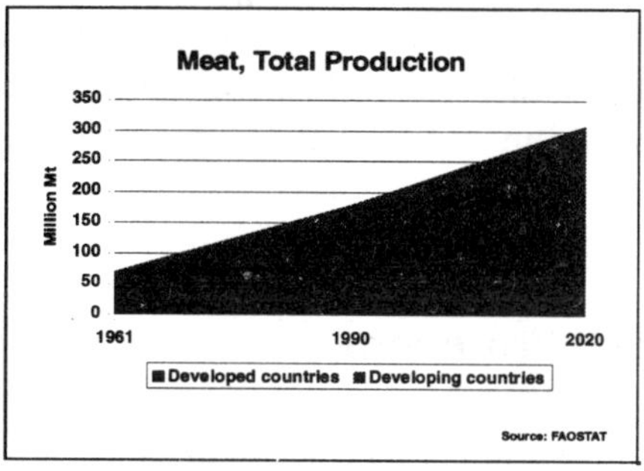

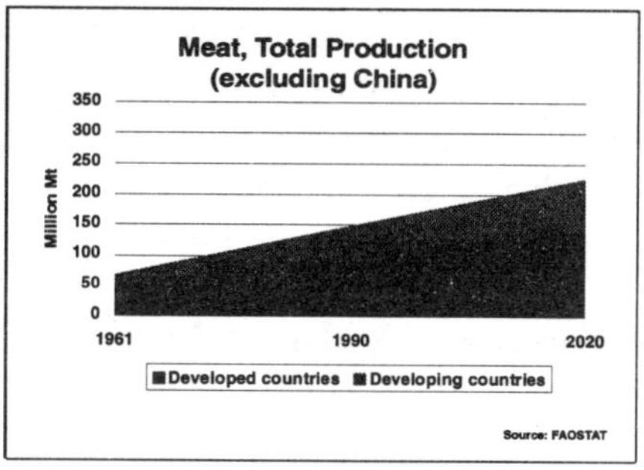

The shortage of livestock products in certain areas in developing countries is a consequence of the traditional marketing systems, where fresh, unprocessed meat is sold at meat markets a few hours after slaughter.

Common Processing Methods Employed in Meat Sector

Method	Products
Carcass dressing at the abattoir level	Dressed whole carcasses, carcass sides or quarters Edible animal by-products (internal organs)
Carcass/meat cutting	Wholesale and retail meat cuts for frying, cooking, roasting, etc.; meat for further processing
Drying of unsalted pieces of meat	Dried meat for preparation of meat dishes
Salting and curing of meat cuts with or without heat treatment	Cooked ham, cooked cured beef, bacon, raw ham and other raw cured products such as dried cured products
Coarse comminuting of raw meat	Hamburger-type products
Coarse or fine comminuting of raw meat with subsequent fermentation	Fermented sausages of the salami type, fermented Asian products (e.g. naem)
Fine comminuting of raw meat with subsequent heat treatment	Emulsion-type sat/sages, e.g. frankfurters, meatballs, emulsion type meat pies
Fine comminuting of precooked meat together with internal organs, blood, etc.	Liver pate, liver sausage, blood sausage, meat pies.

Meat Processing

Meat is sold primarily as whole carcasses or as bone-in cubes. Value is added to raw chilled meat by changing the form or utility. The ISI standards for fresh sheep, goat describes standard cutting practices for merchandising of uniform goat meat cuts. Meat properties and preservation are changed by many single or combination unit-processing operations. Primary processing operations include tenderization, grinding, flaking, freezing, and case-ready fabrication and packaging, whereas examples of further processing are curing, smoking, marinating, injection, emulsifying, forming, and cooking (Pearson and Gillett, 1996). Value-added product areas also would include irradiated products for microbial safety, precooked products for convenience, portioned and institutional items for uniformity, and nutritionally enhanced meat for healthfulness.

Avenues of Value Addition in Meat Processing

Inclusion of Non-meat Binders and Extenders

Incorporation of non-meat binders and extenders such as textured soy, soy protein concentrate, refined wheat, gram and baked lentil flours, corn starch, tapioca starch, wheat semolina, milk solids, milk co-precipitates, mashed potato, green peas, wheat and maize germs, whole egg liquid and egg albumen, etc. in comminuted meat products has resulted in improvement of their nutritional and organoleptic quality as well as reduction in production cost. Isolated soy protein is used for moisture and fat binding, emulsifying, and as an emulsion-stabilizing agent in finely comminuted meat products (Rhee, 1994). Sodium caseinate, soy protein isolate, whey protein concentrate or wheat germ flour can increase cooking yield of frankfurters (Atughonu *et al.*, 1998).

Soy Proteins

Soy products as additives can improve the nutrition (Kamal et al., 1988), flavor, and valuable functional properties to the meat preparations. When used as emulsifiers, binders, moisture retainers, and stabilizers, soy proteins can alter or improve the appearance, taste, and texture of the finished product. Soy proteins can also be used to extend or simulate meats. On the basis of protein content, soy proteins are the lowest cost functional proteins commercially available (Joseph Rakosky, 1970). Important factors in the use of textured soy protein are cost control, improved functionality, fat control, and nutrition (Dean Wilding, 1974).

Starch

All the starches significantly reduce cooking loss and modified tapioca showed the best improvement in cook yield (Zhang and Barbut1 2005). Incorporation of modified tapioca starch in frankfurters improves water and fat binding (Marita Ruusunen *et al.*, 2003).

Liquid Whey

Liquid whey could be substituted for up to 100% of the ice in frankfurter formulations (Yetim *et al.*, 2006) .Rajesh Kumar et al. (2007) reported that the added water can be upto 20% level with frozen liquid whey without affecting the quality of pork nuggets. In addition, the product can also be stored upto 14 days of refrigerated storage without affecting much of its shelf life qualities. By incorporating liquid whey in the comminuted meat products, the most desirable quality pork nuggets can be developed with additional advantage of fortification, excluding the drying expenses to produce whey protein concentrates and elimination of environmental pollution due to draining of surplus liquid whey. Hence, the utilization of liquid whey in comminuted meat products offers an enormous economic advantage and potential to fortify and improve functionality of processed meat products.

Dried Eggs

Incorporation of dried eggs and liquid eggs improves the quality of chicken patties made with spent hen meat and byproducts viz., Skin, Heart meat, Gizzard meat in terms of emulsion stability, product yield, shear force and sensory attributes (Kalaikannan et al., 2007). During the refrigerated storage, the incorporation of egg proteins in chicken patties was beneficial in terms of physico-chemical, microbiological and sensory characteristics. Thus the spent hen meat and byproducts which will fetch very minimal profit to the farmers can be made into this type of communited value added products for better returns.

Hence, by keeping the lean meat levels from 50%, 60 or 70% with remaining edible offals and non-meat extenders and seasonings, etc. to produce economy, choice and prime grade emulsion-type poultry products can be produced to suit the pocket of different income-group consumers.

Utilization of Low Value Meat and Edible Offals

Research efforts have also enabled utilization of less preferred spent/culled breeder meats and carcass components viz. skin, gizzard, heart, ova yolk and adipose tissue as raw materials for production of premier value-added comminuted meat products. Toughness of spent hen meat is no obstacle to the production of a variety of comminuted products. But the by-products as skin, gizzard, heart, ova yolk and fat can be utilized in emulsion-based chicken and other meat products with limited extend as it causes emulsion break down due to its high collagen content. Hence, to overcome this short coming these products can be improved by addition of phosphates, dried eggs, soya proteins and other similar non meat.

Cut up parts

Consumers are willing to pay for the convenience of buying just the parts of the chicken carcass they want. If the breast meat is all that is desired, then they could buy a package of just breasts without the wings and legs. Hence, by diverting the less preferred meat cuts to product making processes the processor can get better profit by marketing these high value products in the fresh market. Further, Improved deboning techniques and modern methods of meat processing are capable of enhancing the quality and value of meat from spent hens leading to greater demand and better financial returns (Kondaiah and Panda, 1992).

Chicken Meat Spread and Emulsion Based Products

The chicken meat spread from spent (culled) hen meat with desirable spreadability and acceptability can be produced by addition of 0.5% sodium tripolyphosphate (STP) and 100ppm á-tocopherol acetate in the formulation. In addition to improved emulsion stability, it inhibited lipid oxidation as measured by TBA assay. With the inclusion of vacuum-packaging with PFP laminate these could be stored for 12 and 60 days under refrigerated (4-5°C) and frozen (-18°C) storage, respectively.

Chicken Skin Meat Cutlet

Incorporation of alcoholic extract of garlic at 2% level (w/v) in unspiced minced meat base containing processed chicken skin at 10% level resulted in delicious chicken skin-meat cutlet. The finished product remained acceptable till 14 and 28 days under refrigerated (4±1°C) and frozen (-18±1°C) storage, respectively.

Poultry Meat Preservation by Phyto-products

Aqueous garlic extract at 4.0% (w/v) or ethyl alcoholic extracts of cinnamon and clove at 0.4% (w/w) and 0.15% (w/w) respectively, in combination with Lactobacillus acidophilus enhanced the shelf-life of minced chicken meat up to 7-8 days.

Chevon

Chevon samosa

Chevon samosa in semi fried and deep fat fried conditions using spent goat meat, can be stored at –20°C up to four months. These products will have most of the microbial quality within the permissible limits.

Chevon nuggets

When compared to spent goat meat nugget the lamb meat nuggets revealed high yield (87.18%). In order to get the optimum yield the combination of lamb meat and spent goat meat at 50:50 levels were prepared. By using this combination the cholesterol level of the nugget was reduced from 136.25 mg to 120.11 mg. Thus by preparing these types of novelty products such as Chevon samosa, Chevon nuggets, Chevon kofta, Chevon patties the spent goat meat were effectively used to get the high value meat products and can be easily marketed without having drawback of tenderness problem on the spent goat meat.

Enrobed Products

Enrobing/coating of meat products with edible materials improves their cooking yield and provides better protection against oxidative and microbiological deterioration. It can also significantly enhance the sensory qualities of meat products and could be an effective method of value addition with better consumer acceptability. Enrobed cutlets can be successfully added to the list of value-added convenience meat-based products.

Enrobing is the process of making "further processed products" by applying edible coating to the products. It includes two distinct steps, i.e., breading and battering. Enrobing brings several advantages to meat products such as value addition, versatility to consumers and improvement of nutritive value as well as eating and microbial qualities of the products (Richardson 1989). Battered and breaded or coated systems have the benefits of versatility and familiarity because they enhance the flavor and texture of processed food products (Mandava and Hoogen Kamp, 1999; Biswas et al., 2004). Coating or enrobing improves the texture of the product, remarkably reducing the product cost, making it an avenue for value addition and better consumer acceptability. Enrobing also contributes other advantages like preserving the nutritive value, reducing moisture and weight loss, and improving juiciness and tenderness. These improvements are brought about by coating ingredients that act as sealants and also prevent high oil uptake during frying of the product (Cunningham, 1989).

Constraints of Meat Industry

Poor market outlet for the country's livestock has been disincentive for investment to exploit the potential. Factors that have contributed towards the inability of our livestock and livestock products to access international markets:

- Marketing infrastructure,
- Conventional dietary habits of consumers,
- Inadequate advertising and campaign to highlight the virtues of processed meat products,
- Consumer's lack of faith in the quality of such products in the absence of stringent
- quality assurance programs for domestic market,
- Non-awareness about availability
- High price of processed products
- Inadequate infrastructures like processing equipments, cold-chain including refrigerated transport and marketing network
- Inefficient production that make our products not competitive in the internal market.
- Stringent sanitary and phytosanitary measures imposed by importing countries as technical barrier to trade.

- Unfair competition in the world market from highly subsidized production from developed world.

Needs of Our Meat Industry

- Development of quality abattoirs for meat processing particularly in the production areas.
- Focus on enhanced productivity and profitability by producers to produce according to market demands and agro-processing initiative to be in accordance to standards set.
- The promotion of private/public sector and trader/producer partnerships. This will ensure access to markets of value added livestock products.
- Strengthen animal health services for increased productivity and high quality products. Establishment of disease-free zone to facilitate access of the E.U. market for animal and animal products.
- Provision of credit facilities.
- Value addition in various forms and packaging.

Imperatives for Value Addition

To increase value addition, the following steps are crucial:

- Providing appropriate incentives for the establishment of agro-industries in the rural areas.
- Focused research on value addition especially processing, storage and packaging of the meat products.
- Promoting partnership between small hold farmers and agribusiness.
- Encouraging the development of supportive infrastructure services such as rural access roads, electricity, water and telecommunications.
- Capacity building of farmers and farmer organizations in value addition and agro-processing.
- Despite revolution in food packaging, leading to availability of several rigid, semi-rigid, and flexible monolayer, co-extruded and laminated retortable and non-retortable packaging materials of high barrier and mechanical strength and innovative packaging systems (MAP, CAP, aseptic, retort packaging etc.), efficient packaging for many of the value-added products developed indigenously has not received much attention.

Scaling-up of the processes/products developed and their familiarization to processing industry and consumers are the need of hour for commercial exploitation and popularization among the masses.

Government Initiatives for Value Addition in Food Sector

- Food Policy is proposed to support projects for reduction in post-harvest losses, enhance the level of processing, encourage value addition, employment generation and ensure remunerative price to farmers. In the Government of India Budget 2006-2007, Food Processing has been placed as a priority sector for bank credit.
- National Bank for Agriculture and Rural Development (NABARD) has set aside Rs. 1,000 crore, especially for agro-processing infrastructure and market development.
- Excise duties have been significantly reduced. For instance, on Dairy Processing equipment, excise duty has been reduced from 16% to zero, food grade hexane from 32 to 16%, and meat, poultry and fish from 16 to 8%, packaging machinery from 15 to 5% and so on.
- Excise duty has been reduced to zero in case of condensed milk, ice cream, and preparations of meat, fish and poultry, pectins, pasta and yeast.
- Presently, food and food processing is regulated by 13 different laws and regulatory orders. Efforts are being made to integrate them.
- The Ministry of Food Processing Industries has already approved establishment of 47 Food Parks in different parts of the country having facilities for cold storing, IQF, packaging and so on at one place.

Future Prospects

Diversity in tradition and culture of India has resulted in different culinary practices for the preparation of a variety of traditional poultry products, some of which (tandoori, curried chicken and kababs) have become popular abroad. However, non-availability of standard processing methods to develop products of uniform quality has met with limited success to commercialize traditional products. Hardly about a dozen chicken preparations are available to Indian consumers in sharp contrast to more than a hundred varieties of value-added poultry products being marketed in some developed countries. Hence, concerted efforts should be made to develop process for some newer products to suit the palate and pocket of the consumers. Similarly, development of more and more comminuted, restructured, enrobed and marinated flavored meat products as ready-meals, snacks, carry-home meals and full or supplementary military pack rations in accordance with the requirements of end-users would enhance the consumption of such products. In this respect, emphasis should be laid on developing shelf-stable products based on Indian cuisine for which there is an enormous potential in both domestic and export markets.

Conclusion and Recommendations

Value can be added at many points in the meat system of production, distribution, processing, and sale of sheep and goat meat products. Younger, leaner, and more heavily muscled animals are more valuable regardless of breed and diet, whereas increased animal age generally decreases meat tenderness and sensory properties. Most meat processing and preservation technologies can be used to produce meat products, with improved product consistency through uniform cutting and fabrication practices and sorting of raw materials. Sheep and goat meat provides improvements in emulsification, textural, and flavor properties that would be advantageous in lower fat or processed meat products, but oxidation of cooked meat and the specific product type influence storage and display shelf-life. Acceptability of sheep and goat meat and meat products is highly dependent on consumer culture and desires. The availability of sheep and goat meat and more convenient product forms would add value in the marketing channels.

REFERENCES

Atughonu, A.G., Zayas, J.F., Herald, T.J. & Harbers, L.H. 1998. Thermo-rheology, quality characteristics, and microstructure of frankfurters prepared with selected plant and milk additives. *Journal of Food Quality* 21, 223-238.

Biswas, A.K., Keshri, R.C. and Bisht, G.S. 2004. Effect of enrobing and antioxidants on quality characteristics of precooked pork patties under chilled and frozen storage conditions. *Meat Science, 66*(3), 733–741.

Cheeke, P.R. 2004. *In* Contemporary Issues in Animal Agriculture. 3rd ed. Pearson Prentice Hall, Upper Saddle River, NJ. pp.253–292.

Cunningham, F.E. 1989. Developments in enrobed products. *In* Processing of Poultry (G.C. Mead, ed, Elsevier Applied Science, London, U.K.) pp. 325–359.

Dawkins, N.L., Phelps, O., McMillin, K.W.and Forrester, I.T. 1999. Composition and physicochemical properties of chevon patties containing oat bran. *Journal of Food Science* 64, 597–600.

Dean Wilding,M. 1974. Textured proteins in meats and meat-like products. *Journal of the American Oil Chemists Society.* 51,1.

FAO. 1994a. *FAO production yearbook.* Vol. 47. Rome, FAO.

FAO. 1994b. *Medium-term prospects for agricultural commodities: projections to the year 2000.* FAO Economic and Social Development Paper No. 120. Rome, FAO.

FAO. 1994c. Agro-industries in FAO. Consultancy report.

FAO. 1995. Development and promotion of value added meat products. *Project document.*

Joseph Rakosky, Jr., 1970 Soy Products for the Meat Industry. *Journal of Agriculture and Food Chemistry,* 18, 6.

Kalaikannan, A, Anjaneyulu A.S.R and Santhi D. 2007. Effect of egg proteins on the quality and refrigerated storage life of chicken patties made with broiler-spent hen meat and by-products. *International Journal of Food Science and Technology,* 42, 579–586

Kamal, G., Mekhael Nargis, S., Bassily, Malak, M. and El-Shafei, Amin K. Said. 1988. Nutritive value of meat-soy mixture Food / Nahrung. 32, 729 -735.

Karmas, E. 1976. Processed meat technology. New Jersey, USA, Noyes Data Corporation.

Kondaiah, N. and Panda, B. 1992 Processing and utilization of spent hens. *World's Poultry Science Journal*, 48(3), 255-268.

Mandava, R. and Hoogen Kamp, H. 1999. The role of processed products in the poultry meat industry. *In* Poultry Meat Science, Poultry Science Symposium Series Vol. 25 (R.J. Richardson and G.C. Mead, eds.), pp. 390–425, Cabi Publishing, Oxon, U.K.

Marita Ruusunen, Jukka Vainionpaa, Eero Puolanne, Marika Lyly, Liisa Lahteenmaki, Markku Niemisto and Raija Ahvenainen.2003. Physical and sensory properties of low-salt phosphate-free frankfurters composed with various ingredients. *Meat Science*, 63(1): 9-16.

Pearson, A.M. and Gillette. T.A. 1996. *In* Processed Meats. 3rd ed. Chapman and Hall, New York, NY. pp.210–241

Rajesh Kumar, B, Kalaikannan, A and Radhakrishnan, K.T. 2007. Studies on processing and shelf life of pork nuggets with liquid whey as a replacer for added water. *American Journal of Food Technology*. 2 (1), 38-43.

Rhee, K.C. (1994). Functionality of soy proteins. *Ch. 10 in Protein Functionality in Food Systems*.. Marcel Dekker. Inc., New York. pp. 311-324.

Richardson, R.J. 1989. Utilization of turkey meat in further processed products. In *Processing of Poultry* (G.C. Mead, ed.) pp. 283-324, Elsevier Applied Science, London.

Sloan, A.E. 2005. Healthy vending and other emerging trends. *Food Technology* 59(2), 26–28, 30, 32–35.

USDA. 2004. Announcement of value-added producer grant application deadlines and funding levels. USDA Rural Business-Cooperative Service.

Yetim, H, Muller, W.D., Dogan, M. and Klettner, P.G. 2006. Using fluid whey in comminuted meat products: effects on textural properties of frankfurter-type sausages *Journal of Muscle Foods* 17, 354-366.

Zhang, L and Barbut, S.2005 Effects of Regular and Modified Starches on Cooked Pale, Soft,and Exudative; Normal; and Dry, Firm, and Dark Breast Meat Batters. 2005. *Poultry Science* 84, 789–796.

□□□

34

Pelt Processing Technologies: Recent Trends

R.S. DHILLON

Arid Region Campus, Central Sheep & Wool Research Institute,
Bikaner 334 006, Rajasthan

Introduction

Pelt processing technology is, perhaps, one of the oldest crafts devised and developed by man. The prehistoric man started using raw, crude animal skins to cover his body to face the vagaries of cold and arctic weather; and from those earliest days of human existence, the animal skins have played an important role in providing clothing to the human race. In the past, the skins of trapped wild fur animals and game animals constituted substantial part of raw material of the fur trade. But of late, there has been marked decrease in the number of wild animal skins owing to increased awareness of environment protection and ecological balance. The governments the world over, has become more and more concerned about wild life conservation and the laws governing hunting have been made more stringent. This has caused acute shortage of fur-skins of wild animals and as a result, the fur industry of India has to depend upon the rabbit pelts, a large part of which is imported from France. The ranching/farming of fur-skin animals is yet to take off in India. Among the most important and hence the costliest furs are Mink, Martin, Sable, Silver Fox etc. The chief fur animals recognized by the fur industry are listed in Table 1. The fur-bearing animals that can be introduced in India are given in Table 2.

Table 1. Principal furs recognized by the industry.

Badger	Marten Sable	Persian lamb (Karakul)
Beaver	Mink wild	Raccoon
Cheetah	Mink ranched	Seal (Fur)
Chinchilla	Musk rat	Seal Hair)
Ermine (weasel)	Nutria	Skunk
Fisher	Ocelot	Squirrel
Jaguar-leopard	Opossum American	Wolf
Lynx	Opossum Australian	Wolverine
Marmot	Otter	

Prentice (1976)

Table 2. Fur animals which can be introduced in India.

S. No.	Species	Age at maturity (months)	Gestation period (days)	Average litter size (No.)	No. of litters per year	Age at slaug-hter (months)	Feeding habits	Climate required
1	Mink	10-12	51-55	5-6	1	8-10	Carnivorous	Temperate
2	Nutria	8-10	120-130	5-8	1-2	8-10	Herbivorous	Tropical/ Temperate
3	Chinch -illa	10	110- 115	3-4	1-2	8-12	Herbivorous	Tropical/ Temperate
4	Marten	16	220-240	3-5	1	10	Carnivorous	Temperate
5	Fox	10-12	54-60	6-8	1	7	Carnivorous	Temperate

Singh (1985)

Raw animal skins, which are a highly perishable material, can be converted into a stable and durable product only by adroitly employing adequate and skilful methods of technological processing. By virtue of continuous improvement in processing technologies, the fur clothing which in ancient times, was one of the basic necessities of the primitive man, has become a sophisticated article of wear today. Proper processing adds value to a raw pelt so much that the sale price of the finished product increases many-fold in comparison to its cost of production. Pelt processing industry in India, is mainly concentrated in the state of Jammu and Kashmir. Around 3.5 m fur-skins are imported in India every year. Out of these, about 0.8 m are rabbit pelts, a substantial part of which is imported from France. The processed fur-skins, lamb pelts and rabbit pelts have great potential for exports as such, as well as, in the form of various articles. Rabbit pelts and fur-skins are used for making fur garments, toys, shoes, gloves, lining material, fancy items like jewellery-box covers, lady purses and many other profitable attractive objects for sale in elite markets in the country, as well as, for exports. The lamb pelts are utilized in making various products viz. caps, ladies and children coats etc.

Processing technologies have, so far, been by and large, traditional and the primary motive of most processes developed has been aimed at the addition of the desired end-property rather than at the environmental and economic sustainability. But the latest trends indicate that huge changes are taking place in pelt processing technologies so as to reduce the pollutant load on the effluent treatment plants to the extent possible. New technologies being adopted in this direction have been discussed in this paper.

Lamb pelts

Karakul, a fat-tailed breed of sheep is known for production of quality lamb pelt. This breed is very well adapted to extreme climatic conditions and sparse vegetation resources. Karakul pelt production is an important facet of sheep production system in many countries like Soviet Russia, Uzbekistan, Afghanistan, South Africa, South–west Africa and Iran. The characters like ornament, type of curls, size of curls, density of curls, colour of the hair-cover and weight determine the quality of the pelt.

Fur-skins and rabbit pelts

Fur-skins are obtained from rare and small fur-bearing wild animals like Mink, Silver Fox, Nutria and Chinchilla etc. These are known in the industry as real fur animals (table-1). Although the rabbit fur is comparatively thin, uneven in hairy coat and less stable and hence not true fur, yet the rabbit is the most promising animal for fur-pelt production. The rabbit fur-pelts constitute about 50 % of the fur-skin trade in the international market. Since a rabbit producer is mainly interested in meat production, fur is a byproduct which gives him additional income. The fur-pelts of rabbits of Rex, Satin, Lilac and Polish breeds are extensively used as substitutes of costly pelts from real fur animals, as they possess a very wide range of colours, shapes and designs. Most of the rabbit production is basically for meat. The rabbits are, therefore, slaughtered at an early date and majority of the skins are immature. Such pelts are used for production of gelatin, fats and manufacture of semi-synthetic fibre. France is the only country where the rabbits are slaughtered on complete development of fur-coat. This is because the feeding cost is so high that after attaining the age of 12 to 14 weeks the rabbit meat cannot be sold at competitive rates. Good profit can be earned from the fur-skins if the slaughter age is extended to 16 to 18 weeks without affecting the quality of meat. This can be achieved by regulating the quality and quantity of feed. The effect of slaughter age on fur quality is given in Table-3.

Fur quality characteristics

1.Colour

Colour is one of the most important characters of a fur-pelt. Though many varieties of colour strains and breeds of rabbits have been evolved during the last century, yet white furs are preferred universally because they can be dyed in any desired shade. New Zealand White and White Giant breeds yield good quality white fur-skins. Among coloured breeds Chinchilla rabbits are preferred especially due to their pattern of colour bands in hair and denseness of hair-coat.

Table 3. Effect of slaughter-age on quality of processed pelts.

S. No.	Breed	Quality characters	Age at slaughter		
			12 Weeks	24 Weeks	> 24 Weeks
1	Soviet Chinchilla	Pelt weight (g)	60.00	122.50	132.50
		Pelt size (cm²)	1077.00	1513.00	2059.75
		Grade	C	B	A
2	Grey giant	Pelt weight (g)	67.50	125.00	130.00
		Pelt size (cm²)	1029.50	1653.00	2158.00
		Grade	C	B	A
3	White giant	Pelt weight (g)	60.00	125.00	137.50
		Pelt size (cm²)	963.25	1530.00	1988.00
		Grade	C	B	A
4	New Zealand white	Pelt weight (g)	40.00	90.00	115.00
		Pelt size (cm²)	821.00	1172.00	1648.75
		Grade	C	B	B

Singh (1985)

2. Pelt size and weight

Heavier and large sized pelts are more valuable than the light weight and small sized ones. These characters may vary with breed age at slaughter, season of feeding and other environmental factors. Autumn and winter pelts are heavier than spring and summer pelts. Large size pelts yield more surface area and are, therefore, suitable for making bigger articles.

3. Density and uniformity of hair-coat

A dense and strong fur with a sheared look will have a velvet like appearance. By placing a hand with palm down on the animal, one can feel a cushion of dense fur and it is practically impossible to see any flesh if such a fur is blown into. Hair-coat density is a highly heritable trait. In Soviet Chinchilla, the heritability of hair-coat density has been estimated to be 0.41 to 0.64. Autumn and winter coats are denser than summer and spring coats.

4. Resilience

This inherent property in fur is responsible for giving strength and elasticity to the fibres. On brushing the hair in opposite direction it should shape back and if it lies flat, it is devoid of life and reflects wooly character. This does not make a fur of good marketable quality.

Pelt processing technologies

Preservation of raw pelts

Curing

Raw pelts are constituted of proteins (30-33%) and water (60-70 %) which makes highly putrefiable. The putrefying bacteria which are present in the skin or get access from the air develop proteolytic enzymes - the proteases and peptidases. These enzymes diffuse into the cell walls and disintegrate the moleculer structure of the skin proteins. In the live animal the metabolic defences do not permit the micro-organisms to cause any harm, but after death these defences cease to function and putrefaction starts. Curing reduces the moisture content to as low a level as 12 %, at which the bacterial decomposition and biochemical changes do not take place.

Curing methods

1. Drying

(a) *Air-drying* : Flayed skins are air-dried in shade at room temperature. Though this method is eco-friendly and economical but does not yield good results because many a times the drying is not deep in the skin and incomplete drying can induce bacterial action around the papilla and hair roots which may cause massive loss of fur (hair-slip) in the subsequent soaking (wetting) treatment.

(b) *Sun-drying* : The raw 'cased skins' (fur inside and flesh outside) are spread on the floor and allowed to dry in the sun. Drying in the direct sun rays can damage the pelt by coagulating the inter-fibrillary cementing substance in the matrix making wetting of the fur-skin difficult in the subsequent operation of soaking. This may also cause blistering (crumpling) in the skins because the surface layer of the skin dries rapidly and prevents the escape of moisture from the interior of the pelt.

2. Salting

(a) *Dry-salting* : Dry common salt is applied on the flesh side of each skin and the skins are then piled one over the other, flesh to flesh, in stacks of 10 to 20. After a week or so the skins are removed from the stacks, dusted gently to remove excess salt and air-dried.

(b) *Wet-salting* : Raw pelts are washed with water to remove blood, dirt and other extraneous matter. After draining out the excess water from the skins, dry common salt is spread on the flesh side of each skin. Salt is applied in three installments. The they are piled one ther the other, the flesh side up, in stacks of 10 to 20 pelts. The salt should be of medium grain, its particle size varying between 1 to 2 mm.

3. Brine-curing

The raw pelts are washed well and cleaned thoroughly in running water. Then they are dipped in saturated solution of common salt (Brine) containing 1.0 % zinc chloride on the weight of solution. After about 10 minutes, the skins are taken out and air-dried in shade on wire stretchers. Skins preserved by this method can be kept for longer periods without any sign of hair-slip.

4. Solvent dehydration

Solvent dehydration is a new method of curing. This is preferred to the conventional methods because it eliminates the use of common salt - the chief source of salt pollution in tanneries. The organic solvent acetone is used in this method as a dehydrating agent which has many advantages over other systems of preservation. The acetone cured skins have approximately one half the shipping weight of the salt-cured skins. This considerably reduces the cost of transportation. Secondly, such skins take very less time (2 to 3 hours) for rehydration (wetting) in comparison to 16 to 20 hours required for dry-salted skins. Acetone preserved skins are defleshed readily and easily. However, this system of preservation involves a little higher cost, but this can be considerably reduced by redistilling the used acetone. Experiments have shown that almost 80% of acetone can be recovered.

Processing

Cured pelts are subjected to a number of chemical treatments before they are converted into the end-product.. These are:

1. Soaking

The pelts are soaked in water containing salt and wetting agents. The objective of soaking is (i) rehydration i.e. wetting, (ii) cleansing and (iii) swelling of the skins to some extent. Soaking facilitates the impregnation of the chemicals into the body of the skin, in the treatments that follow.

Recipe

Weight of raw skins	:	W kg.
Water	:	30x W litres.
Common salt	:	10 g/l
Soda Ash	:	1g/l
Sod. Trichlorophenate	:	0.1g/l
Nonionic detergent	:	1 g/l
Time	:	over-night

Soaked skins are then defleshed and scoured as follows.

2. *Defleshing*

The adipose tissue, adhered tenaciously on the flesh side of the skin does not allow free entry of the chemicals into the body of the skin. Removal of this tissue and allied flesh is called 'Defleshing'. Defleshing should be done very carefully because the hair roots, at this stage, are in a loosened state and depilation causing 'hair-slip' can take place easily.

3. *Scouring*

Scouring removes grease (wool wax) from the wool/hair part and fat of the skin part of the pelt.

Recipe

Water	:	20xW litres.
Soda Ash	:	1-2 g/l
Noigen L.S.	:	1 g/l
Sod. lauryl sulphate	:	3 g/l
Time	:	3 - 4 hrs.

Handle with a wooden stick. Wash thoroughly for at least three changes of water.

4. *Pickling*

The objective of Picking process is to prepare the scoured pelts for tanning by opening up their structure (i.e. fibre texture) through swelling and to bring them into acidic condition (pH 3 - 4) so that the tanning chemicals impregnate into the skins easily and completely.

Recipe

Water	:	15xW litres
Common salt	:	50 g/l
Sul;phuric acid	:	1g/l
Leave for over-night. Next add	:	
Formic acid	:	1 g/l
Total duration	:	40 - 48 hrs.

Stir well. Put the pilcked skins on a wooden horse and allow the pickle solution to drain out completely. Then tan as under without washing:

5. *Tanning*

Recipe

Water	:	15xW litres
Common salt	:	50 g/l

Formaldehyde	:	10 g/l
Sodium bicarbonate	:	1g/l
Basyntan DI	:	1 g/l

The pickled skins are put in the bath containing only common salt, handed for 10 minutes and then formaldehyde is added in two installments at half hour interval while handling thoroughly. Then sodium bicarbonate is added and skins are stirred for another 1 hr. Then Basyntan is added and the skins are left in this bath for over-night. Next day the skins are piled flesh to flesh for a period of 24 hrs. Then the skins are fat- liquored in semidry condition.

6. *Fat-liquoring*

Recipe

Lipoderm liquor 2 FB (BASF)	:	1 part
Water (60 °C)	:	2 parts

The emulsion is applied on the flesh side of each skin with the help of a brush. Skins are piled flesh to flesh for over-night and hanged to dry the next day.

7. *Sawdust drumming*

The dried skins are drummed with the moist sawdust for half an hour, caged and hand- staked. If found greasy, drummed again with saw-dust containing mineral turpentine @ 2-3% on the weight of saw-dust for 1 hr. Caged and staked again. Then the skins are buffed on the flesh side using emery cloth of grit No. 120 or 140.

8. *Bleaching*

Bleaching is not necessary if the skins are naturally coloured or they are to be dyed. Only off-white or pale yellow skins need bleaching.

Recipe

Water	:	15xW litres
Soda ash	:	0.50 g/l
Sandozin NIS liquid	:	0.50g/l
Sodium hydrosulphite	:	3-4 g/l
Temperature	:	35 °C
Stir for 2 hrs. Then add	:	
Sulphuric acid (98%)	:	1 g/l

Stir for another one hour. Then take out and wash thoroughly. Add Leucophor WSI liq. @ 0.1 g/l to the final wash. Handle for 15 minutes. Then take out the skins and allow to drain out the water. Re-fatliquor, if necessary. Dry, hand-stake and finally buff and trim. The skins processed as above have a bright, silky, lustrous and dense fur. They are very soft, supple and strong.

Recent trends

Latest trends in processing technologies, indicate optimization of processing systems so as to reduce the pollutant load to the minimum level in the tannery effluents. Changes are taking place at large scale, to make the technologies of pelt processing eco-friendly. The most attractive strategy is to audit the processes, identify the sources of pollution and generate methods of pollution reduction. These are based on three principles - (i) pollution avoidance, (ii) pollution reduction through process changes and (iii) use of recovery/reuse measures.

Commonly identified sources of pollution in process-houses are from dissolved salts, chromium and sulphide. Almost half the amount of salt entering a process-house is through the skin in the form of salt used for curing. A skin carries with it as much as 30% by its weight of sodium chloride. The management practices of better curing options are being exercised. Replacement of 'salt-curing' method by 'solvent dehydration' technique, for preservation of raw pelts, is one of them. This works on the principle of pollution avoidance rather than effective control or recovery and reuse because of cost factors. For the additional amount of salt used in processing, the pollution reduction measures such as recovery and reuse are being explored and employed.

For chromium (III) salts used in tanning, all the three options of pollution reduction namely pollution avoidance, control and recovery-reuse are available. The chrome-free tannages, using organic tanning agents (i.e. synthetic tanning agents known as 'syntans') has made complete eradication of chromium salts possible and such proven processing options are available today which give better end-products with desired end-properties than the old conventional methods of processing.

Conclusions and Recommendations

Owing to growing concern about wild life preservation, the availability of fur-skins of wild animals have become almost impossible. The import of rabbit pelts from France and other countries has, therefore, escalated many-fold. Thus, in the changed scenario, it has become extremely necessary that organized farms and ranches are established and developed where, besides rabbits, the tamable wild fur-bearing animals are reared and propagated on extensive production system. This will go a long way in sustaining the pelt processing industry of India.

The impact of environment protection awareness is being felt on processing technologies also. Recent trends indicate the availability of attractive strategies for pollution reduction to the maximum extent through process changes and/or recovery and reuse of pollutants. Solvent dehydration method should be used in

Table 4. Effect of processing method on quality of processed pelts

S. No	Process	Quality Characters	White giant		Grey giant		Soviet Chinchilla	
			Raw pelt	Processed pelt	Raw pelt	Processed pelt	Raw pelt	Processed pelt
1.	Basified Chrome Method	Weight (g)	187.50	152.50	140.00	105.00	120.00	97.50
		Weight loss (%)	-	18.70	-	25.00	-	18.80
		Area (cm²)	1637.00	1354.50	1397.10	1217.00	1170.50	1119.10
		Area shrinkage (%)	-	17.30	-	12.90	-	4.40
		Grade	-	A	-	A	-	A
2.	Alum + Potassium Nitrate Method	Weight (g)	237.50	195.00	117.50	91.90	170.00	133.80
		Weight loss (%)	-	17.90	-	21.80	-	21.30
		Area (cm²)	1856.90	1451.40	1501.80	1340.10	1555.40	1363.30
		Area shrinkage (%)	-	21.80	-	10.80	-	12.40
		Grade	-	A-B	-	A	-	A

Singh (1985)

place of salt curing method because it reduces not only the salt pollution in a process-house to half of its quantity but also the transportation cost of the cured pelts. It also helps in retention of natural properties of a raw pelt better than any other method of curing. For reduction of chromium, alternative methods made available recently are being used extensively as no compromise with regard to the quality of processed pelts has to be made. Enzymes have replaced the sulphide treatments thereby reducing its pollution to a large extent. Reduction of pollution in process-house has become a technical feasibility today.

REFERENCES

Annual Report. 1983. Central Sheep and Wool Research Institute, Avikanagar

Bangaruswamy, S. 1982. Development of Indian fur industry – A case for fur animal breeding farmers.

Paper in: 1st National Seminar on *"Fur Animal Breeding"* April 24 -26, North Temperate Regional Station (CSWRI), Garsa (Kullu) H.P.

Cheeke et al. 1982. *"Rabbit Production"*: International printers and publishers, Illinois, USA.

Jyoti Kaul & Wali, B.K. 1988. Preservation of furs-skins. Paper in: Indo-British Workshop on "Rabbit Fur Technology", Oct. 10-15, Regional Research Laboratory (C.S.I.R.), Srinagar (J & K).

Leach, I.B. & Barrett, J.C.1984. Rabbit skins-problems and prospects for less developed countries. *Journal of Applied Rabbit Research* 7, 96.

Mahajan, J.M. 1982. Some fur and wool technology studies. Paper in: 1st National Seminar on *"Fur Animal*

Breeding" April 24 -26, North Temperate Regional Station (CSWRI), Garsa (Kullu) H.P.

Prentice, A.C. 1976. "A candid view of the fur industry": Clay Publishing Co., Ontario, Canada.

Ramaswami, T. 1997. Strategy for reducing tannery pollution. *The Tanner* 12I, 8.

Singh, R.N. 1985. Breeding strategies for fur animal production in India. Paper in: National Seminar on "*Fur Animal Production and Utilization*", Oct. 15 -17, Regional Research Laboratory (C.S.I.R.) Srinagar (J&K).

❑❑❑

35

Socio-economic Status of Small Ruminant Producers and Sustainability of Small Ruminant Production in India

D. C. GUPTA, SURESH A., J.S. MANN

Central Sheep and Wool Research Institute, Avikanagar- 304 501, Rajasthan

Introduction

In India, there are about 124 million goats and 61 million sheep according to livestock census 2003. Sheep and goats contribute greatly to the agrarian economy through their produce and play important role in the livelihood of a large percentage of small, marginal and landless farmers. These are important species among livestock mainly on account of their short generation intervals, high prolificacy and multifaceted utility as producer of wool, meat, skin, milk and manure with its added advantage of survival in low vegetation. They are reared mostly on common degraded grazing lands (Singh *et al.* 2005). It is estimated that sheep contributed 0.239 million tonnes of mutton and 51.4 million kg of wool accounting for Rs. 3770 million and goat contributed 2.7 million tonnes of milk and 0.475 million tonnes of meat accounting for Rs 9350 million in the Indian economy in the year 2004. The contribution of agriculture sector to national GDP is 25% and the share of livestock in agricultural GDP is 23%, of which small ruminants (SR) contribute about 10% of the total value of livestock sector. At the national level, sheep and goats account for 14% of the meat output, 4% of milk output, and 15% of hides and skin production in the country. Small ruminant rearing provides significant proportion of self- employment opportunities. The main role of ovines in the arid, semi-arid and hilly regions are to provide food security, generate income and employment, reduce poverty in the rural areas and act as an insurance against natural vagaries like drought. In this paper an attempt has been made to look into the socio-economic status of small ruminant producers and to discuss the strength, weakness, opportunities and threat to the small ruminant production.

Distribution of Small Ruminants in India

The country has 42 sheep breeds and 22 goat breeds. Nearly 40 per cent of livestock in the country are small ruminants. The distribution of goats in the country is more homogeneous than sheep, as the later are mostly confined to relatively dryer and arid regions (Table 1). Goats are widely distributed across all agroclimatic regions of the country. More than 68 per cent of goats are concentrated mainly in the states of Bihar, Madhya Pradesh, Maharashtra, Rajasthan, Tamilnadu, UP and West Bengal. But, sheep are reared mainly in arid, semiarid and hilly/ mountain regions of the country and nearly 86 per cent of that are concentrated in seven states namely Andhra Pradesh, Gujarat, Jammu & Kashmir, Karnataka, Maharashtra, Rajasthan and Tamilnadu.

Table 1. Share of small ruminant in livestock across the States (India) in the year 2003.

S. No.	Major contributor States	Livestock population (%)				Total livestock '000'
		Bovine	Sheep	Goat	Small ruminant	
1	Andhra Pradesh	41.35	44.35	13.02	57.37	48195
2	Assam	65.93	1.23	21.60	22.83	13830
3	Bihar	60.65	1.41	34.94	36.35	27161
4	Chhatisgarh	77.68	0.90	17.31	18.21	13492
5	Gujarat	67.25	9.52	20.97	30.49	21655
6	Haryana	85.26	7.12	5.18	12.30	8885
7	Himachal Pradesh	58.82	18.10	21.98	40.08	5117
8	Jammu & Kashmir	41.65	34.45	20.76	55.21	9900
9	Jharkhand	56.88	4.30	31.79	36.09	15826
10	Karnataka	52.81	28.32	17.50	45.82	25621
11	Kerala	62.84	0.11	34.86	34.97	3480
12	Madhya Pradesh	74.37	1.53	22.86	24.39	35617
13	Maharashtra	61.06	8.42	29.06	37.48	36763
14	Orissa	65.39	6.92	24.81	31.73	23392

Contd....

Table 1. Contd....

15	Punjab	93.34	2.56	3.23	5.79	8607
16	Rajasthan	43.28	20.46	34.21	54.67	49138
17	Tamilnadu	43.30	22.42	32.78	55.20	24941
18	Uttar Pradesh	70.84	2.45	22.11	24.56	58531
19	Uttaranchal	69.11	5.99	23.43	29.42	4943
20	West Bengal	48.06	3.66	45.11	48.77	41613
21	All India	60.37 (283103)	13.11 (61469)	26.52 (124358)	39.63 (185827)	(468930)

Source: Livestock census 2003.
Within parenthesis number of heads in '000'.

Distribution of Land and Livestock Holding

Based on National Sample Survey Organization estimates (1991-92), more than 84 per cent farmers in the country are either landless or in possession of less than 2.0 ha. of land (Table 2). The marginal and small farmers were 45.3 and 14.2 per cent of the total farmers and their average size of holding was 0.35 and 1.41 ha respectively. Small, marginal and landless households reared nearly 71 per cent population of small ruminant This implies that the marginal and small landholders derive a considerable proportion of their income from small ruminants (Birthal *et. al.* 2002). It reflects that the small ruminants farming are traditional occupation of weaker segment of the society throughout the country.

Table 2. Size and distribution of land and livestock holdings in India, 1991-92

Size class of land holding	Percent of households	Size of land holding (ha.)	No. of Sheep and Goat/households
Landless	21.8	0.0	0.20 (5.1)
<1.0 ha	48.3	0.35 (15.5)	0.81 (46.2)
1 to 2 ha	14.2	1.41 (18.6)	1.15 (19.3)
2 to 4 ha	9.7	2.69 (24.2)	1.31 (15.0)
4 to 10 ha	4.9	5.78 (26.4)	1.66 (9.7)
> 10 ha	1.1	15.44 (15.3)	3.75 (4.7)
Overall	100	1.08 (100)	0.85 (100)

Figures in parentheses are shares in total population of a species.
Source: National Sample Survey Organization, Ministry of programme planning and Implementation, Govt. of India.

Who Rears Small Ruminant?

More than 70 per cent households own livestock and majority of livestock owning households are small, marginal and landless. Land scarce poor households at large kept small ruminants because of their low initial investment and operational costs (CALPI, 2005). A study on the livelihood systems of the rearers in selected villages in Andhra Pradesh indicated that large percentages of small ruminant rearers are classified as poor. However, in some villages well-to-do households are in greater percentage among the small ruminant holders (Katticaren, Rebecca. 2006). The traditional rearers belong to the Yadava community, which is socially backward and illiterate.

Sheep and goat species are predominantly maintained under extensive management on community rangeland, cropped land after harvesting of standing crops and forestland in mixed grazing with other livestock species. Ovine farming can be considered as low input and marginal output system of livestock production. The rearers are not much input intensive and therefore the small ruminant production system mainly depends on common property resources (CPR) for their entire nutrition. Majority of small ruminant farmers are illiterate, having large family size, dominated by other backward classes viz. Jat, Gujjar and Yadav (Riyazuddin et. al. 2001). He reported that the average family size was 10 with family composition 52 % males and 48% females. The average age of the head of household was 50 years and the distribution by caste indicated that 54.8 % farmers were other backward classes, 33.8 % general, 9.7 % scheduled caste and 1.7% scheduled tribes.

The survey conducted by CSWRI, Avikanagar in Tonk district revealed that the average age of the farmer was more than 44 years, males dominated large family size of 10 and Literacy 37.4 per cent (Suresh, *et al.* 2007). Literacy in females was negligible. About 53.3 per cent joint families were observed. About 82.1 per cent farmers were from other backward communities followed by SC (9.4 %) and ST (6.5 %). The involvement of general caste is generally less in raising the small ruminants due to low prestige attached with this profession in the society. Both Hindu and Muslim raise the small ruminants in the state. The religion is not the constraint in sheep and goat rearing. More than 95 per cent Hindu families are raising the small ruminants.

In a socio-economic survey covering 2309 households, the average family size of Mecheri sheep farmers was 4 with literacy rate 51.7 per cent belonged to backward, most backward and other communities in Tamil Nadu state (Thiruvenkadan, A. K. 2004).

Small Ruminant Production Systems

A survey of 5992 households in Rajasthan state under 'National Agricultural Technology Project' was conducted in the year 2001-02 by CSWRI, Avikanagar considering all the agro-climatic zones of the state. The sampled households were essentially keeping the livestock. More than 53 per cent farmers are either landless or marginal or small category. Most of the land is un-irrigated in the arid and semi-arid zones of the state. A total of 54.36 per cent farmers were rearing sheep and goats (Table 3). Small ruminants are reared in combination with the large ruminants. Sheep rearing alone was not observed. A total of 5.24 per cent households were rearing sheep in combination with goat, cattle & goat, buffalo & goat, bovine & goat and cattle, goat & camel. Goat rearing alone among the small ruminants is one of the major livestock production systems in the state. The goat is reared by all categories of farmers and in all the agro-climatic zones of the state. The proportion of goat farmers was highest (64.75 %) in landless category, which declined with the increase in land holding size and reduced to 41.38 % in large category of farmers. In case of sheep, the proportion was highest (15.92) in the large category of farmers, which reduced to 7.9 per cent in landless and lowest (1.39%) in marginal farmers.

Table 3. Distribution of sheep and goat farmers across the household categories in Rajasthan

Category of farmer's	No. of households	No. of farmers rearing (%)	
		Sheep	Goat
Landless	417 (6.96)	7.9	64.75
Marginal	1290 (21.53)	1.39	58.52
Small	1468 (24.52)	2.45	55.73
Semi-medium	1563 (26.08)	5.69	52.71
Medium	720 (12.02)	7.79	51.26
Large	534 (8.91)	15.92	41.38
overall	5992 (100.0)	5.24	54.36

Source: A report on ' Complete enumeration of households survey-(2001-02) under NATP on 'Livestock crop production systems analysis for sustainable production' of CSWRI, Avikanagar unit.

Economics of Small Ruminant Production

The small ruminants are generally penned under the open sky, which is fenced with thorny hedge and are grazed in the outskirts of the villages and on stubbles in the fields of harvested Kharif and Rabi crops. The sheep and goat rearing is based on low input and marginal output system of production. The major input required to these two species is mainly farmer's own labour and veterinary care. The existing production system of small ruminants does not require specialized skills, expensive infrastructure, organized breeding plan, feeding management, husbandry practices, health coverage and marketing procedures as the farmers adopt age old traditional skills of sheep & goat rearing passed over the generations. The variable cost components viz. feed, fodder, labour, veterinary charges, interest on working capital etc. and miscellaneous charges viz. electricity and water will be considered towards the cost of sheep and goat production. The fixed cost components like depreciation on assets, interest on fixed capital need not be taken in to account, as the assets are heritable in nature (Gupta, D. C. 2004).

The benefits derived from small ruminants are meat, wool, milk, hides & skins and manure. The farmers to supplement the wage earned by him, keep the small ruminants to meet the day to day and an emergent type of expenditures. The main advantage of sheep and goats in livestock-cop production system is their distinguished ability to utilize natural grasses, bushes, shrubs, crop stubbles and tree leaves to produce saleable products. Sheep is important species mainly on account of their short generation intervals and multifaceted utility as producer of wool, meat, skin manure and milk to some extent. The mutton is the first priority of sheep rearing and wool, hide & skins are the next priority. Goat is also very important species of the country, mainly on account of their short generation intervals, higher rates of prolificacy and the ease with which the goats and their products can be marketed. The first priority of goat rearing is meat and next is milk, hides & skins and manure. Manure is also very important return from sheep and goats.

Average cost per sheep and goat reared in the extensive system of management in semi-arid region of Rajasthan was Rs. 23.59 and Rs. 21.13 respectively (Rawat, *et al.*, 1998). The cost components do not include the labour cost on account of own family labour employed in raising the small ruminants. The component-wise income from small ruminant showed the maximum return through sale of live animals followed by milk, wool and manure respectively. The gross average return per sheep and goat had been Rs.227.81 and Rs. 280.71 respectively. Therefore, the 'Benefit- Cost ratio' in goat and sheep was 12.28 and 9.66 respectively. The B/C ratio is an indicator of the economic viability. The goat is economic more viable as compared to sheep.

Riyazuddin *et. al.*(2001) reported that the small ruminant rearing cost decreases with the increase in flock size. The sheep and goat rearing annual average cost per ewe/ doe in Rajasthan for a flock size more than 50 was observed Rs. 10.53 and Rs. 30.40 respectively . The small ruminant farmers employed relatively longer hours of grazing in all the seasons. The returns from sheep and goat per animal are higher in small flocks as compared to larger flocks. The returns from sale of kids born, milk and manure were Rs 1076, Rs 736 and Rs 69 respectively in Rajasthan. The total return to the farmer amounted to Rs 1891per doe per year. The sheep farmers of Rajasthan realized on an average Rs 684, Rs 74.00 and Rs 51.00 from sale of lambs born, wool and manure respectively with a total return of Rs. 809.00 /ewe/ year.

SWOT Analysis

In order to critically examine the future prospects and sustainability of small ruminant production system a SWOT analysis was undertaken and is described below.

Strength

1 Small ruminant plays an important role in livelihood security and economic sustenance of landless, marginal and small farmers living in rural areas. Sheep and goat rearing are low input ventures wherein farmer raised these species on free grazing land employing own family labour. Both species can be maintained even in harsh climatic condition and survived on low vegetation mostly waste lands where crop cultivation is not possible or uneconomical, harvested fields, fallow/ barren lands, stubbles, bushes, shrubs and tree leaves along road side.

2 Goats are reared mainly for meat and milk where, sheep reared for mutton and wool. The goat meat and mutton are preferred in the diet of non-vegetarian people. In addition to that the sheep & goat milk is providing nutritional and food security to the families raising them under extensive system of management.

3 There is mixed livestock farming system wherein the small ruminant easily fit into the smallholder production system, as they need low initial investment and low operational cost. They also give quick returns because of short generation interval and high prolificacy.

4 The small ruminant owner sells animals very easily when he require money to meet an emergent need like wedding, funeral, to buy inputs such as vaccine & other preventive medicines for the protection of his livestock, seed, fertilizer and insecticide for crop protection. Hence, the liquidity of ovine assets is easy and quick without any role of compulsion factors in the market.

5 The ovine meat from India will be preferred in national and international markets because of organic nature of production

6 The cost of SR production is very low due to which the consumer's price is not rising so high in comparison to other livestock products. Hence the producers of SR will always find an easy market.

7 The milk from small ruminants is a good source for meeting the nutritional requirement of family members and source of regular income and employment to the owner.

Weakness

1 The majority of small ruminant owners belong to SC, ST, OBC and Muslims and their socio-economic condition is poor. It is the major constraint due to which they do not put their efforts to get literate and therefore the implementation of development programs is very slow.

2 There is a decline in the area under common property resources from 1950 onwards due to expansion of agricultural land, industrial land, encroachment, privatization and other land reform measures (CALPI, 2005).

The reduction in CPR due to decrease in barren and uncultivable land, miscellaneous tree crops, cultivable waste land and fallow land and increase in forest area has serious implications on livelihood of smallholder livestock production. Further, the increase in human and livestock population is exerting pressure on natural resources due to increase in demand. It leads to damage the valuable plant species, biomass production, soil fertility, water resources etc.

3 Inadequate veterinary health services restrict desired growth in SR sector. It receives low priority possibly due to lack of infrastructure facilities required for SR, shortage of trained personnel, non-availability of medicines/ vaccines and lack of disease information on ovines etc.

4 The genetic development in small ruminants is slow and difficult due to various reasons i.e. (i) migration of SR in search of fodder and drinking water, (ii) maintained under free grazing land and on low plane of nutrition, (iii) males are sold relatively at early age by smallholders making selection difficult.

5 Absence of active supportive organization for the mobilization of the rearers is attributing to lack of development of ovine rearers.

6 Majority of ovine rearers are illiterate and do not have easy access to avail the benefit of credit and insurance policies.

7 The SR rearers seldom have a direct access to the markets for live animals, meat, milk and wool. However, the wool market is semi organized and other markets are un-organized operating under the clutches of a nexus of small traders, butchers, market agents and middlemen. The small rearers find it difficult to penetrate these markets because of the nexus. They also do not have the access to market information on market prices, margins, cost of operations etc. at the local or distant market.

8 The development programmes pertaining to ovines involve support of various organizations / agencies to make it successful. But due to lack of institutional coordination, the programmes do not achieve success.

Opportunities

1. Superior indigenous breeding males in the identified breeding tracts should be selected from the elite flocks. A sheep / goat breeder society or cooperative can be formed and may be regulated and the high producing animals shall be pooled to form the nucleus herd, which will produce proven bucks / rams, which can be sold to farmers.

   In case the proven bucks / rams are not sufficient to meet the requirement in the field, the A.I. technology may be used in the farmer's flock to enhance the lamb / kid production of high yield.

2. A focused programme of extension and training shall be an integral part of capacity building of SR rearers. It should focus to educate and train the SR rearers on awareness of human and constitutional rights, awareness of networking and communication, awareness of scientific management of flocks, awareness on better husbandry practices and preventive health care including first aid.

3. Village committees / village Panchayat can take the responsibility of managing the land and provide preferential access to landless, small and marginal farmers.

   Develop a multi-tire silvi pasture in waste / fallow / barren / forest lands. It will provide shrubs and tree lopping for goats and grass for sheep. Emphasis on multi purpose fodder trees should be propagated to receive fodder, fuel, medicinal values etc.

   Controlled grazing in CPR through community controlled customary rationing mechanism is a must for sustainable use of generated resources. Social norms and codes of conduct are to be fixed by the communities for rotational grazing. Special attention should be given to make sure that the deprived segment of the community is mostly benefited through this arrangement.

4. In order to reduce the dependence on CPR, the sheep and goat rearers should be encouraged to adopt semi-intensive system of raising the sheep / goats by showing the institutional flocks performing better than those under extensive system of management. It will reduce the hours of open grazing which could lesser the pressure on CPRs.

5. Diagnostic services need to be speedy, effective and accessible. Greater coverage should be provided for vaccination against diseases like enterotoxaemia, sheep pox, foot & mouth disease, etc. The veterinary care to migratory flocks may include advice on deworming and anthelmatic use, protection against endemic diseases prevalent in the area where the animals are migrating to and prevention of spread of disease through these animals.

6. The short term credit facility need to be extended to small ruminant sector which should be linked with insurance for covering the risk of loss of production and life. Ovine rearers may be given bank loans to establish pastures and growing fodder trees on their own lands.

7. Complete feed blocks prepared from roughage and concentrate mixture in varying proportions will be useful for intensive feeding system of management. Feed blocks can be stored for 6-8 months without loosing the quality. These feed blocks can be fed to animals during drought and famine periods.

8. The demand of organic meat is growing up in the country. The use of herbal and biological drugs for treatment of diseased animals would reduce the harmful chemicals in the animal produce.
9. The Animal Husbandry Development, N.G.O's and rearers association should take initiative to co-ordinate with concerned department in the recipient state for fixing migratory routs, norms of migration, code of conduct, services during migration and other related issues faced by the rearers during the course of migration.

Threat

1. The farmers in the region are changing breed composition by inter mixing the established breeds for short term gains. Efforts should be made to maintain the purity of established breeds and exploit full production potential udder optimum feeding.
2. Small ruminant population is on increase and the grazing land is decreasing every year. As a result the grazing pressure is increasing on grazing land. Under the circumstances, the population of small ruminant will decline and sustainability of farmers on ovines will be threatened.
3. The poor prestige associated with the small ruminant production is weaning away the younger generation from this profession.
4. Intensification of agriculture in the traditionally rainfed regions drives the small ruminants out.
5. The decline/stagnation in wool price is affecting farmers' interest in production of wool. In long term this would affect the wool production.

Conclusions and Recommendations

Small ruminant development is an important intervention in poverty reduction. But the small ruminant production in India is facing challenges in the form of reduction in the grazing resources, unorganized marketing system, poor veterinary care and lack of policy intervention towards the development of the sector. Poor attention towards this sector in terms of inadequate institutional changes and policy interventions may hamper the livelihood generation capacity of the poor and downtrodden in the rural areas. For sustaining the small ruminant production, change is needed in the organization of the production system, in the form of adoption of an intensive or semi-intensive feeding system with increased use of veterinary inputs. Active intervention of farmers, non-governmental organization and policy support from the Government can make the small ruminant production in India sustainable.

REFERENCES

Birthal, P.S., Joshi, P. K. and Kumar , A. 2002. Assessment of research priorities for livestock sector in India. Policy paper 15. published by National Centre for Agricultural Economics and Policy Research, New Delhi.

CALPI, New Delhi. 2005. Sustainable livelihood through small ruminant production: Critical issues and approaches, draft for stakeholder review.

Government of India. 2003. Agricultural Statistics at a Glance, Agricultural Statistics Division, Directorate of Economics, Statistics, Department of Agriculture and Co-operation, Ministry of Agriculture, New Delhi.

Gupta, D.C. 2004. Economics of sheep and goat farming in comparison to other livestock. Advances in Sheep, Goat and rabbit and their Products Utilization, CSWRI, Avikanagar, Rajasthan, pp. 176 - 183.

Katticaren, Rebecca. 2006. A hollastic approach for the development of goat and sheep farming: Evidences from the field. Lead paper in: National Workshop-cum-Seminar on "Commercial Goat & Sheep Farming and Marketing: Farmer-Industry-Research Interface", March4-5, Central Institute for Research on Goats, Makhdoom, pp. 159 - 166.

Nayyar, S.S. 2006. Export of meat and meat products. Lead paper in: National Workshop-cum-Seminar on "Commercial Goat & Sheep Farming and Marketing: Farmer-Industry-Research Interface", March4-5, Central Institute for Research on Goats, Makhdoom, pp. 127 - 130.

Rawat, P.S. , Riyazuddin, Kumar Abbhay and singh, R.N.1998. A report on socio-economic status of sheep and goat breeders and agriculture farmers covered under IVLP/ TOT programme. Published by Statistics, Economics and Marketing Section of CSW RI, Avikanagar.

Riyazuddin, Kumar Abbhay, Jayashankar, j. and Rawat, P. S. 2001. Final report of the ad-hoc project on 'Development of small ruminant sector in India' Published by CSWRI, Avikanagar

Rathore, M.S. 2006. Status and strategies for marketing of goat and sheep, their meat and meat products. Lead paper in: National Workshop-cum-Seminar on "Commercial Goat & Sheep Farming and Marketing: Farmer-Industry-Research Interface", March4-5, Central Institute for Research on Goats, Makhdoom, pp. 97-108.

Suresh.A., Gupta, D.C., Mann, J.S. and Singh V.K. 2007. Sheep Production in Semi-Arid Zones: Management and Economics, Research Bulletin, Central Sheep and Wool Research Institute, Avikanagar.

Singh, V.K., Suresh, A.., Gupta, D.C. and Jakhmola, R.C. 2005. Common property resources rural livelihood and small ruminants in India: A review. Indian Journal of Animal Sciences 75, 1027 - 1036.

Thiruvenkadan, A.K. 2004. Socio-economic status of the mecheri sheep farmers and economics of rearing under farmer's management system. *Indian Journal of Small Ruminants* 10, 1-6.

❑❑❑

36

Economic Fat Lamb Production Technologies for Field Applications

O.H. CHATURVEDI, B. SRINIVAS, R. MALIK
Central Sheep and Wool Research Institute, Avikanagar-304 501, Rajasthan

Introduction

The crop production as well as animal husbandry has been the backbone of rural economy in our country since ages. Today, uncertainty in the crop production in most of the regions, particularly in arid and semi-arid regions of India is a impeccable fact due to low and erratic rainfall. The farmers of these areas have, therefore, diversified from crop production to livestock production to counter the risks of crop failures as the livestock in general are more tolerant to harsh climatic conditions compared to crops. Such necessity leads to gradient increase in the contribution of livestock in agricultural sector that was 14 % in 1980-81, 23 % in 1997-98 and 35 % in 2001-02 (Sharma, 2004). Livestock sector is important for the livelihood of small farmers who are generally constrained in the availability of land for cultivation. Share of dairying in household income was 53% in case of landless farmers, 30% in case of marginal farmers and only 19% in case of large farmers (Shukla and Brahmankar, 1999). Cattle, Buffalo, Camel, Sheep, Goats, Yak and Mithun are the preferred ruminant livestock reared in various parts of our country owing to their adaptability. About 10 % of the total land area of the country is arid and 30 % semi-arid. The arid zone is mostly confined to the States of Rajasthan and Gujarat while almost all the states in Central and Peninsular India and part of Rajasthan, Gujarat, Punjab and Haryana come under semi-arid region. Rajasthan has the maximum (73.60%) of the total arid and semi-arid area followed by Gujarat (29.50 %) and Andhra Pradesh (21.50 %). The annual rainfall in the arid region varies from 10 to 40 cm, quite often erratic, so much so that the entire rainfall of the year may fall on few days and rest of the year may go dry. The summer temperatures may be as high as 49 ^{0}C during day and fall to less than 20 ^{0}C during night. In winters, the day temperatures are higher but the night temperatures may be near freezing point. The annual rainfall in the semi-arid region varies from 50 to 60 cm. In semi-arid regions, the summer and winter temperatures are not as extreme as in the arid region but may reach 45 ^{0}C during summers and 8 ^{0}C during winters. The egalitarian role of livestock in rural economy is well appreciated in the literature (Taneja and Birthal, 2005). The small ruminants (sheep and goat) contribute to the rural economy in terms of meat, milk, wool, hair and dung and accounts for 4.2% of the total milk production and 15 % of the total meat production. The small ruminant rearing is commonly undertaken by economically weaker section of the society who is generally not able to undertake farming and rearing of large ruminants due to high operational costs.

The role of livestock particularly small ruminants more pronounced in the arid and semiarid zone of country, where the risk and uncertainty of crop failure is high due to low availability of water and frequent natural vagaries like drought. Livestock sector acts as a cushion in substantiating the farmers livelihood during financial distress. Sheep is well adapted with arid and semi arid regions even under very scarce water and fodder availability conditions. They also act as a means of asset retention with high liquidity and help in absorbing unemployed family labour (Suresh *et al.*, 2007).

Land and Livestock Holdings and Utilization Pattern

The total land area of the country is around 304 million hectares. Of the total area, about 22.5 % is under forest cover, 3.6 % is under permanent pasture, 1.2 % is under tree crops and groves, 6.25 % is barren and uncultivable land, 4.5 % is under waste lands and 8.0 % is under fallow lands. The total area available for livestock grazing worked out to be about 46 % of the total land in the country. Permanent pastures occupy over 5.0 %, forest cover over 14 % and wastelands over 12 % in the semi-arid regions of the country. In the semi-arid area of Rajasthan, average land holding of the marginal, small, medium and large size farmers was 0.4, 3.6, 6.8 and 20.8 acre with percentage distribution of 0.2, 5.0, 19.1 and 75.7, respectively (Chaturvedi *et al.*, 2002). The irrigated land was 27.2 % in the semi-arid region of Rajasthan. The proportion of irrigated land was slightly lower in the area as compared to the national figure (35%) because the area is typically semi-arid and mostly rain fed. The average family size was 9.7 in semi arid Rajasthan and it was higher in case of large farmers (11.9) followed by marginal (8.5), small (8.0) and medium (7.8) farmers. The number of persons engaged and dependent on agriculture was higher in case of large farmers followed by medium, small and marginal farmers. However, due to shortage of resources like land and agricultural inputs, marginal and small farmers engaged themselves in other jobs including hired agricultural labourers for their livelihood. The human to land ratio in the semi-arid region of Uttar Pradesh and Rajasthan was about 1.4 (Saran *et al.*, 2000; Chaturvedi *et al.*, 2002). It was highest in case of marginal farmers (9.3) followed by small 2.3), medium (1.1) and large (0.6) farmers. Marginal farmers who constitute 48% of the farming population in India accounted for 16% of the arable land, 42.2% small farmers accounted for 19% of the arable land and 6% large farmers accounted for 42% of the arable land whereas their share in the total milch animal was 49, 22 and 11 %, respectively (Taneja and Birthal, 2005). The arable land under fodder cultivation was only 2.4 %. Most of the arable land is used for cultivation of maize, sorghum, pearl millet, cowpea, groundnut, sesamum, pigeon pea, green gram and black gram during kharif and mustard, chickpea, wheat, barley and vegetable crops during rabi season. Major fodder crops grown were M.P. chari, sorghum, pearl millet and cowpea in kharif and lucerne, berseem, kasni, senji and oats in rabi season.

The livestock holding in terms of number and adult cattle unit (ACU) in semi arid region of Rajasthan is 39.2 and 12.7, respectively. The livestock holding in terms of ACU was higher in case of large farmers (15.9) followed by marginal (12.6), medium (12.5) and small (4.6) farmers (Saran *et al.*, 2000; Chaturvedi et al., 2002). The livestock: land (ACU/acre) ratio was highest in case of marginal farmers followed by medium, small and large farmers. Thus, large farmers having larger family size maintained more livestock compared to small, marginal and medium farmers because larger families having more manpower could manage larger livestock herds. Further, an acre of land with a marginal farmer sustained more livestock and human lives compared to that of large farmers. The proportion of cattle, buffaloes, sheep and goats among the livestock was 22.5, 17.1, 31.4 and 29.4 %, respectively in semi arid region. The proportion of large (cattle and buffaloes) and small ruminants (sheep and goats) was 39.6 and 60.4 % indicating that among the livestock, sheep and goats have major contribution in the economy of the farmers of the arid region. Among the livestock, sheep husbandry fits in better with economical and climatologically choice of the farmers in Rajasthan (Sirohi and Rawat, 2000). The proportion of small ruminants was higher in case of marginal farmers followed by medium, large and small farmers (Chaturvedi *et al.*, 2002).

Feed and Fodder Resources

The major feed resources in our country are grasses, grazing, crop residues, cultivated fodders, edible weeds, tree leaves and agro-industrial by-products. Ranjhan (1994), Kumar and Mathur (1996), Ramachandra *et al.* (2001) and Ranjhan (2003) have reported potential availability and requirements of livestock feeds and fodders in the country. Crop residues and by-products constitute the main feeds accounting for 40 % of the total consumption of different livestock. Green fodders contribute 26 %, the concentrates 3 % and the rest is coming from grazing (Mathur, 2004). A large gap exists between the requirement and the availability of feeds and fodders in the country as a whole. Surveys conducted by the National Institute of Animal Nutrition and Physiology, Banglore, have revealed 45 % shortage of dry roughages, 44 % shortage of concentrates and 38 % shortage of green fodders in the country (Singh and Ramachandra, 2004). The situation in arid and semi-arid regions is much more serious. The feed scarcity is mostly due to increasing human and livestock population, deterioration of common grazing lands both in quality and quantity, lack of adoption of feed and fodder production and processing technologies and low priorities given for identifying, improving and utilizing the newer feed and fodder resources (Pradhan, 2003). The feed resources available for ruminants in arid region include seasonal and perennial grasses, stubbles, shrubs, bushes and forbs while grazing on the community range lands, roadsides, canal banks and harvested fields. Top feeds, crop residues, agro-industrial by-products and agricultural waste materials are also fed to the ruminants depending on their availability. The natural/ community grazing lands are under poor to very poor condition mainly due to high stocking density and round the year grazing. These lands have never been protected, harrowed, reseeded, irrigated and properly managed. The biomass yield of these community rangelands is 2, 4.5 and 6.5 quintal DM/hectare during summer, winter and monsoon season, respectively (Santra *et al.*, 2007). The annual biomass yield of community rangeland (4.3 quintal DM/h) is sufficient to maintain at the best 1 ACU (6 sheep or goats) round the year. However, the stocking density (2 to 50 ACU per hectare) on these lands far exceeds its carrying capacity. The natural rangelands are undergoing extensive degradation and ecological changes due to prevailing grazing pressure. Inadequate energy supply round the year and protein supply half the year is the major limitation of natural rangelands and pastures in arid region. During the lean period from January to June the problem of energy supply is acute as the intake is lower than even maintenance requirement of animals. The top feed resources (shrubs and trees) are lopped and fed to ruminants during lean period to overcome the feed shortage. The top feed resources invariably contain some anti-nutritional factors. Farmers apply the dilution technique for feeding of top feeds to their ruminants as they never maintain their animals on sole feeding of tree leaves. Further, with the passing of winter during feed scarcity in arid region of the country, farmers migrate with their sheep to neighbouring states, seeking good pasture and return back to their native tract in monsoon season. Similarly farmers with their sheep and goats migrate from high altitude sub-temperate and temperate ranges to foot hills of Himalayas during winter and return back to their native tract at onset of summer in northern parts of our country (Santra *et al.*, 2007).

Feeding Practices

Feeding practices of livestock are variable throughout the World and these are mainly dependent on climate, feed and fodder availability, socio-economic status of farmers, type of species and breed, level of production etc. in a particular area. Three feeding management system viz. extensive range management, intensive and semi-intensive are in vogue in Indian conditions. The extensive system is principally low resource use and a low level of productivity emerging from poor nutritional availability. There is marked fluctuation in feed availability and its nutritive value in different regions, years and seasons. The greatest limitation in Indian rangelands and natural pastures is on the availability of adequate energy throughout the year and adequate amounts of protein for more than half the year (Singh and Chaturvedi, 2004). During the period from January to June the problem of energy is very acute and the intake is much less than even the maintenance requirements. Although the grazing on the rangelands is considered as the cheapest method of livestock production, the over grazing/ over stocking on available lands has caused serious problems of vegetative destruction, soil erosion and land degradation. Due to long dry summers, erratic rains and light textured soils deficient in organic matter content having tendency to salinity, the natural potential of these lands rendered shortage of the nutritional requirements of livestock. The adult body weights, reproduction rates and milk production are very low and morbidity and mortality rates are very high. Feeding of adequate roughage and concentrate in a proper ration either as mixed feed or individually in stalls is known as intensive feeding management. A kind of compromise between extensive and intensive systems is known as the semi-intensive system of feeding management. This system is mostly followed in semi-arid regions of India. It is a combination of limited free range grazing on available pasturelands and feeding in the stalls with feed and fodder supplements. Sheep, goats, idle bullocks and non-producing cattle are mostly grazed in mixed grazing on community/public rangelands for about 8-10 h a day (Chaturvedi *et al.*, 2002).

Ruminant production in India is mainly based on highly degraded rangeland pastures and/or stubble grazing. Such pastures produce low quality forages which are ineffective to provide a sustained supply of nutrients to ruminants. Grazing ruminants are supplemented with top feeds during lean season to meet their nutritional requirements. These rangeland are covered with a wide variety of vegetation mainly grasses, bushes, shrubs and trees. The biomass yield of the community rangeland is low and stocking density is high. Sheep grazing on Indian rangelands round the year without rest to pasture in semi-arid regions are underfed for most part of the year. Majority of the sheep farmers in semi-arid region do not supplement their sheep with concentrate mixture even in critical physiological stages (Chaturvedi et al., 2002). However, limited concentrate supplementation in addition to free grazing on community rangeland substantially improved production performance of ewes (Chaturvedi *et al.*, 2001a; 2003) and growth performance of lambs (Chaturvedi *et al.*, 2000; Santra *et al.*, 2002). Supplementation of concentrates not only improved growth (McDonald *et al.*, 1996) but also enhanced reproductive performance (Njoya *et al.*, 2005; Chaturvedi *et al.*, 2006), fertility rate (Molina *et al.*, 1994) and overall sheep productivity (Stephenson and Bird, 1992) on low quality forages. Utilization of poor quality feeds by ruminants can be improved through concentrate supplementation, which increases the digestibility of nutrients (Santra *et al.*, 2002) through stimulating rumen fermentation (Sultan and Loerch, 1992). Digestible organic matter intake improved due to supplementation of maize and soybean meal on wheat straw based diets (Abebe *et al.*, 2004). Similarly supplementation of peanut oil meal (2.5g/kg live weight) resulted in higher average daily gain and was economic for growing kid production (Ott *et al.*, 2004). A supplementation of 250 g concentrate in addition to grazing has been recommended to raise sheep under extensive production system in semi-arid regions of India (Karim *et al.*, 2004). Such situations have increased the pressure on supplemental resources under tropical animal production systems (Shem *et al.*, 2001).

Supplementation of Ewes

The nutrient requirements of ewes are higher during critical physiological stages viz. last quarter of lactation and early part of lactation. In view of poor pasture quality and high stocking density on grazing lands, such ewes remain underfed during gestation and lactation. Healthy lambs of optimum birth weight and their adequate nutrition through dam during pre-weaning phase is important for cost effective mutton production. Healthy lambs of comparatively higher birth weight suckling from optimum fed ewes grow faster and attain finishing weight at an early age. Therefore, the supplementation of ewes grazing on pasture is important for economic fat lamb production.

During Gestation

The growth of fetus during first three and half months of pregnancy becomes slow and dam becomes able to maintain at normal level of feeding. In contrast, the growth of fetus becomes faster during later part of gestation (last six weeks of pregnancy). Therefore the nutrient requirement during later part of gestation with respect to DM, ME, DCP, Ca and P, respectively becomes 65, 90, 150, 160 and 165 per cent higher than the maintenance. On the other hand, the biomass yield of community rangeland in most of the part of our country is low and insufficient to meet the nutrient requirement of ewes during later part of gestation. Besides the grazing on good pasture, concentrate supplementation at the rate of 1 % of body weight to ewes during last quarter of gestation enhanced their production performance, general condition as well as birth weight and growth rate of lambs. The birth weight of lambs born from concentrate supplemented ewes was higher from 800 to 850 g as compared to those not supplemented (Chaturvedi et al., 2001a; 2003). The subsequent daily gain of these lambs was also higher (15-45 g/d) during pre-weaning phase because the ewes reared on concentrate supplementation yielded more milk (150-200 g/day) after parturition (Chaturvedi et al., 2001a; 2003). Higher milk yield from concentrate supplemented ewes provided better nutrition to lambs during pre weaning phase.

During Early Part of Lactation

The nutrient requirement of lactating ewes during first two months of lactation becomes twice of the maintenance requirement because mobilization nutrients towards milk production. In lactating ruminants, a complex interaction exists among the major substrates for milk synthesis i.e. acetate, glucose, amino acids and fatty acids and a multitude of regulatory hormones. The regulatory hormones those exert their influence on the galactopoietic process in epithelial cells of the mammary gland in maintaining milk secretion are insulin, prolactin, growth hormone (GH), thyroid hormone and glucocorticoides. Of these, insulin and GH are the key hormones in the homeorhetic control of lactation and are involved in the coordination of metabolism in body tissues. In fact, GH plays a significant role in the partitioning of nutrients away from tissue deposition and towards mammary gland for milk synthesis. Besides the grazing on good pasture, concentrate supplementation at the rate of 1 % of body weight to ewes during early part of lactation enhanced their production performance, general condition as well as growth rate of lambs. The daily gain of lambs nourished by supplemented ewes increased (25 g/d) due to higher daily milk yield by these ewes from 150-200 g (Chaturvedi et al., 2003; 2007a). The lambs suckling from supplemented ewes got adequate nutrition during pre- weaning phase.

During Gestation and Early Part of Lactation

The nutrient requirements of ewes during last quarter of pregnancy and early quarter of lactation are higher due to the obvious reasons. Like the supplementation during late gestation or early lactation, concentrate supplementation at the rate of 1 % of body weight to ewes from last quarter of gestation to early part of lactation in addition to grazing also enhanced production performance and general condition of ewes as well as growth rate of lambs. The birth weight of lambs born from concentrate supplemented ewes was higher 1000 g as compared to those not supplemented (Chaturvedi *et al.*, 2003). The growth rate of lambs nourished by supplemented ewes increased (55 g/d) due to higher daily milk yield by the ewes from 200-250 g (Chaturvedi *et al.*, 2003; 2007a). There exists a positive relationship between milk yield of the dams and pre weaning growth because the lambs meet their nutrient requirement from dam's milk (Singh *et al.* 1973).

Feeding of Lambs

Optimum feeding of lambs during pre and post weaning results in desired finishing weight at 5-6 months age for economic mutton production. The feeding methods of lambs during pre and post weaning are described below.

During Pre-weaning Phase

It is established that nutrition of lactating ewes affects its milk yield which in turn influences the pre weaning growth. The efficiency of milk utilization for growth during pre weaning phase ranges from 17 to over 30 % in different studies (Sharma, 1998). Feeing management of lambs during pre weaning phase is a prerequisite for economic fat lamb production because it affects the finishing weight of fat lambs. Because of optimum biomass availability on the pastures, the desirable growth rate of pre weaner lambs is not achieved under field conditions. The lambs in the famers' flock achieve 68-78 g average daily gain (ADG). Under organized farm feeding management on ad libitum concentrate and free suckling, the pre weaning growth rate increased up to 172 g ADG in male Mutton synthetic lambs and 168 g ADG in female Malpura lambs (Arora *et al.*, 1986). Average daily gain of Avivastra lambs maintained on free suckling and ad libitum creep mixtures varying in CP levels was 140 g with a feed conversion ratio of 1:3.4 during pre weaning phase (Santra and Karim, 1999). In another study, creep mixture containing 10 % CP and 2.7 Mcal DE/kg feed was adequate to support 124 g ADG in pre weaner lambs (Karim et al., 2001). An ADG of 156 g was achieved in pre weaning lambs maintained on free suckling to their dams and *ad libitum* feeding of creep mixture and *Zizyphus nummularia* (pala) leaves as source of roughage (Tripathi et al., 2007). Thus besides the free suckling and good quality leguminous hay or dried tree leaves, creep mixture of 16 % CP and 60 % TDN should be offered *ad libitum* to lambs for their optimum growth during pre weaning phase. Creep mixture should be prepared by using ground maize, groundnut oil meal, sunflower oil meal, milk powder, wheat bran, mineral mixture and common salt 50, 15, 10, 5, 18, 1 and 1, respectively in parts. Besides, vitamin blend at the rate of 25 g per quintal and quality animal protein supplement should be included in creep mixture to balance the amino acids.

During Post-weaning Phase

Feeding of weaner lambs is done under three feeding systems viz. extensive, semi intensive and intensive. The three systems are being discussed as under.

Extensive Feeding System

Under field conditions, most of the farmers maintain their lambs under this system i.e. the lambs are maintained exclusively under grazing on degraded rangelands for 8-10 hours. However due to inadequate nutrient supply through grazing alone, the lambs achieve 18-20 kg finishing weight at the age of 5-6 moths (Chaturvedi *et al.*, 2000; Chaturvedi *et al.*, 2001b). Weaner lambs reared for mutton production should attain the finishing weight of 25-30 kg at 5-6 months of age otherwise it seems uneconomical in terms of benefit cost ratio. Therefore, economic fat lamb production does not permit to adopt the extensive feeding of lambs under field conditions.

Semi-intensive Feeding System

Under this system of feeding management, fat lambs are grazed on rangelands for 8-10 hours and thereafter, concentrate mixture at the rate of 1.5 to 2.5 % of the body weight is offered to fat lambs. This system of fat lamb feeding is followed by few farmers under field conditions wherein little of concentrates (barley, *dal churi*, bran) in the form of supplement is fed. However, the quantities of concentrates offered by the farmers remain low in terms of nutrient supply to exploit the production potential of lambs. Therefore, it is recommended that besides the free grazing of weaner lambs on rangeland, concentrate mixture at the rate of 1.5 to 2.0% of their body weight would provide 25 to 30 kg finishing weight at 6 and 9 months of age, respectively (Karim *et al.*, 2004). Grazing with supplementation on established pasture has still higher production potential since established Cenchrus (*Cenchrus ciliaris*) pasture stocked as per carrying capacity with concentrate supplementation at 1.5 % of body weight provided 27.3 kg finishing body weight in fat lambs at 6 months age (Shinde *et al.*, 1995). In other studies, free grazing of weaner lambs on reseeded Cenchrus pasture with concentrate supplementation of 300 g/lamb/day provided 23-27.7 kg finishing weight and 80-106 g ADG with 9-12% feed conversion efficiency (Karim and Verma, 2000; Chaturvedi *et al.*, 2007b). The feed cost in terms of Rupee per kg gain ranged from 28 to 30 under semi-intensive feeding of lambs. Under field conditions, weaner fat lambs grazing on community rangeland with supplementation of concentrate mixture at the rate of 1 to 1.5% of their body weight achieved 21.6 to 24.2 kg finishing weight with 54 to 127 g ADG (Chaturvedi et al., 2000; Chaturvedi et al., 2005). The cost benefit ratio due to concentrate supplementation under farmers' field ranged from 1.45 to 1.75 indicating that farmers get additional profit of Rs. 1.45 to 1.75 on per Rupee spent. In general farmers spent Rupees 135 to 150 per fat lamb in terms of concentrate supplement and they get Rupees 300 to 400 more on a supplemented fat lamb. Thus, net profit per fat lamb ranges from Rupees 165 to 250 on selling of fat lambs. Concentrate mixture containing 10 % DCP and 65 % TDN should be used in supplementation of weaner fat lambs under semi-intensive feeding system. This concentrate mixture can be prepared by using maize, barley, groundnut oil meal, sunflower oil meal, wheat bran, mineral mixture and common salt 25, 25, 15,15, 18,1 and 1 parts respectively and vitamin blend (20 g per quintal). Further, the concentrate mixture can be made cost effective by using animal feed grade damaged wheat at the rate of 10 kg /quintal.

Intensive Feeding System

Under this system, the male fat lambs of two to three months age are stall fed *ad libitum* a complete feed containing 50 % roughage and 50 % concentrates for a period of 90 days. The rumen of two months old lambs is developed and thus intensive feeding of complete feed can be started at two months age to harvest early economic return. Under field conditions, farmers do not adopt intensive feeding for fat lamb production because of high input and extra management. However, this system can be adopted for commercial mutton production around the mega cities where due to non availability of pastureland, semi-intensive system of feeding can not adopted, while the demand of mutton is regular even at comparatively higher price. The weaner lambs prior to initiation in feed lot should be de-wormed and vaccinated against prevalent contagious disease particularly enterotoxaemia. Locally available good quality tree/shrub leaves, leguminous hay and grasses can be used as roughage portion of complete feed used as intensive ration. Maize, barley, pearl millet (*bajra*) and wheat as energy supplements and groundnut cake, mustard cake, till cake, cottonseed cake and guar korma as protein supplements can be used in least cost ration formulation for intensive mutton production. The ration can be made cheaper by replacing 50% cereal grains (maize or barley) with animal feed grade damaged wheat. The average daily gain (ADG) in Malpura weaner lambs fed complete feed (concentrate and roughage, 50:50) ranged from 95 to 151 g with feed to gain ratio of 7.3 to 9.1 (Karim and Rawat, 1996; Santra and Karim, 2002). The cost of feed input per kg gain during 3-6 month ranged from Rs. 34.4 to 40.0 (Karim *et al.*, 2004; Chaturvedi *et al.*, 2007b).

Conclusion and Recommendations

For economic fat lamb production under field conditions, the care should be taken from the fetus development to finishing weight at 5-6 months age. The ewes should be supplemented with concentrate mixture during late gestation and early lactation to achieve at least 3-4 kg birth weight and 12 -13 kg weaning weight. The proper feeding of creep mixture and roughage should be done during pre-weaning phase to get higher finishing weight. During post weaning in addition to grazing concentrate supplementation (semi-intensive system) reduces the risk and enhances economical gain over intensive feeding. Therefore, in addition to free grazing, concentrate mixture (10 % DCP and 60 % TDN) should be offered to lambs at the rate of 1.5 to 2.5 % of their body weight to harvest the required finishing weight (25 kg). Concentrate should be offered to the lambs after grazing. For commercial fat lamb production, the feed conversion efficiency should be 17-18% and cost of feed input/kg gain in body weight should be Rs. 25-28.

REFERENCES

Abebe, G., Merkel, R.C., Animut, G., Sahlu, T. and Goetsch, A.L. 2004. Effect of ammoniation of wheat straw and supplementation with soybean meal or broiler litter on feed intake and digestion in yearling Spanish goat wethers. *Small Ruminant Research* 51, 37-46.

Arora, A.L., Prasad , V.S.S., Gupta, D.C. Maru, A., Gupta, N. and Mahajan, J.M. 1986 . *All India coordinated research project on sheep breeding for mutton. Annual Report,* Central Sheep and Wool Research Institute, Avikanagar Rajasthan, India.

Chaturvedi, O.H., Bhatta, R., Verma, D.L. and Jakhmola, R.C. 2001b. Effect of supplementary feeding on growth performance of lambs grazing on community rangeland. In: *Proc. of IV National Seminar of Indian Society for Sheep and Goat production and Utilization (ISSGPU),* November 9-11, Coimbatore, India. pp.41.

Chaturvedi, O. H., Bhatta, R., Verma, D. L. and Singh, N. P. 2006. Effect of flushing on nutrient utilization and reproductive performance of ewes grazing on community rangeland. *Asian-Australasian Journal of Animal Sciences* 19, 521-525

Chaturvedi, O.H., Bhatta, R., Santra, A., Mishra, A.S. and Mann, J.S. 2003. Effect of supplementary feeding of concentrate on nutrient utilization and production performance of ewes grazing on community rangeland during late gestation and early lactation. *Asian-Australasian Journal of Animal Sciences* 16, 983-987.

Chaturvedi, O.H., Kumar, S., Mishra, A.K., Arora, A.L. and Karim, S.A. 2007b. Effect of complete feed or grazing and supplementation of lambs on performance, nutrient utilization and feed cost of production. In: *Proc. of International Tropical Animal Nutrition Conference Volume II*, October 4-7, National Dairy Research Institute, Karnal (Haryana), India. pp.247.

Chaturvedi, O.H., Mishra, A.S., Karim, S.A. and Jakhmola, R.C. 2000. Effect of supplementary feeding on growth performance of lambs under field condition. *Indian Journal of Small Ruminants* 6, 110-112.

Chaturvedi, O.H., Mishra, A.S., Santra, A., Karim, S.A. and Jakhmola, R.. 2001a. Effect of supplementary feeding during late gestation on production performance of ewes grazing on community rangeland. *Indian Journal of Animal Sciences* 71, 714-717.

Chaturvedi, O.H., Tripathi, M.K., Mishra, A.S., Verma, D.L., Rawat, P.S. and Jakhmola, R.C. 2002. Land as well as livestock holding pattern and feeding practices of livestock in Malpura taluk of semiarid eastern Rajasthan. *Indian Journal of Small Ruminants* 8,143-146.

Chaturvedi, O.H., Verma, D.L. and Mann, J.S. 2005. Benefit cost ratio and impact of concentrate supplementation to weaner lambs during active phase of growth. In: Compendium (Lead papers and Abstracts): National Seminar on "*Extension Methodological Issues in Impact Assessment of Agricultural and Rural Development Programmes*", January 21-23, University of Agricultural Sciences, Bangalore, India. pp. 318-319.

Chaturvedi, O.H., Verma, D.L., Singh, N.P. and Mann, J.S. 2007a. Effect of concentrate supplementation on production performance of ewes grazing on community rangeland. *Indian Journal of Animal Sciences* 77, in press.

Karim, S.A., Santra, A. and Singh, V.K. 2004. *Lamb feeding for mutton production.* Central Sheep and Wool Research Institute, Avikanagar (Rajasthan), India.

Karim, S.A. and Rawat, P.S. 1996. Growth performance of native and crossbred weaner lambs under intensive feeding. *Indian Journal of Animal Sciences* 66, 830-832.

Karim, S.A. and Verma, D.L. 2000. Growth performance of weaner lambs under protocol of intensive feeding or grazing with supplementation. *Indian Journal of Animal Sciences* 71, 586-588.

Karim, S.A., Santra, A. and Sharma, V.K. 2001. Preweaning growth response of lambs fed creep mixtures with varying levels of energy and protein. *Small Ruminant Research* 39,137-144.

Kumar, P. and Mathur, V.C. 1996. Agriculture in future: demand and supply perspective for the Ninth five-year plan. *Economic and Political Weekly* 31(39), 131-139.

Mathur, B.K. 2004. Drought proofing through utilization of low-grade roughages and non-conventional feed resources for small ruminants. In: Proc. of the National Seminar on "*Opportunities and Challenges in Nutrition and feeding Management of Sheep, Goat and Rabbit for Sustainable Production*", February 10-12, Central Sheep and Wool Research Institute, Avikanagar, pp.191-201.

McDonald, P., Edwards, R.A., Greenhalgh, J.F.D. and Morgan, C.A. 1996. Animal Nutrition, Longman Scientific and Technical, Group UK Limited, Longman House, Burnt Mill, Harlow, Essex CM20 2JE, England.

Molina, A., Gallego, L., Torres, A. and Vergra, H. 1994. Effect of mating season and level of body reserves on fertility and prolificacy of Machege ewes. *Small Ruminant Research* 14, 209-217.

Njoya, A., Awa, D.N. and Chupamon, J. 2005. The effect of a strategic supplementation and prophylaxis on the reproductive performance of primiparous Fulbe ewes in semi-arid zone of Cameroon. *Small Ruminant Research* 56, 21-29.

Ott, J.P., Muir, J.P., Brown, T.F. and Wittie, R.D. 2004. Peanut meal supplementation for growing doe kids on woodland range. *Small Ruminant Research* 52, 63-74.

Pradhan, K. 2003. Feed resources and technological interventions for enhanced livestock productivity. In: Proc. of Workshop on "Current Status of Feed Processing and Future Strategies", July 22-23, Acharya N G Ranga Agricultural University, Hyderabad, India.

Ramachandra, K.S., Anandan, S. and Raju, S.S. 2001. Livestock feed resources availability in India and need for a national database. In*: Proc. of 10th Animal Nutrition Research Workers Conference*, November 9-11, National Dairy Research Institute, Karnal, Haryana, India. pp. 134-144.

Ranjhan, S.K. 1994. Consultancy Report on availability and requirement of feed and fodder for livestock and poultry. Department of Animal Husbandry and Dairying, Ministery of Agriculture, Government of India, New Delhi.

Ranjhan, S.K. 2003. Livestock and feed situation in India. *In: Sustainable Animal Production* Edited by R. C. Jakhmola and R. K. Jain. Pointers Publishers, S.M.S. Highway, Jaipur 302 003 (Rajathsn), India.

Santra, A. and Karim, S.A. 1999. Effect of protein levels in creep mixture on nutrient utilization and growth performance of pre weaner lambs. *Small Ruminant Research* 33, 131-136.

Santra, A. and Karim, S.A. 2002. Nutrient utilization and growth performance of defaunated and faunated lambs maintained on complete diets containing varying proportion of roughage and concentrate. *Animal Feed Science and Technology* 101, 87-99.

Santra, A. Karim, S.A. and Singh, V.K. 2007. *Defaunation for improving mutton and wool production in sheep.* Central Sheep and Wool Research Institute, Avikanagar (Rajasthan), India.

Santra, A., Chaturvedi, O.H. and Karim, S.A. 2002. Effect of concentrate supplementation on nutrient intake and performance of lambs of two genotypes grazing a semiarid rangeland. *Small Ruminant Research* 44, 37-45.

Saran, S., Singh, R.A., Singh R, Rani, S.I. and Singh, K.K. 2000. Feed resources for rearing livestock in Bundelkhand region of Uttar Pradesh. *Indian Journal of Animal Sciences* 70, 526-529.

Sharma, V.K. 1998. Influence of pre weaning nutrition on post weaning gain of lambs. *Ph. D. Thesis*. Chaudhary Charan Singh University, Meerut (U. P.), India.

Sharma, V.P. 2004. Livestock economy of India: current status, emerging issues and long term prospects. *Indian Journal of Agricultural Economics* 59, 512-554.

Shem, M.N., Kakeni, A.M., Mtengeti, E.J. and Otsyina, R. 2001. Performance of grazing dairy cattle in semi-arid area of Western Tanzania and the marginal productivity of *Leucaena leucocephala* leaf meal supplement. *Agroforestry Systems Research* 52, 73-82.

Shinde, A.K., Karim, S.A., Singh, N.P. and Patnayak, B.C. 1995. Growth performance of weaner lambs and kids under intensive and semi-intensive feeding management. *Indian Journal Animal Sciences* 65, 830-833.

Shukla, R.K. and Brahmankar, S.D. 1999. Impact evaluation of Operational Flood on the rural dairy sector. National Council of Applied economic Research, New Delhi.

Singh, K. and Ramchandra, K.S. 2004. Feed and fodder availability, requirement and strategy for meeting the shortage in different regions for small ruminants. In: Proc. of the National Seminar on "Opportunities and Challenges in Nutrition and feeding Management of Sheep, Goat and Rabbit for Sustainable Production", February 10-12, Central Sheep and Wool Research Institute, Avikanagar, pp.1-9.

Singh, N.P. and Chaturvedi, O.H. 2004. Feed Resources and Their Development for Raising Buffaloes in Arid Region of the Country. In: *Buffalo Production Under Different Climatic Regions (Editors)* Kundu S S, Misra, A.K. and Pathak P.S. International Book Distributing Company, Charbagh, Lucknow, India. pp. 329-349.

Singh, V.K., Tiwari, S.B., Singh, L.B. and Homode, J. 1973. Efficiency of milk production and its conversion into lamb weight in Malpura, Chokla and crossbred ewes. *Indian Veterinary Journal* 50, 1199-1204.

Sirohi, Smita and Rawat, P.S. 2000. Resource use efficiency in sheep farming: A case study of district Tonk, Rajasthan. *Indian Journal of Small Ruminants* 6, 42-47.

Stephenson, R.G.A. and Bird, A. 1992.Response to protein plus energy supplementations of pregnant ewes eating mature grass diets. *Australian Journal of Experimental Agriculture Animal Husbandry* 32, 157-162.

Sultan, J.I. and Loerch, S.C. 1992. Effect of protein and energy supplementation of wheat straw based diets on side of nutrient digestion and nitrogen metabolism of lamb. *Journal of Animal Science* 70, 2228-2234.

Suresh, A., Gupta, D.C., Mann, J.S. and Singh, V.K. 2007. Sheep production in semi-arid zones- management and economics. Central Sheep and Wool Research Institute, Avikanagar (Rajasthan), India.

Taneja, V.K. and Birthal, P.S. 2005. Small holder dairying in India: Experiences and development prospects-A review. *Indian Journal of Animal Sciences* 75, 1020-1026.

Tripathi, M.K., Karim, S.A., Chaturvedi, O.H and Verma, D.L. 2007. Effect of different liquid cultures of live yeast on performance, ruminal fermentation and microbial protein synthesis in lambs. *Journal of Animal Physiology and Animal Nutrition*, in press.

□□□

37

Socioeconomic Status of Sheep Farmers and Marketing Structure for Sheep Products in India with Particular Emphasis on Semi-arid Regions of Rajasthan

SURESH. A., D.C. GUPTA, J.S. MANN
Central Sheep & Wool Research Institute, Avikanagar - 304 501, Rajasthan

Introduction

Livestock sector assumes key role in providing livelihood to the rural poor in arid and semiarid regions of India. These regions are prone to prolonged water scarcity and resultant crop failure. During drought periods-towards which these regions are prone-the crops generally fail and the livestock sector is a pillar, which support the livelihood of the farmers. During the severe drought of 2002, crop failure was almost 80 percent in *kharif* season and 60 percent in *rabi* season in the arid parts of Rajasthan (Narain and Kar, 2005). Sheep is one of the major sources of livelihood generation in these regions of India as sheep can adapt with the harsh environment and thrive on the sparse vegetation. Besides their contribution to household nutritional security in terms of milk and meat, they act as a source of livelihood and as an asset of high liquidity. During drought sheep are the animals liquidated first compared to large ruminants or other assets of fixed nature.

India ranks sixth in world sheep population. Rajasthan was the state holding highest sheep population till recently. As per 2003 livestock census, Rajasthan had 10.0 million sheep accounting for 16 per cent of total sheep population of the country (GoI, 2003). The state produced nearly 20 million kg of wool as on 2006 accounting for 40 per cent of total wool production. Most of the Indian sheep breeds are of dual purpose i.e. for mutton and carpet wool. Chokla and Nali found in Rajasthan are the best carpet wool producing sheep breeds. The wool produced in the state is highly preferred for carpet manufacturing. From the farmers' point of view, mutton is the most important product of sheep. The domestic demand for non-vegetarian food in general and that for mutton in particular is highly income elastic. The demand for non-vegetarian food products is on increase and it will be so owing to high per capita income growth, urbanisation and changes in the taste and preference of the consumers (Birthal and Rao, 2004). As of 1999-2000 the total meat consumption in India was to the tune of 3.1 million tonnes and it would be to the tune of 8.0-9.0 million tones by 2020, of which the contribution of mutton would be substantial (Birthal and Taneja, 2006). Further, the small ruminant meat from India is highly preferred in the export market due to lean meat and organic nature of production.

Despite this fabulous contribution towards the economy, sheep sector rests upon very precarious survival base due to shrinking grazing area in terms uncultivable land, fallows and other grazing land and the land under forest. The sheep farmers are generally of poor economic background, socially backward, migratory and left out of decision-making process. The socioeconomic status of the farmers can be improved through technological intervention and measures to augment technical efficiency. The Central Sheep and Wool Research Institute (CSWRI), Avikanagar undertakes applied and basic research on sheep management as well as tests and disseminates its technology through Transfer of Technology (TOT) Programme. Understanding the socioeconomic background of the farmers, their sheep management practices and technology adoption status are important to formulate strategies on research priorities and transfer of the technologies to the farmers. The post production management of sheep products-wool and mutton- is equally important in ensuring better income to the farmers. Marketing is the most important post-production activity. But the markets for wool and mutton are unorganized and the trade practices mostly go against the interest of farmers. With this background the present paper discusses the socio-economic background of sheep farmers, farming practices followed by them, the cost and return in sheep farming and marketing of wool and live sheep with particular emphasis on semi-arid Rajasthan.

Methodology

The data used for the paper is culled out from the study on socio-economic status of sheep farmers and economics of sheep farming conducted as a study for the Transfer of Technology (TOT) programme of Central Sheep and Wool Research Institute (CSWRI), Avikanagar. The data was collected from 107 sheep farmers belonging to 20 villages of Malpura and Toda Rai Singh Tahsils of Tonk district of Rajasthan during March, 2005. The villages were selected purposively as they were identified to be covered under TOT programme. Simple statistical methods were used in estimation.

Findings and Discussion

Socio-economic status

The farmers generally maintain sheep in flock size less than 100 sheep. Of the total respondents 27 per cent had small sheep. The average age of the farmers varied between 43 and 47 years. Agriculture is the main occupation for 51 percent sheep farmers followed by animal husbandry (48 per cent). This is also supported by relatively higher size of operational holding. On an average the farmers possessed an operational holding of 5 ha varying from 2.58 ha in case of the small category sheep farmers to 6.97 ha in case of large category sheep farmers. Nearly 37 per cent farmers had operational holding above 4 ha. Joint family system is the dominant settlement (53 per cent) and it has more members so that they could manage larger flocks. Joint management of flocks was also possible in the joint family systems. Among the large flocks 72 per cent belonged to the joint family set up, where as it was only 38 per cent in case of small flocks. The average family size was 9.54 varying between 6.9 in case of small flock farmers to 12.7 in case of large sheep farmers. Illiteracy is a main hindrance in the social development of the sheep farmers. Only 37 per cent of farmers were literate, that too a very minimal education. The case of the female literacy was much worse, accounting for less than 10 per cent.

Livestock holding

The sheep farmers possessed other livestock like goat and bovines (Table 2). The overall livestock possession expressed in the form of ACU is 19.3 ranging from 9.7 in case of the small sheep farmers to 31.2 in case of large sheep farmers. The share of the small ruminants in the total ACU was 67 per cent. The average flock size per small, medium and large category sheep farmers were 24, 43 and 91 respectively with overall average flock size of 54. The stocking rate indicated that the ACU per ha of operational holding was coming down consistently from 16.0 in case of marginal category to 2.2 in case of large farms giving an average number of ACU of 7.2 per ha.

The popular belief that the sheep rearing is the job of very poor is gradually changing in the rural areas. Nearly 64 per cent semi-medium, medium and large farmers accounted for 70 per cent of the sheep where as 36 per cent landless, marginal and small farmers possessed only 30 per cent of the total sheep population. However, sheep rearing is a major livelihood for the resource poor farmers. Such result has been reported by some other studies also (Pasha, 1991and Birthal and Ravishankar, 1999).

Table 1. Socioeconomic status of sheep farmers in semi-arid region of Rajasthan

S. No	Particulars	Small (<31)	Medium (31-60)	Large (>60)	Overall
1.	No of farmers	29	42	36	107
2.	Average age of the farmer (Years)	44.4	42.8	46.5	44.5
3.	Size of operational holding (OH) (ha)	2.58	4.98	6.97	5.00
4	Percentage of farmers based on OH				
	a. Landless	6.9	2.4	5.6	4.7
	b. Marginal	17.2	14.3	0	10.3
	c. Small	24.1	21.4	19.4	21.5
	d. Semi-Medium	31.2	26.2	22.2	26.2
	e. Medium	17.2	19.0	27.8	21.5
	f. Large	3.4	16.7	25	15.8
5.	Family Details (no)				
	a. Males	3.33	4.48	6.42	4.81
	b. Females	3.57	4.19	6.27	4.73
	c. Total	6.90	8.67	12.69	9.54
6.	Joint family (per cent to the total)	37.9	47.6	72.2	53.3
7.	Literacy status of the farmer (per cent)	20.7	54.8	30.6	37.4

Table 2. Distribution of livestock across flock categories

Livestock	Small	Medium	Large	Overall
Sheep	24	43.4	91.4	54.3
Goat	6.7	4.8	8.9	6.7
Cattle	1.4	2.3	3.8	2.6
Buffalo	1.6	2.8	5.5	3.4
Total Livestock	33.7	53.3	109.5	66.9
Adult Cattle Unit (ACU	9.7	15.7	31.2	19.3

The traditional sheep breed of the locality is *Malpura*. But, of late, there is a tendency to shift towards *Kheri* breed (a non-descript breed which is a cross between *Malpura* and *Marwari* breeds). The *Kheri* breeds are better in body weight and walking capacity. The latter attribute is becoming very important in the selection of breed as the rainfall in the locality is getting scantier and a persistent drought condition prevails. Due to this the availability of fodder is becoming a major issue and farmers are traveling greater distance in search of greener pastures. In our survey, almost three fourth of farmers were found to possess *kheri* breed compared to the *Malpura* breed.

The sheep, goat, cattle and buffalo accounted for 81, 10, 4 and 5 % of the total livestock in the study area (Table 3). The sheep and goat accounted for 90 and 8 % of the total livestock of landless category and 88 and 10 % of that of marginal farmers. The share of sheep in the total livestock composition generally reduced as the land size increased. The trend was opposite in case of cattle and buffalo. Buffalo accounted for 1 % in case of the landless farmers, but increased to 9 % in case of large farmers.

The data given in Table 4 shows the share of different land size groups in total livestock. The large farmers who accounted for 46 per cent of the land also possessed highest share of livestock, especially buffalo and cattle. However, farmers falling in landless to semi-medium categories hold greater share in livestock than in land indicating the importance of livestock in the livelihood of small farmers. This also highlight the more equitable distribution of livestock compared to land.

Table 3. Composition of livestock owned by farmer categories

Category	Sheep (%)	Goat (%)	Cattle (%)	Buffalo (%)	Total livestock (no)
Landless	90.30	7.80	0.92	0.92	218
Marginal	88.10	10.05	0.70	1.17	428
Small	86.01	7.94	2.93	3.22	1400
S-Medium	78.42	14.21	3.68	3.68	1710
Medium	81.37	8.51	4.85	5.26	1691
Large	76.88	9.26	4.80	9.14	1708
Overall	81.28	10.02	3.82	5.04	7155

Table 4. Distribution of livestock across farmer category

Particulars	Land less	Marginal	Small	S Med	Med	Large	Total
Sheep	3.41	6.52	20.81	23.20	23.80	22.70	100
Goat	2.37	6.00	15.51	33.95	20.12	22.07	100
Cattle	0.73	1.10	15.02	23.08	30.04	30.04	100
Buffalo	0.56	1.39	12.50	17.50	24.72	43.3	100
All livestock	3.05	5.99	19.60	23.94	23.67	23.91	100
Land	0	1.21	5.93	14.71	31.96	46.16	100

Feeding Management

The various sources of green fodder during different seasons were the ordinary pasture, harvested cropland, fodder tree and *babool* pods. The major source is the ordinary pasture in all the three seasons of winter, summer and rainy. These lands are under the common property regimes and as such the productivity is low. The availability of pasturelands in the state is becoming scarce due to conversion of common lands for various other purposes. This is increasingly becoming a major limiting factor for sheep rearing. Further, the stocking rate of the common pastures far exceeds the carrying capacity. This negatively affects the regeneration capacity as well as productivity of the pastures. The traditional institutional mechanisms to manage the pasture productivity are gradually disappearing in the villages (Jodha, 1985). It needs deliberate attempt from the stakeholders like Government, NGOs and the farmers to regulate the access to the pastureland and to keep the pasture productivity at optimal level. The productivity of the pastureland is dwindling due to the erratic rainfall also. This compels the farmers to travel long distances in search of fodder. These lands have to withstand high grazing pressure from the large ruminants also. Poor conservation efforts and high stocking rate deplete fodder productivity and quality. Therefore the sheep farmers have to depend on other sources particularly during summer season. Nearly 60 per cent farmers depend on fodder trees (*ardu, neem* etc) during summer season. During the summer season the farmers even hire fodder trees from other farmers.

Due to scarcity of fodder, sheep farmers in this locality resorted to temporary migration along with the flock. Sheep migration is generally considered as a coping mechanism to mitigate fodder shortage as well as an intermediate stage between settled life and nomadic life style. Nearly one-fifth sheep farmers (22 %) resorted to temporary migration to meet fodder requirement. This includes 7 % from the small category, 17 % from the medium category and 39 % from the large category. The migration was undertaken mostly by the family members, and some times with hired labour also. Generally, migration starts at the end of winter season and returns back at the onset of monsoon. The place of migration varies from neighbouring districts to other states (mostly Uttar Pradesh in case of sheep farmers of Tonk district of Rajasthan). It is observed that the larger flocks are more probable to migrate compared to smaller flocks. Similarly, the farmers who were better adopters of technology were found to migrate more than the low adopters.

Veterinary status

Disease is a major constraint faced by sheep farmers. The high rate of morbidity and mortality of the sheep significantly reduces the net returns mainly due to the increase in cost of treatment and physical loss of body weight. The data regarding the morbidity and mortality are given in Table 5.While interpretation, leverage may be given to the data inadequacy due to the one time data collection and the accuracy of farmers identifying the disease. Nevertheless, the data helps as an indicator of the disease profile of sheep. The morbidity was nearly 87%. Due to the overlapping of data of the animals reported under the different diseases, it is not possible to spell out exactly how many sheep were affected. Among the reported cases, FMD and ET accounted for 39 and 18 % cases. The overall mortality was 14 %, the highest being attributed to ET and FMD (4.9 and 4.7 % respectively). Farmers reported that nearly 86 % of the diseased animals are treated in one way or other. Dependence of the farmers on various medical sources is indicated in Figure 1. Nearly 41% farmers depended on Government veterinary clinic for the treatment of sheep. Treatment by self (34 %) is also widely prevalent in the locality. The farmers also availed treatment from private veterinary practitioners. But the charges were exorbitant and the qualified persons were also not available. The average cost of treatment for flock was Rs

1004, varying from Rs. 558 for small, Rs. 876 for the medium and Rs 1512 for the large flocks respectively.

Table 5. Disease wise morbidity and mortality in sheep flocks (%)

Disease	Morbidity	Mortality
Sheep pox	5.11	0.79
Enterotoxaemia	18.16	4.94
Pneumonia	7.59	0.43
Foot and Mouth Disease	38.90	4.75
Parasites	11.81	1.86
Others	5.49	1.09
Total	87.06	13.86

Labour utilisation

The overall average labour employed was 581 man days per annum. This comprises of 80% male labour, 12 % female labour and 8 % child labour all calculated in terms of mandays. The labour utilisation in small, medium and large flocks was 549, 582 and 604 mandays respectively. The role of the female labour was higher in small flock group, who generally had low family size. The contribution of the child labour was the highest in terms of absolute and the percentage share for the large category farmers. Around 99% of the labour required is absorbed from the family itself. Only an insignificant part (4.9 mandays) of the total labour requirement is contributed by hired labour, that too male labour (except few cases of child labour). The overall imputed value of the labour cost was Rs 28,780 per annum. The labour utilisation per flock per annum was Rs. 29123 giving rise to per animal labour cost of Rs 539 per year. The imputed value of labour for family and hired labour, all calculated at the prevailing wage rate by the small, medium and large flock was of Rs 27494, Rs. 29162 and Rs 30377 respectively.

The major labour absorbing activities were feeding, watering, shelter management, milking, grazing, health management, marketing, breeding and shearing activities. Of the 581 mandays employed in sheep husbandry, three-fourth was used for grazing. The role of men is generally high in grazing the livestock. Female labour is mostly engaged in shelter management and managing the sheep in the vicinity of the household.

Cost and returns

The economics of sheep farming was worked out and is given in Table 6. Only the variable costs were considered for the analysis, as fixed cost is heritable in nature from year to year. The imputed value of family labour also was not included in the analysis. The major costs involved in sheep rearing were feed and fodder cost, veterinary care, hired labour and interest on variable cost. The overall average variable cost was Rs. 3520 of which the feed and fodder cost accounted for 56 per cent followed by medicine and veterinary care (29%). The percentage share of expenditure for the different sheep classes was similar. The major return items were animal sale, milk of the sheep, wool and manure together giving a return of Rs. 28252 for a flock of 54. The percentage shares of these heads were 81, 3, 9 and 7 respectively. The return over variable cost (net return/ profit) was Rs. 24732 giving per animal return of Rs. 458 per annum. The net returns for the small, medium and large sheep flocks were 10280, 19087 and 42961 respectively with the corresponding per animal net return of Rs. 428, 440 and 470 respectively.

Table 6. Cost and Return in Sheep Production

Item	Category of farmers			Overall n=54
	Small n=24	Medium n=43	Large n=91	
Expenditure				
a. Fodder and feed (%)	62.50	58.57	52.69	56.08
b. Medicine (%)	26.30	32.03	27.18	28.52
c. Hired labour (%)	5.48	3.73	14.48	9.74
d. Interest (%)	5.63	5.67	5.66	5.65
e. Total variable cost (Rs.)	2115	2735	5567	3520
Returns				
a. Animal sale (%)	77.35	82.31	81.69	81.36
b. Milk (%)	4.87	2.63	3.26	3.26
c. Wool (%)	11.00	8.10	8.46	8.65
d. Manure (%)	6.87	6.96	6.59	6.72
e. Total Returns (Rs.)	12395	21822	48528	28252
Return over variable cost (ROVC) (Rs.)	10280	19087	42961	24732
ROVC per animal (Rs.)	428	440	470	456

The overall average income from the livestock was Rs 61349 of which the sheep accounted for 46 %, followed by buffalo (32%), cattle (12%) and goat (10%) (Table 7). The income from livestock increased with the increase in land holding. For the landless sheep farmers, major source of income was sheep (88%) followed by goat (12 %). They were earning more income from the livestock as compared to the marginal farmers. The share of the sheep in total income from livestock came down from 88 % in case of the landless farmers to 35% in case of the large farmers and the opposite is true in case of contribution of buffaloes and cattle whose share increased as the land category increased. The goat did not show any significant trend.

The contribution of various livestock products towards the total income from livestock is given in Figure 2. We can clearly observe that the share of live animal sale in the total income generated declined consistently as the land size increased. This was accompanied by corresponding increase of the share of milk. The major income sources were milk, wool, dung and sale of live animals. Of the average income of Rs. 61349, milk accounted for 48 %, animal sale for 44%, wool for 4 % and dung for 3.4 %. The share of animal sale was 65 % in case of landless farmers and it reduced to 34 % in case of the large farmers. Correspondingly the share of milk increased from 23 % in case of the landless farmers to 59 % in case of large farmers. This is mainly due to the contribution of the buffaloes in the income generation. The data shows only the gross return. The net return is not considered as the cost of rearing other livestock was not considered in the study. Wool accounted for 6 % in case of the marginal farmers and its share was merely 4 % in case of large farmers.

Table 7. Distribution of income generated from livestock group across the categories of farmers) (Rs)

Category	Sources of Income				
	Goat	Sheep	Cattle	Buffalo	Total
Landless	11.55	88.45	Neg	Neg	28483
Marginal	14.23	69.58	Neg.	16.19	25934
Small	9.88	59.38	11.43	19.31	46633
Se-Medium	11.50	44.82	13.74	29.94	53770
Medium	7.14	43.27	14.27	35.32	70465
Large	8.95	35.38	13.13	42.54	113990
Overall	9.53	46,05	12.42	32.00	61349

Marketing of Sheep Products

Sheep is characterized by multiple products with joint nature of production. The substitution of the two products varies across breeds. Therefore the selection of the breeds also depends on the ratio of substitution of these products along with the absolute yield. As far as the farmers' income portfolio is concerned, as already discussed, farmers derive nearly 80 per cent of their income from sale of live sheep for slaughter purpose. Though the importance of wool is not so pronounced as that of live animal sale as far as the income portfolio of the farmers is concerned, its importance is significant as far as the wool utilization in India, is concerned. The wool produces from India, particularly carpets, are of high economic demand in the international markets. Further the wool processing sector provides direct employment to more than 4 lakh people. The export of carpets earned more than Rs. 2500 crores as on 2005. However, the profits earned at the higher level is not transmitted to the lower section of the value chain particularly to the wool producers. The reduction in custom duty of the wool in India during the post liberalization phase led to sudden import surge from other wool producing countries like New Zealand and China. The primary market structure of the wool is important as far as the income of the farmers is concerned. In this section, we examine the market and marketing structure of wool and live sheep in Rajasthan.

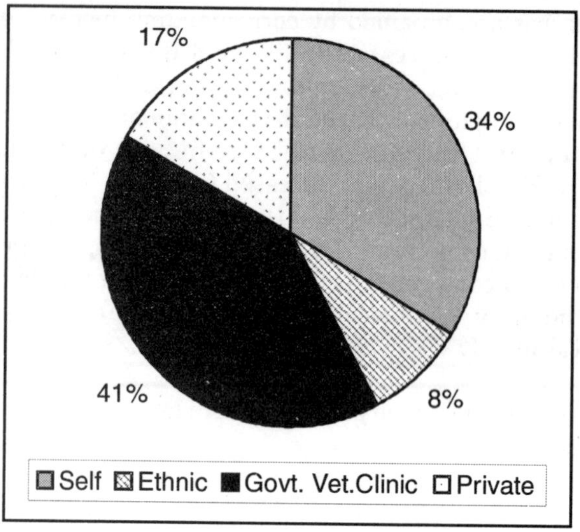

Fig. 1. Source of veterinary service

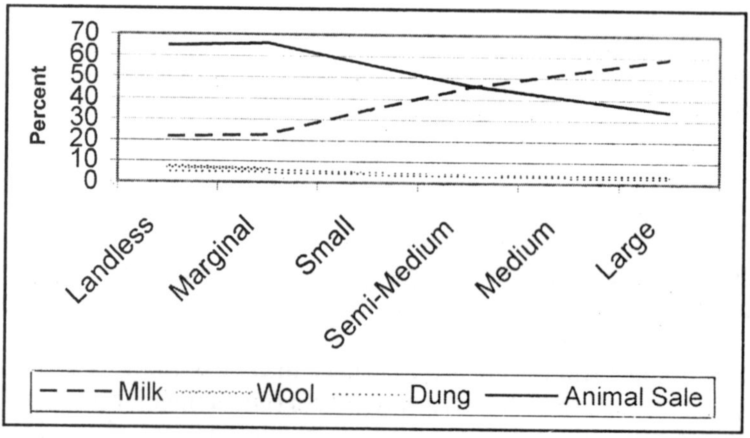

Fig. 2. Contribution of livestock products to income from livestock origin

Wool marketing in Rajasthan

The wool availability in Rajasthan is seasonal coinciding with the shearing season of sheep. Usually sheep are sheared three times a year, during March, July and November. Nearly 80 percent of wool is produced during March and November seasons. The marketing of raw wool in Rajasthan is embedded in multiple exchange relations, that the market continues to be old relying on traditional relations which still persists among wool producers in many parts of the state. The process of price formation in case of wool goes beyond the operation of demand and supply factors and is ruled by the contracts, tradition and age-old institutions (Ray, 1997).

The major wool markets are Bikaner market (Bikaner district), Beawer and Kekri market (in Ajmer district), Jodhpur (Jodhpur district), Shigganj (Pali district) Lunkaaransar (Ganganagar district), Sikar (Sikar district) and Barmer (Barmer district). Bikaner, Beawer, Jodhpur and Kekri markets together accounts for more than 90 per cent of total wool traded in the state. These markets handle even wool produced in other states. The wool markets in Rajasthan are integrated both vertically and horizontally. Many traders in the markets operate yarn manufacturing units and/or even small-scale industries of carpet manufacturing.

Import of wool to India

The raw wool production in India is not sufficient to meet the domestic demand, the least for apparel purpose. However, with the liberalising of the economy and with reduction in the import duty for the wool above 32 micron, import of wool to India increased phenomenally. The major sources of wool import are Australia (mainly for apparel purpose) and New Zealand and China (mainly for carpet purpose). Globally there is reduction in price of wool due to slump in the major apparel and carpet consuming economies. Import of wool along with a host of other conditions is keeping the price of domestic wool at low level. Even though some of the wool types produced in India are the best of its kind in the world for carpet production (like Bikaner Chokla of Rajasthan), there is bias against Indian wool over the imported carpet wool from New Zealand even by the domestic buyers on the pretext of low yield from Indian wool. The wool brought to the market by our farmers is generally not graded and cleaned. It contains high proportion of extraneous matter and impurities. The price of wool is determined by these factors also. Other important parameters affecting the wool price are staple length, lustre and colour. Simple practices like washing the wool (scouring) and grading can improve the price received by the farmers. The carpet manufacturers use mix of imported New Zealand wool and Indian wool to produce carpet as the Indian wool could impart sufficient resilience to the imported wool. However, realising better price for the domestic wool depends on the bargaining strength of the local farmers or farmers' organisation. However, marketing of wool by collectives are conspicuous by its absence.

Market channels and Marketing cost

Wool marketing channels in Rajasthan are highly unorganised and traditional. Those who directly sell the wool to the markets constitute only a minority of the farmers who are mainly located nearby these markets. The wool produced by the farmers in the remote areas of the villages is sold to the village merchants (middlemen) who are traditionally operating in the village. In some cases, the agents of some of the wool traders visit the potential wool suppliers and collect the wool directly from them, but their proportion remains smaller compared to the middlemen who are linked with the commission agents operating in the market.

The linkages between the middlemen and the farmers determine the price spread of raw wool. The farmers avail credit from the middlemen (many times at high interest rate than prevailing in the market) and the farmers are in turn compelled to sell the wool to the middlemen even at lower price. Thus a kind of coercion also exists in the market for selling the wool. Often a kind of monopolistic element also exists in the price prevailing in the market mainly due to the mercantile power overcoming the competitive market forces. In some of the markets even the wool trade is undertaken by a few families who undertake the processing aspects also. The existing marketing practices like under cover method of price fixation, standard deduction and weighing methods betray the spirit of competitive price formation against the interests of the farmers (Ray, 1997).

For further clarity of the price changes, we have analysed the micro-level data from *Kekri* wool market in Rajasthan for the two periods. The data was available for the period of 1990-91 to 1993-94 and for the period of 2000-01 to 2005-06 (Table 8). The data clearly indicate decline in the wool price in recent times, both in terms of the nominal price and real price. As of 2005-06, the nominal price of wool was Rs 4764 per quintal comparable to the level of the price in 1992-93. In real terms, the price declined to Rs. 981/ q from Rs 1823/q in 1993-94.

The decline in price of wool has resulted in farmers gradually getting disinterested in proper post shearing management of wool. Often, the low price should be attributed not only to the quantum jump in import, but also to the flawed market practices. Given the fact that the Indian wool is needed in some proportion to mix with imported wool for imparting resilience for carpet grade yarn, it is imperative that this sector cannot be discarded.

Marketing of live sheep

Live sheep marketing is one of the major sources of income to the farmers. However, the market is not organized and the farmers become prey of the oligopolistic buyers who are limited in nature. The marketing channels don't vary much among the small ruminants.

The collection of the animals and assembling them for sale is an age old practice. The market structure, conduct and performance of livestock marketing and trade depend largely on these traditions and customs which are followed for generations. The sheep are generally reared in large numbers compared to goats. The marketed surplus of the sheep consists mostly of the young lambs of the age group of 6-8 months along with the culled animals. However, there is a mismatch between the marketed surplus and marketable surplus. This mismatch depends on the push and pull factors arising due to variations in market, environment and the domestic/household financial urgency. For example, during times of urgent financial liquidity like marriage or death or other functions, the immediate asset of liquidation is the sheep. During such time even the productive stock may be marketed. Similar is in case of drought. At these times the availability of fodder and feed are limited and the farmers are compelled to leave the animal for whatever money he is getting rather inviting financial loses by keeping them in poor nutrition condition. Higher market price is observed during some particular seasons of the year like Ramdan. Thus we can observe seasonality in the supply of animal as well as the price of the live animal.

Marketing channels

In our discussion, we are not concerned with the sale and purchase of the rams/ ewes between two households to add to the existing flock as they are not major sheep marketing activity. But, it may be remembered that the price fetched in such transactions are higher than that observed during trading for slaughtering as it carries an element of productive value of the animal.

The prevalence of the marketing channels depends on the location of the villages. The main factors involved in the location aspects are the prevalence of sheep farming in the locality determined by the availability of common pastures and the distance from the major consumer market. As the cities are the main consumer centres for mutton, involvement of the *dalal* and *mandi* is a major peculiarity of the live animal sale. Since Rajasthan is dominantly vegetarian in diet habit, the live animals are taken from the state to other markets, mainly to Delhi.

The most important live sheep marketing channels are

1. Breeder- Butcher-Consumer
2. Breeder- Primary Trader-Bakra Mandi Dalal- Meat Shop Keeper-Consumer

In the first marketing channel, the butchers visit the villages on a regular basis and purchase the animal from the prospective sellers. Here intensity of the visit of the butcher depends on the seasonality. The price of the animal is determined by the visual observation of the animal over and above the bargaining power of the farmer. However, the number of animals purchased by the butcher for daily slaughter is less as it is circumvented by the daily off take of mutton from the shop. Mostly the traders or butchers are having a caste basis. Only *Khatiks* or *kasais* participate in the live animal sale or trade as it is the traditional occupation of these communities (Rathore, 1993). The popularity of a particular channel probably depends on the services the channel provide, the demand in the major consuming centres and the profit margins each agent in the channel accrue. Here the commission agents purchase the animals and store them in the premises of their households. He has to bear the cost of keeping them healthy.

Bakra mandies

The *bakra* mandi is a goat and sheep trading centre, where goat and sheep are assembled at a single geographical location in large number to be sold on a private treaty basis and which possess facilities for receiving, caring, handling and selling the animal (Rathore, 1993). These can be called secondary animal markets. A *bakra* mandi plays an important role in mobilising the sheep from the rural areas to supplying it to the consumers. The size, operations and performance of the mandies vary across *mandi* to *mandi*. Generally auction takes places on fixed days and on the remaining days transaction takes place in terms of transporting or sending the animals to other markets. In each *mandi* the number of functionaries varies and the charge for the functionaries also varies.

Marketing margin and Price spread

Marketing Margin and price spread are considered as a measure of marketing efficiency. Marketing margin is defined as the difference between what the consumer pays for a product and what the farmer receives. Price spread is the difference between the retail price per unit and the farm value of an equivalent amount of product sold by the farmer. The farmers share is widely considered as a measure of the fairness of farm price and the efficiency of product marketing. The price spread and the marketing margin varies according to the marketing channel and the number and type of functions involved. Calculation of marketing channel in case of animal marketing is difficult because of various problems like (1) the traders purchase the animal on individual price basis, but are sold in the secondary market of mandies on a group basis (2)mixing of various age and qualities of animals including females which are not highly priced (3) mixing of the meat of male and female animals at the local meat shop level (4) presence of other economically important parts like skin, hide, hoof, blood, intestine etc which vary widely.

Price of live animal

The price of the animal depends mainly on the age, sex, body weight and health of the animal, of which the body weight of the animal carries important role. The male sheep always fetch superior price compared to the female one as the female sheep are slaughtered only when they are culled from the productive stock or debilitated. The market price of sheep per body weight is observed to be higher during the age group of 5-8 months that is upto a period before achieving sexual maturity. The price of sheep at farmer level is determined by prevailing price of mutton at the major consuming centres. The price at the village level is more of a sellers market (in normal circumstances) as the cost of rearing the animal till getting better price (say a few weeks) is less to the farmers as the sheep is reared based on a social cost and the farmers has to spend only some labour for keeping it. Therefore, it won't cost much to keep the animal at his possession till he gets better prices, unless he is not in financial urgency.

Slaughter houses

The slaughtering of animals takes place either in the registered *mandies* (registered with the local bodies, which requires furnishing of some documents for minimum required facilities) or in unregistered markets. The figures shown for the amount slaughtered in the registered *mandies* is far less compared to the total meat consumption indicating widespread prevalence of unregistered slaughter houses. The condition of the slaughter house is far less than the recommended one as they does not keep the required level of hygiene and safety standards. The facilities for cleaning, processing, preservation and storing of meat are not well developed in the state. In the health point of view of the consumers, it is urgently needed to improve the hygienic status of the slaughter houses.

Table 8. Nominal and real price of raw wool *kekri* mandi

Year	Nominal Price	Real price (1982-83 constant)
1990-91	3500	1915.70
1991-92	3233	1555..8
1992-93	4727	2066.9
1993-94	4517	1822.8
1999-00	4260	1182.3
2000-01	4173	1081.6
2001-02	4833	1209.2
2002-03	4838	1170.5
2003-04	5884	1349.9
2004-05	6158	1326.8
2005-06	4764	980.9

Conclusion and Recommendations

The sheep farming is an important occupation of weaker section of the society in the semi-arid regions of Rajasthan. Sheep farming is unorganized and is extensive in nature and it depends on the fodder availability in the pasture lands. The productivity of the sheep depends to a great extent on the production and productivity of the pasture lands. Therefore, it is imperative that the productivity of the pastureland is to be improved sustain sheep production. Next area of concern is availability and accessibility to veterinary facilities in the villages where the farmers still depend on the self treatment. Marketing of wool and the live sheep is unorganized. The farmers are at the receiving end, particularly so in case of wool. Innovations in the institutional set-up, particularly introduction of breeders association will work a great way in solving these problems.

REFERENCES

Birthal, P.S and Rao, P. 2004. Intensification of livestock production in India: pattern, trends and determinants, *Indian Journal of Agricultural Economics* 59, 555-565

Birthal, P.S. and Ravishankar,A. (1999). Sources of growth in the livestock sector of Karnataka. In (Eds.). Birthal, P.S., Anjani Kumar, A.Ravishankar and U.K.Pandey. Sources of Growth in the Livestock Sector, Policy Paper 9. NCAP, New Delhi

Birthal, P.S. and.Taneja. V.K. 2006, "Livestock Sector in India: Opportunities and Challenges for Small Holders" Paper presented in the International Workshop on Smallholder Livestock Production in India: Opportunities and Challenges, Indian Council of Agricultural Research, International Livestock Research

Institute and National Centre for Agricultural Economics and Policy Research. 31 January- 1 Feb, New Delhi

Government of India, 2003. Livestock Census.

Jodha, N.S. 1985a. Population growth and the decline of common property resource in Rajasthan, India, *Population Development Review*, 11(2)

Narain, Prathap and Amal Kar (Ed.) 2005a. Drought in Western Rajasthan. Impact, Coping Mechanism and Management Strategies. Central Arid Zone Research Institute, Jodhpur

Pasha, S.A. (1991) Sustainability and viability of small and large farmers: animal husbandry and common property resources. *Economic and Political Weekly* 26, A 27-A30.

Rathore, M.S. 1993. Marketing of Goats in Rajasthan. Institute of Development Studies, Rajasthan

Ray, Sunil. 1997. Natural Resources, Organization and Technology Linkages. The case of wool based Industies. Rawat Publications, Jaipur.

❑❑❑

38

Rural Employment and Business Opportunities in Value Chain of Small Ruminant Production: National and International Trend

MAYA KANT AWASTHI

Indian Institute of Management, Lucknow, Uttar Pradesh-226013

Introduction

The poverty is a serious problem throughout the world. Though both rural and urban areas are suffering with the problem of poverty but across the geographic regions intensity of poverty is more in rural areas. In rural areas mainly small, marginal and landless farmers along with fringe social groups such as nomadic and tribal groups are major poverty ridden groups. As majority of them do not have sufficient land resource, majority of them rear some types of animals primarily to augment and support their family income. These animals are critical in livelihood support system and food security of these groups and very important source of animal proteins for farm family members. These animals also ensure year round flow of cash income to resource poor farmers. Like any developing country in India, about 75% of farmers are small and marginal farmers. Small and marginal farmers, own 56% of the large ruminants and 62% of the small ruminants (World Bank 1996). Interestingly same livestock ownership pattern we can observe in other part of the world too. Livestock development therefore could be very important vehicle for increasing income and employment in the rural areas where a large number of people are still trapped in poverty and malnutrition situations.

Small Ruminant Production Systems

Small ruminant animal provide very important role in livelihood support system. In fact small ruminants play a vital role in generating income for small marginal and landless, farmers. Such system also help in augmenting nutritional support for the economically weaker section of society. Small ruminant animal based farming systems have very strong common property resources and ecological component. Farmers with poor resource base who keep small ruminant animal such as goats and sheep are forced to extract all the fodder specially in the dry season from common property resources primarily through grazing. This over extraction of natural resources from already fragile and degraded CPR further accelerate the degradation process. Though in some cases these small ruminant animals also help in maintaining soil fertility and in semi arid and arid regions it is an effective means of soil nutrient replenishment.

Rising Problem of Rural Employment

Problem of unemployment is universal problem therefore providing gainful employment to small and marginal farmers who do not have necessary productive resource base have been a challenge for policy planners through out the world. This problem has serious Socio-economic and political ramifications. Solving or addressing this problem has therefore become very important policy goal of various state and central governments.

There are two approaches which are being adopted to tackle this problem of unemployment. In first approach we try to generate employment through various infrastructural/developmental schemes such as Rastrya Gramin Rojgar Yojna, Pradhan Mantri Grameen Sadak Yojna etc. Marginal and small farmers are then provided employed in these schemes. In second approach these resource poor farmers are facilitated in creation of productive resource at family and community level. Through use of these productive resources, employment gets generated which can be than utilized by the resource poor marginal and small farmers. Both these approaches of employment generation are depicted in figure-1 below.

First approach of generating employment addresses the problem from short term perspective. Through this approach though immediate employment demand can be fulfilled but it does not provides long teams solution to the problem rural unemployment. Second approach on the other hand attempt to create employment on sustained basis but the problem with this approach of employment generation is that creation of productive asset it self does not generate the employment. To generate employment from this productive asset say for example milch animal we need further investment on sustained basis on variable cost items such as cost of green fodder, concentrate and cost of veterinary services etc. for any farmer who do not have sustained cash flow and is resource poor, it is very difficult to arrange such sustained investment on various variable cost items. In such situations farmers are unable to realize full employment generation potential from these schemes. Not only this often these productive assets are created on part of loan and part of subsidy model and therefore, need repayment of loan installments at regular intervals. Any default on repayment of loan further drag them deep in red financial situation (Fig 2).

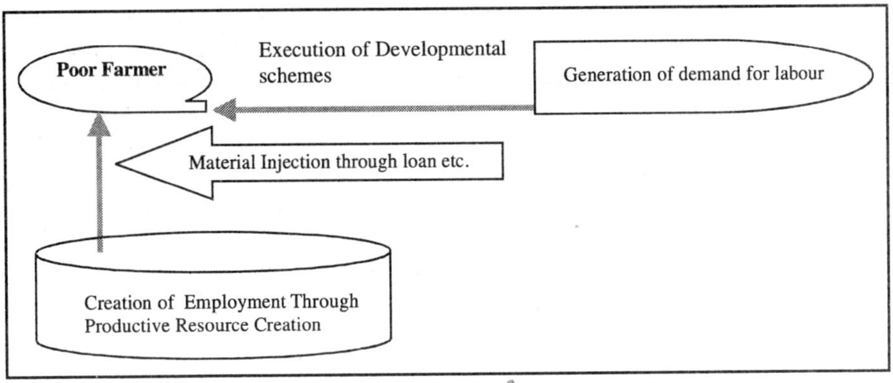

Fig. 1. Employment generation approaches

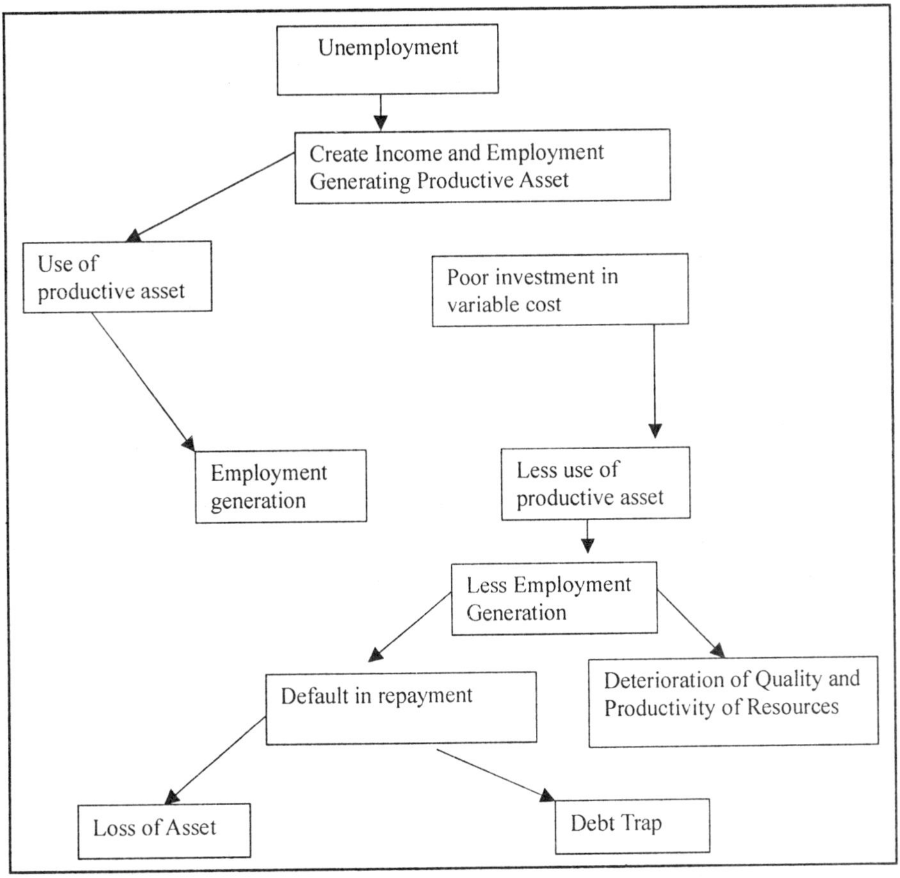

Fig. 2. Employment Generation Process

Small Ruminant Animals Can Offer Robust Solution to Unemployment Problem

In order to solve rural unemployment problem we need to identify a productive resource which is within the buying power of resource poor farmers and does not require market cost/value based resources in their process of employment / income generation. Small ruminant animals provide ideal solution to this paradoxical situation. As small ruminant animal such as sheep and goat does not require heavy investment on variable cost as they can be reared on agricultural waste which does not have much market value and can be very easily integrated in any farming system and give excellent return on investment on sustained basis. Such animals also help in providing nutritional support to farm family and soil. Because of these qualities of small ruminant animals they are becoming very popular through out the would. Table below provide a comparative analysis of small and large ruminant animals.

Value Optimization Points of Value Chain of Small and Ruminant Animals

Small ruminant animal have a very high potential value chain. This value chain has maximum linkages with the other farming systems as evident from figure 3. Each these value chain points provide excellent opportunities of value maximization. For example manure received form small ruminant animal affluent can be further processed, packed and sold in the market. Such value optimization will lead to maximization of employment creation and increase in family income. Therefore, our focus should be for tapping such value maximization / employment generation potential at each of these points of value chain. Across the globe there is increasing trend of capturing such values at all such potential points of value chain.

Table 1. Comparison of characteristics of large and small ruminant animals

Characteristics	Animal category	
	Small ruminant	Large ruminant
Initial investment	Low	High
Regular investment in variable costs	Negligible	High
Sunsitility to poor variable cost investment	Low	High
Gestation period	Low	High
Integration capacity with farming system	High	Low

Global and National Trend

Population Growth of Small Ruminant Animals

It is interesting to note that though small ruminants provide main livelihood support to resource poor farmers and is one of the most important sources of employment for them but over the period of time we observe only marginal increase in absolute population of these animals. Table 2 below demonstrates the population trend of small ruminant and large ruminant animals.

Table 2. Livestock population in India (In millions)

Year	Large Ruminant		Small Ruminant	
	Cattle	Buffaloes	Sheep	Goat
1982	192.5	69.8	48.8	95.3
1992	204.6	84.2	50.8	115.3
1997	198.9	89.9	57.5	122.7
2003	185.2	97.9	61.5	124.4

Source: Jabir Ali, 2007

A close look of population growth pattern of small ruminant animal on the other hand present even more bleak picture (Table 3). The population growth pattern of small ruminants indicates that population growth of small ruminant animals such as goat population growth is slipping constantly since 1982. However, decline have been more steep from 1997 onward when it came down from 1.3 percent to as low as 0.2 percent. During his period, sheep population has declined from 2.5% to 1.2%. When we compare these population growth figures with the growth figures of large ruminant such as buffalo, comparison becomes even more revealing. During this period though cattle population growth continued to decline but buffalo population registered only marginal decline (1.3% to 1.2%).

Table 3. Animal growth rate of large and small ruminant animals

Period	Cattle	Buffalo	Sheep	Goat
1982-92	0.6	1.9	0.4	1.9
1992-97	-0.6	1.3	2.5	1.3
1997-2003	-1.0	1.25	1.2	0.2

Production Trend

Actual impact of large and small ruminant population on production of livestock products can be traced back from the perspective of production patterns in various livestock products (Table-4). Results presented in table 4 reveal that though in general production growth rate is declining for products of all ruminant animals but production decline is more sharp in case of small ruminant animals.

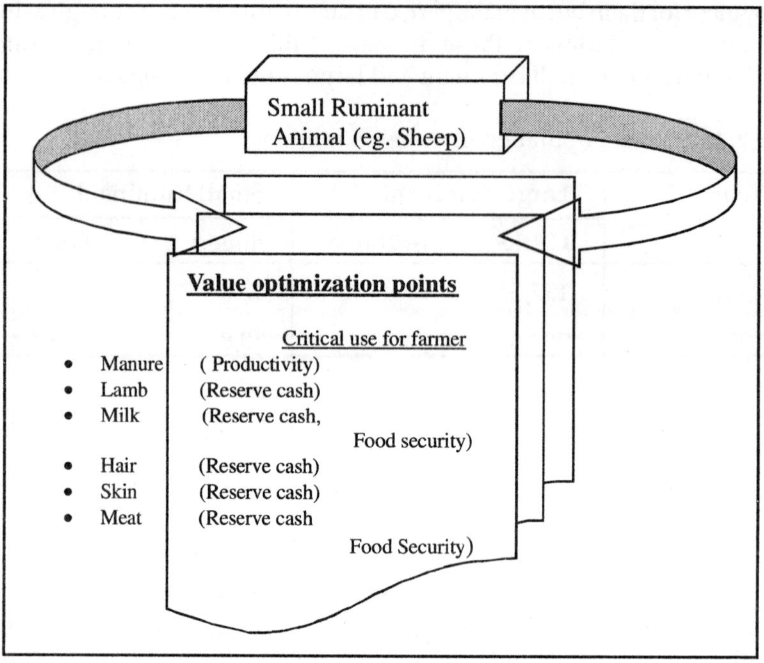

Fig 3. Value optimization points in small ruminant animal value chain

Table 4. Production of livestock products in India

Commodity	Production			Annual Growth (%)	
	1980.-82	1990.92	2001.2003	1980-90	1990.2003
Milk					
Cow	14.2	23.7	36.7	5.2	4.1
Buffalo	18.6	29.7	47.9	5.0	4.7
Goat	1.1	2.5	1.6	6.8	0.2
Meat					
Beaf & Veal	0.88	1.28	1.47	1.2	1.6
Buffalo	0.87	1.19	1.45	2.5	2.3
Goat	0.31	0.43	0.47	2.9	3.2
Mutton	0.17	0.19	0.23		
Wool	33.2	40.5	49.5	2.7	2.0

Source: Jabir Ali, 2007

Trend in Value of Livestock Production

Though in most of the developing countries including India contribution of livestock sector in gross value of agricultural sector is increasing over period of time but in majority of cases rate of growth is declining (Table 5). This is despite the fact that is general purchasing power of consumer is rising and demand and price various livestock product is also rising. This trend clearly indicates presence of problem of value creation and value capturing at various locations of value chain of livestock sector.

Table 5. Value creation in livestock sector in India.

Commodity	Value at 1993-94 prices			Annual Growth rate (%)	
	1980-82	1990-92	2001-2003	1980-90	1990-2003
Beaf and buffalo	7.70	12.5	14.6	4.82	1.45
Goat meat/ mutton	21.1	38.2	46.8	5.11	1.42

Source: Jabir Ali, 2007

Owns Small Ruminant Animals

Small ruminant animals require relatively less initial investment and need virtually negligible variable cost investment. Because of these advantages, these animals were believed to be very popular among landless, marginal and small farmers. However, a close look of data presented in table-6 below indicates a contrary picture. There are not too many landless farmers who are rearing small ruminant animals to augment their family income and engage surplus family labour. Squeeze is common property resources and degradation of CPR appears two important reasons for this besides several other socio- economic effectors influencing in this process.

Table 6. Distribution of large and small ruminant animal

Category of animal	Proportion of Households		Size of livestock flock (animals/Family)	
	Bovine	Ovine	Bovine	Ovine
Landless (<0.002 ha)	2.5	5.1	23	20
Marginal (0.002-1.0 ha)	43.8	46.2	180	81
Small (1-2 ha)	23.3	19.3	324	115
Medium (2-4 ha)	17.7	15.0	361	131
Large (<4.0 ha)	12.7	14.4	418	203

Source: NSS Report 408, 1991-92, GOI

Blue Print of Small Ruminant Animals

In whole South Asia, crop and animal production is well developed and well integrated in the farming systems. It is specially strong in case of smallholder agriculture. In this region, there are different types of interaction and integration due to a shift in some component of the system (Devendra et al., 1997). Tribal farming systems, which involve vertical or horizontal migration, are common in the arid/semi-arid regions of India and Pakistan and in some of the cool tropics of the Himalayas. In the annual crop - livestock systems animals use local community pastures and crop residues for feeding. In certain situations animals are also stall-fed and allowed to graze. Figure 3 below indicate a typical blue print of value chain of small ruminant animals.

Illustrative value chain for sheep rearing in semi arid region of Rajasthan state of India (Fig 4) really indicate the vertical and horizontal expansion of value chain of small and ruminant animals such as sheep. This value chain also explains great value creation potential at various points of value chain. However, in this process key issue will be not only to create and capture value from the value chain, rather more critical issue would be to ensure that the person who has created value should also reap that value. Created value should not be taken away by some other person.

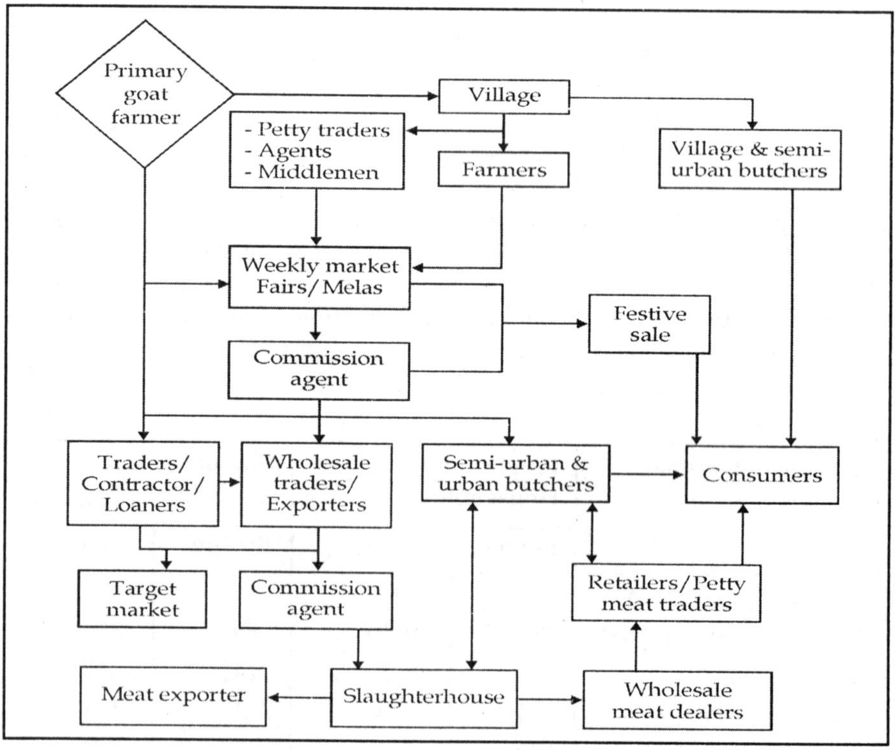

Source: Suresh A, DC Gupta and JS Mann. 2007

Fig. 4. Typical value chain of small and ruminant animals

Key Researchable Issues

To get maximum out of employment generation potential of small ruminant animals there are number of research issue, which we have to tackle. Some of these research issues have been listed below.

- Development of a new generation of vaccines and diagnostics, diagnostic technologies/ methodologies, and immunobiologicals against important diseases.
- Strengthening of monitoring and surveillance systems to develop an animal disease database, and the evolution of strategies for the control of animal diseases
- At the national level increased research on various exotic diseases to tackle their control by specialised laboratories.
- Further research on environmental pollutants and toxicants, mycotoxins and mycotoxicosis.
- Research on identification of value optimization points, value creation and value capturing in value chain of small ruminant animals.

Justification for Research

The justification for research on small ruminant based crop–animal systems is based on following factors:

- There is widespread presence of poverty and concerns about equity and food security. A very high proportion of the human population (49%) currently lives below the poverty line. Furthermore, the percentage of rural poor as a percentage of the total poor is also very high.
- In future, rising populations, higher incomes, urbanisation and changing consumer preferences will fuel an increased demand for animal products.
- In Asia, productivity of these animals in both the rained and irrigated areas needs to be increased to meet future demands for animal proteins, notably milk, beef, goat meat and mutton
- The arid/semi-arid regions are fragile environments that require protection. Environmental pollution is a problem in peri-urban and urban production systems.

Increased resource allocation for research on small ruminant based animal systems is clearly necessary in the context of the need for more food, greater equity and also environmental considerations.

Conclusion and Recommendations

In rural employment generation process small ruminants can play a vital role. Especially in livestock integrated farming systems of South Asia region, generating income for small, marginal and landless farmers through small ruminant animal has very high potential. Such small ruminant based farming system would be useful in providing proteins for the poorer sections of society too. We also observe a strong link between these systems and environmental issues. Sedentary farmers, who keep goats and sheep, are forced into migration in the dry season in a search for feed as they are dependent on common property resources for grazing. As a natural consequence of this, we have to face overgrazing and erosion induced ecological problems. During the course of these migrations, small ruminants also play an important role in the maintenance of soil fertility. Therefore, existing small ruminant farming systems need to be thoroughly studied and the constraints/opportunities needs to be identified for better integration of small ruminants with annual cropping systems in order to realize full employment and income generation potential. For this we need to assess environmental impacts of the traditional systems along with the effects of changes resulting from the introduction of new or available technologies. Finally, there is a strong linkage between value optimization and value capturing issues in value chain of small ruminant products with employment and income generation potential from such systems.

REFERENCES

Ali Jabir . 2007. Designing Vertical Coordination for Indian Meat Industry. In Bhaskaran S. and Mohanty Suchitra (Eds.), Marketing of Livestock and Livestock Products in India. The Icfai University Press, Hyderabad

Devendra, C. 1996. Opportunities for increasing the economic contribution of small ruminants in Asia. ACIAR Proceedings 74:27-32. ACIAR (Australian Centre for International Agricultural Research), Canberra, Australia.

Devendra C., Thomas D., Jabbar M.A. and Kudo H. 1997. Improvement of Livestock Production in Crop-Animal Systems in Rainfed Agro-ecolological Zones of South-East Asia. ILRI (International Livestock Research Institute), Nairobi, Kenya. 116 pp.

Ishaque S.M. 1993. Sheep management systems. In: Mackintosh J.B. (ed), Sheep Production in Pakistan. PARC (Pakistan Agricultural Research Council), Islamabad, Pakistan. pp. 257-291.

Seré, C., Steinfeld, H. and Groenewold, J. 1995. World livestock production systems: Current status, issues and trends. In: Gardiner P. and Devendra C. (eds), Global Agenda for Livestock Research. Proceedings of a Consultation, ILRI, Nairobi, Kenya, 18-20 January 1995. ILRI (International Livestock Research Institute), Nairobi, Kenya. pp. 11-39.

Singh K. 1995. Research priorities for improving animal agriculture by agro-ecological zone in India. In: Devendra C. and Gardiner P. (eds), Global Agenda for Livestock Research. Proceedings of the Consultation for the South Asia Region, ICRISAT Asia Center, Patancheru, India, 6-8 June 1995. ILRI (International Livestock Research Institute), Nairobi, Kenya. pp. 77-83.

Suresh, A, Gupta, D.C. and Mann, J.S. 2007. Contribution of Sheep towards the Livelihood of Rural Poor and Marketing of Wool in Rajasthan. In Bhaskaran S. and Mohanty

Suchitra (Eds.), Marketing of Livestock and Livestock Products in India. The ICFAI University Press, Hyderabad

World Bank. 1993. Implementing the World Bank's Strategy to Reduce Poverty: Progress and Challenges. The World Bank, Washington, DC, USA. (Mimeograph).

World Bank. 1996. India Livestock Sector Review: Enhancing Growth and Development. Report 14522-IN. The World Bank, Washington, DC, USA. 33 pp.

❑❑❑

39

Organized Health Care Technologies for Reducing Morbidity and Mortality in Sheep

D. SINGH, C.P. SWARNKAR

Division of Animal Health, Central Sheep & Wool Research Institute, Avikanagar

Introduction

Sheep is probably the first farm animal species to be domesticated by man around 9000 years ago. Sheep owning to their multiple utility for wool, meat, milk and manure plays a pivotal role in the rural agrarian economy of arid to semi-arid zones of India. Millions of small and marginal farmers and landless labourers are solely dependent on sheep production for earning their livelihood. The demand of meet and fiber has been growing consistently due to ever increasing human population, but the supplies do not commensurate with the requirements both in the domestic as well as overseas markets.

Although sheep is well adapted to the harsh climate and inadequate grazing conditions, they are subjected to frequent outbreaks, which are tacking heavy toll causing extensive financial losses to the farmers. Poor exploitation of genetic potential of indigenous animals, inadequate feed and fodder resources, in sufficient health cover, inadequate marketing and credit support are basic reason behind the low productivity of sheep. The productivity of livestock is greatly hampered by various infectious and/ or non-infectious diseases, thus keeping healthy status of flock is a major concern of the rural families. The losses due to disease includes slow growth rate, decreased production of wool, deterioration in skin quality, reduced fertility rate, decline in stock replacement rate, expenditure on treatment and control, increased labor and management cost and mortality losses.

Management, nutrition and availability of efficient health care delivery systems are vital to keep the sheep flock healthy and are the key to a successful sheep enterprise. The farmer accepting an improver breed of livestock would need an accompanying package of animal management and disease preventive practices in order to exploit the full productive potential of the breed. The management schedule will differ according to:

- *Type of management / rearing system* (Intensive / Semi-intensive / Extensive; Stationary / Migratory)
- *Geographical area* (Plains / Hills / Humid / Arid / Semi-arid)
- *Aim / goal of rearing* (Commercial / Research / Extension)

In a commercial flock, management programme should be economics and tailored towards net-worth, while in research and extension flocks, breeding values will allow more elaborate programme. In all sorts of management / animal production system diseases cause losses through mortality, inefficiency of production, retarded growth, decrease in stock replacement rate and expenditure on treatment and control. *Prevention is better than Cure* is basic philosophy. Depending on the type of management, diseases, vaccination background of the flock and / or location, one may be able to eliminate or add some of the health practices suggested in this topic and able to develop a sound and economical flock health programme for particular area.

Basic Steps Involved in Preventive Health Care

Preventive flock health management means avoidance of diseases in flock or reducing their incidence as well as losses caused by them through proper *planning* and *involving* Veterinarians in all matters related to disease diagnosis and prevention. The steps involved in formulating preventive health care programme are:

a) *Develop a concept of disease*: Disease is any biological abnormality in individual resulting in failure to reproduce, thrive, grow, perform, survive etc.

b) *Appreciate cause / origin of disease:* A variety of infectious and non-infectious agents causes diseases, it may be more important to know the source and nature of causal agent as well as how it might get entrance to the flock.

c) *Know the nature of disease:* By recording and monitoring the disease prevalence, pathogenicity, losses due to disease, age and sex affected, flock dynamics.

d) *Apply knowledge at flock level*: With the help of diagnostic laboratory, utilize expertise to evolve preventive practices for particular diseases.

Origin of Disease in Flock

Usually four major aspects were considered to play role in occurrence of a disease in flock.

i) *Genetic*: eg. Dwarfism, cerebellar atrophy, entropion etc. These disorders are rare and well controlled by breeders.

ii) *Management:* eg. starvation / inanition, debility, hypothermia, heatstroke, dehydration, pica, fracture, foot wound etc. These entities can be corrected by regular and frequent observation of flocks.

iii) *Feed and water:* eg. Nutritional deficiency / excesses, intoxication, chemical residue etc. Monitoring of quality and quantity of feed and water and maintaining sufficient amount of feed and water will eliminate related problem in the flock.

iv) *Infectious:* In comparison to non-infectious diseases, infectious diseases are important for flock productivity. These diseases are caused by a variety of organisms like viruses, bacteria, fungi, protozoan and metazoan parasites. To formulate preventive care system to these diseases, it is necessary for us to know:

- The reservoir or source of infection (food. water, air, vectors, carriers)
- The mode of transmission (mechanical, vector-borne)
- The precipitating factors (nutritional, physical, environmental stress)
- The virulence of infectious agent

  i. *High*: produce severe diseased and override all defense mechanisms.

  ii. *Moderate*: cause mild to moderate disease

  iii. *Low*: cause disease only in highly susceptible sheep.

- The manifestation of disease

  i) Death of sheep

  ii) Obvious system malfunction: respiratory sign, diarrhoea, nervous disorders, urogenital disorder, skin / musculoskeletal abnormality.

  iii) Sub-clinical disease: disease is there but not severe enough to show, even minor stress will trigger serious signs.

It should be kept in mind that after reaching the animals on new farm, the activity that you can take on the first day would have a long term affect upon the health and welfare of flock. As many of the diseases that affect sheep can not be detected by external examination of sheep, one need to assume that one or more of the new animals could be bringing diseases in to flock and therefore following points must be considered:

- Quarantine all sheep for 3 weeks. Restocking provides an ideal opportunity to exclude many of the diseases that affect sheep and therefore improve the health, welfare and profitability of flock.
- Treat for ectoparasites.
- Treat for anthelmintic resistant gastrointestinal nematodes.
- Vaccinate against clostridial infection, pox and PPR.
- Examine all the feet of all the sheep and if necessary give foot bath.

Basic Preventive Health Care Approaches

Disease control measures are used to curtail the morbidity and mortality in a flock. Preventive care is the first line of defense against diseases. Following practices must be followed by farm owner / manager to keep flock in healthy status:

- *Eradication* of infectious agents from nature (regional extinction): It depends upon factors responsible for prevalence of disease viz.,
    - Knowledge of the etiological agent
    - Disease transmission pattern
    - Type of host affected (age and sex)
    - Relationship between host and pathogen
    - Diagnostic facility and veterinary interventions
    - Replacement stock
    - Opinion of society and cooperation of farmers
    - Public health consideration
- *Quarantine:* It is the oldest tool of the preventive medicine and is based on system of enclosure in which an infectious agent is limited to its habitat and thus prevented from getting out. The extent of quarantine may be international, interstate or local. The period of quarantine depends upon the incubation period of agent and varied from 2-3 weeks. It is useful but its effectiveness is dubious when vector / carrier are involved.
- *Slaughter* of chronically infected or carrier animal is recommended to eliminate the source of infection. It's effectiveness is influenced by public awareness and cooperation, financial compensation, public health importance and supporting legislation.
- *Improved managmental practices*: It involves grazing, nutritional, breeding and housing management. It is essential that one should watch the flock closely and sort off thin ewes and offer extra feed as well as treat sick sheep. If sheep dies in a flock, find out why? (*Often the most important sheep in a flock is the first one to die. If you do not know why it died, you can't correct the problem*). The other regular managemental practices are:

- Check for abortion.
- Proper disinfection of naval cord after trimming with 7% Iodine solution / tr. Iodine to prevent naval ill and associated problems in new born lambs.
- Early treatment and precaution for exposure-starvation syndrome, pneumonia and enteritis is necessary for lamb's survival.
- Observe ewes for vaginal discharge and udder infection.
- Regular culling of broken mouth sheep and ewes with bad udder, chronic wasting diseases.

- *Use of chemotherapeutics and biologicals*: A typical flock health chart must take care of the farm operations, age group at risk and local agro-climatic conditions. Programme for an organized farm/ field flock will vary from place to place. Thus, health maintenance programme is a flexible and dynamic process which can be modified considering changing situations.

The common infectious and non-infectious diseases require due attention to keep flock in healthy conditions are as under:

Infectious diseases: Sheep pox, Pestes des Petitis Ruminants (PPR), bluetongue (BT), Foot and Mouth disease (FMD), Enterotoxaemia (ET), Johne's disease (JD), Pneumonia (Pasteurellosis), Foot Rot, Anthrax, Contagious ecthyma, Parasitic gastroenteritis (PGE), Scabies, Coccidiosis, Nasal Myiasis and ectoparasitism.

Non-infectious: Among non-specific causes, diseases of respiratory system took the heaviest toll on the flock. Two distinct form of pneumonia, one affecting neonatal lambs and other involving animals of all age group (suppurative pneumonia) are common and could be due to prevalence of ovine adenovirus (Dubey *et al.*, 1985) or other viruses (Lonkar *et al.*, 1982). Inadequate transfer of maternal immunity, through colostrums and low birth weight of lambs are precipitating factors (Maru *et al*, 1993). In a flock suckling lambs mortality rate was nearly four times that of other age group. Retrospective study on pattern and causes of lambs death (Maru *et al.*, 1987) exhibited that first seven days after birth of lambs are crucial for lamb losses to be curtailed. Strategic health cover of pregnant ewes and neonates around lambing may be a major contributor to the reduction in mortality. Prior to lambing, body condition is corrected by supplementing especially with concentrate (Speedy and Clark, 1981). The first two days in life of lambs need adequate protection against exposure and starvation. Feeding of colostrums to the lambs in the first few hours after birth may be important for survival of lambs (Mellor, 1990). The other common non specific manifestations responsible for morbidity and mortality in a flock are bloat, anaemia, vitamin and mineral deficiency, hypothermia, exposure-Inanition syndrome, hyperthermia, debility and metabolic diseases.

The best way to reduce morbidity and mortality losses in flock is through implementation of preventive flock health approaches that incorporate strategic vaccination of prevalent diseases, deworming for endoparasites, vigilance over seasonal diseases and management adopted.

Sheep pox: It ranks first among pox diseases of animals and continues to be most serious economic problem of sheep industry in India as it inflicts losses due to high mortality in lambs, reduction in body weight and wool yield, damaged skin, abortions and mastitis (sharma et al., 1986). It is caused by DNA virus of *Poxviridae* family. High rise of temperature, nasal and ocular discharge and cutaneous pox lesions are main clinical signs. The package of practices adopted to prevent occurrence of sheep pox in flock includes:

- Annual vaccination of all adult animals (during last trimester of pregnancy in major lambing season)
- Vaccination of new born lambs at 2-3 month of age.

To curtail the losses during outbreak of sheep pox following practices are being recommended:

- Segregation of affected and in-contact animals with limiting movement of specified person.
- Restricting the grazing area for infected flock.
- Proper disposal of carcasses.
- Disinfection of premises and farm equipments.

Pestes des Petitis Ruminants (PPR) is a highly contagious disease caused by RNA virus of *Paramyxoviridae* family. High rise of body temperature, mucopurulant nasal and ocular discharge, ulceration in oral cavity and severe diarrhoea are commonly observed. Zebra stripping in caecum, colon and rectum are very prominent on post-mortem. The incorporation of strategic vaccination with cell culture vaccine of all the adult stock (once in three year) and first vaccination of lambs at 3-4 month of age is successful in preventing the occurrence of disease in flock. In face of outbreak quarantine, disinfection and symptomatic treatment of affected animal is recommended.

Bluetongue: It is a non-contagious insect born viral disease caused by RNA virus of family *Reoviridae* and is transmitted by biting midges (*Culicoides* spp.). Fever, oedema of muzzle, lips and face, ulcer in oral mucosae and dental pad, myositis and coronitis, congenital abnormalities, abortion and loss of ram fertility are common manifestation of disease (Mathur *et al.*, 1989; Parsonson, 1990; Sharma *et al.*, 1986; Srivastava *et al.*, 1989). Prevention of diseases is complicated due to prevalence of multiple serotypes with no amount of cross protection between them. In planned flock health approach, to curtail the losses and control of diseases following points should be considered:

- Destruction of breeding sites of midges through smoking or spraying with insecticides.

- Avoid grazing of flock on wet low land where midges abound.
- Adjustment of shearing timing to allow wool growth before the onset of midge's activity.
- Reschedule breeding of ewes if cases of lamb / foetal deformity increase.
- Symptomatic treatment of affected animals.

Foot and Mouth Disease: It is a highly contagious viral disease (RNA virus of *Picornaviridae* family) of cloven hooved animals, which can shatter the livestock economy and close market for export. The agroclimatic conditions, socio-economic and mixed animal husbandry practices provide a conducive ecological niche to the FMD virus. Clinical signs are often unapparent in sheep. The vaccination at 6 monthly intervals is recommended to prevent occurrence of FMD in flocks of endemic area, however existing cost of vaccine is not economical to farmer.

Enterotoxaemia: Clostridial infections are a major risk to flock where a sudden change in feeding practice is expected. To prevent the losses due to ET in flock following package of practice should be implemented:

- Vaccination of adult at 6 monthly intervals.
- Primary vaccination of lambs at 2-3 months of age and booster vaccination 15 days later.
- Care should be taken while sudden change of feed (quality and quantity).
- Avoid excess amount of starchy feed.

Pasteurellosis: Both *Pasteurella haemolytica* and *P. multicida* found to cause disease in sheep flocks particularly during sudden change in climate. The clinical manifestations include dyspnoea, high temperature, muscular tremors, ocular and nasal discharge containing blood and sudden death in young animals (Dimri et al., 1994, Das et al., 1997). To prevent the precipitation of disease in flock following practices are useful:

- Flock management should be designed to reduce the environmental stress and nutritional stress.
- Avoid over crowding and exposure to inclement weather.

Foot rot: It is prevalent in wet, marshy and badly drained pasture. Foot injuries, moisture and warmth are essential for organism (*Fusiformis necrophorus* and *F. nodosus*) to cause disease. There is generalized lameness, pain, loss of body weight and exfoliation of hooves. To prevent occurrence of foot rot in flock following measures should be practiced:

- Footbath of all animals (4-8% $CuSO_4$ solution).
- Regular trimming of hooves after first rains every year.
- Maintenance of good cleanliness and hygiene of corrals and pasture.
- Do not buy animal from infected flock.

Johne's disease: A chronic wasting disease mostly occurs in adult sheep, caused by *Mycobactrium paratuberculosis* and manifested by persistent diarrhea, gradual emaciation and no response to treatment. The following practices should include as an integral part of flock management to reduce the incidence of JD in flock (Maru et al., 1993):

- Regular screening for JD through faecal smear examination and elimination of positive cases through slaughter.
- Regular lime dusting, turning of soil and reduce per stocking rate.
- Infected premises should be left vacant for 12 months.

Parasitic Diseases

It is one problem limiting the animal productivity throughout the country, but the effect of parasitism may be more in areas of high moisture and rainfall and where permanent pasture situation exists and intensive and semi intensive grazing systems are practiced.

Fascioliosis: It is caused by *Fasciola gigantica* and *F. hepatica* (in hilly areas) and transmitted by snail (*Lymnaea auricularia*). The affected animal exhibit anemia, anorexia, oedema, emaciation and wool loss. The disease also caused losses through condemnation of liver.

Amphistomosis: Usually two peaks of infection were manifested, first in February-March and second in July-September. Immature amphistomes are more pathogenic and clinical manifestation includes diarrhea, straining, unthriftiness, dehydration, weakness and loss of body weight.

Schistosomiosis: *Schistosoma indicum* and *S. spindale* are recorded species of blood flukes. These are elongate, sexually differentiated flukes which live in the circulatory system of their host. Like other trematodes, these flukes require an aquatic or amphibious snail as an intermediate host in order to complete their life cycle. Infection occurs either via skin penetration or by penetration of the GIT after ingestion of cercariae with water. The immature flukes migrate through the lungs and the liver to the mesenteric veins, where they mature. Around two month post infection animal develops profuse some time bloody diarrhea, dehydration, loss of appetite and anaemia. The eggs become lodged in the liver and an intense immunological response results in to formation of granuloma and disruption of liver function.

In endemic area following measures should be adopted to reduce the incidence of flukes and curtail the losses caused by them:

- Use of flukicides at strategic point based on epidemiology.
- Avoidance of contaminated fodder.
- Proper drainage of lands.

Gastrointestinal Nematodes: The economical pathogenic nematode species affecting sheep productivity are *Haemonchus contortus, Trichostrongylus colubriformis, T. oxei* and *Oesophagostomum columbianum.* Besides the other nematodes like *Bunostomum, Chabertia, Trichuris* and *Strongyloides papillosus* are also recorded (Singh et al., 1997). The clinical signs in general are anaemia, diarrhea, submandibular oedema, debility and pica.

In temperate and sub temperate region an increasing trend in GIN has been observed during spring (March - May) and from summer (June) to autumn (October- November). The onset of monsoon by middle of June with moderate temperature makes the environment favorable for development and survival of pre-parasitic stages leading to availability of larvae on pasture. The favourable environmental conditions exist during monsoon (June to October) causing high incidence of GIN with peak intensity during monsoon and post monsoon periods (Swarnkar *et al.,* 1997).

In planned flock health approach the control of GIN involve integrated worm management practices. Keeping in view the epidemiology of most pathogenic species, environmental conditions, management (breeding, lambing, grazing) of flock, availability and use pattern of fodder resources, status of anthelmintic resistance, occurrence of peri-parturient rise in faecal egg count, hypobiosis, a strategic deworming plan should be formulated (Singh and Swarnkar, 2007). To manage the worm intensity at acceptable level and to delay the emergence of anthelmintic resistance in nematode parasites following practice should be adopted:

- Deworming at strategic point (1-2 time/year) base on epidemiological knowledge, particularly when parasitic population on strategic is high.
- Use of proper drug (after establishing the prevalence of species) at proper dose rate (based on the heaviest animal of group).
- Keep class of anthelmintic in rotation (at least annually).
- Provision of safe pasture for newborn lambs.
- Optimum nutrition during stress (particularly around lambing)
- Regular monitoring of flocks for faecal egg count.
- Do not buy or introduce resistant worms in flock (deworm the new animals with effective anthelmintic in quarantine).

Now in the face of high prevalence of anthelmintic resistance in parasitic nematodes, the role of refugia in maintaining the susceptibility of parasites against anthelmintics is considered by taking the advantage of over-dispersion phenomenon in faecal egg count. The main focus of anthelmintic research conducted outside the pharmaceutical industry is on the ways of ensuring efficacy of anthelmintic by reducing uses, targeting treatment to those stock that require them. In stead of mass drenching of flock targeted selective treatment (TST) is useful. TST seems to represent one of the most realistic method for restraining worm resistance and an opportunity for management by resource-poor farmers. TST offer the possibility at minimal cost to treat and utilize the relatively small proportion of individuals unable to withstand parasitic challenge[122]. As wit most parasites, burdens of nematodes infecting farm animals are grossly over-dispersed with vast majority of worms occurring in a small percentage of hosts (Hoste et al., 2002). The detailed epidemiological studies in sheep flocks of Rajasthan under All India Network Programme on Gastrointestinal Parasitism revealed that even during favourable monsoon season (July - September) around 20-30% and 50% of animals in field and farm conditions, respectively had faecal egg count (FEC) to the tune of >1000 epg. Other wise in rest of seasons, majority of animals had either < 200 epg or free of infection (Singh *et al.*,2007). In recent times, it has been recognized that the phenomenon of parasite over-dispersion could be put to good use if those animals suffering from level of infection sufficient to cause considerable production loss or health effects, can be identified and treated individually. Following the same TST principle and based on strong correlation that exist between the coloration of the ocular mucous membrane of sheep and degree of aneamia caused by haematophagous parasite (*H. contortus*), the FAMACHA© (Faffa Malan Chart) system has been developed and validated by South African Scientists (Nari, 2005; Van Wyk and Bath, 2002). With this approach not only cost of anthelmintic and labor will be reduced but it would help to maintain population in refugia which in turn should help to conserve the genes for susceptibility within parasite population.

Nasal myiasis: Oestrus ovis larvae are obligatory parasite of nasal and sinusal cavities of sheep and goats (Hall and Wall, 1995). Female flies seriously annoy sheep since they deposit first instar larvae on the nostril. The infestation of larvae of *Oestrus ovis* is characterized by shaking of head purulent nasal discharge, dysponea, sneezing sometimes circling movement and loss of weight. Attempt should be made to break the life cycle of the fly using rafoxnide / ivermectin in the first week of June i.e. during pre-monsoon period (Sanyal et al., 1986).

Mange: Both type of mange, viz. facial mange (*Sarcoptes scabie*) and body mange (*Psoroptes ovis*) is commonly found in sheep. The disease is more prominent in winter and group instinct in sheep lead to quick spread of infection. In organized farm the extent of losses caused by mange could be prevented by:

- Dipping with acaricide / insecticide 15 days after shearing.
- Regular inspection of flock to isolate and treat the infected animals.

In addition to above diseases, it is essential for farm manager or Veterinarian to remain watchful for following diseases/entities in flocks:

- Maggot wounds - during monsoon, after shearing.

- Naval ill in lambs could be controlled by implementing proper disinfection of naval cord after birth.
- Hypothermia (Exposure and Inanition Syndrome) in lambs.
- Hyperthermia
- Dehydration due to lamb scour.

Thus, complex problems involved in implementation of disease control programme are probably more difficult than any other production phases. Mass inoculation to control the epidemic of killer disease is the easiest and most effective intervention. For preventive flock health programme, it is important for the Veterinarians to set up a guideline for the procedure, indicating what and how to do and what to use. A region specific flock health calendar must consider farm operations, age group and agro climatic condition in relation to their role in disease incidence. It has been observed that implementation of planned flock health programme cost around Rs. 30.00 per sheep per year (Rs. 10 - 15 for prophylactic and Rs. 15 - 20 curative medicines).

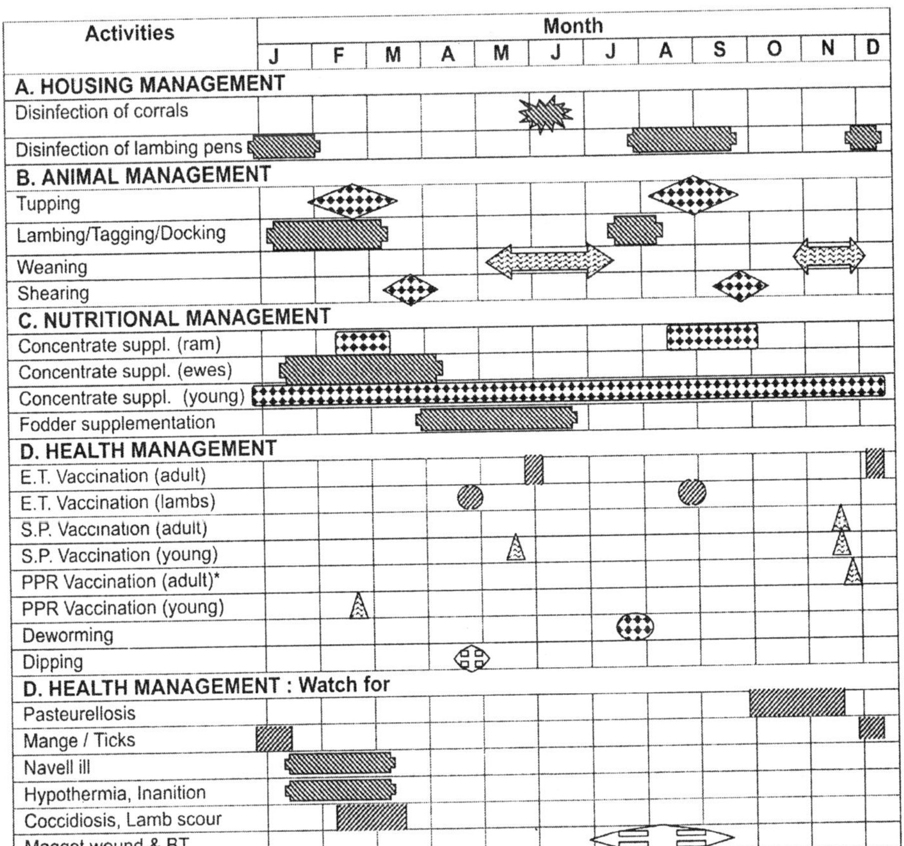

(* Every 3 year)

Model Prophylactic Health Calendar for Sheep Flock

For preventive flock health programme it is important for the veterinarians to set up a guideline for the procedure, indicating what and how to do, when to do and what to use. A model flock health calendar must consider farm operation, age groups and local agroclimatic conditions in relation to their role in disease incidence. On the basis of epidemiology of disease, following annual flock health programme was framed and implemented with success in semi-arid conditions.

REFERENCES

Das, S.C., Mishra, S., Mishra, N. and Bhagwan, P.S.K. 1997. Comparision of competitive ELISA, indirect ELISA and agar fel precipitation test for the detection of bluetongue virus antibodies in sheep. *Indian Journal of Animal Science* 67, 1033-1035.

Dimri, U., Lonkar, P.S., Srivastava, C.P. and Bhagwan, P.S.K. 1994. An outbreak of pneumonia due to *Pasteurella multocida* in sheep. *Indian Veterinary Journal* 71, 1163-1167.

Dubey, S.C., Joshi, D.V. and Sharma, S.N. 1985. Serological evidence for ovine adeno virus antibodies in sheep and goat of semi-arid Rajasthan. *Indian Journal of Comparative Microbiology and Immunology* 6, 167-169.

Hall, M. and Wall, R. 1995. Myiasis of human and domestic animals. *Advances in Parasitology* 35, 257-334.

Hoste, H., Chartier, C. and Le Frileux, Y. 2002. Control of gastrointestinal parasitism with nematodes in dairy goats by treating the host category at risk. *Veterinary Research* 33, 531-545.

Lonkar, P.S., Sharma, M.M., Uppal, P.K. and Mathur, P.B. 1982. Some observations on pneumonia in sheep at an organized farm in semi-arid part of Rajasthan. In: *Proceeding of National Seminr on sheep and goat diseases*, CSWRI, Avikanagar, 28-30 May, pp. 166-171.

Maru, A., Lonkar, P.S., Srivastava, C.P. and Dubey, S.C. 1987. Pattern and causes of lamb death in housed sheep flocks of different breeds. *Indian Veterinary Medical Journal* 11, 160-164.

Maru, A., Srivastava, C.P., Dubey, S.C. and Lonkar, P.S. 1993. Disease profile of a sheep flock in semi-arid India. *Preventive Veterinary Medicine* 15, 1-17.

Mathur, A.K., Srivastava, R.S. and Kalra, D.B. 1989. Influence of bluetongue disease on the semen qualities of exotic rams. *Indian Journal of Animal Reproduction* 11, 38-40.

Mellor, D. 1990. Meeting colostrums needs of newborn lambs. *In Practice, Veterinary Record* (Suppl) 12, 239-244.

Nari, A. 2005. The FAMACHA assay as a component of integrated pest management. In: *Proc. FAO Symposium on "Integrated Animal Parasite Management: From Acedemic Interest to Reality"* pp. 95 - 102.

Parsonson, I.M. 1990. Pathology and pathogenesis of bluetongue infection. *Current Topics in Microbilology and Immunology* 162, 119-141.

Sanyal, P.K., Maru, A. and Gupta, S.K. 1986. The use of injectible rafoxanide against ovine nasal myiasis. *Veterinary Parasitology* 19, 127-131.

Sharma, M.M., Lonkar, P.S., Srivastava, C.P. and Kalra, D.B. 1986. Congenital anomalies in lambs as a sequelae to bluetongue infection of ewes. *Indian Journal of Animal Science* 56, 1055-1059.

Singh, D. and Swarnkar, C.P. 2007. Alternative methods of control of gastrointestinal nematodosis in small ruminants. In: *Small Ruminants Emerging Diseases, Containment Under WTO Regime* (Eds. V.S. Vihan, Ashok Kumar, V.K. Gupta and N.P. Singh), Satish Serial Publishing House, Delhi, pp. 267-303.

Singh, D., Swarnkar, C.P., Khan, F.A., Srivastava, C.P. and Bhagwan, P.S.K. 1997. Epidemiology of ovine gastrointestinal nematodes at an organized farm in Rajasthan, India. *Small Ruminant Research* 26, 31-37.

Singh D, Swarnkar C P and Khan F A. 2007. Annual Progress Report on *All India Network Programme on Gastrointestinal Parasitism.* Central Sheep and Wool Research Institute, Avikanagar (Rajasthan) India.

Swarnkar, C.P., Singh, D., Khan, F.A., Srivastava, C.P. and Bhagwan, P.S.K. 1997. Development and survival of *Haemonchus contortus* larvae on pasture under semi-arid conditions of Rajasthan. *Journal of Veterinary Parasitology* 11, 37-41.

Srivatava, C.P., Maru, A. and Dubey, S.C. 1989. Epidemiology of congenital defects in foetus / lambs in bluetongue endemic flock. *Indian Journal of Animal Science* 59, 14-20.

Speedy, A.W. and Clark, C.F.S. 1981. Lamb land sheep; the nutritional management cycle. *Veterinary Record* 108, 493-498.

van Wyk, J.A. and Bath, G F. 2002. The FAMACHA© system for managing haemonchosis in sheep and goats by clinically identifying individual animal for treatment. *Veterinary Research* 33, 509-529.

❑❑❑

40

Alternative Approaches for Worm Management with Emphasis on Use of Phyto-anthelmintics

C.P. SWARNKAR, D. SINGH

Division of Animal Health, Central Sheep & Wool Research Inst., Avikanagar

Introduction

In developing countries like India, livestock, in general, and small ruminants in particular, are mainly reared by marginal and small farmers representing poor strata of the society. Parasitic infection, particularly, internal parasitism is one of the major disease syndrome of pastoral ruminants simply because environment which is responsible for pastoralism also favor development, survival and dissemination of parasites. Since long, the menace of gastrointestinal parasitism (GIP) in livestock is being managed by frequent use of anthelmintics which results in emergence of anthelmintic resistance strains of parasites. The wide-spread prevalence of anthelmintic resistance in nematode parasites has led to realization that unless the pharmaceutical industry provides an endless supply of new molecules, intensive chemoprophylaxis in the way that is currently practiced; it is not a suitable option for the control of GIP (Jackson and Miller, 2006). In addition, the end of the anthelmintic spectrum, high cost and tendency of consumer to prefer products from livestock managed with minimal chemical interventions led to uniform agreement among parasitologists throughout the world that conventional paradigm of chemical based worm control must be changed from one of maximum to one of the optimal worm management compatible with sustainability.

Nature is full of medicinal plants having tremendous ability to cure and prevent various ailments when used with wisdom. Herbal medicines are part and parcel of society to combat diseases in human as well as livestock population, from the dawn of civilization. As per WHO estimate around 80 % of world population in developing countries relies on the traditional medicines for primary health care (De Smet, 1997). The world market for herbal medicine has reached US$ 60 billion with annual growth rate between 5- 15 %. The present communiqué elaborates the alternate method of worm control with emphasis on potential of plant based approach.

Spectrum of Parasites

Throughout India, among gastrointestinal nematodes, the prevalent predominant species in small ruminants is *Haemonchus contortus* followed by *Trichostrongylus* spp., *Oesophagostomum* spp., *Strongyloides papillosus* and *Trichuris* spp (Singh *et al.*, 1997). *Oestertagia, Bunostomum* and *Gaigeria* spp. were also on record from hilly parts of the country. In general, the life cycle of gastrointestinal nematode is consisting of free-living and parasitic phase. The stages in the free-living phase are known as "Supra-population" consisting of eggs and different stages of larvae (1- 3) on pasture. Their development, survival and dissemination are mainly governed by the climatic factors and status of pasture. The stages in parasitic phase are known as "Infra-population" consisting of L_4, immature and mature worms residing in the host. Over 90 % of the populations of various stages of nematode parasites live outside the animal as supra-population. Conventional control program (based entirely on chemotherapeutics) are targeted only against infra-population (i.e. only against 10 % of worm population).

Impact of Parasites

In a flock or herd, the effects of internal parasitism vary between wide extremes. At one end, death and total economic loss can occur while on other end, effect may be undetectable and control measures un-warranted. Economically major consequences of GIP occur at individual and flock/herd level along with impact on human welfare (Poppi *et al.*, 1990; Swarnkar and Singh, 2005).

a) Impact at individual level

- Increased mortality,
- Reduced in live weight gain (35-40 %) and lower market value,
- Reduced in wool yield (35-40 %) and deterioration in quality,
- Increased slaughter house condemnation of carcasses or organs for human consumption,
- Decreased milk yield (10-15 %),
- Reduced feed intake and feed conversion efficiency,
- Reduced draught power and dung out put (fuel and fertilizer).

b) Impact at flock or herd level

- Reduced productive life span of flock/herd,
- Hamper the effort for genetic selection,
- Increased immuno-suppression and susceptibility to other diseases.

c) Impact on human welfare

- Reduced supply of animal protein and other products,
- Adverse effect on community development,
- Reduced traction and dung out put.

Present Worm Control Approach and Associated Problems

The control of GI nematodes is mainly achieved by the use of chemotherapeutic agents popularly known as anthelmintics. Role of highly effective anthelmintics in managing this menace is well established. History of modern anthelmintics dates from the 1920's when carbon tetrachloride was introduced for the control of fasciolosis and haemonchosis in sheep. The advent of phenothiazine in late 1930's marked the first broad spectrum anthelmintic. Thiabendazole, the first benzimidazole anthelmintic introduced in the market in 1960 - 61, further improved the control of helminths as it had a broad spectrum of activity, high efficacy and wide margin of safety. A large number of anthelmintic compounds were launched during 1960 - 1980 which has a high potency at mg/kg body weight instead of gram doses. In 1981, a major step forward was made with the launch of avermectins, having excellent broad spectrum activity. Since the advent of anthelmintics, their use in controlling GIP is mainly tactical and the animals are treated without considering the epidemiological profile and economic significance of parasites in particular agroclimatic area. This sole dependence on chemotherapeutics coupled with their extensive use, improper dosage and faulty worm management practices unfortunately led to emergence of anthelmintic resistant strains of parasites thus posses serious limitation on the use of anthelmintic and challenging the ability of animals to produce and maintain high level of productivity in the flock.

The problem become even more severe if new generation of anthelmintics with unique mode of action are not developed and/or replaced in future. The renewed hope of a lasting solution with introduction of broad spectrum anthelmintics in early 1960's was short- lived as several report of anthelmintic resistance attracted our attention (Singh *et al.*, 2002). At global level anthelmintic resistance has emerged as the single most important problem confronting the control of parasites in livestock. A survey of resistance to parasiticides carried out by the OIE and FAO in 77 out of 151 OIE member countries revealed that 54.5 % of countries are affected by anthelmintic resistance (Nari and Hanson, 1999). In India the reports on the occurrence of anthelmintic failure and emergence of anthelmintic resistance have shown increasing trend since 1990 and it has been reported from farm and field in all the major sheep rearing zones of country (Singh *et al.*, 2002).

Issues in Worm Management

The usual mode of control of GIN has up to now based on the repeated use of anthelmintics. However, the higher prevalence of anthelmintic resistance, consumer's concern regarding residue in animal products and environment and negative effect of anthelmintics on the development of natural immunity against worms makes it mandatory to reconsider the present method of worm control and look for alternative solutions which address both the management of worms and anthelmintic resistance (Hoste *et al.*, 2002). The other issues which arise in worm management are:

- Can pharmaceutical industry provide an endless supply of new molecule?
- Intensive anthelmintic intervention and the way it is currently practiced is sustainable?
- How rising cost of anthelmintic intervention and increasing demand of animal products free of chemical residue will justify farmer and public concern in the era of liberalization, privatization and globalization?
- How the environmental issue will be addressed with the use of anthelmintics?

Alternative Approaches in Worm Management

Unlike anthelmintics, alternative approaches in worm management usually do not have direct effect on parasitic infection in the host. Instead, the mode of action is based on improving the immunity of the hosts or decreasing exposure of the hosts to the parasites (Waller, 1999; Singh and Swarnkar, 2007). A closer look on the control strategies alternative to chemotherapy reveals that some of them are in the technological state while others are still under research and development. Based on type of targeted population the alternative approaches are as follows:

A. Approaches Targeted Against Infra-Population

- Chemotherapeutic management (targeted selective treatment - TST),
- Nutritional management,
- Immunological control (Worm vaccine),
- Genetic control (Breed resistance),
- Bioactive forages and phyto-medicines.

B. Approaches Targeted Against Supra-Population

- Grazing and pasture management,
- Biological control.

As per the preview of the present paper, all the approaches mentioned above could not be described here and detailed relevance of bioactive forages and phyto-anthelmintics in worm management is being discussed below.

Bioactive Forage and Role of Condensed Tannin

Beneficial effect on the ability of ruminants to withstand internal parasitic infection has been found with tannin-containing plants (Hoskin *et al.*, 1999). Condensed tannins (CTs) are polyphenolic secondary plant metabolites, well recognized for their ability to form complex with soluble protein at normal ruminal pH (when fed at concentration less than 10 % of diet). This restrict the normal break down of protein to ammonia in the rumen and allowing more dietary pre formed protein to pass unchanged into the small intestine, acting as by-pass protein. The increased protein supply to animals leads to increased resistance to infection (Molan *et al.*, 2000; Paolini *et al.*, 2003) as evident by suppression of growth and fecundity of worms, reduced worm burden and FEC and improved status of protein and phosphorus (Athanasiadou *et al.*, 2001; Butter *et al.*, 2000). Recently, emphasis has been given to the potential anthelmintics properties of certain range forages. Forages like *Lotus, Hedysarum* and *Onobrychis* species are highly nutritive and rich in CT and other secondary metabolites. Sheep grazing on sulla (*Hedysarum coronarium*) found to maintain high level of productivity while carrying a substantial burden of *Teladorsagia circumcincta* and *Trichostrongylus colubriformis* (Niezan *et al.*, 1995) compared to sheep grazed on low CT forage (*Medicago sativa*).

On *in vitro* trials, it has been observed that CTs significantly inhibited the development of eggs to infective larvae. The study conducted at CSWRI, Avikanagar revealed significant reduction in larval development rate in feces from CT fed animals (Swarnkar *et al.*, 2007a). CTs could bind to the protein available in the nutrient media for larvae leading to larval starvation and death as well as developing larvae can ingest CT, which bind to the intestinal mucosa and cause autolysis. CTs may also inactivate hatching enzymes in worm eggs (causing reduction in egg hatching) and can paralyze the body musculature (pharyngeal muscles) of larvae (leading to reduced viability of larvae). Hence, sufficient evidences are available suggesting that CTs can disrupt the life cycle of nematodes and help in worm management. On *in vivo* study, up to 50 % reduction in worm burden was observed in lambs fed with sulla (*Hedysarum coronarium*) or great trefoil (*Lotus pedunculatus*) leaves containing CTs (Niezen *et al.*, 1998a, b). Around 40- 60 % reduction in FEC with marginal rise in haemoglobulin and packed cell volume were observed in sheep (infected with GIN) fed CTs (@ 5 % of DM) either as leaves of Khejri (*Prosopis, cineraria*) or as extract (acetone: water) of leaves compared to those offered diet having no CT (Swarnkar *et al.*, 2007). The production losses caused by GIN were found to be alleviated by feeding the animals on White clover (Niezen *et al.*, 2002).

The consumption of CTs may not result in economic gain because of the negative effect of excessive CT consumption on the performance of host in term of reduced feed intake, nutrient digestibility, impairment of rumen metabolism and mucosal toxicity (Dawson *et al.*, 1999). The economic benefits of CTs were more pronounced in grazing studies with parasitized animals, however it is not certain whether the improvement in performance was the direct consequence of a reduced larval intake, an anthelmintic like effect or grazing of high quality fodder. The letter might have been the case as the majority of tanniferous forages are leguminous and usually are of higher nutritive value than conventional range land vegetation. It is therefore possible that improved performance observed in animals grazing such forages is a consequence of improved resilience rather than the consequence of improved resistance.

Phyto (herbal) Anthelmintics

These have been used for eons by farmers and traditional healers to treat parasitism and improve livestock performance and productivity, but most of information on phyto-medicine are anecdotal and their use is on "trial and error" basis or based on conventional wisdom. Recently there has been an increasing trend and global resurgence in ITK of herbal drugs due to reasons like minimal or no side effect, eco-friendly, cost effective with least problem of drug resistance. The increasing interest in the research and development (scientific validation) of phyto- medicines paved the way for the formulation of a number of preparations for treating various diseases and enhance production potential of livestock. Through improved understanding of nematode parasites and diseases they cause, new opportunities are being created to develop anthelmintic therapies using natural molecules (small peptides, enzymes etc) to interfere with natural processes in worm (metabolism, egg production, larval development etc). In many cases the active compounds in the herbal remedies have not been fully identified. Plant enzymes or secondary metabolites such as alkaloids, glycosides and tannins have shown dose- dependent anti-parasitic properties Traditional societies use number of herbal remedies deemed suitable for a number of parasitic diseases. For example, seeds or the foliage of plant as garlic, onion, mint, walnut or parsley have been used to control GIP while cucumber and pumpkin seeds have been associated with expulsion of tape worms from GIT (Akhtar *et al.*, 2000; Guarrera, 1999). The administration of extracts from *Acasia* and *Artemisia* spp in animals infected with haemoprotozoan parasites (*Trypanosoma, Plasmodium*) have resulted in a reduction of the number of blood parasites (Adewunmi *et al.*, 2001; Deharo *et al.*, 2001; Lans *et al.*,2000).

The mechanism whereby the consumption of certain plants or their extracts can affect parasite viability, motility and fecundity both *in vitro* and *in vivo* are largely unknown. For some plants it has been suggested that their consumption could be associated with an enhanced immune response of the host towards the parasites as a result of nutrient supplementation and improved nutrition (Athanasiadou and Kyriazakis, 2004) while many of the plants have been reported to have direct activity against parasites.

In India, under NATP a large number of crude extracts from different plants were evaluated for antiparasitic activities and it has been found that *Malia azedarach* (Bakain), *Butea frondosa* (Palas), *Artemesia martima* (Kirmala), *Mallotus philippinensis* (Kamala), *Azadirachta indica* (Neem), *Embelia ribes* (Baberang), *Tagetes patula* (Gaatakadi), *Andrographis peniculata* (Kalmegh), *Vernonia anthelmintica* (Kaljiri), *Vitex negundo* (Bana), *Xanthium stramarium* (Chotta Gokhru) etc posses low to high level of antiparasitic activity against different stages of GI nematodes (Anon, 2003; Sharma *et al.*, 2003; Swarnkar *et al.*, 2004, 2007b; Vihan *et al.*, 2002 a, b). The results of *in vitro* and *in vivo* studies are as under.

Effect on egg embryonation: Out of 51 plant extracts tested, a total of 11 extracts were found to cause significant inhibition of embryonation in *H. contortus* eggs, revealing existence of embryonicidal activity. At the concentration of 1.25 mg/ml, the corrected mean % egg embryonation ranged from 1.71±0.59 (cold alcoholic extract of *Pongamia glabra* seed) to 49.90±2.91 (aqueous extract of *Melia azedarach* fruit). The plant extract which showed embryonicidal activity were aqueous extracts of *Andrographis peniculata* (leaves), *Azadirachta indica* (leaves), *Bauhimia variegata* (leaves), *Melia azedarach* (leaves), *Melia azedarach* (fruit), *Vernonia anthelmintica* (seed), ether extracts of *Butea frondosa* (seed), *Pongamia glabra* (seed) and alcoholic extract of *Tagetes patula* (flower) with ED_{50} value to the tune of 0.915, 1.115, 0.913, 0.467, 1.262, 1.129, 0.993, 0.753, 0.390 and 0.823 mg/ml, respectively.

Effects on egg hatching: Out of 51 plant extracts evaluated, a total of 27 plant extracts were found to cause significant inhibition of hatching of *H. contortus* eggs exhibiting potent to moderate ovicidal property. The corrected mean % egg hatching (@1.25 mg/ml) ranged from 0.07±0.07 (hot aqueous extract of *M. azedarach* leaves) to 47.48±11.47 (aqueous extract of *C. camphora* leaves). The aqueous extracts found to possess ovicidal property belonged to *A. peniculata* (leaves), *A. indica* (leaves), *B. frondosa* (seed), *B. variegata* (leaves), *C. camphora* (leaves), *E. ribes* (seed), *M. azedarach* (fruit and leaves), *T. patula* (flower and leaves), *V. anthelmintica* (seed) and *V. negundo* (leaves). The alcoholic extracts of *B. frondosa* (seed), *C. camphora* (leaves), *M. azedarach* (leaves and fruits), *P. glabra* (seed), *T. patula* (flower) and *Tribulus terrestris* (seed), ether extracts of *B. frondosa* (seed), *P. glabra* (seed) and *V. anthelmintica* (seed) and chloroform extracts of *C. camphora* leaves also exhibited significant ovicidal activity against *H. contortus* eggs. The ED_{50} values of extracts required to inhibit hatching in 50% of eggs ranged from 0.194 mg/ ml for alcoholic extract of *T. patula* flower to 1.078 mg/ml for alcoholic extract of *M. azedarach* fruit (Table 1).

Table 1. ED_{50} (mg/ml) values for plant extracts required to inhibit hatching of *H. contortus* eggs.

Plant	Part of plant	Type of extract	ED_{50}	95%F. limits
A. peniculata	Leaves	Aqueous	0.431	0.283 – 0.655
A. indica	Leaves	Aqueous	0.705	0.541 – 0.724
B. frondosa	Seed	Alcoholic	0.399	0.282 – 0.564
		Chloroform	0.761	0.696 – 0.832
		Aqueous	0.623	0.609 – 0.637
		Ether	0.648	0.627 – 0.669
B. variegata	Leaves	Aqueous	0.332	0.188 – 0.552
C. camphora	Leaves	Chloroform	0.917	0.838 – 1.003
		Alcoholic	0.672	0.444 – 1.027
		Aqueous	0.800	0.416 – 1.538
E. ribes	Seed	Aqueous	0.692	0.682 – 0.702
M. azedarach	Fruit	Hot aqueous	0.669	0.649 – 0.689
		Hot Alcoholic	1.078	0.981 – 1.185
	Leaves	Cold Aqueous	0.306	0.233 – 0.402
		Hot Aqueous	0.515	0.497 – 0.553
		Cold Alcoholic	0.436	0.380 – 0.500
		Hot Alcoholic	0.327	0.245 – 0.436
P. glabra	Seed	Cold Alcoholic	0.267	0.145 – 0.493
		Cold Ether	0.418	0.304 – 0.576
T. patula	Leaves	Aqueous	0.272	0.170 – 0.434
	Flower	Aqueous	0.260	0.164 – 0.413
		Alcoholic	0.194	0.144 – 0.262
T. terrestris	Seed	Hot Alcoholic	0.278	0.213 - 0.363
V. anthelmintica	Seed	Aqueous	0.672	0.661 – 0.695
		Ether	0.914	0.837 – 0.998
V. negundo	Leaves	Aqueous	0.339	0.203 – 0.565

Effect on infective larvae of H. contortus: On larval mortality assay, aqueous extract of *B. frondosa* seed (@20.0 mg/ml) caused maximum mortality (90.73 %) in infective larvae of *H. contortus* after 72 hrs incubation. The other extracts those exhibited moderate larvicidal activity (30.97 to 67.26%) were aqueous extracts of *A. peniculata* (leaves), *A. indica* (leaves), *V. anthelmintica* (seed), alcoholic extracts of *M. azedarach* (leaves), *P. glabra* (seed), *V. anthelmintica* (seed) and ether extracts of *A. indica* (leaves), *B. frondosa* (seed) and *P. glabra* seed. A total of 19 extracts from

11 plants showed low level of larvicidal (10- 30 %) property against infective larvae of *H. contortus* while 20 extracts from 12 plants failed to produce any evident effect on larval mortality. The extracts possessing low larvicidal activity were aqueous extracts of *E. ribes, M. azedarach, S. officinalis, T. patula,* alcoholic extracts of *A. indica, B. frondosa, C. camphora, E. ribes, M. azedarach, P. glabra, S. officinalis, T. patula, T. terrestris,* and *V. negundo*. In addition chloroform extracts of *A. indica* and *B. frondosa* also showed low larvicidal activity whereas ether extracts of plants did not produce any effect on larval mortality.

*In vivo evaluation of plant extracts:*On the basis of *in vitro* results, 31 extracts from 16 plants were evaluated for *in vivo* efficacy in sheep experimentally infected with *H. contortus*. It was observed that on day 10th post- treatment four plant extracts (@ 50.0 mg/kg body weight) caused marked reduction in fecal egg count in sheep. The plant extracts showing good efficacy were aqueous extract of *T. patula* flower, alcoholic extracts of *A. indica* bark and *B. frondosa* seed with corrected faecal egg count reduction to the tune of 74.8, 56.9, 51.0 and 78.9 %, respectively. Similarly alcoholic extract of *E. ribes* seed (@ 6.0 ml/sheep) caused 58.5 % reduction in fecal egg counts on day 10th post-treatment. The plant extracts showing moderate efficacy (30- 50 %) in sheep experimentally infected with *H. contortus* were aqueous extracts of *E. ribes* (seed), *M. phillippinensis* (leaves), *M. azedarach* (leaves and fruits), *P. persica* (leaves), alcoholic extracts of *B. variegta* (leaves), *P. persica* (leaves), *T. terrestris* (seed) and ether extract of *P. persica* (leaves), *V. anthelmintica* (seed). The extracts failed to cause any appreciable reduction in fecal egg count in sheep experimentally infected with *H. contortus* were *B. frondosa* seed (ether), *E. ribes* seed (ether and chloroform), *M. phillippinensis* (chloroform), *M. azedarach* leaves (hot aqueous), *T. patula* flower (alcoholic), *T. terrestris* seed (chloroform) and *V. anthelmintica* seed (aqueous, alcoholic, chloroform).

The plants or active principle isolated from plants having antiparasitic activity in different parts of world are listed in table 2 and 3. Garlic does not found to prevent production of eggs but prevents the eggs of certain parasites from developing in to larvae (Bastidae, 1969). Common mugwart (*Artemisia vulgaris*) is effective against lung worms and hook worms. Santonin derived Eurasian worm wood (*A. cina*) is effective against *Echinococcus* and *Ascaris* spp. In North-Africa, *A. herba-alba* proved to be highly effective against *H. contortus* in sheep (Idris *et al.*, 1982).

Kembewa (1999) reported that *Tephrosia vagelli* (@ 200 gm of dried leaves) is effective in treating the parasitic gastroenteritis in livestock. In Cameroon, Ndi *et al.* (1999) observed reduction in fecal egg count to the tune of 34.4, 52.4 and 55.8 % for GIN with *Solanum aculeastrum, Khaya anthotheca* and *Veronia amyglandina,* respectively. Mostafa *et al.* (1996) observed an anthelmintic efficacy of 35.53 % against natural GIN infection using seeds of *A. indica*. A reduction to the tune of 80- 100 % was noticed with *Carica papaya* latex in animal infected with *Ascaris suum* (Satrija *et al.*, 1994).

Table 2. Plants or their parts showing anthelmintic activity

Plant	Part	Target parasites	Host	Reference
Allium sativum (Garlic)	Cloves	Nematodes	Ruminants	Cabaret et al., 2002
Azadirachta indica (Neem)	Leaf, Bark, Seed	GIN	Ruminants	Costa et al., 2006
Areca catechu	Nut	Taenia	Cattle, Dog, Goat	Roepke, 1996
Adhatoda vesica	Seed	Nematodes	Sheep	Lateef et al., 2003
Butea frondosa (Palas)	Flower	Ascarids		Lal et al., 1978
Carica papaya (Papita)	Seed	Ascarids		Stepek et al., 2004
Caesalpinias crista	Seed	Fasciola	Buffalo	Kailani et al., 1995
Cucurbita pepo (Pumpkin)	Seed	GIN	Sheep	Duval, 1994
Nigella sativa (Kalonji)	Seed	Fasciola	Buffalo	Kailani et al., 1995
Punica granatum (Anar)	Fruit	GIN	Sheep	Akhtar and Riffat, 1985
Mimosa pudica	Stem	Haemonchus	Goat	Fernandez, 1991
Melia azedarach (Bakain)		GIN	Goat	Akhtar and Riffat, 1984
Khaya anthotheca		GIN	Calves	Ndi et al., 1999

Table 3. Active molecules isolated from plants and possessing anthelmintic activity

Plant	Part	Active molecule	Reference
Allium sativum (Garlic)	Bulbs	Allicin	Vieira et al., 1999
Annona squamosa	Leaves	Anthraquinone, Terpenoids	Vieira et al., 1999 Jovellanos, 1997
Azadirachta indica (Neem)	Seed	Azadirachtin	Katker and Katker, 1984
Areca catechu	Nut	Arecolin	Roepke, 1996
Adhatoda vesica	Root	Alkaloides, Glycosides	Lateef et al., 2003
Artemisia herba-alba	Shorts	Santonin	Idris et al., 1982
Albizia anthelmintica	Root, Bark	Sesquiterpene, Kosotoxins	Gothuma et al., 2004 Gakuya, 2001
Ananas comosus	Leaves	Bromelain	Hordegen et al., 2003
Butea frondosa (Palas)	Flower	Palasonin	Lal et al., 1978
Carica papaya (Papita)	Seed	Benzyle isothiocynate	Vieira et al., 1999
Calotropis procera	Leaves	Triterpenoids, Anthocyanins	Al-Qarawi et al., 2001
Chenopodium ambrosioides	Seed	Ascaridole	Ketiz et al., 2002
Cucurbita pepo (Pumpkin)	Seed	Cocurbitacin	Sharma et al., 1971

In Bangladesh, Mostofa *et al.* (1983) observed some degree of anthelmintic activity in Pineapple, Bet, Chirata and Haritaki against GIN in cattle. *A. indica* contain azadirachtin, vepaol, isovepaol, nimbin, nimbidin and vanillic acid (Katkar and Katker, 1984) and oil from seed is used as anthelmintic and parasiticides. The efficacy of neem seed against natural GIN ranged from 40.00 to 52.77 % in goats (Mostafa *et al.*, 1995) and from 35.29 to 40.00 % in sheep (Ahmed *et al.*, 1994). *In vitro* studies with aqueous extract of *A. indica* leaves exhibited potent embrynocidal and ovicidal activity and on *in vivo* studies with alcoholic extract of bark, a reduction in the faecal egg count to the tune of 48.2 % was observed in artificially infected sheep (Anon, 2003). In Philippines, plants which have been used by the farmers to control parasites in goats (Fernandez, 1995) include *Quisqualis indica, Momordica charantia, Clitorea ternatea, Bixa orellana, Lansium domesticum, Karicia papaya, Leucaena leucocephala, Cajanus cajan, Magnifera indica, Manihot esculenta, Tamarindus indica, Moringa oleifera, Artemisia vulagaris, Mimosa pudica, Chrysoyllum caimito* and *Tinospora rumphii*. All of these plants showed an efficacy of > 50 % against *H. contortus*.

Conclusion and Recommendations

Despite ample evidence of antiparasitic properties of several plants, it has been felt that scientific evidences on the antiparasitic effect of most plant products are limited, regardless of their wide ethno-veterinary usages. Scientific validation of the antiparasitic activities and possible side effects of phyto-products in ruminants is necessary prior to their adoption as an alternate mean of worm control (Pessoa *et al.*, 2002). The question of efficacy is the main issue for herbal de wormer. In most cases the active material has to be extracted from plants and *in vitro* conditions and concentrations used are not always comparable to those *in vivo*, thus often result can differ in the two assays. One may question the relationship of *in vitro* studies with *in vivo* efficacy. Considerable physiological differences are expected to be present between the *in vitro* conditions and in the predilection site of worm within their animal hosts including factors those affect bioavailability of the active compounds. The antiparasitic and anti- nutritional effects of plant secondary metabolites ingestion on infected host performance should be considered and evaluated simultaneously.

While based on traditional methods, automation of screening has greatly increased the number of biological extracts which can be screened, potentially leading to more rapid identification of promising compounds. A variety of other technical breakthroughs have enhanced the ability of researchers to develop biological anthelmintics (Hein *et al.*, 2001). These include powerful "phage display" strategies to screen a huge variety of peptides or recombinant antibodies for those capable of binding to specific targets (Cwiria *et al.*, 1990). Mixtures of bacteriophage displaying different peptides on their surface are exposed to a biological target and phage that binds are selectively enriched by affinity. Since the phage contains the DNA that encodes the binding peptide, this peptide can be characterized simply by sequencing the DNA. If desired, the peptide or antibody can be re-expressed in a different recombinant protein expression system. Such peptides could be used to identify exposed proteins or available binding sites, present on complex biological targets such as a worm surface. Alternatively, the peptides could become biological anthelmintics as targeting sequences for bioactive molecules such as protein toxins or as receptor binding agonists or antagonists.

REFERENCES

Adewunmi, C.O., Agbedahunsi, J.M., Adebajo, A.C., Aladesanmi, A.J., Murphy, N. and Wando, J. 2001. Ethno-veterinary medicine: screening of Nigerian medicinal plants for trypanocidal properties. *Journal of Ethnopharmacology,* 77, 19-24.

Akhtar, M.S., and Riffat, S. 1984. Efficacy of *Melia azedarach,* Linn. (Bahain) and morantel against naturally acquired gastrointestinal nematodes in goats. *Pakistan Veterinary Journal,* 4, 176 – 179.

Akhtar, M.S., and Riffat, S. 1985. Efficacy of *Punica granatum* fruit rinds against naturally acquired nematodal and cestodal infection. *Journal of Pharmacology,* Punjab University, Lahore, 6, 296-297.

Akthar, M.S, Iqbal, Z., Khan, M.N. and Lateef, M. 2000. Anthelmintic activity of medicinal plants with particular reference to their use in animals in the Indo-Pakistan subcontinent. *Small Ruminant Research,* 38, 99-107.

Ahmed, N.A., Mostofa, M., Awal, M.A. and Alam, M.N. 1994. Comparative efficacy of modern anthelmintics with that of neem seeds against gastrointestinal nematodiasis in sheep. *Bangladesh Veterinary Journal,* 28, 21-23.

Al-Qarawi, A.A., Mahmoud, O.M., Sabaih Haroun, E.M. and Adam, S.E. 2001. A preliminary study on the anthelmintic activity of *Calotropis procera* latex against *Haemonchus contortus* infection in Najdi sheep. *Veterinary Research Communication,* 25, 61-70.

Anon. 2003. Final Report on NATP "*Identification and Evaluation of Medicinal Plants for Control of Parasitic Diseases on Livestock*", Central Sheep and Wool Research Institute, Avikanagar (Rajasthan) pp. 1-136.

Athanasiadou, S. and Kyriazakis, I. 2004. Plant secondary metabolites: antiparasitic effects and their role in ruminant production systems. *Proceeding of Nutritional Society,* 63, 631-639.

Athanasiadou, S., Kyriazakis, F., Jackson, F. and Coop, R. L. 2001. Direct anthelmintic effects of condensed tannin towards different gastrointestinal nematodes of sheep: *in-vitro* and *in-vivo* studies. *Veterinary Parasitology,* 99, 205-219.

Bastidae, G.J. 1969. Effect of ingested garlic on *Necator americanus* and *Ancylostoma caninum. American Journal of Tropical Medicine and Hygiene,* 18, 920-923.

Butter, N.L., Dawson, J.M., Wakelin, D. and Buttery, P. J. 2000. Effect of dietary tannin and protein concentration on nematode infection (*Trichostrongylus colubriformis*) in lambs. *Journal of Agricultural Science,* 134, 89-99.

Cabaret, J., Bouilhol, M. and Moga, C. 2002. Managing helminths in ruminants in organic farming. *Veterinary Research,* 33, 625-640.

Costa, C.T., Bevilaqua, C.M., Maciel, M.V., Camurea-Vasconcelos, A.L. Morais, S.M., Monteiro, M.V., Farias, V.M., de Silva, M.V. and Souza, M.M. 2006. Anthelmintic activity of *Azadirachta indica* A Juss against gastrointestinal nematodes. *Veterinary Parasitology*, 137, 306-316.

Cwiria, S.E., Peters, E.A., Barrett, R.W. and Dower, W.J. 1990. Peptides on phage: a vast library of peptides for identifying ligands. *Proc. The National Academy of Sciences*, 87, 6378-6382.

Dawson, J. M., Buttery, P. J., Jenkins, D., Wood, C. D. and Gill, M. 1999. Effects of dietary Quebracho tannin on nutrient utilization and tissue metabolism in sheep and rats. *Journal of Science for Food and Agriculture*, 79, 1423-1430.

Deharo, E., Bourdy, G., Quenev, O.C., Munoz, V., Ruiz, G. and Sauvain, M. 2001. A search for natural bioactive compounds in Bolivia through a multidisciplinary approach Part V. Evaluation of the antmalarial activity of plants used by the T. acana Indiana. *Journal of Ethnopharmacology*, 77, 91-98.

De Smet, P.A.G.M. 1997. Adverse effects of herbal drugs. Springer–Verlog, Berlin Heidelberg, New York, 3.

Duval, J. 1994. The control of internal parasites in ruminants. www.eap.mcgill.ca/AgrpBio/ab 370-04e.htm.

Fernando, S. 1982. Herbal Food and Medicine in Sri lanka. National NCO Council with support of FAO, Colombo, Sri lanka, 114.

Fernandez, T.J. 1991. Local plants having anthelmintic activity. *ASEAN Journal of Science and Technology Development*, 8, 115-119.

Fernandez, T.J. 1995. Anthelmintic value of *Tinospora rumphii*. In: PSAS, Proceedings of the 32nd Philippine Society of Animal Science Annual Convention, Volume I. Manila, Philippine, pp. 478-483.

Gakuya, D.W. 2001. Pharmacological and clinical evaluation of anthelmintic activity of *Albizia anthelmintica* Bragan, *Maeura edulis* dewolf and *Maerua subcordata* Dewolf plant extracts in sheep and mice. Ph.D. Theses, University of Nairobi.

Gothuma, J.M., Mbaria, J.M., Wanyama, J., Kaburia, H.F.A., Mpoke, L and Mwangi, J.N. 2004. Efficacy of *Myrsine apicana, Albizia anthelmintica* and *Hildebrandtia sealosa* herbal remedies against mixed natural sheep helminthosis in Samburu district, Kenya. *Journal of Ethnopharmacology*, 91, 7-12.

Guarrera, P. M. 1999. Traditional anthelmintic, antiparasitics and repellants uses of plants in Central Italy. *Journal of Ethnopharmacology*, 68, 183-192.

Hein, W.R., Shoemaker, C.B. and Heath, A.C.G. 2001. Future technologies for control of nematodes of sheep. *New Zealand Veterinary Journal*, 49, 247-251.

Hordegen, P., Hertberg, H., Heilmann, J., Langhans, W. and Maurer, V. 2003. The anthelmintic efficacy of five plant products against gastrointestinal trichostrongylids in artificially infected lambs. *Veterinary Parasitology*, 117, 51-60.

Hoskin, S.A., Barry, T.N., Wilson, P.R., Charleston, W.A.G. and Kemp, P. P. 1999. Growth and carcass production of young farmed deer grazing sulla, chicory or perennial ryegrass/white clover pasture in New Zealand. *New Zealand Journal of Agricultural Research*, 42, 83- 92.

Hoste, H., Chartier, C. and Le Frileux, Y. 2002. Control of gastrointestinal parasitism with nematodes in dairy goats by treating the host category at risk. *Veterinary Research,* 33, 531-545.

Idris, U.M. E.I.A.A., Adam, S.E.A. and Tartour, G. 1982. The anthelmintic efficacy of *Artemisia herba-alba* against *Haemonchus contortus* in goats. National Institute of Animal Health, *Quarterly,* Japan, 22, 138-143.

Jackson, F. and Miller, J.M. 2006. Alternative approaches to control - Quo Vadit? *Veterinary Parasitology,* 139, 371-384.

Jovellanos, J.M.M. 1997. Efficacy of three selected herbal plants on gastrointestinal parasites of cattle. DVM, University of Philippines, Los Banos, Philippines.

Kailani, S.U.R., Akhtar, M.S. and Asraf, M. 1995. Anti-fasciolic efficacy of indigenous plant drugs, kalongi, shahterah and karanjwa in buffalo. *Pakistan Journal Pharmacological Science,* 8, 17-27.

Katkar, C.M. and katker, M.S. 1984. Potential of Neem oil and Neem cake production in India. Research Planning Workshop on Botanical Pest Controlo Project (IRRI, Los Banos), Philippines.

Kembewa, B.M.D. 1999. Integration of indigenous veterinary remedies in farming system in Malawi. In: *Proceeding of International Seminar,* Kozhikode, 4-6 February.

Ketiz, J.K., Taylor, A., Bowman, D.P., Brown, D.L., Warnick, L.D. and Erb, H.N. 2002. *Chenopodium ambrosioides* and its essential oil as treatment for *Haemonchus contortus* and mixed adult nematode infections in goats. *Small Ruminant Research,* 44, 193-200.

Lal, J., Chandra, S. and Sabir, M. 1978. Modified methods for isolation of palasonin - the anthelmintic principle of *Butea frondosa. Indian Journal of Pharmacological Science,*40, 97-98.

Lans, C, Harper, T., Georges, K. and Bridgewater, E. 2000. Medicinal plants used for dogs in Trinidad and Tobago. *Preventive Veterinary Medicine,* 45, 201-220.

Lateef, M., Icbal, Z., Khan M.N., Akhtar, M.S. and Jabbar, A. 2003. An anthelmintic activity of *Adhotoda visica* root. *International Journal of Agricultutre and Biology,* 5, 86-90.

Molan A I, Waghorn G C, Min Br and Mc Nabb W C. 2000. The effect of condensed tannins from seven herbage on *Trichostrongylus colubriformis* larval migration *in-vitro. Folio Parasitology,* 47, 39-44.

Mostofa, M., Hasan, Q. and Sobhan, M.A. 1983. Efficacy of some indigenous medicinal plants against gastrointestinal nematodiasis in cattle and their comparative activities with that of nemafax. *Bangladesh Veterinary Journal,* 17, 1-4.

Mostofa, M., Alam, M.N., Ahmed, N.U. and Awal, M.A. 1995. Comparative efficacy of modern anthelmintics and neem seeds against natural infection of gastrointestinal nematodes in goats. *Bangladesh Agricultural University Research Progress,* 9.

Mostafa, M., Mckellar, M. and Alam, M.N. 1996. Epidemiology of gastrointestinal helminth parasites in small ruminants in Bangladesh and their anthelmintic therapy. In: *Sustainable Parasite Control in Small Ruminants,* (Eds. L.F. Le Jambre and M.R. Knox) ACIAR Proceedings, 74, 105-108.

Nari, A., and Hanson, J.W. 1999. Resistance of ecto- and endo-parasites: current and future solutions, 67th General Session, International committee, OIE, Paris, 17-21 May.

Ndi, A.N., Bayemi, C., Njwe, P.H., Tchoumboul, R., Njakoi, J., Moopoi, H., Njakoi, N. and Sali-Django, M. 1999. The anthelmintic efficacy of some indigenous plants in the Nort-west Province of Cameroon. *Rev. Elev. Med. Vet. Tropicaux,* 52, 103-106.

Niezen, J. H., Wahgorn, T. S., Charleston, W. A. G. and Waghorn, G. C. 1995. Growth and gastrointestinal nematode parasitism in lambs grazing lucern (*Medicago sativa)* or sulla (*Hedysarum coronarium)* which contasins condensed tannins. *Journal of Agricultural Science,* 125, 281-289.

Niezen, J.H., Charleston, W.A.G., Hodgson, J., Miller, C.M., Wahgorn, T.S. and Robertson, H.A. 1998 a. Effect of plant species on the larvae of gastrointestinal nematodes which parasitize sheep. *International Journal for Parasitology,* 28, 791-803.

Niezen, J.H., Robertson, H.A., Wahgorn, G.C. and Charleston, W.A.G. 1998 b. Production, faecal egg counts and worm burdens of ewe lambs which grazed six contrasting forages. *Veterinary Parasitology,* 80, 15-27.

Niezen, J. H., Robertson, H. A., Sidex, A. and Wilson, S. R. 2002. The effect of pasture species on parasitism and performance of lambs grazing one of three grass-white clover pasture swards. *Veterinary Parasitology,* 105, 303-315.

Paolini, V., Bergaud, J.P., Grisez, C., Prevot, F., Drochies, P. and Hoste, H. 2003. Effects of contensed tannins on goats experimentally infected with *Haemonchus contortus. Veterinary Parasitology,* 113, 253-261.

Pessoa, L. M., Morais, S. M, Bevilaqua, C. M. L. and Luciano, J. H. S. 2002. Anthelmintic activity of essential oil of *Ocimum gratissimum* Linn. And euganol against *Haemonchus contortus. Veterinary Parasitology,* 109, 59 – 63.

Poppi, D.P., Sykes, A.R. and Dynes, R.A. 1990. The effects of parasitism on host nutrition - the implications for nutrient manipulation. *Proc. New Zealand Society for Animal Production,* 50, 237-243.

Roepke, D.A.1996. Traditional and reapplied veterinary medicine in East Africa, Ethno-veterinary research and development, International Technology Publication, London.

Satrija, F., Nansen, P., Bjorn, H., Murtini, S. and He, S. 1994. Effects of papaya latex against *Ascaris suum* in naturally infected pigs. *Journal of Helminthology,* 68, 343-346.

Sharma, L.D., Bahga, H.S. and Srivastava, P.S. 1971. *In-vitro* anthelmintic screening of indigenous plants against *Haemonchus contortus* of sheep and goats. *Indian Journal of Animal Research,* 5, 33-38.

Sharma, S.R., Singh, D., Khan, F. ., Swarnkar, C.P. and Bhagwan, P.S.K. 2003. *In* vitro anthelmintic potential of *Xanthium strumarium* leaves against *Haemonchus contortus* of sheep. *Indian Journal of Animal Science,* 73, 342-344.

Singh, D. and Swarnkar, C.P. 2007. Alternative methods of control of gastrointestinal nematodosis in small ruminants. In: *Small Ruminants Emerging Diseases,*

Containment Under WTO Regime (Eds. V.S. Vihan, Ashok Kumar, V.K. Gupta and N.P. Singh), Satish Serial Publishing House, Delhi, pp. 267-303.

Singh, D,. Swarnkar, C.P. and Khan, F.A. 2002. Anthelmintic resistance in gastrointestinal nematodes of livestock in India. *Journal of Veterinary Parasitology,* 16, 115-130

Singh, D., Swarnkar, C.P., Khan, F.A., Srivastava, C.P. and Bhagwan, P.S.K. 1997. Epidemiology of ovine gastrointestinal nematodes at an organised farm in Rajasthan, India. *Small Ruminant Research,* 26 : 31-37.

Stepek, G., Behnke, J.M., Buttle, D.J. and Duce, I.R. 2004. Natural plant cystein proteinases as anthelmintics. *Trends in Parasiotology,* 20,

Swarnkar, C.P. and Singh, D. 2005. Nutritional intervention in management of gastrointestinal nematodes in small ruminants. In: *Proc. FAO Symposium on "Integrated Animal Parasite Management: From Acedemic Interest to Reality"* pp. 41 - 54.

Swarnkar, C.P., Singh, D., Khan, F.A., Bhagwan, P.S.K. and Dubey, S.C. 2004. *In-vitro* evaluation of *Melia azedarach* (Bakain) against pre-parasitic stages of *Haemonchus contortus. Indian Journal of Animal Science,* 74, 940 - 943.

Swarnkar, C. P., Singh, D., Khan, F. A., and Vaithiyanathan, S. 2007a. Potential role of tanniferous plants (*Prosopis cineraria*) in managing *Haemonchus contortus* infection in sheep. In: *Proceedings of XVIII National Congress of Veterinary Parasitology,* SKUAST-J, Jammu, 7 - 9 September.

Swarnkar, C.P., Singh, D., Khan, F.A., Kumar, M., Bhagwan, P.S.K. and Dubey, S.C. 2007 b. *In-vitro* ovicidal and larvicidal activity of *Butea frondosa* (Palas) seeds extracts on *Haemonchus contortus. Journal of Veterinary Parasitology,* (In-Press)

Vieira, L. ., Cavalcante, A.C.R., Pereira, M,F., Dantas, I.B.A. and Ximenes, L.J.F. 1999. Evaluation of anthelmintic efficacy of plants available in Geara state, North-Cost Brazil, for the control of goat gastrointestinal nematodes. *Revue De Medecine Veterinaire.* 150, 447-452.

Vihan, V.S., Kumar, A. and Arora, N. 2002a. Evaluation of anthelmintic activity of *Melia azedarach* (Bakain) against *Haemonchus contortus* infection in goats. In: *X International Congress of Indian Association of Animal Production,* New Delhi.

Vihan, V.S., Arora, N., Kumar, A. and Sharma, D.K. 2002b. Evaluation of anthelmintic activity of *Butea frondosa* (Palas) against *Haemonchus contortus* in goats. In: *National Symposium on Recent Trends in Diagnostics and Therapeutics of Animal Diseases,* Rajasthan Agricultural University, Bikaner.

Waller, P.J. 1999. International approaches to the concept on integrated control of parasites in livestock. *International Journal for Parasitology,* 29, 155-164.

❑❑❑

41

Relevance of Nematophagous Fungi in Control of Gastrointestinal Nematodes of Small Ruminants

F.A. KHAN
Central Sheep and Wool Research Institute, Avikanagar

Introduction

Gastrointestinal (GI) nematodes compromise welfare and cause serious production losses in small ruminants on pasture throughout the world. Anthelmintic resistance in GI nematode populations constitutes a major problem to grazing live stock globally. In order to avoid continued threat of anthelmintic resistance, work is being taken up throughout world to find new supplements or alternatives to chemotherapeutics. Restrictive legislation regarding the chemical control of parasites in organic live stock production together with a wish to eliminate drug residues in agricultural products are additional driving forces for the interest in alternative parasite control methods. The available alternate control approaches today are breeding for animals with a higher resistance to GI nematodes, nutritional interventions, vaccines, herbal anthelmintics and biological control (BC). BC using nematode destroying fungi is a potentially useful strategy that may become available for use in an integrated approach of controlling GI nematodes in near future.

Principles and Background for Biological Control

Nematophagous fungi have been known for a long time and are the natural enemies of nematodes. They have been found in all regions of the world, from tropics to Antartica and have been reported from agricultural, garden and forest soils, and are especially abundant in soils rich in organic materials. They comprise more than 200 species of taxonomically diverse fungi which utilize nematodes either as the main source of nutrients, or supplementary to a saprophytic existence (Barron, 1997).

These have several characteristics including highly specific parasitism to nematodes, not to plants and higher animals, and growth at suitable pH and temperature ranges on natural or synthetic media. These have been placed into the class Fungi Imperfecti (Deuteromycotina) and have been divided into three major groups-

- *Nematode- trapping (predacious) fungi,* which produce trapping devices on growing hyphae, such as adhesive networks, knobs and constricting and non-constricting rings.
- *Endoparasitic fungi* producing spores, which either stick to or penetrate through the cuticle of the nematode or passively enter the oesophagous of feeding nematodes.
- *Egg parasitic fungi* that attack eggs with their hyphal tips.

These BC agents rarely eliminate the larvae, but reduce the larval numbers to acceptable levels. In contrast to chemical control of nematode parasites which is directed towards the adult nematodes within the host, BC focuses on the free living larvae on pastures leading to reduction in their number. This reduction in infective larvae on herbage subsequently will prevent the build-up of worm burdens in hosts, which otherwise would cause sub clinical or clinical infections, particularly in young animals. Since any BC agent does not eliminate the number of infective larvae to zero, the grazing small ruminants will constantly get a small quantity of larvae and as a result should be able to develop a natural immune response.

To successfully develop a BC product a number of criteria have to be met, such as survival through GI tract, able to grow and be active in faeces, cheap to produce, and safe to handle. Of these, the gut survival ability has been the prime goal for the selection of candidates among the nematophagous fungi (Larsen *et al.*, 1991, 1992, 1994; Waller and Faedo, 1993; Waller *et al.*, 1994; Faedo *et al.*, 1997; Chandrawathani *et al.*, 2002; Sanyal, 2001; Khan *et al.*, 2002). Fungi are naturally picked up by grazing ruminants and subsequently excreted in the voided faeces (Hashmi and Connan, 1989; Larsen *et al.*, 1994; Manueli *et al.*, 1999; Saumell *et al.*, 1999; Chandrawathani *et al.*, 2002).

To date, *Duddingtonia flagrans* (DF) is the only nematophagous fungus having prospect of success against GI nematode larvae. Morphologically DF has short, erect and septate conidiophore. Conidia are formed at the conidiophore apex and slightly below without being borne on pronounced denticles (Fig.1). Conidia are ellipsoidal, truncate and 1-2 septate (Fig. 2). There is enormous production of chlamydospores, which are thick walled and red brown (Fig. 3). Mature chlamydospores have warty protuberances on their surfaces.

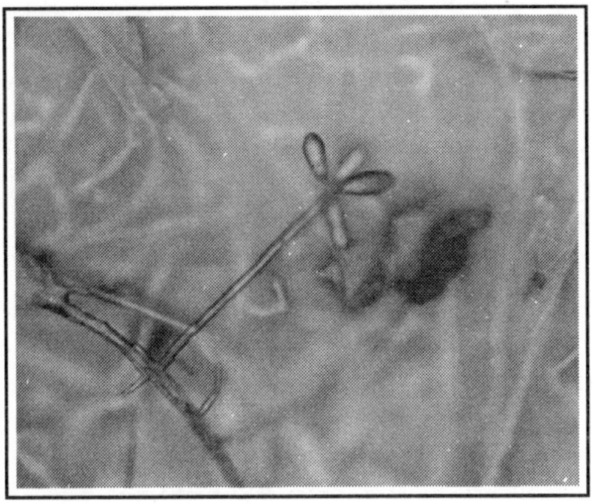

Fig 1. Conidiophore of *Duddingtonia flagrans* bearing conidia (400X)

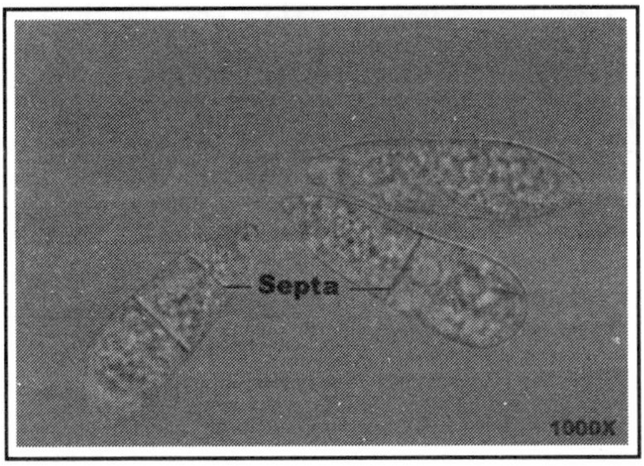

Fig 2. Conidia of *Duddingtonia flagrans* (1000X)

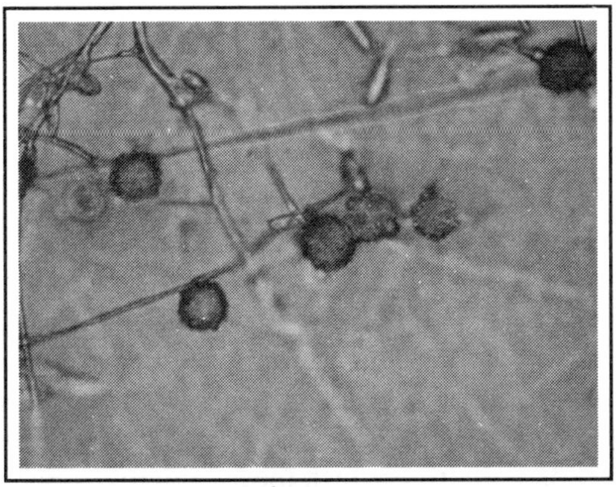

Fig 3. Chlamydospores of *Duddingtonia flagrans* (400X)

Duddingtonia flagrans is a natural part of the soil micro flora and may be isolated from different sources such as soil, grass, old faecal pellets/ pats and fresh faeces. However, fresh faecal samples are ideal source as the isolation of fungus through this way will be indicative of its gut survival. In this way we can save unnecessary labor at a later stage, because for deployment gut survival ability of the fungus is necessary so that the fungus will come in the faeces and will trap the developing larvae in faeces. It is very important to isolate fungus locally because of the reasons that laboratory maintained stocks may loose their nematode trapping ability and importation of such fungi from other countries has become very difficult these days. For isolating fungus from faeces, a large number of small samples *per rectum* should be collected from the grazing ruminants. A number of isolates of *D. flagrans* have been isolated from different sources from different parts of world (Table1).

Table 1. *Duddingtonia flagrans* isolated from different countries

Source of isolation	Country	No. of isolates	Reference
Compost and soil samples	Denmark	8	Larsen *et al.*,1991
Faeces from sheep, cattle, horse	Australia	16	Larsen *et al.*, 1994
Faeces from sheep, buffalo	India	2	Sanyal, 2000a
Faeces from sheep	India	1	Khan *et al.*, 2001
Faeces from sheep	Malaysia	1	Chandrawathani *et al.*, 2002
Faeces from sheep	India	1	Anon., 2006

Duddingtonia flagrans has been selected on the basis for its ability to survive passage through the GI tract of ruminants and subsequently to trap the developing parasite larvae in the deposited faeces (Larsen *et al.*, 1991, 1992; Sanyal, 2001; Khan *et al.*, 2002). This is because of the enormous production of chlamydospores. These chlamydospores are resistant to the stress of the GI tract. After passing through the animal, the spores germinate in faeces, forming specialized, 3-dimensional sticky networks which trap the developing larvae. The net formation is induced by physical contact with migrating nematodes. The trap formation starts by formation of a lateral branch. This branch grows out curls back and fuses with the parent hyphae. From this primary loop, many loops may be formed and as a result a sticky three- dimensional network is formed (Fig. 4 & 5). A mucous substance covers the trapping nets and is sticky to nematodes. Lectins are found in adhesive layer, which has affinity to the carbohydrate of the nematode cuticle. Extra cellular polymers (sugars, uronic acids and proteins) mediate the adhesion and represent the site for enzymes, which are important for the penetration of the nematode larvae. After penetration of the cuticle there is formation of infection bulb from which the assimilative (trophic) hyphae grow into the body of larvae.

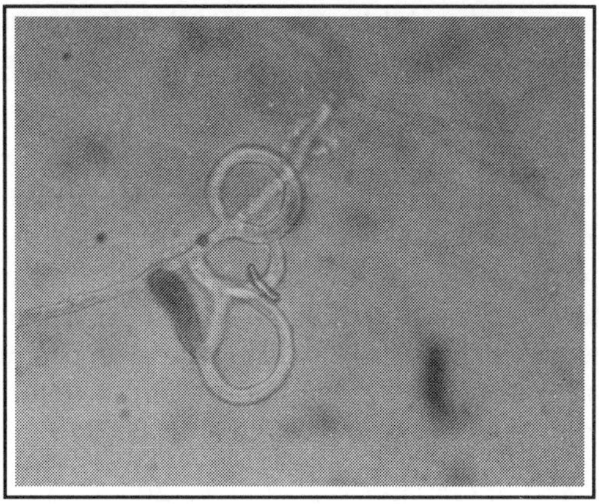

Fig 4. Three dimensional hyphal loop formed by *Duddingtonia flagrans* (400 X)

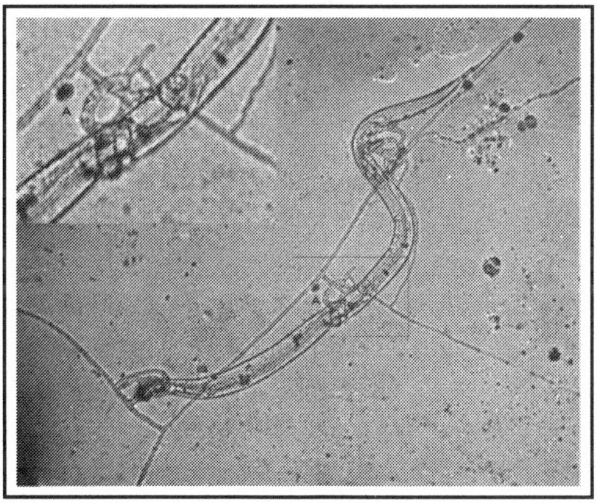

Fig 5. Infective larva trapped by *Duddingtonia flagrans* (200X)

Research results are now steadily accumulating to show parasitological as well as animal performance benefits by the strategic use of DF against GI nematodes of ruminants. Work has moved from concept phase and there are now companies actively involved in the technology in some countries. We can assess the relevance of DF in controlling GI nematodes of SR by looking into the results of research done in different parts of the world having different climatic conditions:

Temperate Regions

In Denmark the fungus has been shown to work well in reducing trichostrongyle infections in lambs at the dose rate of 1 million spores of DF/kg body weights daily mixed in 100 g of barley (Githigia *et al.*, 1997). From 2002 to 2004, three field trials were carried out to test the potential effect of daily fungus treatment to grazing ewes with lambs at foot through mixing spores with supplement (Larsen *et al.*, 2005). These trials demonstrated a positive effect on production with increased weight gains in lambs after feeding fungal spores daily to ewes and lambs for 2 months in the first half of the grazing season under Danish conditions.

Waghorn *et* al. (2003) evaluated the ability of DF to reduce number of infective larvae of three species of GI nematodes (*Haemonchus contortus, Teladorsagia circumcincta* and *Trichostrongylus colubriformis* in dung of both sheep and goats in New Zealand. The fungus could pass equally well through both species of ruminants and was equally effective against those GI nematodes. The fungus has been found effective in reducing larval numbers on pasture grazed by goats fed a daily dose at the rate of 5x 10^7 chlamydospores per animal per day for the 26 day grazing period in New Zealand (Wright *et al.*, 2003).

In Swedish island of Gotland Waller *et al.* (2004) conducted a study on 2 commercial farms, where a daily feed supplement of DF was provided for 6 weeks to ewes after lambing. Tracer tests showed that *T. circumcincta* and *Trichostrongylus* levels were significantly less in the fungal treatment group.

In Louisinia (USA) daily feeding of fungus to grazing ewes reduced larval population in faecal material, which reduced pasture infectivity and subsequently nematode burden in tracer animals (Fontenot *et al.*, 2003). The study showed that DF fed daily to grazing ewes was an effective BC agent in reducing a predominantly *H. contortus* population on pasture.

Burke *et al.* (2005) evaluated the efficacy of combine treatment of copper wire particles (COWP) and DF against GI nematodes in lambs in USA. They found no adverse effect of COWP on the ability of DF to inhibit residual larval development after COWP treatment. There was a beneficial effect of treating lambs with both COWP and DF. With fewer eggs being excreted due to the effect of copper and the additional larval reducing effect of DF, much lower larval challenge on pasture would be anticipated.

In Spain Gomez-Rincon *et al.*, (2006) evaluated the efficacy of DF on mountain pastures and concluded that fungal spores fed to sheep at critical times with regard to the epidemiology of parasites could have a significant effect on the infective larvae present on pasture, which could further improve lamb performance.

Eysker *et al.* (2006) evaluated the combined effect of evasive grazing and application of DF in lambs in Netherlands and found no clear effect of fungus on weight gain, faecal egg counts, worm counts, tracer worm counts and pasture larval counts. However, the yield of infective larvae in faecal cultures clearly demonstrated an effect of fungal application.

Tropical/ Subtropical Regions

In Australia single and sustained dosing experiments with locally isolated DF established that a dose rate between 5 x 10^5 and 10^6 chlamydospores per day given to sheep resulted in > 80% reduction in the number of infective larvae derived from nematode eggs in faeces (Larsen *et al.*, 1998). During a pasture plot study, where faeces collected from sheep given chlamydospores at a rate of 5 million/sheep/day was deposited on pasture plots, Faedo *et al.* (1998) found significantly lower numbers of larvae recovered from faeces and pasture from sheep dosed with fungus than on plots from control sheep. This study demonstrated that supplementary feeding or blocks administration offer potential fungal deployment options. Knox and Faedo (2001) conducted a field study to determine the effect of feeding 2 million spores of DF/animal daily for 6 months to lambs on pasture. Feeding fungus reduced the faecal egg counts and caused improvement in live weight gain. Successful attempts have been made to control the pre-parasitic stages of sheep nematodes using DF incorporated into nutrient supplement blocks (Waller *et al.*, 2001a) or a fungal controlled-release device (Waller *et al.*, 2001b). In blocks DF survived the manufacturing process and resulted into significant reductions in larval numbers. Chlamydospores were found capable of surviving pressures of several tones when incorporated into matrices and pressed into tablets for developing intraruminal controlled release devices. Chlamydospores released could reduce the number of infective larvae that developed in cultures of faeces collected from *H. contortus* infected sheep. Although such slow release device has been found successful in controlled experiments, there are still many unanswered questions to attend before this becomes an economical feasible option.

Pena *et al.* (2002) found a >94% reduction in number of developing larvae in cultures from lambs fed about 250,000 spores of DF/kg of body weight. There was >80% reduction in larval development in faeces from goat fed with 1 lac spores of DF/kg of body weight (Terril *et al.*, 2004).

In South Africa Maingi *et al.*, 2006 evaluated the effectiveness of selective anthelmintic treatments and feeding of DF (at the dose rate of 0.5 million chlamydospores/kg body weight daily) in reducing levels of GI nematodes in goats. Although the results did not show the improvement in parasite control by the deployment of fungus, there were fewer animals which required the anthelmintic treatment in fungus fed group.

In Malaysia pen feeding trials were conducted using sheep naturally and artificially infected with *H. contortus* and local isolate of fungus either in a grain supplement or incorporated into a feed block (Chandrawathani *et al.*, 2002). Fungus survived the gut passage in sheep and at about 1X 10^6 spores/animal/day, reduced the percentage of infective larvae developing in faecal cultures by more than 90%. Chandrawathani *et al.* (2003) found satisfactory results while assessing the control of nematode parasites of small ruminants using DF given as a daily feed supplement or incorporated into feed blocks in a small scale field trial. Young sheep receiving the fungus through both the methods did not suffer clinical haemonchosis, which necessitated salvage treatment in the control animals.

Further, long term field studies were conducted on two government managed sheep farms located in different agro climatic regions in Malaysia by Chandrawathani *et al.* (2004). The results have shown a benefit of using DF as a daily feed supplement which could improve the level of parasite control of sheep along with rapid rotational grazing.

In India work on BC of GI nematodes was initiated in 1998 with attempts to isolate local strains of these fungi from fresh faecal samples of ruminants. During a survey of fresh faecal samples from ruminants conducted at Anand (Gujarat) a total of 6 isolates (4 isolates of *Arthrobotrys oligospora* and 2 of *Duddingtonia flagrans*) were obtained (Sanyal, 2000a). In another study Khan *et al.* (2001) obtained 1 isolate each of *A. oligospora* and *D. flagrans* after examining 100 fresh faecal samples of sheep at Central Sheep and Wool Research Institute, Avikanagar (Rajasthan).

Laboratory experiments have been performed to evaluate the *in- vitro* trapping efficiency of these isolates grown in petri dishes containing the nutrient medium as well as in faecal cultures from ruminants infected with *H. contortus.* Sanyal (2000b) screened *D. flagrans* isolates obtained from the faeces of grazing livestock in western India for their suitability as bio control agents. In the experiment on predatory activity, it was observed that there was little trapping of infective *H. contortus* larvae during the initial 48- 72 hour post inoculation (PI), followed by a surge of predator (> 70%) in the following 72- 96 hour. While studying the trapping efficacy of local isolate of the fungus obtained from sheep faeces Khan *et al* (2000) found the isolate very effective as it trapped all *H. contortus* larvae within 48- 72 hours PI. Khan *et al.* (2001) reported that for the success as a nematode destroying fungus the concentration of chlamydospores of *D. flagrans* was important in faeces. About 92% reduction in larval numbers in faecal cultures was observed at the concentration of 2,000 chlamydospores/ g of faeces. Further the isolate was subjected to the investigations to examine its potential to survive passage through the gastrointestinal tract of sheep and subsequently to destroy *H. contortus* larvae in faecal culture (Khan *et al.*, 2002). Two different protocols of fungal dosing were tried. A single oral dose of 1.5 million chlamydospores was given to one sheep infected with *H. contortus.* Another *H. contortus* infected sheep received a total of 3 million chlamydospores in two equal doses 2 hour apart. The larval reduction capability of the fungus was evaluated by faecal culture set up from the fungus infused sheep. A clear reduction in the number of larvae 24-hour post dosing of both the treatment protocols was observed. Sanyal (2004) demonstrated that larval recovery from coprocultures decreased with increase in chlamydospores concentration added to the faecal cultures.

Now efforts are on to find out a suitable and cheap method for delivering fungus, which could be integrated with existing animal husbandry practices without causing problems to the farmers. The occurrence of night housing of animals in India provides the practical option of deploying fungus spores in a feed supplement. The incorporation of chlamydospores of *Duddingtonia flagrans* in concentrate feed pellets has been tried at National Dairy Development Board, Anand, however, with a very poor shelf life of spores (Anon., 2001). The fairly long shelf life of desiccated chlamydospores led to be used for top dressing of concentrate feeds. Feeding trials were conducted by Sanyal (2002) with stall- fed sheep parasitized with *H. contortus*. Animals were offered concentrate feed which had been top-dressed with desiccated chlamydospores for 10 days. It was observed that spore level of 1×10^6 / kg body weight virtually eliminated larvae from both the pasture and the faecal cultures.

At Avikanagar barley grains supporting the growth of *D. flagrans* have been incorporated into straw based complete feed blocks (CFB) to develop a delivery system for biological control of gastro intestinal nematodes. CFB have been developed as an alternative to grazing and supplementation feeding strategy for sheep production, especially during adverse conditions such as drought, which are common phenomena in arid and semi- arid regions (Raghuvansi *et al*. 2007). The fungus survived well the block making process and had 4 to 5 weeks of shelf life. A paddock trial was conducted for 6 weeks during monsoon season (August and September) when the environment is most conducive for development and survival of pre-parasitic stages of GIN on pasture. From the day of turnout and over the following 6 weeks animals of fungus treated group were fed once a day with chlamydospores of *D. flagrans* through feed blocks. The fungus feeding resulted in significant reductions in larval numbers in faecal cultures set up during the feeding period (Table 2). Larval recovery from the herbage samples started from 2nd week of the experiment and from then on larval counts increased markedly on the plot grazed by the control group. On the other hand low levels of herbal larval counts were maintained in the plot grazed by fungus treated group up to the 4th week of experiment (Figure 4). After that no difference could be observed in herbage larval counts from both the pastures. It is obvious as the shelf life of chlamydospores in the block was observed only for 4 to 5 weeks. Therefore, in order to have continued effect preparation of fresh batch of CFB is required after every 4 weeks. The study demonstrated that block feeding might prove a potential means of delivery of *D. flagrans* for successful biological control of *Haemonchus contortus* (Anon, 2004).

There are few important things which should be considered for potential use of DF. The most important one is the timing of deployment of fungus. It should be fed at the time when there is maximum availability of infective larvae on pasture. Therefore, one very important feature in implementing biological control with DF is to know the detailed epidemiology of GI nematodes in question in the system in question. In order to get best results timing of treatment is also very important and it should start just before an expected increase in herbage larval number appears. The treatment period should be long enough to cover most of the anticipated average life expectancy (months) of adult parasites. By using right dose and time it should be possible to significantly reduce the number of infective larvae on pasture resulting into the safe environment for young stock from heavy infection and disease. The treatment should be given daily otherwise it will result in an uneven shedding of fungal spores in the faeces of animals which will allow some eggs to develop to infective larvae in a faecal environment with sub optimal concentration of DF.

Table 2. Mean faecal egg count (FEC), mean infective larvae (per gram of faeces) and percent reduction of infective larvae from faecal cultures of control and fungus- fed sheep over an 6- week grazing experiment.

Days	FEC (eggs per gram)		Faecal culture L_3		
	Control	Fungus - fed	Control	Fungus - fed	Reduction (%)
1	400	700	42.9	Nil	100
4	333	533	65.0	10.3	84.2
7	433	1067	70.8	15.3	78.4
10	2767	2733	41.2	8.3	80.0
13	2700	2600	63.8	11.8	81.5
16	2667	3233	53.6	5.6	90.0
19	1867	4300	52.6	3.2	93.9
22	1733	7133	49.6	1.4	97.2
25	9350	4050	55.2	10.2	81.5
28	9350	9000	71.3	28.0	60.7
31	20100	13650	53.6	32.6	39.2
34	22300	13800	60.2	18.4	69.4
37	15100	5800	55.3	21.8	60.6
40	14450	3150	60.1	60.2	0
43	11150	2800	65.3	56.1	14.1

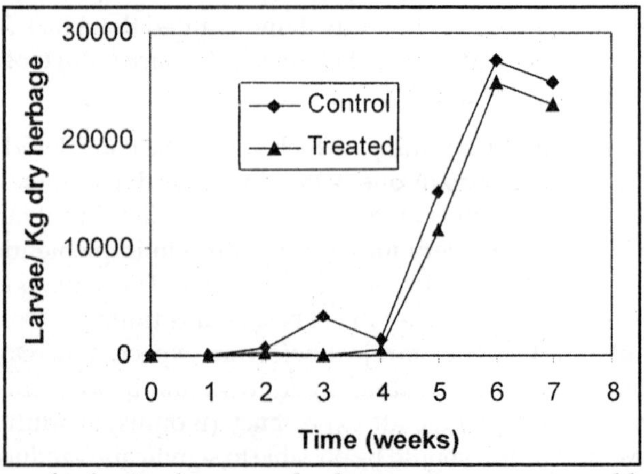

Fig 6. Number of infective larvae on herbage on pasture grazed by fungus fed sheep and unfed control sheep

Conclusion and Recommendation

From the above discussion it is apparent that there is substantial evidence supporting the use of DF against the free-living larval stages of many of the most important GI nematodes of ruminants. Today, research and trials are either on going or being planned in many countries in Europe, North and South America, Southeast Asia and in the pacific region. With the progress, continuing interest and investment in nematophagous fungi as BC agents, the future looks very promising and it is likely that a control strategy of GIN, alternative or adjunct to chaemotherapy, will surface for use in the near future. BC is advantageous in the way that it could be applicable to a range of worm parasites not only within, but also between livestock species. The approach of BC is preventive in nature. BC may be used as a supplement or adjunct to chaemotherapy in the attempt to control widespread anthelmintic resistance within nematode populations of livestock. Besides, such non-chemical alternative control options become more important with regard to control of parasitic nematodes of livestock on organic farms. One major task in the future will be to test the effect of DF in integrated parasite control strategies involving different potential elements such as restricted use of anthelmintics, exploitation of resistant animals and implementation of various grazing managements. In both traditional and organic production systems diet may also influence the level of parasitism and needs attention.

REFERENCES

Anon., 2001. Biotechnology Laboratory Report 2000-2001. National Dairy Development Board, Anand, Gujarat, 15-16.

Anon., 2004. Annual Progress Report 2002-2003. All India Network Programme on Gastro intestinal Parasitism, Central Sheep and Wool Research Institute, Avikanagar, Rajasthan, 44-45.

Anon., 2006. Annual Progress Report 2002-2003. All India Network Program on Gastro intestinal Parasitism, Central Sheep and Wool Research Institute, Avikanagar, Rajasthan, 37-38.

Barron, G.L. 1977. The nematode- Destroying Fungi. Canadian Biological Publications, pp 1-140, Guelph, Ontario, Canada.

Burke, J.M., Miller, J.E., Larsen, M. and Terril, T.H. 2005. Interaction between copper oxide wire particles and *Duddingtonia flagrans* in lambs. *Veterinary Parasitology* 134, 141-146.

Chadrawathani, P., Omar, J., Waller, P.J., Hoglund, J., Larsen, M. and Zahar, W. M. 2002. Nematophagous fungi as a biological control agent for nematode parasites of small ruminants in Malaysia: a special emphasis on *Duddingtonia flagrans. Veterinary Research* 33, 685-696.

Chandrawathani, P. Jamnah, O. Waller, P.J. Larsen, M. Gillespie, A.T. and Zahari, W.M. 2003. Biological control of nematode parasites of small ruminants in Malaysia using the nematophagous fungus *Duddingtonia flagrans. Veterinary Parasitology* 117, 173-183.

Chandrawathani, P. Jamnah, O., Adnan, M., Waller, P.J., Larsen, M. and Gillespie, A.T. 2004. Field studies on the biological control of nematode parasites in the tropics, using the microfungus *Duddingtonia flagrans*. *Veterinary Parasitology* 120,177-187.

Eysker, M., Bakker, N., Kooyman, F.N.J., Olthus, S.O. and Ploeger, H.W. 2006. Effect of biological control through the daily application of spores of *Duddingtonia flagrans* in lambs kept under an evasive grazing system in the Netherlands. *Veterinary Parasitology* 140, 312-320.

Faedo, M., Larsen, M. and Waller, P.J. 1997. The potential of nematophagous fungi to control the free-living stages of nematode parasites of sheep: Comparison between Australian isolate of *Arthrobotrys* spp. and *Duddingtonia flagrans*. *Veterinary Parasitology* 72, 149-155.

Faedo, M., Warnes,E. H., Dobson, R.J. and Waller, P.J. 1998. The potential of nematophagous fungi to control the free- living stages of nematode parasites of sheep: Pasture plot study with *Duddingtonia flagrans*. *Veterinary Parasitology* 76, 129-135.

Fontenot, M. E., Miller, J. E., Pena, M. T., Larsen, M. and Gillespie, A. 2003. Efficiency of feeding *Duddingtonia flagrans* chlamydospores to grazing ewes on reducing availability of parasitic nematode larvae on pasture. *Veterinary Parasitology* 118, 203-213.

Githiga, S.M, Thamsborg, S.M., Larsen, M., Kyvsgaard, N.C. and Nansen, P. 1997. The preventive effect of the fungus *Duddingtonia flagrans* on trichostrongyles infections of lambs on pasture. *International Journal for Parasitology* 27, 931-939.

Gomez-Rincon,C., Uriarte, J. and Valderbano, J. 2006. Efficiency of *Duddingtonia flagrans* against richostrongyle infections of sheep on mountain pastures. *Veterinary Parasitology* 141, 84-90.

Hashmi, H.A. and Connan R.M. 1989. Biological control of ruminant trichostrongylids by *Arthrobotrys oligospora*, a predacious fungus. *Parasitology Today* 5, 28-30.

Khan, F.A., Swarnkar, C.P., Singh, D., Bhagwan, P.S.K. and Dubey, S.C. 2001. Nematophagous fungi isolated from sheep faeces: A promising biological control of *Haemonchus contortus*. *Indian Journal of Animal Sciences* 71, 834-836.

Khan, F.A., Swarnkar, C.P., Singh, D., Bhagwan, P.S.K. and Dubey, S.C. 2002. Predacious activity of *Duddingtonia* flagrans against *Haemonchus contortus* larvae in faeces after passage through the gastro-intestinal tract of sheep. *Indian Journal of Animal Sciences* 72, 559-560.

Knox, M. R. and Faedo, M. 2001. Biological control of field infections of nematode parasites of young sheep with *Duddingtonia flagrans* and effects of spore intake on efficacy. *Veterinary Parasitology* 101, 155-160.

Larsen, M., Wollstrop, J., Henriksen, S.A., Dackman, C., Gronvold, J. and Nansen, P. 1991. *In -vitro* stress selection of nematophagous fungi for biocontrol of parasitic nematodes in ruminants. *Journal of Helminthology* 65, 193-200.

Larsen, M., Wollstrop, J., Henriksen,S. A., Gronvold, J. and Nansen, P. 1992. *In- vivo* passage through calves of nematophagous fungi selected for biocontrol of parasitic nematodes. *Journal of Helminthology* 66, 137-141.

Larsen, M., Faedo, M and Waller, P.J. 1994. The potential of nematophagous fungi to control the free living stages of nematode parasites of sheep: Survey for the presence of fungi in fresh faeces of grazing livestock in Australia. *Veterinary Parasitology* 53, 275-281.

Larsen, M., Faedo, M, Waller, P.J. and Hennessy, D. R. 1998. The potential of nematophagous fungi to control the free living stages of nematode parasites of sheeo: Studies with *Duddingtonia flagrans. Veterinary Parasitology* 76, 121-128.

Larsen, M., Thamsborg, S. M. and Mejer, H. 2005. Experiences from Danish field trials to evaluate the effect of the biological control agent Duddingtonia flagrans, fed to ewe with lambs at foot. Addendum at Novel Approaches to the Control of Helminth Parasites of Livestock, 4th International Conference, Merida, Mexico, 10-12 January, 2005.

Maingi, N., Krecek, R. C. and Biljon, N. van. 2006. Control of gastrointestinal nematodes in goats on pastures in South Africa using nematophagous fungi *Duddingtonia flagrans* and selective anthelmintic treatments. *Veterinary Parasitology* 138, 328-336.

Manueli, P.R., Waller, P.J., Faedo, M. and Mahommed, F. 1999. Biological control of nematode parasites of livestock in Fiji: screening of fresh dung of small ruminants for the presence of nematophagous fungi. *Veterinary Parasitology* 91, 39-45.

Pena, M.T., Fontenot, M. E., Miller, J. E., Gillespie, A. and Larsen, M. 2002. Evaluation of of the efficacy of *Duddingtonia flagrans* in reducing infective larvae of *Haemonchus contortus* in faeces of sheep. *Veterinary Parasitology* 103, 259-265.

Raghuvansi S K S, Prasad R, Tripathi M K, Mishra A S, Chaturvedi O H, Misra A K, and Jakhmola R C. 2007. Effect of complete feed blocks or grazing and supplementation of lambs on performance, nutrient utilization, rumen fermentation and rumen microbial enzymes. *Animal* 1, 221-226.

Sanyal, P.K. 2000a. Presence of predacious fungi in fresh faeces of ruminants from western India. *Journal of Veterinary Parasitology*, 14, 133-135.

Sanyal, P.K. 2000b. Screening for Indian isolates of predacious fungi for use in biological control against nematode parasites of ruminants. Veterinary Research Communications, 24: 55-62.

Sanyal, P.K. 2001. Biological control of *Haemonchus contortus* in sheep by nematophagous fungus, *Duddingtonia flagrans. Indian Veterinary Journal* 78, 1091-1094.

Sanyal, P.K. and Mukhopadhyaya, P.N. 2003. Top dressing of feed with desiccated chlamydospores of *Duddingtonia flagrans* for biological control of the pre-parasitic stages of ovine *Haemonchus contortus. Veterinary Research Communications* 27, 381-390.

Sanyal, P.K. 2004. Density dependent nematophagous behaviour of *Duddingtonia flagrans. Indian Veterinary Journal* 81, 16-19.

Saumell, C.A., Padilha, T. C., Santos, P. de. and Roque, M.V.C. 1999. Nematophagous fungi in fresh faeces of cattle in the Mata region of Minas Gerais State, Brazil. *Veterinary Parasitology* 82, 217-220.

Waghorn, T. S., Leathwick, D. M., Chen, L. Y. and Skipp, R. A. 2003. Efficacy of the nematode- trapping fungus *Duddingtonia flagrans* against three species of gastrointestinal nematodes in laboratory fecal cultures from sheep and goats. *Veterinary Parasitology* 118, 227-234.

Waller, P.J. and Faedo, M. 1993. The potential of nematophagous fungi to control the free-living stages of nematode parasites of sheep: Screening studies. *Veterinary Parasitology* 49, 285-297.

Waller, P.J., Larsen, M. Faedo, M. and Hennessy, D.R. 1994. The potential of nematophagous fungi to control the free-living stages of nematode parasites of sheep: *in-vitro* and *in-vivo* studies. *Veterinary Parasitology* 51, 289-299.

Waller, P. J., Knox, M. R. and Faedo, M. 2001. The potential of nematophagous fungi to control the free living stages of nematode parasites of sheep: feeding and block studies with *Duddingtonia flagrans*. *Veterinary Parasitology* 102, 321-330.

Waller, P. J., Faedo, M. and Ellis, K. 2001. The potential of nematophagous fungi to control the free living stages of nematode parasites of sheep: towards the development of a fungal controlled release device. *Veterinary Parasitology* 102, 299-308.

Waller, P. J., Schwan, O., Ljungstrom, B. L., Rydzik, A. and Yeates, G. W. 2004. Evaluation of biological control of sheep parasites using *Duddingtonia flagrans* under commercial farming conditions on the island of Gotland, Sweden. *Veterinary Parasitology* 126, 299-315.

Wright, D.A., Mc Anulty, R.W., Noonan, M.J. and Stankiewicz, M. 2003. The effect of *Duddingtonia flagrans* on trichostrongyles infections of Saanen goats on pasture. *Veterinary Parasitology* 118, 61-69.

❑❑❑

42

Grassland and Pasture Management Systems in Arid and Semiarid Region An Over View

R.K. JAIN, R.P. NAGAR

Regional Research Station, Indian Grassland & Fodder Research Institute, Avikanagar - 304 501, Rajasthan

Introduction

In arid and semiarid regions of India, animal husbandry occupies an important role, increasing with the aridity. The grasslands and grazing lands serve as major forage resource base for livestock in such regions. Such areas were considered very important since ancient times as is evident from Vedic literature in which these have been projected as the ecosystems i.e. the plants, the animals they support, and the soil supporting them. Like in other parts of the world, the major roles of such areas in the country are: (i) a primary feed resource for livestock and (ii) an effective means for maintaining soil fertility. These two roles are closely linked through grazing animals. From the ecological viewpoint these act as a natural protective screen to protect arable lands from invasion of desert in dry regions and to prevent lands from water erosion in wet and mountainous regions. However, because of misuse and neglect their present condition is far from satisfactory besides shrinkage in area on account of land pressure for other activities. So today, these are generally considered as mere assemblage of vegetation that offers grazing and browsing. In fact such lands are in an advanced stage of deterioration due to many factors such as large bovine population, free range grazing, lack of management practices and deforestation for development activities. Also, the arid and semiarid regions experience several natural constraints *viz.*, extremes of temperature, steepness of slope, irregular rainfall and lack of water. However, the most important constraint affecting these areas is their extreme fragility.

Thus, protection and re-vegetation of range/grazing lands in the country has become a challenge. Such aspects, in context of arid and semiarid regions, have received attention of Indian Council of Agricultural Research (ICAR) through its institutes *viz.*, Indian Grassland and Fodder Research Institute (IGFRI), Jhansi; Central Arid Zone Research Institute (CAZRI), Jodhpur and Central Soil Water Conservation Training and Research Institute (CSWCR&TI), Dehra Dun besides several traditional and state agricultural universities in the country. Consequently, a good amount of information is now available on improvement of grasslands/grazing lands through various ecological approaches, including soil and water conservation and appropriate range management practices.

Arid and semiarid areas are essentially water deficit zones. In India such areas constitute over 40 per cent of total land area. The arid zone has comparatively less coverage (10 % of total land area), largely confined to western states *viz.*, Rajasthan and Gujarat. The semiarid zone has more coverage (> 30 % of total land area), largely confined to southern states and western parts of north-western states (Table 1). Rajasthan has highest per cent of area (73.6 %) as arid and semiarid followed by Gujrat (29.5 %) and Andhra Pradesh (21.5 %).

Table 1. Extent of arid and semiarid zones in different states of India.

State	Area (million km²)			% Area of state
	Arid	Semiarid	Total	
Andhra Pradesh	0.022	0.138	0.182	21.5
Gujarat	0.064	0.090	0.154	29.5
Haryana	0.013	0.027	0.040	6.8
Karnataka	0.009	0.139	0.148	17.6
Madhya Pradesh	-	0.059	0.059	6.2
Maharashtra	0.001	0.189	0.190	20.2
Punjab	0.016	0.032	0.048	8.3
Rajasthan	0.195	0.121	0.316	73.6
Tamil Nadu	-	0.095	0.095	10.0
Uttar Pradesh	-	0.064	0.064	6.7

Livestock Population *vs* Grazing Area

The grazing pressure (ACU/ha) calculated on the basis of area available for grazing indicate that it is as high as 25.1 in Punjab (due to very less area for grazing) to 1.18 in Madhya Pradesh and 1.30 in Karnataka. The experimental results show that in arid region, the carrying capacity is hardly in the range of 0.2-0.5 ACU/ha whereas in semiarid region it is 1 ACU/ha when range condition is classified as good (Raheja, 1966; Singh and Roy, 1998). Table 2 depicts information related to livestock population, calculated as Adult Cattle Units (ACU), and availability of grazing area.

Table 2. Livestock population and grazing area in arid/semiarid region of nine states in India.

State	Livestock Population (ACU) (x 1000)	Total Grazing Area (x 1000 ha)	Grazing Intensity (ACU/ha)	Grazing Area/ACU (ha)
Andhra Pradesh	20108	8943	2.25	0.44
Gujrat	13012	2916	4.46	0.22
Haryana	4111	338	12.16	0.08
Karnataka	18415	14179	1.30	0.77
Madhya Pradesh	6739	5729	1.18	0.85
Maharashtra	13329	6495	2.05	0.49
Punjab	4982	198	25.1	0.04
Rajasthan	36197	5465	6.63	0.15
Uttar Pradesh	14123	1131	12.48	0.08

Grazing Systems and Pastoralism

A variety of grazing systems are practiced in different climatic zones of the world where pastoralism is the way of life. The community grazing activity is mainly dependent on the availability of grazing resources from pastures and other grazing lands.

In India, there are nearly 30 pastoral communities, located particularly in the northern and western parts of the country covering the states of Jammu & Kashmir, Himachal Pradesh, Uttar Pradesh, Haryana, Rajasthan and Gujrat. These communities are also found, in relatively less concentration, in the south *viz.*, Andhra Pradesh, Karnataka and Tamil Nadu. In other states *viz.*, Maharastra and Madhya Pradesh only a few communities are there.

The "Kharak" system of grazing in Uttar Pradesh Hills and Gols" system of grazing in Rajasthan are the examples of a grazing practice in which tribal move forward or change the site after using it.

In case of migration, nomadic tribes move in accordance with a definite scheme of periodic/seasonal migration on the basis of which four grazing systems have been recognised in India *viz.*, (i) Total Nomadism (ii) Semi Nomadism (iii) Transhumance (iv) Partial Nomadism (Tyagi and Singh, 1988).

Total Nomadism

It covers systems in which livestock owners do not have a permanent place of residence and their families move with herds in search of forage and water.

Their movement tends to follow repetitive pattern to and from fairly well defined areas. Now, number of such tribes is very low. The sheep breeders of Madhya Pradesh and "Banjaras" of Bikaner (Rajasthan) are still nomadic tribes. To have a separate identity of sheep breeders of Madhya Pradesh, the state government have introduced a higher grazing licence fee for them. Similarly, the "Banjaras" of Bikaner are totally nomadic. They move in desert area with their herds particularly cattle of "Rathi" breed.

Semi-Nomadism

It covers systems in which livestock owners have a permanent place of residence near which supplementary cultivation is practiced. However, for greater part of the year, they travel with their herds to distant grazing areas. There are still large number of pastoral tribes who are semi-nomadic such as "Gujjars" of Punjab and Haryana; "Raikas" and "Rabari" of Marwar tract in Rajasthan; "Bhrawad" and "Rabari" of Gujrat; "Dhangar" of Maharashtra; "Golla", Khurba" and Lambada" of Andhra Pradesh and Karnataka.

Transhumance

It covers systems in which farmers have a permanent place of residence and their herd are tended by a herdsman for long periods of time in distant grazing areas, including forest grazing areas. Rajasthan has a large tribal population, especially in the hilly areas in south where they practice transhumance. The herdsmen take the animals of rich farmers to grazing areas in other states also. The tribes of Madhya Pradesh also act as herdsmen and graze the animals in forest grazing areas.

Partial Nomadism

It covers systems in which farmers have a permanent place of residence and keep herds in the vicinity. Village grazing systems may also be put in this category in which grazing animals remain in the vicinity. The "Gadariya", "Ahirs" and "Gujjars" of Uttar Pradesh, Madhya Pradesh and Rajasthan; "Ahirs" of Bihar, Gujrat and Maharastra; "Bhils" of Madhya Pradesh; "Toda" of Nilgiris in Tamil Nadu and "Gauda" of Orissa are the examples in this category. Village grazing system may also be put in this category in which grazing animals remain in the vicinity. In Bundelkhand region villagers send their animals to the nearby upland and hilly areas where during rainy season natural vegetation is at its peak. In Rajasthan, the majority of "Gujjars" are cultivators, but a large number of them living in hills forests are graziers.

In most cases, nomadic migration is confined to their respective states. The frequent draught and famine conditions in western Rajasthan lead to a large scale migration of the human and livestock population to Haryana, Punjab, Madhya Pradesh and Gujrat. However, the problem of nomadic animal husbandry is not so acute in Rajasthan when compared to the Himalayan region. The selective grazing and overstocking along grazing routes are the main reasons of decline in range health with the hazards like soil erosion and weed invasion in the Himalaya (Tyagi and Shankar, 1988).

Grass cover

In India (8°-35° N and 60°-97.5° E), there is a great variation in climate, soil and physiography and consequently the rangelands are also very heterogeneous in nature. Dabadghao and Shankarnarayan (1973) have identified five major grass covers of India in their monumental book - Grass Covers of India. This work has included the ICAR survey work conducted during 1954-1962 at 507 sites spread over the whole country and also the work of Whyte (1964). The grass covers of arid and semiarid zones are mainly *Sehima-Dichanthium* and *Dichanthium-Cenchrus-Lasiurus* types. The *Sehima-Dichanthium* covers the whole of peninsular India, Satpura and the Vindhyan area below Jhansi in the south and up to southern Bihar in east spreading over 1.74 million km^2. The floristic list includes 24 perennial grasses and 129 other herbaceous species of which 56 are legumes. *Sehima* is more prevalent on gravel and the cover may be 87 per cent of the ground. *Dichanthium* flourishes on level soils and the cover may be 80 per cent of the ground.The *Dichanthium-Cenchrus-Lasiurus* cover is restricted to north-western arid regions having an area of 0.436 million km^2. The floristic list includes 11 perennial grasses and 45 other herbaceous species of which 19 are legumes. Highest development of cover is *Dichanthium* type on heavier soils, *Cenchrus* type on loamy soil to sandy loam soils and *Laisurus* type on sandy soils. Species wise major grasses in arid and semiarid zones include *Sehima nervosum, Dichanthium annulatum, Cenchrus* Sp., *Lasiurus sindicus* , *Desmostachya bipinnata* and *Sporobolus marginatus*. The details for these species are presented in Table 3.

Grassland Restoration

As the primary functions of rangelands vary, the objectives of restoration of a degraded rangeland will vary as per the intended major use. For example, if the major emphasis is placed on livestock productivity, the output per animal and the productivity per unit of land are to be increased. Even for normal grazing use provision for drinking water, stands for salt licks and shade trees for grazing animals are required. Basically, the range restoration techniques are based on ecological approaches e.g. protection for vegetation recovery, soil and water conservation, reseeding, improved range management techniques and grazing management.

Protection for Vegetation Recovery

The rangeland degradation due to overgrazing leads to a situation where nutritionally low grade and unpalatable species dominate the scene. The vegetation becomes sparse and soil working in such situation is likely to further accentuate the erosion process. It has been found that just by mere protection through fencing a lot of benefit in terms of vegetation recovery can be obtained. Rangelands can be protected through barbed/woven/chicken wires or chain links supported by angle iron/cement/ stone/wooden poles or through thorn fencing/live hedges/cattle proof ditches/bunds etc. (Kanodia and Patil, 1982). Studies conducted on protection of grasslands at IGFRI, Jhansi revealed increase in herbage yield from 0.1 t/ha - 3.5 t/ha within three years. The plant population of desired perennial grasses increased from 11 plants to 397 plants/m^2 and undesirable forbs decreased from 1444 to 33 plants/ m^2. (Trivedi and Kanodia, 1982).Once the desired level of vegetation recovery has been achieved, the rangelands should be released for grazing in the deferred rotational grazing management system by compartmentalizing with the help of inner fencing into paddocks (Shankar, 1986).

Table 3. The major grass species encountered in arid and semiarid regions of India.

Grass Species	Habitat and Fodder Value	Major Associated Species
Cenchrus Sp. (*C. ciliaris, C. setigerus*)	On well drained sandy alluvial soils (sandy to sandy loams). Good fodder value.	*Aristida funiculate, A. adscensionis, Eragrostis ciliaris, Cenchrus biflorus, Eleusine compressa, Dactyloctenium sindicum, Urochloa panicoides, Indigofera linifolia, Tephrosia purpurea, Phaseolus trilobus, Rhynchosia minima*
Desmostachya bipinnata	On young alluvium. The river terraces liable to inundation during monsoon support this cover. It is deep rooted drought resistant, tussocky perennial grazed only when young. It is some times an obnoxious weed in cultivated fields which get inundated during monsoon.	-
Dichanthium annulatum	On old alluvium – alluvial flats with sandy clay loam to clay soils. Good fodder value	*Brachiaria ramosa, Heteropogon contortus, Eragrostis* Sp., *Indigofera* Sp., *Alysicarpus* Sp.
Lasiurus sindicus	On loose sandy soils in low rainfall regions. Species highly nutritious	-
Sehima nervosum	On hills and piedmonts where soil and moisture conditions are favourable. It does not occur on heavy soils and very wet ground. Good fodder value	*Aristida funiculate, A. adscensionis, Cympogon jawarancusa, C. martini, Heteropogon contortus* and *Eremopogon faveolatus*
Sporobolus marginatus	On low lying heavy saline soils– in depressions where water stagnates and some salinity is there due to impeded drainage. It is a drought resistant tussocky perennial in lightly grazed areas.	*Dichanthium annulatum* in patches where salinity is reduced by natural leaching.
Panicum antidotale	On sandy soils – an excellent soil binder. It is a deep rooted drought resistant grass.	-

Soil and Water Conservation

Measures related to soil and water conservation in range and pasture management in dry areas are of crucial importance. The measures are especially important on degraded sloppy lands. Most useful such measures are described below.

Bunding : Along the contour lines and periphery bunds are made, either manually or using tractor mounted implements. The details *viz.*, cross section and interval of location of bunds are decided according to topography. In dry degraded areas, bunds are required to be stabilized by sodding with suitable grass species *viz.*, *Dichanthium annulatum*, *Cenchrus ciliaris* and *Panicum antidotale*. The excess runoff may be safely dispersed off through grassed waterways or pipe outlets (Srinivasan et al.,1962).

Water Storage/Discharge Structures: The construction of water storage/ discharge structures *viz.*, checkdams, anne-cuts and embankments, of desired engineering specifications, in areas like ravines and watersheds etc. are required to be constructed for efficient conservation and utilization of already scarce water resources in dry areas. In some ravinous hilly areas gully plugging may also require erecting structure of various engineering specifications.

Biological measures: Various grass and legume species are known to play an important role in soil and moisture conservation. The runoff and soil loss under such systems is reduced considerably in every agro-climatic region. In excessive windy areas a combination of trees may be planted as wind/shelter belt, in specific designs, to check the wind impact. The spreading of dry grasses or shrubby growth on bare ground along the contours or terraces have the potential to check soil and water loss. Various chemicals may also be used for mulching effect and soil stabilization. Some examples are as under:

(1) Contour furrows (60 cm wide and 25 cm deep) at a distance of 10-15 m across the slope increased forage production of *Cenchrus ciliaris* considerably (130 per cent over non-furrow plots) (Singh, 1990a).

(2) In IRWH (Inter Row Water Harvesting) system (30 cm wide ditch alternated with 70 cm wide raised beds), the yield of pasture grasses and legumes raised on the edge of ditch was higher by 124 and 66 percent respectively over the conventional system of planting (Singh, 1990a).

(3) Pitting in arid lands is a good device for effective moisture conservation. Use of Staggered Pitter Discer, a tractor attached device developed at CAZRI, Jodhpur has significantly increased forage yield (80-115 per cent in natural pasture and 50-70 per cent in seeded pasture respectively) (Das and Yadav, 1979).

(4) At IGFRI, Jhansi in *Sehima-Dichanthium* grassland contour furrows made at an interval of 6 m was found to be most effective in soil moisture conservation when compared to 3 and 12 m spaced furrows (Shankar and Gupta, 1993).

Reseeding: The grasslands in dry region are usually of poor quality, containing less than seven percent CP for most of the time. Also, because of excessive grazing pressure on most of these grasslands, the productivity of such lands has deteriorated to considerable extent. In order to increase the pasture productivity in dry areas, it is essential to replace low yielding annual grasses with high yielding perennial grasses that are adaptable to the prevailing condition of that region. Similarly, as the use of fertilizers in these lands is of little practical value on account of high costs, introduction of suitable pasture legumes may be one of the best ways to improve the quality.

Most useful such measure/reseeding techniques are as follows:

(1) The important species identified for reseeding is provided in Table 4. Selection of such species depends on the site conditions. *Dichanthium annulatum* and *Cenchrus ciliaris* have wide range of climate and soil in areas having rainfall up to 750 mm. In such situation suitable legumes include *Alysicarpus rugosus, Stylosanthes hamata, Styzolobium deeringianum, Indigofera endecaphylla* and *Lablab purpureus.*

(2) Pre monsoon broadcasting of *Stylosanthes* seeds in *Sehima-Heteropogon* grasslands was found best in terms of legume establishment. In case of *S. hamata* broadcasting of seeds after harrowing gave maximum establishment as compared to broadcasting alone or line sowing at 50 cm apart. Also, use of pelleted seeds gave better establishment (Rai and Patil, 1985). However, in arid situations pelleting of seed have not proved to be of any specific advantage except in case of *Laisurus sindicus* .

(3) Significant increase in nitrogen fixation is possible by appropriate selection of *Rhizobium* strains and by appropriate seed pelleting practice. By using a seed rate of 4 - 6 kg/ha after required scarification pretreatment and premonsoon sowing/ broadcasting have provided the nitrogen equivalence to the tune of 40-60 kg/ha in natural grassland (Rai *et al.*,1980).

Improved Management Practices

Bush cleaning: The undesirable bushes are required to be removed, either manually or by mechanical felling and removal of stumps or application of selective weedicides on the cut stamps to kill them to stop coppicing. The leaf fodder yielding shrubs may, however, be retained in appropriate proportion. It has been found that in rangelands of Indian deserts medium stock density level of *Zizyphus nummularia* (14 % of land area covered by shrub canopy) was optimum for high forage as well as leaf fodder production (Shankar and Gupta, 1993).

Application of fertilizers: Application of fertilizers on productivity of grasslands has been extensively studied in arid and semi-arid regions. The response is maximum in case of nitrogen doses followed by phosphorus, almost no response with potassium. The timing of fertilizer application is also important, maximum response has been obtained when applied about 30 days after the start of monsoon. Similarly, application of 60 kg N/ha in three split doses at 25 days interval was found to be superior (Rai and Kanodia, 1981). The forage yield increases were in the range of 40-70 per cent over control.

Table 4. Important grass species for reseeding in arid and semiarid regions.

Condition	Species	Strains/Varieties	Seed rate (kg/ha)
Extreme arid situations	*Lasiurus sindicus*	CAZRI-317, 319, 351; M 30-7, 20-5 and 295	4-5
Sand dunes	*Panicum antodotale*	CAZRI-331, 333, 379, 347	2-3
Arid and Semi-arid - sandy loam soils	*Cenchrus ciliaris*	CAZRI-358; Marwar Anjan (75); IGFRI-3108	4-5
Arid and Semi-arid - sandy loam soils	*Cenchrus setigerus*	Marwar Dhaman (CAZRI-76, 296, 175, 415); Pusa Yellow Anjan	4-5
Arid and Semi-aridred gravelly soil/sloppy lands	*Chrysopogon fulvus*	IGFRI-1	3-4
Semi-arid clay soil with higher moisture status	*Iseilema laxum* *Panicum maximum*	Hamil, PGG-1, PGG-9	3-4 5-6
Arid and Semi-arid - loam soils with higher moisture status	*Dichanthium annulatum*	CAZRI-490, 485; IGFRI-495-1; Marvel-8	2-3

Controlled burning: Controlled burning, if used judiciously, is a powerful range/pasture management tool. It is preferable to use this tool in combination with moderate grazing management. Burning is recommended during wetter periods with longer intervals, about three years (Shankar and Trivedi, 1988).

Grazing management: In pasture grazing systems, a period of rest is essential for the grasses to recoup and rejuvenate. In *Sehima* dominated grasslands in Jhansi deferred rotational grazing system has been found to be superior in providing more number of animal days (2925) as compared to continuous grazing system (2097 days) with a body weight gain of 191 g/heifer/day, without being detrimental to the range health (Upadhyay *et al.*,1971). In *Lasiurus* dominated arid grassland the same system was superior in terms of average body weight gain and calving of "Tharparker" breed of heifers over continuous grazing system (Shankarnarayan *et al.*,1981). In semiarid conditions, following proper grazing management carrying capacity of different grasslands varied from 6.2–10.5 sheep/ha/year (Table 5).

Table 5. Carrying capacity of different grasslands at Jhansi.

Composition (sheep/ha/year)	Carrying capacity
Cenchrus ciliaris	7.0
Cenchrus setigerus	10.5
Heteropogon contortus	7.0
Dichanthium annulatum	6.2

(Source: IGFRI, 1990)

A mixed pasture of *Cenchrus ciliaris* + *Stylosanthes hamata* was adjudged best for body weight gain (342.5 g/animal/day) in case of calf, while in case of heifers *Cenchrus ciliaris* + *Macroptelium purpureum* pasture was the best. Studies on management practice for "Mandya" sheep indicated that it is be well maintained throughout the year on natural grasslands, if these animals are allowed to graze on stylo - siratro mixed pasture for about two hours in addition to six hours grazing in natural grasslands (IGFRI, 1990).

Improving Grassland Productivity

Grassland and rangeland degradation is the greatest challenge to the rangeland users in India. In order to meet the various demands of the growing population, this process needs to be halted and methodologies developed for their sustainable development and utilization (Singh 1991). In a study on plant-soil-animal interface at IGFRI, Jhansi during (1987-1991), maximum average biomass production was obtained under sown pasture (Table 6).

Table 6. Biomass yield and economics under three different grassland management systems in semiarid region.

System	Yield (DM t/ha)						Return (B/C)
	1987	1988	1989	1990	1991	Average	
Sown pasture	2.7	10.9	8.3	4.3	3.6	6.8	1.27
Improved pasture	2.3	9.7	8.0	3.7	6.1	6.2	1.17
Natural pasture	1.3	6.1	6.8	6.0	3.7	5.1	2.65

The introduction of stylo added substantially (40 %) to the biomass in the second year but reduced to almost negligible in 1990 after second year of grazing. It showed the necessity of introducing stylo every third year. Natural grasslands, however, proved more economical in terms of output per rupee invested. Body weight gain in goats was similar in sown and improved pastures but it decreased in natural grasslands. Sheep showed best performance in sown pastures. Cattle gave better performance in improved pasture (Singh, 1995).

Agroforestry Systems

In India, because of various factors such as competition for land due to intense population pressure, rapidly depleting fuel wood resources, time-scale for orchard operations, and dispersed distribution of benefits from forestry and seasonal shortage of labour; the growing of trees did not find a suitable place in rural economy and rangeland development programmes in the past. However, due to the ecological and social cost of rangeland degradation besides highly adverse land-man and land-livestock ratio that has placed enormous pressure on non-renewable land resource, emphasis is now shifting towards agroforestry based land-use systems.

Agroforestry Based Fodder Production Systems

Agroforestry based production systems are capable of yielding leaf fodder, wood and food at the same time besides conserving and rehabilitating the ecosystem. The woody perennial/tree is one of the main components in such systems (Deb Roy, 1995).

Silvo-pastoral system

Such system integrates pasture and/or animals with trees. Here, woody perennial, preferably of fodder value, are introduced deliberately and systematically and managed scientifically. Under poor soil, water and nutrient situations where cropping is not possible such systems can serve the twin purposes of forage and firewood production and ecosystem conservation. In India such systems have great potential on wastelands, especially in the arid and semi-arid region (Singh and Roy, 1993).

Integration of livestock with the tree + pasture crops is quite an old practice. It is probable that as soon as cattle were domesticated there was some integration. Cattle are grazed in dry forests and savanna areas and during long dry season fodder trees are lopped so that cattle could feed on browse. Three major ways of utilization of silvo-pastoral systems can be identified:

- grazing and /or browsing in natural forests.
- grazing or harvesting pasture grown under planted trees.
- browsing or harvesting of tree foliage.

Arid Region : Khejri (*Prosopis cineraria*), the most important top feed resource in the arid regions, has been extensively studied on aspects related to economic uses, tree growth and productivity, soil and micro-climate amelioration, water use efficiency and compatibility with the pastures (Mann and Saxena, 1980). Among exotic species, *Acacia tortilis* has been studied on similar lines. This species was found to be the best for sand dune areas (Shankarnarayan et al.,1987). In a study conducted by Shankar et al. (1976) at Jodhpur, highest composition of grass cover and herbage yield was recorded under 14-18 year old plantation of *Prosopis cineraria* (Table 7).

Table 7. Herbage yield and vegetation under various trees (14 -18 year old) at Jodhpur.

Vegetation Character	**Tree**				
	1	**2**	**3**	**4**	**5**
Herbage Yield (t/ha)	2.34	0.78	1.32	1.66	0.85
Plant Population (no./m^2)	313	217	250	483	171
Annuals (no.)	26	18	35	22	13
Perennials (no.)	10	9	8	10	5

Note: Tree 1 = *Prosopis cineraria;* Tree 2 = *Acacia senegal;* Tree 3 = *Albizia lebbeck;* Tree 4 = *Tecomella undulata;* Tree 5 = *Prosopis juliflora*)
(Source: Shankar et al.,1976)

Gupta (1983) worked out the net annual returns for the raising of annual crops and livestock, on marginal lands, following a survey of 70 farm households in five villages. It was compared with an alternative silvo-pastoral system involving growing of pasture grasses and multipurpose trees for production of fodder, fuel wood and small timber. Employment increased to 30-75 man days/ ha annually in silvo-pastoral system when compared to with an average of 26 in annual crops. The returns were also higher if full rotation of the tree species is considered, variation in return in different systems was due to tree species and its rotation.

Semiarid Region : This agro-ecosystem, having a variety of soils, climate and vegetation; silvopastures with native tree species viz., *Azadirachta indica, Cassia siamea, Morus alba, Pongamia pinnata* (Syn. *Derris indica*), *Tamarindus indica* and *Acacia* sp. have been practiced in this region since long. However, tree intensity per ha has remained quite low (Singh and Roy 1993). Extensive work on silvopasture system productivity and their management has been done at IGFRI, Jhansi since 1970s. Table 8 provides information regarding productivity of various improved silvipastoral combinations.

The utilization of tree component through lopping management has also been standardized for several top feed species in silvopastoral system. Of six species studied, *Albizia amara* and *Albizia procera* provided highest forage and firewood yields (Table 9).

In Karnataka, yield of few grasses was studied in association with *Acacia nilotica* by Subbayam (1980). In a 7 year period *Cenchrus ciliaris* gave peak green yield of 3.4 t/ha/cut followed by *Dichanthium annulatum* (1.7 t/ha/cut) and *Chloris bournii* (1.0 t/ha/cut).

Table 8. Production from different silvi-pastoral combinations (mean of 5+ year data) in Bundelkhand region.

Tree species	Pasture species	Yield (t DM/ha)
Acacia tortilis	Natural pasture	3.2
	Cenchrus ciliaris	3.4
Albizia lebbeck	*Cenchrus ciliaris*	7.8
	Chrysopogon fulvus	7.8
	Sehima nervosum	7.9
Albizia procera	*Cenchrus ciliaris + Dolichos lablab*	7.9
Dalbergia sissoo	Natural pasture	2.9
Dichrostachys cinerea	*Cenchrus ciliaris*	4.3
	Chrysopogon fulvus	5.2
	Cenchrus ciliaris	4.2
Hardwickia binata	*Cenchrus ciliaris*	4.2
	Chrysopogon fulvus + Sehima nervosum	3.7
Leucaena leucocephala	*Cenchrus ciliaris*	4.1
	Cenchrus setigerus	3.2
Albizia amara	*Cenchrus ciliaris*	4.1
	Chrysopogon fulvus + Sehima nervosum	3.3

Source: Singh and Roy, 1998

Table 9. Forage and firewood production from some multipurpose trees (at 10 years) in Bundelkhand region.

Tree species	Yield (kg/tree/ha)	
	Forage	Firewood
Acacia tortilis	1.1 (0.51)	18.6 (11.21)
Albizia amara	21.2 (11.2)	32.3 (19.70)
Albizia lebbeck	8.5 (4.21)	8.6 (5.11)
Albizia procera	20.7 (11.38)	15.7 (9.32)
Dichrostachys cinerea	4.3 (2.76)	9.2 (5.63)
Hardwickia binata	6.7 (3.67)	8.5 (5.53)

Values in parenthesis are dry weights (Source: Roy, 1991)

Role of top feeds on rangelands: Fodder from trees is mainly available from two parts *viz.*, leaf twigs and pods. This forage is usually rich in proteins, vitamins and minerals like calcium. They are, however, in general low in phosphorus and crude fibre (Singh, 1990b). Such top feeds species play an important role in human food security through their function as animal feed resource. During dry season, forage from trees may constitute the only source of green forage in many situations

(Deb Roy, 1988). The importance of top feeds increases with the severity of drought and progression of drought season. In dry areas it is the winter and summer months (Raghvan, 1990). Besides fodder, such trees have many other functions. They may provide fire wood, poles for building and fencing, live fence, fibre, fruits, spices, fats, medicines, dyes and tannin. Most of these uses are exploitative and compete, either in the short or long run, with the supply of fodder by the trees. Some other uses such as for live fencing are complimentary, as they encourage cultivation of the species and increase the availability of feeds. Of various uses to which fodder trees are put, perhaps the most significant is that dealing with soil fertility maintenance as it raises the chances of cultivation of fodder trees in production systems. Leguminous fodder trees such as *Gliricidia sepium*, *Leucaena leucocephala* and *Acacia albida* are the examples.

The top feeds are considered very important in vegetation stabilization and sustained productivity of rangelands. They may also play important role as windbreaks and by providing shade for the grazing animals (Sincere and Ranjhan, 1989).

List of common tree and shrub specie

A. Tree species

Botanical name	Common name	Botanical name	Common name
Acacia catechu	Khair	*Leucaena leucocephala*	Subabul
Acacia leucophloea	Oringia	*Madhuca indica*	Tut
Acacia nilotica	Babul	*Madhuca longifolia*	Mahua
Acacia Senegal	Kumat	*Melia azedarach*	Bakain
Acacia tortilis	Israeli Babul	*Milletia auriculata*	Ganj
Aegle marmelos	Bel	*Morud indica*	Mulberry
Ailanthus excelsa	Ardu	*Pithecellobium dulce*	Jungle jalebi
Albizia lebbek	Siris	*Pongamia glabra*	Karanji
Azadirachta indica	Neem	*Prosopis cineraria*	Khejri
Bauhinia variegata	Katchnar	*Prosopis juliflora*	Vilayati baul
Capparis spinosa/deciduas	Karil	*Quercus incana*	Banj
Collophospermum mopane	Mopane	*Salvadora oleoides*	Pilu/Jal
Cordia dichotoma	Losoru	*Schleichera oleosa*	Kusum
Dalbergia sissoo	Shisham	*Syzyium cuminii*	Jaman
Dendrocalamus strictus	Bamboo	*Tamarindus indica*	Imli
Ficus bengalensis	Bargad	*Tecomela undulata*	Rohira
Ficus religiosa	Pipal	*Terminalia arjuna*	Arjun
Grewia oppositifolia	Bhimal/Biuk	*Zizyphus jujuba*	Beri
Hardwickia binata	Anjan	*Zizyphus mauritiana*	Ber
Holoptelea integrifolia	Kanju	*Zizyphus rotundifolia*	Ber

B. Shrubs

Botanical name	Common name	Botanical name	Common name
Acacia jaequemontii	Bambli	*Dichrostachys nutans*	Nutan
Balanites aegytica	Balanites	*Leptadenia pyrotechnica*	Leptadenia
Bauhinia racemosa	Zinja	*Lycium barbarum*	Lycium
Calligonum polygonoides	Phog	*Maytenus emarginata*	Maytenus
Calotropis procera	Aak	*Zizyphus nummularia*	Pala
Caparis aphylla	Kair	*Tamaris articulata*	Tamarix

Some precautions for the lopping of the tree

- Fresh leaves should not be lopped as they are often toxic and thereby nature's device for protection.
- Saplings and poles are not to be lopped.
- It is safe to give two season's rest after lopping in order to recover from the injury due to lopping.
- Lopping may be restricted to lower two-third of the crown protecting the upper one-third which can manufacture the food for the plants till new leaves come up.
- At the time of lopping branches having a diameter of over 7.5 cm may be avoided.
- Lopping should not be carried out as far as possible on the eroded area or area prone to erosion hazard.

Agro-silvi-pastoral systems

Such systems integrate crop, pasture and/or animals with trees. Here again woody perennial, preferably of fodder value, are introduced deliberately and systematically and managed scientifically. Such systems can be used for food production and soil conservation besides providing fodder and fuel wood. These systems hold promise especially in highland humid tropics. It may be tree-livestock-crop mix around homestead, woody hedgerows for browse, green manure, soil conservation or an integrated production of pasture, crops, animals and wood (Singh and Roy, 1991).

Horti-pastoral systems

Such systems integrate fruit trees with pastures. The main objective is to obtain fruits. Trees become focus of the system and activities to establish pastures or produce animals become secondary. In such systems it is important to check that introduction of forage system is beneficial or at worst neutral to tree crop. There may be competition between trees and pasture but the extent and effect will depend upon spacing. The list of fruit crops suitable for developing horti-pastures in different situations is given in Table 10.

Table 10. Fruit trees suitable for developing horti-pastures in arid and semi-arid conditions.

Category	Fruit trees	
	Local name	**Botanical name**
Indigenous drought hardy fruits	Ker	*Capparis decidua*
	Pilu	*Salvadora oleoides*
	Gondi	*Cordia gharaf*
	Jhas ber	*Zizyphus nummularia*
Fruit trees for rainfed conditions	Ber	*Zizyphus mauritiana*
	Guava	*Psidium guajava*
	Karonda	*Psidium guajava*
	Pomegranate	*Punica granatum*
	Custard apple	*Annona squamosa*
	Aonla	*Emblica officinalis*
Fruit trees for irrigated conditions	Fig	*Ficus carica*
	Phalsa	*Grewia subinequalis*
	Date palm	*Phoenix dactilifera*
	Papaya	*Carica papaya*
	Lime	*Citrus auronitifolia*
	Sweet orange	Citrus sinensis
	Sweet orange	*Citrus sinensis*

Source: Singh and Roy, 1991

The fodder species selected for this purpose should be adaptable to local agro-climatic conditions, of fast growth rate, of high coppicing ability and should provide higher economic returns. Depending on the site conditions *Gliricidia sepium*, *Erythrina* sp., *Zizyphus mauritiana*, *Prosopis juliflora*, *Dichrostachys cinerea*, *Acacia* sp. and *Albizia* sp. may be used (Westley, 1990).

Forage Conservation / Transportation

In these regions, there are flush periods of forage production in grassland during July - October after which there is scarcity period. Even if there is fodder cultivation it can barely meet the requirements during December - April. Thus scarcity remains during October - December and April - July. Therefore, it is necessary to conserve surplus green fodder for use in lean period. Similarly, there is regional imbalance in forage availability and occurrence of drought in some or the other part of the country. In order to meet such challenges, a system of efficient transportation is also required. In this part the most common technique of hay making and baling are presented.

Hay: For hay making, A - frame, tripod, zig-zag fence and single pole structures are useful. Harvesting of crop should be started when at least 50 per cent of dew-drops are evaporated from the leaves. The harvested material should not be immediately tied into bundles or collected into small heaps, which is the common practice followed, but should be scattered on the ground. After harvest, it should be collected with the help of hand rake and placed on the structure. The uppermost layer needs to be nicely dressed in the form of a thatch roof, which helps in draining the rain water quickly in zig-zag fence. The material may be tied into loose bundles for placing on horizontal bracings.

About 200 kg of green grass may be placed for drying on the A - frame, tripod stand and zig-zag fence at a time. The green material is spread (6-8 cm thick) uniformly on the frame, on both the slopes. It takes usually 4-7 days during the month of September to have safe moisture content for storage. The material should intermittently be examined for moisture content. About 18-20 per cent moisture in long loose hay is quite safe for storage (IGFRI ,1990).

Baling : The baling and densification of hay material, including crop residues serve as important feed reserve for the scarcity period and also for ease in transportation of areas of deficit forage supplies. Several indigenously designed, manual and bullock-operated presses are used in this region. The machine operated baling machines are more efficient, although a bit costly. Compounded herbage feed mixtures including grass, straw and even tree leaves may be densified to the extent of 350-400 kg/m^3 using such machines. Densified fodders are 6-7 times less in volume and are easy and economical to transport (IGFRI, 1990).

Conclusion and Recommendations

Appropriate grassland/pasture land management coupled with appropriate soil, water and range management techniques offer immense possibilities of ameliorating productivity of degraded areas for a variety of situations in arid and semiarid regions of the country. Already a good deal of literature is available on these lines, but a sounder understanding of such system is required for large-scale adoption. There is a need for a multi-disciplinary team approach for designing further refined technologies for specific situations/ habitats.

REFERENCES

Dabadghao, P. M. and Shankarnarayan, K. A. 1973. *The grass cover of India*, ICAR, New Delhi, 713 p.

Das, R. B. and Yadav, R. C. 1979. Development of CAZRI's "Staggered Pitting Discer" for water harvesting and increased pasture productivity in arid zone. *Annals of Arid Zone* 18, 96-100.

Deb Roy, R. 1988. Strategy for increased forage production through silvipastoral system, In: *Forage production in India* (ed. Panjab Singh), RMSI, Jhansi (India), 164-169.

Deb Roy, R. 1995. Agroforestry based forage production system, In: *New vistas in forage production* (eds C. R. Hazra & Bimal Misri), IGFRI, Jhansi, 47-72.

Gupta, T. 1983. The economics of tree crops on marginal agricultural lands with special reference to the hot arid region in Rajasthan, India. *International Tree Crops Journal* 2, 155-194.

IGFRI, 1990. *Forage production technology*, Indian Grassland & Fodder Research Institute, Jhansi, 43 p.

Kanodia, K. C. and Patil, B. D. 1982. Pasture development, In: *Desert resources and technology* (ed. Alam Singh), Scientific Publishers, Jodhpur, 263-305.

Mann, H. S. and Saxena, S. K. 1980. *Khejri (Prosopis cineraria) in the Indian desert*, Monograph No. 11, CAZRI, Jodhpur.

Raghvan, G.V. 1990. Availability and use of shrubs and tree fodder in India, In: *Shrubs and tree fodder for farm animals* (ed. C. Devendra), IDRC, Ottawa (Canada), 196-210.

Rai, P. and Patil, B.D. 1985. Effect of pelleting on establishment and production of forage and seed in *Stylosanthes* species. *Indian Journal of Range Management* 6, 19-25.

Rai, P. and Kanodia, K.C. 1981. Effect of split application of nitrogen on the quality and quantity of *Sehima-Heteropogon* grassland. *Forage Research* 7, 83-85.

Rai, P., Kanodia, K.C., Patil, B.D., Velaydhan, K.C. and Agrawal, R. 1980. Nitrogen equivalence of range legumes introduced in natural grasslands. *Indian Journal of Range Management* 1, 97-101.

Raheja, P. C. 1966. Rajasthan desert can bloom with forage. *Indian Farming* 15: 47.

Roy, M.M. 1991. Some tropical fodder trees for sustained fodder and firewood availability during lean periods. *Journal of Tropical Forage* 1, 196-205.

Shankar, V. 1986. Vegetation recovery under protection, In: *Desert environment: conservation and management* (eds K.A. Shankarnarayan and V. Shankar), CAZRI, Jodhpur, 25-28.

Shankar, V. and Trivedi, B.K. 1988. Grazing burning rotation for the management of *Sehima-Heteropogon* grasslands in Central India, *Proc. 3rd International Rangeland Congress* (Abstract vol. 1) (eds P. Singh, V. Shankar and A.K. Srivastava), RMSI, Jhansi, 382-385.

Shankar, V. and Gupta, J.N. 1993. Restoration of degraded rangelands, In: *Restoration of degraded lands : concepts and strategies* (ed. J. S. Singh). Rastogi Publications, Meerut, 116-155.

Shankar, V., Dadhich, N.K. and Saxena, S.K. 1976. Effect of Khejri tree (*Prosopis cineraria* Macbridge) on the productivity of range grasses growing in its vicinity. *Forage Research* 2, 91-96.

Shankarnarayan, K.A., Harsh, L.N. and Katju, S. 1987. Agroforestry in arid zones of India. *Agroforestry Systems* 5, 69-88.

Shankarnarayan, K.A., Kackar, N.L. and Mertia, R.S. 1981. Productivity of large herbivores, cattle and sheep in arid grazing land ecosystem, *Monograph* No. 19. CAZRI, Jodhpur.

Sincere, M.N. and Ranjhan, S.K. 1989. The place of fodder trees and shrubs in grassland systems, In: Proc. XVI Intl. Grass. Cong. (4-11 October), Association Francaise pour la Production Fourragere, Nice (France), 1761-1768.

Singh, P. 1990a. Grassland and pasture development in the arid region of Rajasthan, *Proc. Rehabilitation of degraded land ecosystem* (UNESCO-ICAR), CAZRI, Jodhpur, 1-8.

Singh, P. 1990b. Agrisilviculture systems in India, In: *Shrubs and tree fodder for farm animals* (ed. C. Devendra), IDRC, Ottawa (Canada), 186-195.

Singh, P. 1991. Research strategies for re-vegetation and management of degraded rangelands in India, *Proc. IVth Intl. Rangeland Cong.* (VOL. II), M. Association Francaise de Pastoralisme, Paris, 805-806.

Singh, P. 1995. Improved methods in rangeland production and utilization systems for sustainable development in India, *Proc. Vth Intl. Rangeland Cong.* (VOL. I), Society for Range Management, Denver, 516-517.

Singh, P. and Roy, M.M. 1991. Forage production through agroforestry : constraints and priorities. *Range Management and Agroforestry* 12, 169-178.

Singh, P. and Roy, M.M. 1993. Silvipastoral systems for ameliorating productivity of degraded lands in India. *Annals of Forage* 1, 61-73.

Singh, P. and Roy, M.M. 1998. Agroforestry and rangeland development, In: *Fifty years of Agronomic Research in India* (eds R. L. Yadav, Panjab Singh, Rajemdra Prasad and I.P.S. Ahlawat), Indian Society of Agronomy, New Delhi, 221-254.

Srinivasan, V., Bonde, W.C. and Tejwani, K.G. 1962. Studies on grass and their suitability to stabilize and maintain bunds in the ravine lands of Gujrat. *Journal of Soil Water Conservation* (India) 10, 72-78.

Trivedi, B.K. and Kanodia, K.C. 1982. Succession of grassland vegetation on semi-rocky substratum at Jhansi. *Indian Journal of Range Management* 3, 1-10.

Tyagi, R.K. and Shankar, V. 1988. Pastoralism and grazing systems in the Central Himalayas. Proc. 3rd International Rangeland Congress (Abstract vol. II) (ed. P. Singh, V. Shankar and A. K. Srivastava), RMSI, Jhansi, 665-668.

Upadhyay, V.S., Dabadghao, P.M. and Shankarnarayan, K.A. 1971. *Annual Progress Report*, Indian Grassland and Fodder Research Institute, Jhansi.

Westley, S.B. 1990. Living fences - an agroforestry technology. *Agroforestry Today* 2, 11.

Whyte, R.O. 1964. Grassland and Fodder Resources of India. ICAR, New Delhi.

❑❑❑

43

Management of CPR, Cultivated Pastures and Siloi-Pasture in Semi-Arid and Arid Environment for Sustainable Small Ruminant Production

J. S. MANN, S.C. SHARMA
Central Sheep & Wool Research Institute Avikanagar
(via:Jaipur) Rajasthan 304 501

Introduction

Agriculture is the main occupation of the rural people and about 70% of the population is engaged in this activity. Livestock rearing is complementary to agriculture. The total area under cultivation is about 169.7 M ha out of 328.726 M ha total geographical area. India has 180.2 m small ruminant (57.6 and 122.5 m, respectively for sheep and goat) population (FAO, 2001), which is one of the densest in terms of small ruminants per unit of cultivated land. A total of 37% of the geographical area of the country (107.43 m ha) is degraded due to various causes of land degradation. Average size of land holding is 0.15 ha of which most of them (59%) are marginal farmers who depended wholly or partially on small ruminant farming and only 4.4 % of the available land is used for fodder production. Sheep and goat production in India heavily depends on traditional feeding methods including most common grazing on waste and community lands. Further, sustained and high population growth rates, combined with limited and rapidly diminishing land holds and land for food grains and cash crops led to shortage of feed and fodder to small ruminants in India, which resulted in overgrazing and degradation of the available land.

As a result, the productivity of small ruminants is not in accordance with their genetic potential. To overcome this problem, efforts are necessary for establishment of proven pasture and silvi-pastoral systems on waste and community lands to get optimum production from sheep and goat through increasing highly nutritious proteinous top fodder and forage production.

Silvi-pastoral systems are defined as growing of ideal combination of grasses, legumes and trees for producing highly nutritious top fodder and forage, fuel wood, timber and optimising land productivity, conserving plants, soil and nutrients etc. on sustainable basis on the same unit of land. This involves replantation, substitution or intervention in the existing vegetation by desirable species.

Present Status of the Forage Production

The forage resource base has undergone a considerable neglect at the hands of researchers, planners and the development agencies. Green herbage availability varied from 4.7 to 29.1 t/ha. The stocking capacity of pastures between 0.4 and 13.3 sheep/ha/annum under natural vegetation. Forage cultivation is restricted to only about 4.4 % of the cultivated area. This is basically because of the preponderance of marginal and small land holdings. Besides grazing and fodder trees, the major local forage resource is the crop residue, which again is too inadequate to sustain the livestock. In India there is a gap between demand and supply of about 35.0 and 57.0% in dry and green forages, respectively.

Constraints

There is acute shortage of fodder especially green nutritious fodder, which is the major cause of low productivity of the livestock (Deb Roy et al, 1989). The main reasons for low productivity are insufficient and low quality fodder and feed including grazing facilities (Deb Roy, 1993). The shortage of fodder supply in the country is attributed to some of the following reasons.

i) Less area under fodder crops

The division of the families has further fragmented the land. At present land holdings are very small and the farmer is always biased in the choice of the crops. Due to these reasons agricultural land ratio does not permit diversion of land from food production to cultivated fodder. Thus the area under fodder crops is meager and is not more than one per cent of the total cultivated land.

ii) Uncontrolled grazing

Grazing pattern clearly indicates a heavy grazing pressure on pasture. This has led to a decline in biomass availability. The grazing pattern has created manifold problems in pastures. Most of the grassland areas are under heavy and indiscriminate grazing. The pasturelands are used for the grazing of local livestock and migratory livestock. The well-developed grasslands are protected only for about two to three months during monsoon and are grazed indiscriminately during rest of the year. Community rangeland/ pastures are grazed in uncontrolled manner particularly at the active period of growth of most of the vegetation. The edible forage species do not get enough time to grow, set seed and proliferate. Obnoxious weeds also invade the pastures. Excessive and continuous grazing has severely damaged these lands. The herbage species found in these lands represent the third or fourth stage of degradation.

iii) Poor Management Practices

The management practices play an important role in determining the productivity of grasslands, yet this has been the last priority of the farmers. Presence of inferior and unproductive grass species, lack of fertilization, absence of legume component, improper cutting and indiscriminate grazing are some of the factors responsible for poor productivity of the grasslands. There exists a wealth of indigenous knowledge for the proper utilization and management of the natural resource base but the farmers because of increasing population pressure and declining land productivity are not using it.

iv) Heavy livestock population

Livestock is the integral component of Indian agriculture, since time immemorial and its contribution to national economy through milk, meat, wool as well as farmyard manure (FYM) is enormous. We have approximately 20% of world's cattle, 50% of buffaloes, more than 120 million goats and 60 million sheep (Deb Roy, 1993). Due to religious belief the population of unproductive cattle is increasing. This huge population and poor fodder availability has widened the gap between demand and supply of forage crops. Considerable fodder resources were wasted on the maintenance of an excessive number of poorly fed and low yielding animals, which contributed to the process of pasture destruction.

v) Fodder tree use

Tree flora of Indian sub-continent is one of the richest on this globe. The country supports number of trees and shrubs of fodder value, yet not more than 10 trees & shrubs are extensively used by the farmers. Tree leaf fodder is the major feed resource during lean periods, particularly the winters. The tree leaf fodder provides 50-90% of the forage demand during lean periods (Negi, 1977). The over exploitation and unscientific management of fodder trees has depleted this resource at huge environmental cost.

Strategies

Keeping in view the constraints in fodder production and in order to overcome the gap between demand and supply, the emphasis need to be given on several steps for augmenting the fodder production. Existing resource utilization pattern needs to be studied in totality according to a system approach.

Fodder production is a component of the farming system and efforts need to be made for increasing the forage production in a farming system approach. The holistic approach of integrated resource management will be based on maintaining the fragile balance between productivity functions and conservation practices for ecological sustainability. The strategies for improvement and conservation of forage resources will have to be dictated by actual customers. Some of the scientific interventions, which could help in improving the productivity of forages, are described here below.

(A) Agronomic management

The herbage production from grasslands and meadows can be enhanced with the adoption of improved technology. Important components of this technology are:

(a) Control of bushes and weeds
(b) Pasture establishment
(c) Introduction of legumes/grasses
(d) Fertilizer application
(e) Cutting and grazing management
(f) Lopping management of fodder trees

(a) Control of bushes and weeds

The bushes and noxious weeds and poor quality grasses may offer severe competition for light and nutrients. The most common obnoxious weeds in the pasture/grasslands are Lantana, Ageratum etc. These weeds can be controlled by cutting and stems treated with herbicides to prevent re-growth. The herbicides like Weedon 64, Picloram, Paraquat and Glyphosate etc @ 1.0-2.0 Kg/ha could be applied around the bush. Sood and Singh (1986) have found that paraquat spray in the 15 cm band @ 0.6 lt./ha reduced the weed incidence in the grasslands and the fresh herbage yield increased by 26.8%.

(b) Pasture establishment

The successful establishment of a pasture requires more skill and care, as compared to other crops. The method of introduction of improved grasses and legumes in the natural grasslands should be cost effective with minimum soil working. The following methods of establishment could be considered.

(i) *Scratching or pitting* : Singh (1995) found that planting Nandi grass and Guinea grass in circular pits was superior over local practice. Similarly Sood and Kumar (1995) has found that pit method of introduction is superior to scrapping.

(ii) *Hoof and teeth method* : The pasture could be heavily grazed; followed by throwing seeds, then allowing the animals to trample the area when soil is wet.

(*iii*) *Line sowing* : In this method grass seeds are placed in opened furrows and pressed by foot during the light showers. It helps in proper weeding in first year of establishment and introduction of annual and perennial legume species for quality improvement of pasture produce.

(c) Introduction of legumes/grasses

Legumes and grass species can be introduced during July by seeding and tussock planting, respectively. A combination of Siratro has been found quite successful for the mid altitude region (Melkania, 1987). Legumes introduced in the pastures generally do not establish well due to ineffective nodulation. Hazra (1998) observed that the *Rhizobium* inoculation of the pasture legumes provides synergistic effect for better establishment and obtained 59% and 72% higher green and dry herbage yield as compared to control. The herbage yield and nutritive value of the hay from grasses-legume mixtures were found five and two times higher, respectively than the hay of local species. It is essential that during the first year of seeding/tussock planting, grazing is restricted in treated sites and the grass cutting is done carefully to help the establishment of introduced fodder species. In order to increase the quality and quantity of pasture, legumes species are introduced, since legumes are rich source of proteins and are palatable to most of the livestock species. Perennial legumes like Dolichos, Clitoria, Siratro, Stylosanthes and Atylosia can be grown successfully to improve pasture quality as well as quantity. Experiment conducted at C.S.W.R.I., Avikanagar showed a significant increase in dry fodder yield with grass legume pasture (Table 1) and it was found that mixed pasture of Cenchrus and Dolichos yielded the highest (59.80 q/ha.).

Table 1. Production of Cenchrus and its mixture with perennial legumes.

Treatments	**Dry forage (q/ha.)**
Cenchrus alone	28.80
Cenchrus + Siratro (*Macroptilius atropurpureum*)	37.50
Cenchrus + Clitoria (*Clitoria ternatea*)	45.00
Cenchrus + Dolichos (*Dolichos lablab*)	59.00

Thus, it indicates that best combination of grass and legume is with Cenchrus and Dolichos sown in 1:1 ratio as alternate strips of 6-8 rows of each. A better method of introducing Clitoria in Cenchrus pasture is through broadcasting of Clitoria seed followed by cultivation to mix it in to soil prior to sowing of Cenchrus in line.

It has been observed that an annual legume like Cowpea, Guar and Moth can also be successfully grown as nurse crops in Cenchrus pasture during first year of establishment. Study conducted at C.S.W.R.I., Avikanangar indicates that total biomass was increased with the addition of annual legumes with pasture

(Table 2). To get the maximum yield and high quality forage from grasses and legumes they should be harvested at 50 % bloom stage.

Table 2. Production of Cenchrus with and without annual legumes.

Treatments	Dry forage (q/ha.)
Cenchrus alone	8.90
Cenchrus + Moth	13.00
Cenchrus + Guar *(Cymopsis tetragonoloba)*	14.90
Cenchrus + Cowpea (*Vigna sinsensis)*	19.20

(d) Fertilizer management

The present poor production potential of pastures could also be attributed to poor fertility of soils. To raise the fertility status and rectify the deficiencies, soil testing coupled with field trials need to be conducted to work out the fertilizer requirement of different pastures. Generally, no fertilizer is added to rangelands except the dropped excreta by animals. Judicious use of fertilizer for pasture can boost the vegetative growth and is also economically feasible. Application of nitrogen fertilizer must be given in split doses for better utilization, whereas phosphorus and potash should be supplied as basal dose in case of grasses. In legumes the full dose of nitrogen, phosphorus and potash should be given as a basal dose in furrows or by broadcasting at the time of sowing. Herbage yield increased significantly with the application of nitrogen @ 60 Kg/ha and phosphorus @30 Kg/ha (Sood and Sharma, 1996). Nitrogen @ 40 kg/ha and Phosphorus @ 30 Kg/ha applied as basal and two splits (onset of Monsoon and 45 days after first application) in natural grassland increased the forage yield significantly. Two splits were significantly superior to single application (Singh, 1995). Using FYM or sheep manure 10 t/ha and 40 kg N, 40 kg P_2O_5 /ha at establishment of pasture and there after 40 kg N/ha every year can sustain the productivity of the pasture for 5-6 years.

(e) Cutting and grazing management

The response to cutting of a forage plant depends upon its seasonal yield of carbohydrate storage, its growth habit and extent of inflorescence development. Frequency of cutting also significantly influences the yield and quality of herbage produced. The areas with high temperatures may require larger interval and low intensity of cutting to build up sufficient carbohydrate storage for re-growth. Cutting grasses twice from natural grasslands recorded higher fresh forage yield (14.54 t/ha) than one cut (12.08 t/ha) and three cuts (13.30 t/ha). The crude protein content was higher with two cuts compared to one cut (Kaul and Sood, 1986). Studies undertaken by Singh (1995) on cutting management of grasslands suggest that the herbage biomass yields can be doubled if harvested twice during July - October.

Controlling the time, duration and intensity of grazing appears to be the key factors in grazing management. Periods of rest allow grazed perennials to replenish leaf area, seed set and store food reserves in their roots (Merrill, 1983; Adams *et al.*, 1991). Continuous or too frequent access by large numbers of cattle to the same range impedes the ability of new growth to store food. The grazing can lead to the disappearance of nutritive species and infestation by less palatable species and weeds. Deferred rotational grazing system was found superior in Sehima dominated grasslands (Upadhyaya *et al.*, 1971), resulting in greater number of animal days as compared to continuous approach. Rotational grazing

Table 3. Scheme of rotational grazing

Year of grazing	Compartment			
	I	II	III	IV
1st Year	July-Sept.	Oct.-Dec.	Jan.-March	April-June
2nd Year	Jan.-March	April-June	July-Sept.	Oct.-Dec.
3rd Year	Oct.-Dec.	Jan.-March	April-June	July-Sept.
4th Year	July-Sept.	Oct.-Dec.	Jan.-March	April-June

Table 4. Scheme of deferred rotational grazing

Year of grazing	Compartment				
	I	II	III	IV	V
1st Year	July-Sept.	Oct.-Dec.	Jan.-March	April-June	Protected
2nd Year	Protected	July-Sept.	Oct.-Dec.	Jan.-March	April-June
3rd Year	April-June	Protected	July-Sept.	Oct.-Dec.	Jan.-March
4th Year	Oct.-Dec.	Jan.-March	April-June	Protected	July-Sept.

Table 5. Effect of stocking rate / grazing resources on wool yield and body weight of sheep

Treatments	Population of sheep in Village	Farmers sheep population	Stocking rate in terms of sheep (Nos/ha)		Body wt. /sheep (kg)	Wool yield (kg per sheep)
			Common rangelands	Farmer's own area		
Over crowded pasture	33,334	237	13.24	28.55	23.13	1.148
Low level of nutrition	8,934	183	5.75	7.12	26.11	1.283
High level of nutrition	10,989	428	13.4	4.01	26.35	1.259

has steadily gained the popularity in last two decades, because it offers better control over livestock distribution and feeding pattern with goals of periodically resting vegetation (Adams *et al.*, 1991).

In over crowded pasture body wt. of sheep decreased. In two-tier silvipastoral system [Ardu (100 trees/ha) + cenchrus pasture] and three-tier silvipastoral system [Ardu + *Dichrostachys nutans* (100 bushes/ha) + Cenchrus pasture] 8 and 12 sheep can easily be allowed for grazing in a hectare pasture in order to sustain its productivity for a longer period.

(f) Lopping management of fodder trees

Trees and shrubs in the silvi-pastoral system, road side avenue trees often contribute substantial amount of leaf fodder especially in arid and semi-arid regions during lean period through lopping / pruning of trees, popularly known as top feeds. The leaf fodder yield per tree varies considerably and it depends on species, initial age, lopping intensity and interval as well as agro-climatic conditions. Ardu can be lopped fully twice a year in the month of December and May-June at any stages of its age. While, fully-grown neem is lopped fully once in a year in the month of December-January. However, young tree of neem can be lopped fully twice a year in Dec.-Jan. and May-June. Reverse is true for Khejri. Young Khejri plant can be lopped fully only once in a year, however, full-grown tree/age old tree can be lopped two-third twice a year in May-June and Nov.-December.

(B) Growing of fodder crops and fodder trees

Agro-forestry is an integrated system in which trees for fodder fuel and timber are grown in association with cultivated crops. Various fodder and grain crops like Cowpea, Guar, Moong, Urd, Moth, Bajra, Sorghum and Maize etc. are grown with fodder trees. Growing of thin canopy leguminous tree like Khejri (*Prosopis cineraria*) and Ardu (*Ailanthus excelsa*) along with cereals and pulses doesn't affect the grain yield adversely and give additional yield of fodder.

Table 6. Grain and dry fodder yield of crops under different Agroforestry-system.

Trees	Grain Yield (q/ha.)					Dry fodder yield (q/ha.)				
	Sesame	Bajra	Guar	Cench.	Cench+ Cowpea	Sesame	Bajra	Guar	Cench.	Cench+ Cowpea
Ardu	0.35	1.95	3.42	2.58	2.51	7.15	24.35	20.97	12.58	13.61
Neem	0.29	1.74	2.74	2.13	1.56	6.18	21.03	18.97	10.96	11.97
Babool	0.38	2.07	3.50	2.72	2.50	7.66	24.21	22.38	12.82	13.85
Siris	0.34	1.74	3.11	2.25	1.65	6.56	20.67	20.52	11.52	12.60
Dichro	0.35	1.20	2.28	1.99	1.49	4.29	13.19	12.19	8.50	7.93

(C) Silvi-pastoral System

Silvipasture implies sustained and combined management of the same land for herbaceous fodder, top feeds and fuel wood, thereby leading to optimization of production. The rangelands exhibited enormous gain in forage production over existing situation due to multi-tier silvi-pasture techniques amalgamated with an adaptable complementary plant species. Silvi-pastoral systems are the most important for increasing fodder production from the marginal, sub-marginal and other wastelands, which comprise about 50% of the total land area. It involves planting of multipurpose trees in the existing pastures/grazing lands or planting such trees on wasteland/denuded lands followed by sowing / planting of grasses and or legumes in between the inter-spaces of trees.

Table 7. Production and cost of cultivation of natural and different silvi-pastoral systems (1987-91)

Treatments	Average dry fodder + green fodder (q/ha)	Average seed (Kg/ha)	Average cost (Rs/ha)
Natural Pasture	16.58	13.75	562.20
Cenchrus pasture alone (single tier)	19.12	93.15	970.97
Ardu + Cenchrus (2-tier system)	20.93+ 6.21	80.15	1364.71
Ardu + Dichrostachys+ Cenchrus (3-tier system)	23.75 + 9.52	86.65	1867.29

Milk yield of mutton synthetic ewes kept on these silvi pastoral systems was recorded higher (365.0 to 467.22) ml/day) as compared to cenchrus alone (310.55 ml/day) and natural pasture (285.55 ml/day. The higher milk yield of eves maintained on silvi pastoral systems could be due to green Ardu leaves supplementation.

Among comparative species under multi tier systems at the age of two years establishment ardu + Mulberry attained the maximum height and basal diameter and the same time provided more biomass (1196.84 g) dry leaf (252.17 g) and fuel wood (321.42 g) per plant as compared to Ardu + *Bahunia raccmosa* (301.53, 71.28 and 39.78 g) respectively.

Among the various crops combinations with ber (*Zizyphus mauritiana*) during 1984-86, maximum dry matter yield was attained with bajra +cenchrus followed by bajra +clitoria. The fertilizer dose: of 40 kg N and 40 kg P_2O_5 per hectare produced more dry matter as compared to 20kg N & 20kg P per hectare.

(D) Agri-silvipastoral system

Under the agri-silvi-cultural system, multipurpose trees (MPTs) including fodder cum fuel trees can be grown in association with crops. Trees are pruned/ lopped annually, yielding fodder as well as fuel wood. In addition to annual pruning, few trees are also cut down in order to allow light penetration and minimization of competition with the crops. Under alley cropping system MPTs like *Leucaena leucocephala* and even perennial pigeon pea etc. are pruned frequently to provide leaf fodder to get better crop production.

(E) Agri-horti-silvicultural system

Under this system besides growing fruit trees and fodder crops, fast growing NFTs like *Leucaena leucocephala* can be lopped two to three times in a year to provide fodder (2.5-3.0 t/ha) and fuelwood (1.8-2.5 t/ha). These fodder trees also provide some protection to the fruit trees during summer and cold winters.

(F) Hortipastoral system

In this system forage are grown in wide inter-row spaces of fruit trees for economic utilization of orchard lands. Forage from hortipasture is consumed fresh and is also conserved as hay for winters. There is considerable area under orchards. Inter spaces between fruit trees could be utilized for the production of fodder by growing perennial grasses and legumes.

Table 8. Yearly fodder production potential of improved crop rotation in various regions.

Regional rotation	Green forage yield (t/ha)
A- Northern Region	
• Napier-pearmillet+Velvet bean- Berseem+Mustard	122.60
• Deenanath grass- Berseem-Maize+Cowpea	200.90
• Napier- Pearlmillet hybrid+Berseem	211.70
B- Central and Western Region	
• Napier- Pearlmillet hybrid+Cowpea-Cowpea-Berseem	255.00
• Sorghum+Cowpea- Berseem+Mustard- Sorghum+Cowpea	168.60
• Napier- Pearlmillet hybrid+Guar- Lucerne	252.90
C- Eastern Region	
• Pearlmillet+CowpeaMaize+Cowpea-Oat	102.60
• Maize+Cowpea- Deenanath grass-Oat	130.80
• Napier- Pearlmillet hybrid alone	106.40
D- Southern Region	
• Napier- Pearlmillt hybrid + Hedge Lucerne	225.00
• Sorghum+Cowpea- Maize+Cowpea	110.70
• Guinea grass alone	135.00

Conclusions and Recommendations

It may be concluded that the forage production situation in the region is very alarming and corrective measures have to be taken to improve the same. Delineation of the area for various agricultural activities should be created and adhered under legislation. A comprehensive grazing policy needs to be formulated for the entire zone. Both grazing and forage cultivation has to be considered complementary to each other and simultaneous efforts are required to improve the both. Fodder tree improvement programmes for higher leaf fodder have to be initiated. In order to improve the grasslands, the grassland management needs to be considered holistically promoting the interaction between grassland, livestock and the grazing communities, so that this vast natural resource can serve human society substantially, more particularly grazing communities of the region. The recommendations are:

- Forage production must be taken up as a first management goal and 25% of the forest area should be put under trees with regulated accessibility to the farmers.
- Growing forage grasses and fodder trees along village roads and panchayat lands
- Growing forage grasses and fodder trees on terrace risers/bunds- a non competitive land use system
- Conservation of native biodiversity for future improvement
- Breeding biotic, abiotic, stress tolerant cultivars of forage species suitable for area not used under arable agriculture
- Participatory techniques to be adopted to identify the problems and to carry out the improvement programme
- In-depth studies on migratory graziers
- Forage based agroforestry systems
- Controlled grazing to maintain the productivity of pasture (grazing should be allowed as per carrying capacity)

REFERENCES

Adams, B.W., Ehlert, G. and Robertson, A. 1991. Grazing systems for public grazing lands. Range notes No. 10 (Alberta forestry lands and wildlife, public lands division, Leth-bridge, Alberta). pp. 1-8.

Deb Roy 1993. Reap more biomass through diversity in forestry. *Intensive Agriculture.* 31, 23-26

Deb Roy, Shankaranarayan, K.A and Pathak R.S 1989. The Fodder Trees of India and their importance. *Indian Forester,* 106, 306 -311.

Hazra, C.R. 1998. Integrated nutrient management for sustainable forage production in pasture and silvipastures. *Fertilizer. News.* 43, 33-45.

Kaul, A.K. and Sood, B.R. 1986. Production and quality of a natural grassland as affected by cutting frequency and nitrogen application. *Indian Journal of Range Management.* 71, 5-9.

Melkania, N.P. (987. Search for an ideal grass-legume introduction for mid-Himalayan grasslands. *Indian Journal of Range Management.* 8:51-53.

Merrill, J. 1983. " The XXX Ranch: Managing Range for Ecology and Economy", Publication No. ys-83-1, Chapter offprint from 1983. Year Book of Agriculture (U.S. Deptt. of Agriculture, Washington, D.C., pp. 87-94).

Negi, S.S. 1977. Fodder Trees in Himachal Pradesh. *Indian Forester* 103, 616-622

Singh Virendra. 1995. Technology for forage production in Hills of Kumaon. In : New Vistas in Forage Production (ed. Harzra, C.R and Misri Bimal). AICRPF (IGFRI). Publication Information Directorate, New Delhi. pp. 197-202.

Sood, B.R and Singh C.M. 1986. Effect of method of introduction and fertilizer management on the production potential and quality of natural grassland of Kangra valley. *Range Management and Agroforestry* 7, 119-126.

Sood, B.R. and Kumar, P. 1995. Production and economics of natural grasslands improvement. *Range Management and Agroforestry* 16, 17 -22.

Sood, B.R. and Sharma, V.K. 1996. Effect of nitrogen and phosphorus levels on the productivity of natural grassland. *Haryana Journal of Agronomy* 12, 68-74.

Upadhyay, V.S., Dabadghao, P.M and Shankarnarayan, K.A. 1971. Annual Report. Indian Grassland and Fodder Research Institute, Jhansi.

□□□

44

Role of Wool in Sheep Husbandry

N.P. GUPTA, R.K. ARORA
Central Sheep and Wool Research Institute, Avikanagar-304501, Rajasthan

Introduction

Sheep husbandry is an important pastorial activity across the globe. It is a driving force of agriculture sector in India by contributing towards the production system. It bears an important role in providing livelihood to the rural masses especially to the socio-economically weaker sections of the society. About 15 lakh families are estimated to the engaged in various activities related to sheep rearing and utilization of their product. The national sheep wealth is about Rs.2400 crores producing an annual income of about 800 crores in the form of wool, meat, skin, milk and manure. Wool and mutton are the two important products which come out of sheep husbandry activity. However, its purposed now a days is primarily meat production and wool production activity gradually losing its importance because of unfavorable price situation.

Wool obtained from sheep is a prominent animal fibre commonly known to human civilization from time immemorial. It has unique properties of warmth, resilience, drape etc., for which it is used as textile fibre throughout the world since ancient times. The notable wool producing countries are Australia, New Zealand, South Africa, former USSR, China etc. Wool plays a vital role in India economy. Wool is serving important purpose by producing fabrics for civilian and defense personnel as well as for export. There is well developed wool based industry in country which is basically an export oriented industry. It consists of organized and decentralized wool fibre processing and product manufacturing sectors but it is tiny and scattered in comparison to jute and cotton industry. The organized sector comprises woolen and worsted spinning and weaving mills whereas decentralized sector mainly consists of hosiery, carpets, shoddy and Khadi processing units and has a dominant presence in the woolen sector of the country.

Woollen sector mainly consumes indigenous wool and produces yarns of coarser variety which are generally used for carpets, blankets and upholstery where worsted sector consumes fine variety imported wool producing finer yarn utilized for apparels and knit goods - Khadi sector utilizing coarser native and imported fine wools produces wide range of products adopting labour intensive processes which serves a vital social role in employing a handsome number of people. Besides the shoddy sector using various types of used woolen products for different end uses mainly blankets.

In the decentralized sector high employment is generated with low capital investment. The carpet sector is mainly export oriented, the woolen knitwear apparel is also contributing to exports. Industry's export contribution has reached to Rs.5000 crores and there is scope to augment this share. However, non-availability of quality wool and inadequate quantity is the major limitation and as a result the industry is becoming highly import dependant.

In the present article it has been tried to compile the status of wool production, performance of developmental strategy adopted for increasing production and improve the quality of wool, export of woolen products.

World Scenario

Wool production

The world wool production on clean basis indicates a declining trend. The world production has declined to 1264 mkg in the year 2003-04 from a peak of 2010 mkg achieved in the year 1989-90. The production has declined sharply in Austria, New Zealand and other countries except in China, Russia and India. It is unlikely that the worsted wool production and supply will show increasing trend in near future as inter fibre price competition is intensifying and mutton price is increasing relative to wool price.

World productivity

The world wool productivity per sheep per year on greasy basis is around 2.3 kg while that of Australia, New Zealand Chine and India are 5.5 kg, 6 kg and 2.2 kg & 0.8 kg respectively. The productivity of greasy wool per sheep per year in India has remained almost constant (Fig. 1) and it is unlikely to show any improvement in near future which indicates that the developmental activities could not have any positive impact despite huge expenditure incurred.

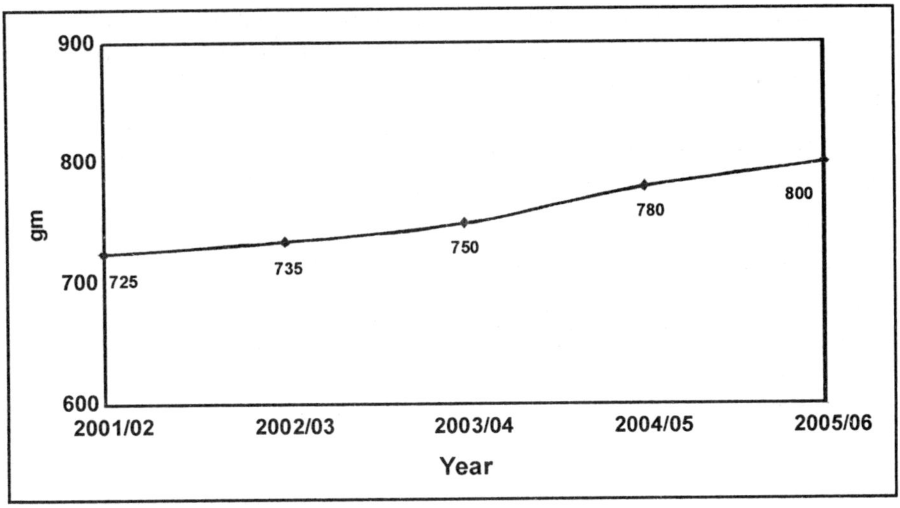

Fig. 1. Productivity per animal per year

Indian Scenario

Production

The annual wool production of the country has increased to 55 mkg greasy for the year 2005/06 (Fig. 2) from 22 mkg in 1950. However the wool production increase is mainly explained as the increase in sheep population.

Most of these wools are of carpet grade wool. Out of total wool about 15 kg wool produced mainly in Southern and Eastern region of country are having very little textile application because of their inherent extreme coarse-ness and

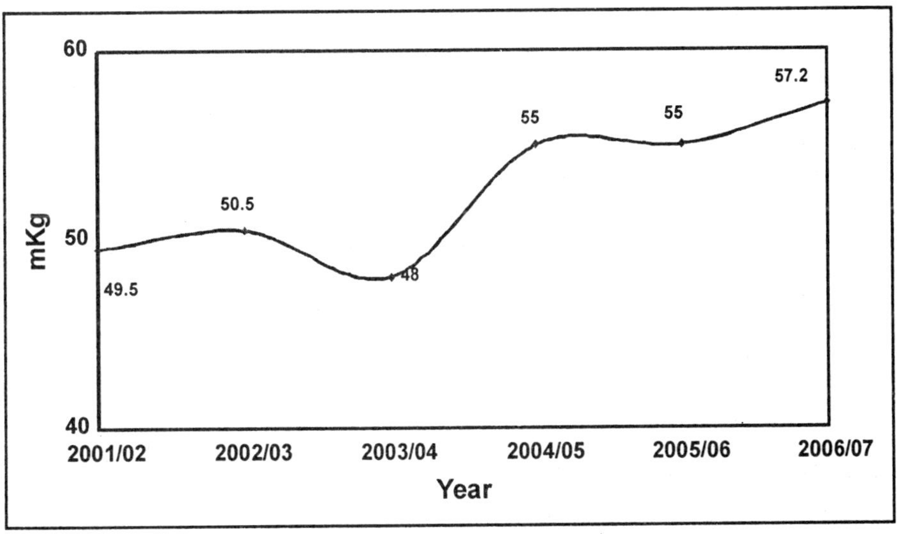

Fig 2. Wool production in India

coloured/pigmented nature. The clean wool production is 36 m kg and it is not sufficient for the country's consumption (Fig. 3) as a result importation is increasing.

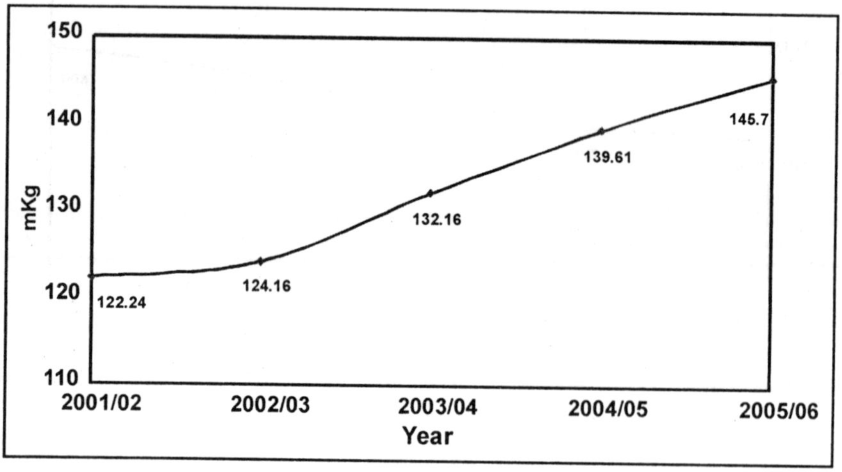

Fig. 3. Consumption of wool

There is hardly any apparel wool production in the country. The wool produced in the hilly region of the country is having adequate fineness but lacks in staple length which makes them unsuitable for finer count spinning on worsted spinning system. As a result worsted industry is almost entirely dependent on imported Australian merino wool.

Wool Quantity & Quality

It may be pointed out here that national sheep & wool developmental programme to increase the annual wool yield and wool quality through mass crossbreeding of indigenous with exotic high wool producing rams were launched throughout the country. A critical analysis of wool production data indicates that these programmes have largely been unsuccessful in improving the fibre out put per animal and to improve wool quality. These have largely contributed in fineness improvement but at the cost of staple length. The length could not be increased through annual shearing. As the increase in length/ growth become stagnant after a certain period. The short staple fine wool find very little application in spinning finer yarns and therefore necessary to restrict fine wool production in hilly belts of J&K, HP and UP only. The North Western plain regions should concentrate to produce quality carpet wool so that carpet of superior quality could be manufacture.

Processing

Processing of wool is done in different systems of processing depending upon their fineness and length. Most of the indigenous wools which are coarser are processed on woolen system of spinning to produce coarse yarn for utilization in carpet making and blanket weaving. Generally 4 Nm yarn is produced for these products. The yarn produced on this system is bulky, processing more amount of entrapped air in the body of yarn helping in retention of warmth of the body when fabrics produced from such yarns are used. Fine yarns are produced on worsted system of processing which are used to produce finer variety of fabrics like shirting and suiting. In this system of processing the fibre lies parallel to the yarn axis and posses least amount of air inside the body of the yarn. Generally 64s variety of wool is processed in this system.

Products

Variety of products like blankets, carpets, felts, shawls, knitwear etc. are produced from wool. Blankets area generally produced from yarns prepared on woolen system. The blankets produced from finer variety of wools are softer and loftier whereas blankets prepared from coarser variety of wools are harsher and rough in feel which needs softening treatment or blending of other softer fibres.

Three types of carpets are produced from wool i.e. hand knotted, hand tufted and handloom woven. The three products do not complete with each other since they altogether different appeal and their use in also limited to certain areas. Carpets having 40 to 400 knots per square inch are generally produced in the country. Tufted carpets are prepared with the help of tufted gun. These carpets are generally used for floor covering purposes. During last two decades hand tufted carpets have found wider use and a potential demand. This system of carpet manufacture has bright future in internal as well as export market because the production cycle is short, adaptability of trendy colours, designs styles and price is competitive with same kind of hand knotted carpets as well as with machine made carpets felts are prepared by hand as well as machine.

The scaly structure of wool and other animal fibres helps in making an irreversible strong structure. Generally felts are produced from wool fibre only. Knitwears are generally produced from 48s to 70s quality wool. Out of the total wool and woolen exports the hosiery export account for about 30%. Shoddy sector utilizes woolen and synthetic rags imported from developed countries and generally cheap blankets are manufactured out of shoddy.

Value Addition

Value addition is most important in every stage of processing and wool is not an exception. In fact value addition at each stage of processing of wool is much higher as compared to other textile fibres. The usual processing routes of wool fibres after harvesting (shearing) is sorting, grading, scouring, early stage processing, yarn spinning, twisting (doubling), yarn dyeing, weaving and chemical processing (finishing) for fabric manufacturing whereas the processing routes for carpet, non woven, knitting and ready made garments are slightly different and a few new processing sequences are involved.

The value addition is comparatively much higher in hand knotted carpet, high quality technical textiles and worsted suiting. It is worth to mention here that out of total value of wool produced annually i.e. Rs.200 crores and imported wool of 1000 crores woolen products of about 5000 crores are exported annually. Moreover woolen products of similar value are being sold annually within the country. It reflects the value addition of Rs.8500 crores for Rs.1200 crores wool i.e. nearly seven time the value of raw wool.

Conclusion and Recommendations

The country has adequate infrastructure of wool based textile industry and there is enormous scope for boasting production and export of carpet, blankets, shawl and technical textile produced from wool along with mass employment generation. Wool based industry is likely to show significant growth as most of the advanced countries are vacating from textile industry because of higher cost, and other associated problems. With the thrust in export and higher demand domestic demand of woolen manufactured items, the demand of wool is increasing for consumption or processing by wool based industry of country. The dream of self sufficiency in wool production in country could not be realized and it is unlikely to be achieved in the near future. On the contrary which has strong position in quality carpet wool production becoming dependant on imports as industrial requirement is growing at higher pace as compared to growth in production.

The country is almost entirely dependant on imports for fine wools requirement for worsted apparel. Currently India imports wool to be extent of 100 mkg costing 1000 crores (2006-07). This includes apparel wool, carpet quality and little quantity of coarser variety of wool. It is essential to supply quality wool and utilize improved technology for increasing production and export. To meet the demand supply gap in wool fibre it is necessary to put thrust for wool productivity maximization with quality improvement. Carpet wool is always in demand. It is very much desired that country should increase the carpet wool production to full fill the demand of the industry. Stress should be to increase the fibre production per animal rather than increasing the number of animals. The medium type carpet wool producing breeds such as Nali, Chokla, Pattanwadi, Magra, Marwari, Jaisalmeri may be further improved through selective breeding. The shoddy products area to be discouraged so that indigenous wool blends can find significant entry in the field of blanket and shawls.

REFERENCES

Gupta, N.P., Pokharna, A.K., and Patni, P.C. 2006. Prospects of production, processing and marketing of sheep and goat fibres. Lead paper in : *National Workshop cum Seminar on "Commercial Goat & Sheep Farming and Marketing"*, March 4-5, Central Institute for Research on Goats, Mukhdoom, pp.79- 86.

Shakyawar, D.B. 2007. Wool and Woolen Industry of India- Its Present Status and Future Prospectus. Lead Paper in: *National Seminar Indian Textiles : Vision 2020"*, September 30 & October 1, Government Central Textile Institute, Kanpur.

Arora, R.K., and Patni, P.C. 2004. Diversification of Product Manufacture from Wools of South India. Lead Paper in : *Training program on "Advances in Sheep, Goat and Rabbit production and their Product Utilization"*, Central Sheep and Wool research Institute, Avikanagar, pp.164- 167.

Anon. 2006. Wool and Woollen Statistics. *Wool Ways*, 16, 22- 40.

❑❑❑

List of Contributors

Dr. A. Santra
Senior Scientist (Animal Nutrition)
Eastern Regional Research Station
National Dairy Research Institute,
Kalyani, West Bengal

Dr. A. Suresh
Scientist (Animal Economics and Statistics)
Central Sheep and Wool
Research Institute
Avikanagar 304 501, Rajasthan

Dr. A.K. Mishra
Senior Scientist
(Animal Genetics and Breeding)
Central Sheep and Wool
Research Institute
Avikanagar 304 501, Rajasthan

Dr. A.K. Shinde
Principal Scientist
(Livestock Production and Management)
Central Sheep and Wool
Research Institute
Avikanagar 304 501, Rajasthan

Dr. A.L. Arora
Head
Animal Genetics and Breeding
Central Sheep and Wool
Research Institute
Avikanagar 304 501, Rajasthan

Dr. A.R. Sen
Senior Scientist
(Livestock Product Technology)
National Research Centre on Meat
Hyderabad- 500 039, AP

Dr. Anil Joshi
Principal Scientist (Biochemistry)
Central Sheep and Wool
Research Institute
Avikanagar 304 501, Rajasthan

Dr. Asim K. Misra
Senior Scientist (Animal Nutrition)
Indian Grassland and
Fodder Research Institute
Jhansi, UP

Dr. B.S. Punia
Director
Central Institute of Research
on Buffalo
Hisar, Haryana

Dr. Bandla Srinivas
Senior Scientist (Animal Nutrition)
Central Sheep and Wool
Research Institute
Avikanagar 304 501, Rajasthan

Dr. C.P. Swarnkar
Senior Scientist (Animal Nutrition)
Central Sheep and Wool
Research Institute
Avikanagar 304 501, Rajasthan

Dr. D. Lal
Principal Scientist
Central Institute of Research on Buffalo
Hisar, Haryana

Dr. D. Santhi
Assistant Professor
(Livestock Product Technology),
Tamil Nadu Veterinary and
Animal Science University
Chennai-7

Dr. D. Singh
Principal Scientist (Animal Health)
Central Sheep and Wool
Research Institute,
Avikanagar 304 501, Rajasthan

Dr. D.C. Gupta
Senior Scientist
(Animal Economics and Statistics)
Central Sheep and Wool
Research Institute
Avikanagar 304 501, Rajasthan

Dr. Davendra Kumar
Senior Scientist (Animal Physiology)
Central Sheep and Wool
Research Institute
Avikanagar 304 501, Rajasthan

Dr. F.A. Khan
Senior Scientist (Animal Health)
Central Sheep and Wool
Research Institute,
Avikanagar 304 501, Rajasthan

Dr J.S. Mann
Principal Scientist
(Forage Agronomy)
Central Sheep and Wool
Research Institute
Avikanagar 304 501, Rajasthan

Dr. K.K. Singh
Senior Scientist (Animal Nutrition)
Indian Grassland and
Fodder Research Institute
Jhansi, UP

Dr. L. Leslie Leo Prince
Scientist (Animal Genetics and Breeding)
Central Sheep and Wool
Research Institute
Avikanagar 304 501, Rajasthan

Dr. M.K. Agnihotri
Joint Commissioner
Department of Animal
Husbandry, Dairying and
Fisheries, Ministry of Agriculture
Government of India, New Delhi-1

Dr. M.K. Awasthi
Professor
Agribusiness centre
Indian Institute of Management
Lucknow, Uttar Pradesh-226013

Dr. M.K. Tripathi
Senior Scientist (Animal Nutrition)
Central Sheep and Wool
Research Institute
Avikanagar 304 501, Rajasthan

Dr. M.M. Das
Senior Scientist (Animal Nutrition)
Indian Grassland and
Fodder Research Institute, Jhansi, UP

Dr. N. Kondaiah
Director
National Research Centre on Meat
Hyderabad- 500 039, AP

Dr. N.M. Soren
Scientist (Animal Nutrition)
Central Sheep and Wool Research
Inst., Avikanagar 304 501, Rajasthan

Dr. N.P. Gupta
Principal Scientist (Wool Science)
Central Sheep and Wool
Research Institute
Avikanagar 304 501, Rajasthan

Dr. N.P. Singh
Director
Central Institute of
Research on Goats
Makhdoom, Farah
Mathura, Uttar Pradesh

Dr. Narayan Dutta
Senior Scientist (Animal Nutrition)
Centre of Advanced Studies in
Animal Nutrition
Indian Veterinary Research Institute
Izatnagar, Bareilly, UP

Dr. Navneet Saxena
Senior Scientist (Organic Chemistry)
Central Institute of
Research on Buffalo
Hisar, Haryana

Dr. O.H. Chaturvedi
Senior Scientist
(Animal Nutrition)
Central Sheep and Wool
Research Institute
Avikanagar 304 501, Rajasthan

Dr. R. Bhatt
Senior Scientist (Animal Nutrition)
Central Sheep and Wool
Research Institute
Avikanagar 304 501, Rajasthan

Dr. R. Gulyani
Principal Scientist
(Animal Physiology)
Central Sheep and Wool
Research Institute
Avikanagar 304 501, Rajasthan

Dr. R.C. Jakhmola
Head
Arid Regional Research Centre
CSWRI, Bikaner - 334 006, Rajasthan

Dr. R.K. Arora
Principal Scientist
(Textile Technology)
Central Sheep and Wool
Research Institute
Avikanagar 304 501, Rajasthan

Dr R.K. Jain
Head
Regional Research Station
Indian Grassland and
Fodder Research Institute
Avikanagar 304 501, Rajasthan

Dr R.P. Nagar
Senior Scientist
(Forage Seed Technology)
Regional Research Station
Indian Grassland and
Fodder Research Institute
Avikanagar 304 501, Rajasthan

Mr. R.S. Dhillon
Technical Officer (Pelt Technology)
Arid Regional Research Centre
CSWRI, Bikaner - 334 006,
Rajasthan

Dr. Raman Malik
Senior Scientist (Animal Nutrition)
Central Sheep and Wool
Research Institute
Avikanagar 304 501, Rajasthan

Dr. S. Saha
Senior Scientist
(Animal Economics and Statistics)
Central Sheep and Wool
Research Institute
Avikanagar 304 501, Rajasthan

Dr. S. Suresh Kumar
Scientist
(Livestock Product Technology)
Southern Regional Research Centre,
Central Sheep and Wool
Research Institute, Mannavanur
Kodaikanal-634 103 Tamil Nadu

Dr. S.A. Karim
Director
Central Sheep and Wool
Research Institute
Avikanagar 304 501, Rajasthan

Dr. S.C. Sharma
Senior Scientist (Forage Agronomy)
Central Sheep and Wool
Research Institute
Avikanagar 304 501, Rajasthan

Dr. S.K. Kaushish
Principal Scientist (Animal Science)
Central Arid zone Research Institute,
Jodhpur, Rajasthan

Dr. S.K. Sankhyan
Senior Scientist (Animal Nutrition)
Central Sheep and Wool
Research Institute
Avikanagar 304 501, Rajasthan

Dr. S.M.K. Naqvi
Head
Animal Physiology and Biochemistry
Central Sheep and Wool
Research Institute
Avikanagar 304 501, Rajasthan

Dr. Sajjan Singh
Senior Scientist (Animal Nutrition)
Central Sheep and Wool
Research Institute
Avikanagar 304 501, Rajasthan

Dr. Tribhuvan Sharma
Head
Animal Nutrition Division,
College of Veterinary Science,
Rajasthan Agriculture University
Bikaner Rajasthan-334 006.

Dr. V. Rajkumar
Scientist (Meat Science)
Department of Meat Science
and Technology
Madras Veterinary College,
Chennai

Dr. V. Sejian
Scientist (Animal Physiology)
Central Sheep and Wool
Research Institute
Avikanagar 304 501, Rajasthan

Dr. V.P. Maurya
Senior Scientist (Animal Physiology)
Central Sheep and Wool
Research Institute
Avikanagar 304 501, Rajasthan

Dr. Y. Babji
Scientist
(Livestock Product Technology)
National Research Centre on Meat
Hyderabad- 500 039, AP

□□□